W9-BAC-331

THIRD EDITION

Mathematical Statistics and Data Analysis

John A. Rice
University of California, Berkeley

BROOKS/COLE
CENGAGE Learning™

Australia • Brazil • Japan • Korea • Mexico • Singapore • Spain • United Kingdom • U

BROOKS/COLE
CENGAGE Learning™

Mathematical Statistics and Data Analysis, Third Edition
John A. Rice

Acquisitions Editor: Carolyn Crockett

Assistant Editor: Ann Day

Editorial Assistant: Elizabeth Gershman

Technology Project Manager: Fiona Chong

Marketing Manager: Joe Rogove

Marketing Assistant: Brian Smith

Marketing Communications Manager:
    Darlene Amidon-Brent

Project Manager, Editorial Production:
    Kelsey McGee

Creative Director: Rob Hugel

Art Director: Lee Friedman

Print Buyer: Karen Hunt

Permissions Editor: Bob Kauser

Production Service: Macmillan
    Publishing Solutions

Text Designer: Roy Neuhaus

Copy Editor: Victoria Thurman

Illustrator: Macmillan Publishing Solutions

Cover Designer: Denise Davidson

Compositor: Macmillan Publishing Solutions

© 2007 Brooks/Cole, Cengage Learning

ALL RIGHTS RESERVED. No part of this work covered by the copyright herein may be reproduced, transmitted, stored, or used in any form or by any means graphic, electronic, or mechanical, including but not limited to photocopying, recording, scanning, digitizing, taping, Web distribution, information networks, or information storage and retrieval systems, except as permitted under Section 107 or 108 of the 1976 United States Copyright Act, without the prior written permission of the publisher.

For product information and technology assistance, contact us at
Cengage Learning Customer & Sales Support, 1-800-354-9706

For permission to use material from this text or product, submit all requests online at **www.cengage.com/permissions**
Further permissions questions can be emailed to
permissionrequest@cengage.com

Library of Congress Control Number: 2005938314

ISBN-13: 978-0-534-39942-9

ISBN-10: 0-534-39942-8

Brooks/Cole
10 Davis Drive
Belmont, CA 94002-3098
USA

Cengage Learning is a leading provider of customized learning solutions with office locations around the globe, including Singapore, the United Kingdom, Australia, Mexico, Brazil, and Japan. Locate your local office at: **international.cengage.com/region**

Cengage Learning products are represented in Canada by Nelson Education, Ltd.

For your course and learning solutions, visit **academic.cengage.com**

Purchase any of our products at your local college store or at our preferred online store **www.ichapters.com**

Printed in the United States of America
3 4 5 6 7 11 10

We must be careful not to confuse data with the abstractions we use to analyze them.

WILLIAM JAMES (1842–1910)

Contents

Preface

Intended Audience

This text is intended for juniors, seniors, or beginning graduate students in statistics, mathematics, natural sciences, and engineering as well as for adequately prepared students in the social sciences and economics. A year of calculus, including Taylor Series and multivariable calculus, and an introductory course in linear algebra are prerequisites.

This Book's Objectives

This book reflects my view of what a first, and for many students a last, course in statistics should be. Such a course should include some traditional topics in mathematical statistics (such as methods based on likelihood), topics in descriptive statistics and data analysis with special attention to graphical displays, aspects of experimental design, and realistic applications of some complexity. It should also reflect the integral role played by computers in statistics. These themes, properly interwoven, can give students a view of the nature of modern statistics. The alternative of teaching two separate courses, one on theory and one on data analysis, seems to me artificial. Furthermore, many students take only one course in statistics and do not have time for two or more.

Analysis of Data and the Practice of Statistics

In order to draw the above themes together, I have endeavored to write a book closely tied to the practice of statistics. It is in the analysis of real data that one sees the roles played by both formal theory and informal data analytic methods. I have organized this book around various kinds of problems that entail the use of statistical methods and have included many real examples to motivate and introduce the theory. Among the advantages of such an approach are that theoretical constructs are presented in meaningful contexts, that they are gradually supplemented and reinforced, and that they are integrated with more informal methods. This is, I think, a fitting approach to statistics, the historical development of which has been spurred on primarily by practical needs rather than by abstract or aesthetic considerations. At the same time, I have not shied away from using the mathematics that the students are supposed to know.

The Third Edition

Eighteen years have passed since the first edition of this book was published and eleven years since the second. Although the basic intent and stucture of the book have not changed, the new editions reflect developments in the discipline of statistics, primarily the computational revolution.

The most significant change in this edition is the treatment of Bayesian inference. I moved the material from the last chapter, a point that was never reached by many instructors, and integrated it into earlier chapters. Bayesian inference is now first previewed in Chapter 3, in the context of conditional distributions. It is then placed side-by-side with frequentist methods in Chapter 8, where it complements the material on maximum likelihood estimation very naturally. The introductory section on hypothesis testing in Chapter 9 now begins with a Bayesian formulation before moving on to the Neyman-Pearson paradigm. One advantage of this is that the fundamental importance of the likelihood ratio is now much more apparent. In applications, I stress uninformative priors and show the similarity of the qualitative conclusions that would be reached by frequentist and Bayesian methods.

Other new material includes the use of examples from genomics and financial statistics in the probability chapters. In addition to its value as topically relevant, this material naturally reinforces basic concepts. For example, the material on copulas underscores the relationships of marginal and joint distributions. Other changes include the introduction of scatterplots and correlation coefficients within the context of exploratory data analysis in Chapter 10 and a brief introduction to nonparametric smoothing via local linear least squares in Chapter 14. There are nearly 100 new problems, mainly in Chapters 7–14, including several new data sets. Some of the data sets are sufficiently substantial to be the basis for computer lab assignments. I also elucidated many passages that were obscure in earlier editions.

Brief Outline

A complete outline can be found, of course, in the Table of Contents. Here I will just highlight some points and indicate various curricular options for the instructor.

The first six chapters contain an introduction to probability theory, particularly those aspects most relevant to statistics. Chapter 1 introduces the basic ingredients of probability theory and elementary combinatorial methods from a non measure theoretic point of view. In this and the other probability chapters, I tried to use real-world examples rather than balls and urns whenever possible.

The concept of a random variable is introduced in Chapter 2. I chose to discuss discrete and continuous random variables together, instead of putting off the continuous case until later. Several common distributions are introduced. An advantage of this approach is that it provides something to work with and develop in later chapters.

Chapter 3 continues the treatment of random variables by going into joint distributions. The instructor may wish to skip lightly over Jacobians; this can be done with little loss of continuity, since they are rarely used in the rest of the book. The material in Section 3.7 on extrema and order statistics can be omitted if the instructor is willing to do a little backtracking later.

Expectation, variance, covariance, conditional expectation, and moment-generating functions are taken up in Chapter 4. The instructor may wish to pass lightly over conditional expectation and prediction, especially if he or she does not plan to cover sufficiency later. The last section of this chapter introduces the δ method, or the method of propagation of error. This method is used several times in the statistics chapters.

The law of large numbers and the central limit theorem are proved in Chapter 5 under fairly strong assumptions.

Chapter 6 is a compendium of the common distributions related to the normal and sampling distributions of statistics computed from the usual normal random sample. I don't spend a lot of time on this material here but do develop the necessary facts as they are needed in the statistics chapters. It is useful for students to have these distributions collected in one place.

Chapter 7 is on survey sampling, an unconventional, but in some ways natural, beginning to the study of statistics. Survey sampling is an area of statistics with which most students have some vague familiarity, and a set of fairly specific, concrete statistical problems can be naturally posed. It is a context in which, historically, many important statistical concepts have developed, and it can be used as a vehicle for introducing concepts and techniques that are developed further in later chapters, for example:

- The idea of an estimate as a random variable with an associated sampling distribution
- The concepts of bias, standard error, and mean squared error
- Confidence intervals and the application of the central limit theorem
- An exposure to notions of experimental design via the study of stratified estimates and the concept of relative efficiency
- Calculation of expectations, variances, and covariances

One of the unattractive aspects of survey sampling is that the calculations are rather grubby. However, there is a certain virtue in this grubbiness, and students are given practice in such calculations. The instructor has quite a lot of flexibility as to how deeply to cover the concepts in this chapter. The sections on ratio estimation and stratification are optional and can be skipped entirely or returned to at a later time without loss of continuity.

Chapter 8 is concerned with parameter estimation, a subject that is motivated and illustrated by the problem of fitting probability laws to data. The method of moments, the method of maximum likelihood, and Bayesian inference are developed. The concept of efficiency is introduced, and the Cramér-Rao Inequality is proved. Section 8.8 introduces the concept of sufficiency and some of its ramifications. The material on the Cramér-Rao lower bound and on sufficiency can be skipped; to my mind, the importance of sufficiency is usually overstated. Section 8.7.1 (the negative binomial distribution) can also be skipped.

Chapter 9 is an introduction to hypothesis testing with particular application to testing for goodness of fit, which ties in with Chapter 8. (This subject is further developed in Chapter 11.) Informal, graphical methods are presented here as well. Several of the last sections of this chapter can be skipped if the instructor is pressed for time. These include Section 9.6 (the Poisson dispersion test), Section 9.7 (hanging rootograms), and Section 9.9 (tests for normality).

A variety of descriptive methods are introduced in Chapter 10. Many of these techniques are used in later chapters. The importance of graphical procedures is stressed, and notions of robustness are introduced. The placement of a chapter on descriptive methods this late in a book may seem strange. I chose to do so because descriptive procedures usually have a stochastic side and, having been through the three chapters preceding this one, students are by now better equipped to study the statistical behavior of various summary statistics (for example, a confidence interval for the median). When I teach the course, I introduce some of this material earlier. For example, I have students make boxplots and histograms from samples drawn in labs on survey sampling. If the instructor wishes, the material on survival and hazard functions can be skipped.

Classical and nonparametric methods for two-sample problems are introduced in Chapter 11. The concepts of hypothesis testing, first introduced in Chapter 9, are further developed. The chapter concludes with some discussion of experimental design and the interpretation of observational studies.

The first eleven chapters are the heart of an introductory course; the theoretical constructs of estimation and hypothesis testing have been developed, graphical and descriptive methods have been introduced, and aspects of experimental design have been discussed.

The instructor has much more freedom in selecting material from Chapters 12 through 14. In particular, it is not necessary to proceed through these chapters in the order in which they are presented.

Chapter 12 treats the one-way and two-way layouts via analysis of variance and nonparametric techniques. The problem of multiple comparisons, first introduced at the end of Chapter 11, is discussed.

Chapter 13 is a rather brief treatment of the analysis of categorical data. Likelihood ratio tests are developed for homogeneity and independence. McNemar's test is presented and finally, estimation of the odds ratio is motivated by a discussion of prospective and retrospective studies.

Chapter 14 concerns linear least squares. Simple linear regression is developed first and is followed by a more general treatment using linear algebra. I chose to employ matrix algebra but keep the level of the discussion as simple and concrete as possible, not going beyond concepts typically taught in an introductory one-quarter course. In particular, I did not develop a geometric analysis of the general linear model or make any attempt to unify regression and analysis of variance. Throughout this chapter, theoretical results are balanced by more qualitative data analytic procedures based on analysis of residuals. At the end of the chapter, I introduce nonparametric regression via local linear least squares.

Computer Use and Problem Solving

Computation is an integral part of contemporary statistics. It is essential for data analysis and can be an aid to clarifying basic concepts. My students use the open-source package R, which they can install on their own computers. Other packages could be used as well but I do not discuss any particular programs in the text. The data in the text are available on the CD that is bound in the U.S. edition or can be downloaded from academic.cengage.com/statistics.

This book contains a large number of problems, ranging from routine reinforcement of basic concepts to some that students will find quite difficult. I think that problem solving, especially of nonroutine problems, is very important.

Acknowledgments

I am indebted to a large number of people who contributed directly and indirectly to the first edition. Earlier versions were used in courses taught by Richard Olshen, Yosi Rinnot, Donald Ylvisaker, Len Haff, and David Lane, who made many helpful comments. Students in their classes and in my own had many constructive comments. Teaching assistants, especially Joan Staniswalis, Roger Johnson, Terri Bittner, and Peter Kim, worked through many of the problems and found numerous errors. Many reviewers provided useful suggestions: Rollin Brant, University of Toronto; George Casella, Cornell University; Howard B. Christensen, Brigham Young University; David Fairley, Ohio State University; Peter Guttorp, University of Washington; Hari Iyer, Colorado State University; Douglas G. Kelly, University of North Carolina; Thomas Leonard, University of Wisconsin; Albert S. Paulson, Rensselaer Polytechnic Institute; Charles Peters, University of Houston, University Park; Andrew Rukhin, University of Massachusetts, Amherst; Robert Schaefer, Miami University; and Ruth Williams, University of California, San Diego. Richard Royall and W. G. Cumberland kindly provided the data sets used in Chapter 7 on survey sampling. Several other data sets were brought to my attention by statisticians at the National Bureau of Standards, where I was fortunate to spend a year while on sabbatical. I deeply appreciate the patience, persistence, and faith of my editor, John Kimmel, in bringing this project to fruition.

The candid comments of many students and faculty who used the first edition of the book were influential in the creation of the second edition. In particular I would like to thank Ian Abramson, Edward Bedrick, Jon Frank, Richard Gill, Roger Johnson, Torgny Lindvall, Michael Martin, Deb Nolan, Roger Pinkham, Yosi Rinott, Philip Stark, and Bin Yu; I apologize to any individuals who have inadvertently been left off this list. Finally, I would like to thank Alex Kugushev for his encouragement and support in carrying out the revision and the work done by Terri Bittner in carefully reading the manuscript for accuracy and in the solutions of the new problems.

Many people contributed to the third edition. I would like to thank the reviewers of this edition: Marten Wegkamp, Yale University; Aparna Huzurbazar, University of New Mexico; Laura Bernhofen, Clark University; Joe Glaz, University of Connecticut; and Michael Minnotte, Utah State University. I deeply appreciate many readers for generously taking the time to point out errors and make suggestions on improving the exposition. In particular, Roger Pinkham sent many helpful email messages and Nick Cox provided a very long list of grammatical errors. Alice Hsiaw made detailed comments on Chapters 7–14. I also wish to thank Ani Adhikari, Paulo Berata, Patrick Brewer, Sang-Hoon Cho Gier Eide, John Einmahl, David Freedman, Roger Johnson, Paul van der Laan, Patrick Lee, Yi Lin, Jim Linnemann, Rasaan Moshesh, Eugene Schuster, Dylan Small, Luis Tenorio, Richard De Veaux, and Ping Zhang. Bob Stine contributed financial data, Diane Cook provided data on Italian olive oils, and Jim Albert provided a baseball data set that nicely illustrates regression toward the mean. Rainer Sachs provided the lovely data on chromatin separations. I thank my editor, Carolyn Crockett, for her graceful persistence and patience in bringing about this revision, and also the energetic production team. I apologize to any others whose names have inadvertently been left off this list.

CHAPTER 1

Probability

1.1 Introduction

The idea of probability, chance, or randomness is quite old, whereas its rigorous axiomatization in mathematical terms occurred relatively recently. Many of the ideas of probability theory originated in the study of games of chance. In this century, the mathematical theory of probability has been applied to a wide variety of phenomena; the following are some representative examples:

- Probability theory has been used in genetics as a model for mutations and ensuing natural variability, and plays a central role in bioinformatics.
- The kinetic theory of gases has an important probabilistic component.
- In designing and analyzing computer operating systems, the lengths of various queues in the system are modeled as random phenomena.
- There are highly developed theories that treat noise in electrical devices and communication systems as random processes.
- Many models of atmospheric turbulence use concepts of probability theory.
- In operations research, the demands on inventories of goods are often modeled as random.
- Actuarial science, which is used by insurance companies, relies heavily on the tools of probability theory.
- Probability theory is used to study complex systems and improve their reliability, such as in modern commercial or military aircraft.
- Probability theory is a cornerstone of the theory of finance.

The list could go on and on.

This book develops the basic ideas of probability and statistics. The first part explores the theory of probability as a mathematical model for chance phenomena. The second part of the book is about statistics, which is essentially concerned with procedures for analyzing data, especially data that in some vague sense have a random character. To comprehend the theory of statistics, you must have a sound background in probability.

1.2 Sample Spaces

Probability theory is concerned with situations in which the outcomes occur randomly. Generically, such situations are called *experiments*, and the set of all possible outcomes is the **sample space** corresponding to an experiment. The sample space is denoted by Ω, and an element of Ω is denoted by ω. The following are some examples.

E X A M P L E   **A**   Driving to work, a commuter passes through a sequence of three intersections with traffic lights. At each light, she either stops, s, or continues, c. The sample space is the set of all possible outcomes:

$$\Omega = \{ccc, ccs, css, csc, sss, ssc, scc, scs\}$$

where csc, for example, denotes the outcome that the commuter continues through the first light, stops at the second light, and continues through the third light. ∎

E X A M P L E   **B**   The number of jobs in a print queue of a mainframe computer may be modeled as random. Here the sample space can be taken as

$$\Omega = \{0, 1, 2, 3, \ldots\}$$

that is, all the nonnegative integers. In practice, there is probably an upper limit, N, on how large the print queue can be, so instead the sample space might be defined as

$$\Omega = \{0, 1, 2, \ldots, N\}$$

∎

E X A M P L E   **C**   Earthquakes exhibit very erratic behavior, which is sometimes modeled as random. For example, the length of time between successive earthquakes in a particular region that are greater in magnitude than a given threshold may be regarded as an experiment. Here Ω is the set of all nonnegative real numbers:

$$\Omega = \{t \mid t \geq 0\}$$

∎

We are often interested in particular subsets of Ω, which in probability language are called **events.** In Example A, the event that the commuter stops at the first light is the subset of Ω denoted by

$$A = \{sss, ssc, scc, scs\}$$

(Events, or subsets, are usually denoted by italic uppercase letters.) In Example B, the event that there are fewer than five jobs in the print queue can be denoted by

$$A = \{0, 1, 2, 3, 4\}$$

The algebra of set theory carries over directly into probability theory. The **union** of two events, A and B, is the event C that either A occurs or B occurs or both occur: $C = A \cup B$. For example, if A is the event that the commuter stops at the first light (listed before), and if B is the event that she stops at the third light,

$$B = \{sss, scs, ccs, css\}$$

then C is the event that she stops at the first light or stops at the third light and consists of the outcomes that are in A or in B or in both:

$$C = \{sss, ssc, scc, scs, ccs, css\}$$

The **intersection** of two events, $C = A \cap B$, is the event that both A and B occur. If A and B are as given previously, then C is the event that the commuter stops at the first light and stops at the third light and thus consists of those outcomes that are common to both A and B:

$$C = \{sss, scs\}$$

The **complement** of an event, A^c, is the event that A does not occur and thus consists of all those elements in the sample space that are not in A. The complement of the event that the commuter stops at the first light is the event that she continues at the first light:

$$A^c = \{ccc, ccs, css, csc\}$$

You may recall from previous exposure to set theory a rather mysterious set called the empty set, usually denoted by $\emptyset$. The **empty set** is the set with no elements; it is the event with no outcomes. For example, if A is the event that the commuter stops at the first light and C is the event that she continues through all three lights, $C = \{ccc\}$, then A and C have no outcomes in common, and we can write

$$A \cap C = \emptyset$$

In such cases, A and C are said to be **disjoint.**

Venn diagrams, such as those in Figure 1.1, are often a useful tool for visualizing set operations.

The following are some laws of set theory.

Commutative Laws:

$$A \cup B = B \cup A$$
$$A \cap B = B \cap A$$

Associative Laws:

$$(A \cup B) \cup C = A \cup (B \cup C)$$
$$(A \cap B) \cap C = A \cap (B \cap C)$$

Distributive Laws:

$$(A \cup B) \cap C = (A \cap C) \cup (B \cap C)$$
$$(A \cap B) \cup C = (A \cup C) \cap (B \cup C)$$

Of these, the distributive laws are the least intuitive, and you may find it instructive to illustrate them with Venn diagrams.

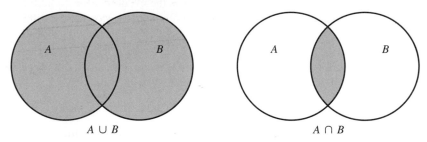

$A \cup B$ $A \cap B$

FIGURE **1.1**   Venn diagrams of $A \cup B$ and $A \cap B$.

1.3 Probability Measures

A **probability measure** on Ω is a function P from subsets of Ω to the real numbers that satisfies the following axioms:

1. $P(\Omega) = 1$.
2. If $A \subset \Omega$, then $P(A) \geq 0$.
3. If A_1 and A_2 are disjoint, then

$$P(A_1 \cup A_2) = P(A_1) + P(A_2).$$

More generally, if $A_1, A_2, \ldots, A_n, \ldots$ are mutually disjoint, then

$$P\left(\bigcup_{i=1}^{\infty} A_i\right) = \sum_{i=1}^{\infty} P(A_i)$$

The first two axioms are obviously desirable. Since Ω consists of all possible outcomes, $P(\Omega) = 1$. The second axiom simply states that a probability is nonnegative. The third axiom states that if A and B are disjoint—that is, have no outcomes in common—then $P(A \cup B) = P(A) + P(B)$ and also that this property extends to limits. For example, the probability that the print queue contains either one or three jobs is equal to the probability that it contains one plus the probability that it contains three.

The following properties of probability measures are consequences of the axioms.

Property A $P(A^c) = 1 - P(A)$. This property follows since A and A^c are disjoint with $A \cup A^c = \Omega$ and thus, by the first and third axioms, $P(A) + P(A^c) = 1$. In words, this property says that the probability that an event does not occur equals one minus the probability that it does occur.

Property B $P(\emptyset) = 0$. This property follows from Property A since $\emptyset = \Omega^c$. In words, this says that the probability that there is no outcome at all is zero.

Property C   If $A \subset B$, then $P(A) \leq P(B)$. This property states that if B occurs whenever A occurs, then $P(A) \leq P(B)$. For example, if whenever it rains (A) it is cloudy (B), then the probability that it rains is less than or equal to the probability that it is cloudy. Formally, it can be proved as follows: B can be expressed as the union of two disjoint sets:

$$B = A \cup (B \cap A^c)$$

Then, from the third axiom,

$$P(B) = P(A) + P(B \cap A^c)$$

and thus

$$P(A) = P(B) - P(B \cap A^c) \leq P(B)$$

Property D   Addition Law $P(A \cup B) = P(A) + P(B) - P(A \cap B)$. This property is easy to see from the Venn diagram in Figure 1.2. If $P(A)$ and $P(B)$ are added together, $P(A \cap B)$ is counted twice. To prove it, we decompose $A \cup B$ into three disjoint subsets, as shown in Figure 1.2:

$$C = A \cap B^c$$
$$D = A \cap B$$
$$E = A^c \cap B$$

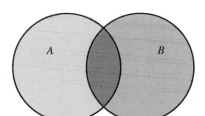
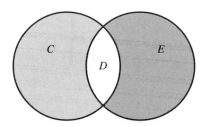

FIGURE **1.2**   Venn diagram illustrating the addition law.

We then have, from the third axiom,

$$P(A \cup B) = P(C) + P(D) + P(E)$$

Also, $A = C \cup D$, and C and D are disjoint; so $P(A) = P(C) + P(D)$. Similarly, $P(B) = P(D) + P(E)$. Putting these results together, we see that

$$P(A) + P(B) = P(C) + P(E) + 2P(D)$$
$$= P(A \cup B) + P(D)$$

or

$$P(A \cup B) = P(A) + P(B) - P(D)$$

E X A M P L E **A**   Suppose that a fair coin is thrown twice. Let A denote the event of heads on the first toss, and let B denote the event of heads on the second toss. The sample space is

$$\Omega = \{hh, ht, th, tt\}$$

We assume that each elementary outcome in Ω is equally likely and has probability $\frac{1}{4}$. $C = A \cup B$ is the event that heads comes up on the first toss or on the second toss. Clearly, $P(C) \neq P(A) + P(B) = 1$. Rather, since $A \cap B$ is the event that heads comes up on the first toss and on the second toss,

$$P(C) = P(A) + P(B) - P(A \cap B) = .5 + .5 - .25 = .75 \qquad \blacksquare$$

E X A M P L E **B**   An article in the *Los Angeles Times* (August 24, 1987) discussed the statistical risks of AIDS infection:

> Several studies of sexual partners of people infected with the virus show that a single act of unprotected vaginal intercourse has a surprisingly low risk of infecting the uninfected partner—perhaps one in 100 to one in 1000. For an average, consider the risk to be one in 500. If there are 100 acts of intercourse with an infected partner, the odds of infection increase to one in five.
>
> Statistically, 500 acts of intercourse with one infected partner or 100 acts with five partners lead to a 100% probability of infection (statistically, not necessarily in reality).

Following this reasoning, 1000 acts of intercourse with one infected partner would lead to a probability of infection equal to 2 (statistically, but not necessarily in reality). To see the flaw in the reasoning that leads to this conclusion, consider two acts of intercourse. Let A_1 denote the event that infection occurs on the first act and let A_2 denote the event that infection occurs on the second act. Then the event that infection occurs is $B = A_1 \cup A_2$ and

$$P(B) = P(A_1) + P(A_2) - P(A_1 \cap A_2) \leq P(A_1) + P(A_2) = \frac{2}{500} \qquad \blacksquare$$

1.4 Computing Probabilities: Counting Methods

Probabilities are especially easy to compute for finite sample spaces. Suppose that $\Omega = \{\omega_1, \omega_2, \ldots, \omega_N\}$ and that $P(\omega_i) = p_i$. To find the probability of an event A, we simply add the probabilities of the ω_i that constitute A.

E X A M P L E **A**   Suppose that a fair coin is thrown twice and the sequence of heads and tails is recorded. The sample space is

$$\Omega = \{hh, ht, th, tt\}$$

As in Example A of the previous section, we assume that each outcome in Ω has probability .25. Let A denote the event that at least one head is thrown. Then $A = \{hh, ht, th\}$, and $P(A) = .75$. ∎

This is a simple example of a fairly common situation. The elements of Ω all have equal probability; so if there are N elements in Ω, each of them has probability $1/N$. If A can occur in any of n mutually exclusive ways, then $P(A) = n/N$, or

$$P(A) = \frac{\text{number of ways } A \text{ can occur}}{\text{total number of outcomes}}$$

Note that this formula holds only if all the outcomes are equally likely. In Example A, if only the number of heads were recorded, then Ω would be $\{0, 1, 2\}$. These outcomes are not equally likely, and $P(A)$ is not $\frac{2}{3}$. ∎

EXAMPLE B   *Simpson's Paradox*

A black urn contains 5 red and 6 green balls, and a white urn contains 3 red and 4 green balls. You are allowed to choose an urn and then choose a ball at random from the urn. If you choose a red ball, you get a prize. Which urn should you choose to draw from? If you draw from the black urn, the probability of choosing a red ball is $\frac{5}{11} = .455$ (the number of ways you can draw a red ball divided by the total number of outcomes). If you choose to draw from the white urn, the probability of choosing a red ball is $\frac{3}{7} = .429$, so you should choose to draw from the black urn.

Now consider another game in which a second black urn has 6 red and 3 green balls, and a second white urn has 9 red and 5 green balls. If you draw from the black urn, the probability of a red ball is $\frac{6}{9} = .667$, whereas if you choose to draw from the white urn, the probability is $\frac{9}{14} = .643$. So, again you should choose to draw from the black urn.

In the final game, the contents of the second black urn are added to the first black urn, and the contents of the second white urn are added to the first white urn. Again, you can choose which urn to draw from. Which should you choose? Intuition says choose the black urn, but let's calculate the probabilities. The black urn now contains 11 red and 9 green balls, so the probability of drawing a red ball from it is $\frac{11}{20} = .55$. The white urn now contains 12 red and 9 green balls, so the probability of drawing a red ball from it is $\frac{12}{21} = .571$. So, you should choose the white urn. This counterintuitive result is an example of *Simpson's paradox*. For an example that occurred in real life, see Section 11.4.7. For more amusing examples, see Gardner (1976). ∎

In the preceding examples, it was easy to count the number of outcomes and calculate probabilities. To compute probabilities for more complex situations, we must develop systematic ways of counting outcomes, which are the subject of the next two sections.

1.4.1 The Multiplication Principle

The following is a statement of the very useful multiplication principle.

MULTIPLICATION PRINCIPLE

If one experiment has m outcomes and another experiment has n outcomes, then there are mn possible outcomes for the two experiments.

Proof

Denote the outcomes of the first experiment by $a_1, \ldots, a_m$ and the outcomes of the second experiment by $b_1, \ldots, b_n$. The outcomes for the two experiments are the ordered pairs (a_i, b_j). These pairs can be exhibited as the entries of an $m \times n$ rectangular array, in which the pair (a_i, b_j) is in the ith row and the jth column. There are mn entries in this array. ∎

EXAMPLE A   Playing cards have 13 face values and 4 suits. There are thus $4 \times 13 = 52$ face-value/suit combinations. ∎

EXAMPLE B   A class has 12 boys and 18 girls. The teacher selects 1 boy and 1 girl to act as representatives to the student government. She can do this in any of $12 \times 18 = 216$ different ways. ∎

EXTENDED MULTIPLICATION PRINCIPLE

If there are p experiments and the first has n_1 possible outcomes, the second $n_2, \ldots$, and the pth n_p possible outcomes, then there are a total of $n_1 \times n_2 \times \cdots \times n_p$ possible outcomes for the p experiments.

Proof

This principle can be proved from the multiplication principle by induction. We saw that it is true for $p = 2$. Assume that it is true for $p = q$—that is, that there are $n_1 \times n_2 \times \cdots \times n_q$ possible outcomes for the first q experiments. To complete the proof by induction, we must show that it follows that the property holds for $p = q + 1$. We apply the multiplication principle, regarding the first q experiments as a single experiment with $n_1 \times \cdots \times n_q$ outcomes, and conclude that there are $(n_1 \times \cdots \times n_q) \times n_{q+1}$ outcomes for the $q + 1$ experiments. ∎

EXAMPLE C   An 8-bit binary word is a sequence of 8 digits, of which each may be either a 0 or a 1. How many different 8-bit words are there?

There are two choices for the first bit, two for the second, etc., and thus there are

$$2 \times 2 \times 2 \times 2 \times 2 \times 2 \times 2 \times 2 = 2^8 = 256$$

such words. ∎

E X A M P L E D A DNA molecule is a sequence of four types of nucleotides, denoted by A, G, C, and T. The molecule can be millions of units long and can thus encode an enormous amount of information. For example, for a molecule 1 million (10^6) units long, there are 4^{10^6} different possible sequences. This is a staggeringly large number having nearly a million digits. An amino acid is coded for by a sequence of three nucleotides; there are $4^3 = 64$ different codes, but there are only 20 amino acids since some of them can be coded for in several ways. A protein molecule is composed of as many as hundreds of amino acid units, and thus there are an incredibly large number of possible proteins. For example, there are 20^{100} different sequences of 100 amino acids. ∎

1.4.2 Permutations and Combinations

A **permutation** is an ordered arrangement of objects. Suppose that from the set $C = \{c_1, c_2, \ldots, c_n\}$ we choose r elements and list them in order. How many ways can we do this? The answer depends on whether we are allowed to duplicate items in the list. If no duplication is allowed, we are **sampling without replacement.** If duplication is allowed, we are **sampling with replacement.** We can think of the problem as that of taking labeled balls from an urn. In the first type of sampling, we are not allowed to put a ball back before choosing the next one, but in the second, we are. In either case, when we are done choosing, we have a list of r balls ordered in the sequence in which they were drawn.

The extended multiplication principle can be used to count the number of different ordered samples possible for a set of n elements. First, suppose that sampling is done with replacement. The first ball can be chosen in any of n ways, the second in any of n ways, etc., so that there are $n \times n \times \cdots \times n = n^r$ samples. Next, suppose that sampling is done without replacement. There are n choices for the first ball, $n - 1$ choices for the second ball, $n - 2$ for the third, $\ldots$, and $n - r + 1$ for the rth. We have just proved the following proposition.

Permutation
=Ordered
Arrangement of
object

PROPOSITION A

For a set of size n and a sample of size r, there are n^r different ordered samples with replacement and $n(n - 1)(n - 2) \cdots (n - r + 1)$ different ordered samples without replacement. ∎

COROLLARY A

The number of orderings of n elements is $n(n - 1)(n - 2) \cdots 1 = n!$. ∎

E X A M P L E **A**   How many ways can five children be lined up?

This corresponds to sampling without replacement. According to Corollary A, there are $5! = 5 \times 4 \times 3 \times 2 \times 1 = 120$ different lines. ∎

E X A M P L E **B**   Suppose that from ten children, five are to be chosen and lined up. How many different lines are possible?

From Proposition A, there are $10 \times 9 \times 8 \times 7 \times 6 = 30{,}240$ different lines. ∎

E X A M P L E **C**   In some states, license plates have six characters: three letters followed by three numbers. How many distinct such plates are possible?

This corresponds to sampling with replacement. There are $26^3 = 17{,}576$ different ways to choose the letters and $10^3 = 1000$ ways to choose the numbers. Using the multiplication principle again, we find there are $17{,}576 \times 1000 = 17{,}576{,}000$ different plates. ∎

E X A M P L E **D**   If all sequences of six characters are equally likely, what is the probability that the license plate for a new car will contain no duplicate letters or numbers?

Call the desired event A; Ω consists of all 17,576,000 possible sequences. Since these are all equally likely, the probability of A is the ratio of the number of ways that A can occur to the total number of possible outcomes. There are 26 choices for the first letter, 25 for the second, 24 for the third, and hence $26 \times 25 \times 24 = 15{,}600$ ways to choose the letters without duplication (doing so corresponds to sampling without replacement), and $10 \times 9 \times 8 = 720$ ways to choose the numbers without duplication. From the multiplication principle, there are $15{,}600 \times 720 = 11{,}232{,}000$ nonrepeating sequences. The probability of A is thus

$$P(A) = \frac{11{,}232{,}000}{17{,}576{,}000} = .64$$ ∎

E X A M P L E **E**   *Birthday Problem*

Suppose that a room contains n people. What is the probability that at least two of them have a common birthday?

This is a famous problem with a counterintuitive answer. Assume that every day of the year is equally likely to be a birthday, disregard leap years, and denote by A the event that at least two people have a common birthday. As is sometimes the case, finding $P(A^c)$ is easier than finding $P(A)$. This is because A can happen in many ways, whereas A^c is much simpler. There are 365^n possible outcomes, and A^c can happen in $365 \times 364 \times \cdots \times (365 - n + 1)$ ways. Thus,

$$P(A^c) = \frac{365 \times 364 \times \cdots \times (365 - n + 1)}{365^n}$$

and

$$P(A) = 1 - \frac{365 \times 364 \times \cdots \times (365 - n + 1)}{365^n}$$

The following table exhibits the latter probabilities for various values of n:

n	$P(A)$
4	.016
16	.284
23	.507
32	.753
40	.891
56	.988

From the table, we see that if there are only 23 people, the probability of at least one match exceeds .5. The probabilities in the table are larger than one might intuitively guess, showing that the coincidence is not unlikely. Try it in your class. ∎

E X A M P L E **F**   How many people must you ask to have a 50 : 50 chance of finding someone who shares your birthday?

Suppose that you ask n people; let A denote the event that someone's birthday is the same as yours. Again, working with A^c is easier. The total number of outcomes is 365^n, and the total number of ways that A^c can happen is 364^n. Thus,

$$P(A^c) = \frac{364^n}{365^n}$$

and

$$P(A) = 1 - \frac{364^n}{365^n}$$

For the latter probability to be .5, n should be 253, which may seem counterintuitive. ∎

We now shift our attention from counting permutations to counting combinations. Here we are no longer interested in ordered samples, but in the constituents of the samples regardless of the order in which they were obtained. In particular, we ask the following question: If r objects are taken from a set of n objects without replacement and disregarding order, how many different samples are possible? From the multiplication principle, the number of ordered samples equals the number of unordered samples multiplied by the number of ways to order each sample. Since the number of ordered samples is $n(n - 1) \cdots (n - r + 1)$, and since a sample of size r can be ordered in $r!$ ways (Corollary A), the number of unordered samples is

$$\frac{n(n - 1) \cdots (n - r + 1)}{r!} = \frac{n!}{(n - r)! r!}$$

This number is also denoted as $\binom{n}{r}$. We have proved the following proposition.

PROPOSITION **B**

The number of unordered samples of r objects selected from n objects without replacement is $\binom{n}{r}$.

The numbers $\binom{n}{k}$, called the **binomial coefficients,** occur in the expansion

$$(a + b)^n = \sum_{k=0}^{n} \binom{n}{k} a^k b^{n-k}$$

In particular,

$$2^n = \sum_{k=0}^{n} \binom{n}{k}$$

This latter result can be interpreted as the number of subsets of a set of n objects. We just add the number of subsets of size 0 (with the usual convention that $0! = 1$), and the number of subsets of size 1, and the number of subsets of size 2, etc. ∎

E X A M P L E **G**   Up until 1991, a player of the California state lottery could win the jackpot prize by choosing the 6 numbers from 1 to 49 that were subsequently chosen at random by the lottery officials. There are $\binom{49}{6} = 13{,}983{,}816$ possible ways to choose 6 numbers from 49, and so the probability of winning was about 1 in 14 million. If there were no winners, the funds thus accumulated were rolled over (carried over) into the next round of play, producing a bigger jackpot. In 1991, the rules were changed so that a winner had to correctly select 6 numbers from 1 to 53. Since $\binom{53}{6} = 22{,}957{,}480$, the probability of winning decreased to about 1 in 23 million. Because of the ensuing rollover, the jackpot accumulated to a record of about \$120 million. This produced a fever of play—people were buying tickets at the rate of between 1 and 2 million per hour and state revenues burgeoned. ∎

E X A M P L E **H**   In the practice of quality control, only a fraction of the output of a manufacturing process is sampled and examined, since it may be too time-consuming and expensive to examine each item, or because sometimes the testing is destructive. Suppose that n items are in a lot and a sample of size r is taken. There are $\binom{n}{r}$ such samples. Now suppose that the lot contains k defective items. What is the probability that the sample contains exactly m defectives?

Clearly, this question is relevant to the efficacy of the sampling scheme, and the most desirable sample size can be determined by computing such probabilities for various values of r. Call the event in question A. The probability of A is the number of ways A can occur divided by the total number of outcomes. To find the number of ways A can occur, we use the multiplication principle. There are $\binom{k}{m}$ ways to choose the m defective items in the sample from the k defectives in the lot, and there are $\binom{n-k}{r-m}$ ways to choose the $r - m$ nondefective items in the sample from the $n - k$ nondefectives in the lot. Therefore, A can occur in $\binom{k}{m}\binom{n-k}{r-m}$ ways. Thus, $P(A)$ is the ratio of the number of ways A can occur to the total number of outcomes, or

$$P(A) = \frac{\binom{k}{m}\binom{n-k}{r-m}}{\binom{n}{r}}$$ ∎

EXAMPLE I *Capture/Recapture Method*
The so-called capture/recapture method is sometimes used to estimate the size of a wildlife population. Suppose that 10 animals are captured, tagged, and released. On a later occasion, 20 animals are captured, and it is found that 4 of them are tagged. How large is the population?

We assume that there are n animals in the population, of which 10 are tagged. If the 20 animals captured later are taken in such a way that all $\binom{n}{20}$ possible groups are equally likely (this is a big assumption), then the probability that 4 of them are tagged is (using the technique of the previous example)

$$\frac{\binom{10}{4}\binom{n-10}{16}}{\binom{n}{20}}$$

Clearly, n cannot be precisely determined from the information at hand, but it can be estimated. One method of estimation, called **maximum likelihood,** is to choose that value of n that makes the observed outcome most probable. (The method of maximum likelihood is one of the main subjects of a later chapter in this text.) The probability of the observed outcome as a function of n is called the **likelihood.** Figure 1.3 shows the likelihood as a function of n; the likelihood is maximized at $n = 50$.

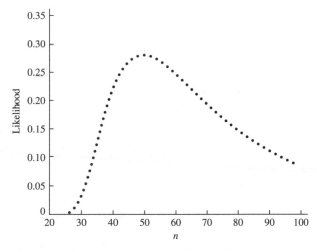

FIGURE **1.3** Likelihood for Example I.

To find the maximum likelihood estimate, suppose that, in general, t animals are tagged. Then, of a second sample of size m, r tagged animals are recaptured. We estimate n by the maximizer of the likelihood

$$L_n = \frac{\binom{t}{r}\binom{n-t}{m-r}}{\binom{n}{m}}$$

To find the value of n that maximizes L_n, consider the ratio of successive terms, which after some algebra is found to be

$$\frac{L_n}{L_{n-1}} = \frac{(n-t)(n-m)}{n(n-t-m+r)}$$

This ratio is greater than 1, i.e., L_n is increasing, if

$$(n-t)(n-m) > n(n-t-m+r)$$
$$n^2 - nm - nt + mt > n^2 - nt - nm + nr$$
$$mt > nr$$
$$\frac{mt}{r} > n$$

Thus, L_n increases for $n < mt/r$ and decreases for $n > mt/r$; so the value of n that maximizes L_n is the greatest integer not exceeding mt/r.

Applying this result to the data given previously, we see that the maximum likelihood estimate of n is $\frac{mt}{r} = \frac{20 \cdot 10}{4} = 50$. This estimate has some intuitive appeal, as it equates the proportion of tagged animals in the second sample to the proportion in the population:

$$\frac{4}{20} = \frac{10}{n} \qquad\qquad ■$$

Proposition B has the following extension.

PROPOSITION C

The number of ways that n objects can be grouped into r classes with n_i in the ith class, $i = 1, \ldots, r$, and $\sum_{i=1}^{r} n_i = n$ is

$$\binom{n}{n_1 n_2 \cdots n_r} = \frac{n!}{n_1! n_2! \cdots n_r!}$$

Proof

This can be seen by using Proposition B and the multiplication principle. (Note that Proposition B is the special case for which $r = 2$.) There are $\binom{n}{n_1}$ ways to choose the objects for the first class. Having done that, there are $\binom{n-n_1}{n_2}$ ways of choosing the objects for the second class. Continuing in this manner, there are

$$\frac{n!}{n_1!(n-n_1)!} \frac{(n-n_1)!}{(n-n_1-n_2)!n_2!} \cdots \frac{(n-n_1-n_2-\cdots-n_{r-1})!}{0!n_r!}$$

choices in all. After cancellation, this yields the desired result.   ■

EXAMPLE **J** A committee of seven members is to be divided into three subcommittees of size three, two, and two. This can be done in

$$\binom{7}{3\,2\,2} = \frac{7!}{3!2!2!} = 210$$

ways. ∎

EXAMPLE **K** In how many ways can the set of nucleotides $\{A, A, G, G, G, G, C, C, C\}$ be arranged in a sequence of nine letters? Proposition C can be applied by realizing that this problem can be cast as determining the number of ways that the nine positions in the sequence can be divided into subgroups of sizes two, four, and three (the locations of the letters A, G, and C):

$$\binom{9}{2\,4\,3} = \frac{9!}{2!4!3!} = 1260$$

∎

EXAMPLE **L** In how many ways can $n = 2m$ people be paired and assigned to m courts for the first round of a tennis tournament?

In this problem, $n_i = 2, i = 1, \ldots, m$, and, according to Proposition C, there are

$$\frac{(2m)!}{2^m}$$

assignments.

One has to be careful with problems such as this one. Suppose we were asked how many ways $2m$ people could be arranged in pairs without assigning the pairs to courts. Since there are $m!$ ways to assign the m pairs to m courts, the preceding result should be divided by $m!$, giving

$$\frac{(2m)!}{m!2^m}$$

pairs in all. ∎

The numbers $\binom{n}{n_1 n_2 \cdots n_r}$ are called **multinomial coefficients.** They occur in the expansion

$$(x_1 + x_2 + \cdots + x_r)^n = \sum \binom{n}{n_1 n_2 \cdots n_r} x_1^{n_1} x_2^{n_2} \cdots x_r^{n_r}$$

where the sum is over all nonnegative integers $n_1, n_2, \ldots, n_r$ such that $n_1 + n_2 + \cdots + n_r = n$.

1.5 Conditional Probability

We introduce the definition and use of conditional probability with an example. Digitalis therapy is often beneficial to patients who have suffered congestive heart failure, but there is the risk of digitalis intoxication, a serious side effect that is difficult to diagnose. To improve the chances of a correct diagnosis, the concentration of digitalis in the blood can be measured. Bellar et al. (1971) conducted a study of the relation of the concentration of digitalis in the blood to digitalis intoxication in 135 patients. Their results are simplified slightly in the following table, where this notation is used:

$$T+ = \text{high blood concentration (positive test)}$$
$$T- = \text{low blood concentration (negative test)}$$
$$D+ = \text{toxicity (disease present)}$$
$$D- = \text{no toxicity (disease absent)}$$

	$D+$	$D-$	Total
$T+$	25	14	39
$T-$	18	78	96
Total	43	92	135

Thus, for example, 25 of the 135 patients had a high blood concentration of digitalis and suffered toxicity.

Assume that the relative frequencies in the study roughly hold in some larger population of patients. (Making inferences about the frequencies in a large population from those observed in a small sample is a statistical problem, which will be taken up in a later chapter of this book.) Converting the frequencies in the preceding table to proportions (relative to 135), which we will regard as probabilities, we obtain the following table:

	$D+$	$D-$	Total
$T+$	.185	.104	.289
$T-$	.133	.578	.711
Total	.318	.682	1.000

From the table, $P(T+) = .289$ and $P(D+) = .318$, for example. Now if a doctor knows that the test was positive (that there was a high blood concentration), what is the probability of disease (toxicity) given this knowledge? We can restrict our attention to the first row of the table, and we see that of the 39 patients who had positive tests, 25 suffered from toxicity. We denote the probability that a patient shows toxicity given that the test is positive by $P(D+ \mid T+)$, which is called the **conditional probability** of $D+$ given $T+$.

$$P(D+ \mid T+) = \frac{25}{39} = .640$$

Equivalently, we can calculate this probability as

$$P(D+ \mid T+) = \frac{P(D+ \cap T+)}{P(T+)}$$

$$= \frac{.185}{.289} = .640$$

In summary, we see that the unconditional probability of $D+$ is .318, whereas the conditional probability $D+$ given $T+$ is .640. Therefore, knowing that the test is positive makes toxicity more than twice as likely. What if the test is negative?

$$P(D- \mid T-) = \frac{.578}{.711} = .848$$

For comparison, $P(D-) = .682$. Two other conditional probabilities from this example are of interest: The probability of a false positive is $P(D- \mid T+) = .360$, and the probability of a false negative is $P(D+ \mid T-) = .187$.

In general, we have the following definition.

DEFINITION

Let A and B be two events with $P(B) \neq 0$. The conditional probability of A given B is defined to be

$$P(A \mid B) = \frac{P(A \cap B)}{P(B)} \qquad P(A \mid B) = \frac{P(A \cap B)}{P(B)} \qquad ∎$$

The idea behind this definition is that if we are given that event B occurred, the relevant sample space becomes B rather than Ω, and conditional probability is a probability measure on B. In the digitalis example, to find $P(D+ \mid T+)$, we restricted our attention to the 39 patients who had positive tests. For this new measure to be a probability measure, it must satisfy the axioms, and this can be shown.

In some situations, $P(A \mid B)$ and $P(B)$ can be found rather easily, and we can then find $P(A \cap B)$.

MULTIPLICATION LAW

Let A and B be events and assume $P(B) \neq 0$. Then

$$P(A \cap B) = P(A \mid B)P(B) \qquad ∎$$

$P(A \cap B) = P(A \mid B)P(B)$

The multiplication law is often useful in finding the probabilities of intersections, as the following examples illustrate.

EXAMPLE A   An urn contains three red balls and one blue ball. Two balls are selected without replacement. What is the probability that they are both red?

Let R_1 and R_2 denote the events that a red ball is drawn on the first trial and on the second trial, respectively. From the multiplication law,

$$P(R_1 \cap R_2) = P(R_1)P(R_2 \mid R_1)$$

$P(R_1)$ is clearly $\frac{3}{4}$, and if a red ball has been removed on the first trial, there are two red balls and one blue ball left. Therefore, $P(R_2 \mid R_1) = \frac{2}{3}$. Thus, $P(R_1 \cap R_2) = \frac{1}{2}$.
∎

EXAMPLE **B**   Suppose that if it is cloudy (B), the probability that it is raining (A) is .3, and that the probability that it is cloudy is $P(B) = .2$ The probability that it is cloudy and raining is

$$P(A \cap B) = P(A \mid B)P(B) = .3 \times .2 = .06$$
∎

Another useful tool for computing probabilities is provided by the following law.

LAW OF TOTAL PROBABILITY

Let $B_1, B_2, \ldots, B_n$ be such that $\bigcup_{i=1}^{n} B_i = \Omega$ and $B_i \cap B_j = \emptyset$ for $i \neq j$, with $P(B_i) > 0$ for all i. Then, for any event A,

$$P(A) = \sum_{i=1}^{n} P(A \mid B_i)P(B_i)$$

Proof

Before going through a formal proof, it is helpful to state the result in words. The B_i are mutually disjoint events whose union is Ω. To find the probability of an event A, we sum the conditional probabilities of A given B_i, weighted by $P(B_i)$. Now, for the proof, we first observe that

$$P(A) = P(A \cap \Omega)$$

$$= P\left(A \cap \left(\bigcup_{i=1}^{n} B_i\right)\right)$$

$$= P\left(\bigcup_{i=1}^{n}(A \cap B_i)\right)$$

Since the events $A \cap B_i$ are disjoint,

$$P\left(\bigcup_{i=1}^{n}(A \cap B_i)\right) = \sum_{i=1}^{n} P(A \cap B_i)$$

$$= \sum_{i=1}^{n} P(A \mid B_i)P(B_i) \qquad ∎$$

The law of total probability is useful in situations where it is not obvious how to calculate $P(A)$ directly but in which $P(A \mid B_i)$ and $P(B_i)$ are more straightforward, such as in the following example.

E X A M P L E **C**   Referring to Example A, what is the probability that a red ball is selected on the second draw?

The answer may or may not be intuitively obvious—that depends on your intuition. On the one hand, you could argue that it is "clear from symmetry" that $P(R_2) = P(R_1) = \frac{3}{4}$. On the other hand, you could say that it is obvious that a red ball is likely to be selected on the first draw, leaving fewer red balls for the second draw, so that $P(R_2) < P(R_1)$. The answer can be derived easily by using the law of total probability:

$$P(R_2) = P(R_2 \mid R_1)P(R_1) + P(R_2 \mid B_1)P(B_1)$$
$$= \frac{2}{3} \times \frac{3}{4} + 1 \times \frac{1}{4} = \frac{3}{4}$$

where B_1 denotes the event that a blue ball is drawn on the first trial.   ■

As another example of the use of conditional probability, we consider a model that has been used for occupational mobility.

E X A M P L E **D**   Suppose that occupations are grouped into upper (U), middle (M), and lower (L) levels. U_1 will denote the event that a father's occupation is upper-level; U_2 will denote the event that a son's occupation is upper-level, etc. (The subscripts index generations.) Glass and Hall (1954) compiled the following statistics on occupational mobility in England and Wales:

	U_2	M_2	L_2
U_1	.45	.48	.07
M_1	.05	.70	.25
L_1	.01	.50	.49

Such a table, which is called a *matrix of transition probabilities*, is to be read in the following way: If a father is in U, the probability that his son is in U is .45, the probability that his son is in M is .48, etc. The table thus gives conditional probabilities: for example, $P(U_2 \mid U_1) = .45$. Examination of the table reveals that there is more upward mobility from L into M than from M into U. Suppose that of the father's generation, 10% are in U, 40% in M, and 50% in L. What is the probability that a son in the next generation is in U?

Applying the law of total probability, we have

$$P(U_2) = P(U_2 \mid U_1)P(U_1) + P(U_2 \mid M_1)P(M_1) + P(U_2 \mid L_1)P(L_1)$$
$$= .45 \times .10 + .05 \times .40 + .01 \times .50 = .07$$

$P(M_2)$ and $P(L_2)$ can be worked out similarly.   ■

Continuing with Example D, suppose we ask a different question: If a son has occupational status U_2, what is the probability that his father had occupational status U_1? Compared to the question asked in Example D, this is an "inverse" problem; we are given an "effect" and are asked to find the probability of a particular "cause." In situations like this, Bayes' rule, which we state shortly, is useful. Before stating the rule, we will see what it amounts to in this particular case.

We wish to find $P(U_1 \mid U_2)$. By definition,

$$
P(U_1 \mid U_2) = \frac{P(U_1 \cap U_2)}{P(U_2)}
$$

$$
= \frac{P(U_2 \mid U_1)P(U_1)}{P(U_2 \mid U_1)P(U_1) + P(U_2 \mid M_1)P(M_1) + P(U_2 \mid L_1)P(L_1)}
$$

Here we used the multiplication law to reexpress the numerator and the law of total probability to restate the denominator. The value of the numerator is $P(U_2 \mid U_1)P(U_1) = .45 \times .10 = .045$, and we calculated the denominator in Example D to be .07, so we find that $P(U_1 \mid U_2) = .64$. In other words, 64% of the sons who are in upper-level occupations have fathers who were in upper-level occupations.

We now state Bayes' rule.

BAYES' RULE

Let A and $B_1, \ldots, B_n$ be events where the B_i are disjoint, $\bigcup_{i=1}^{n} B_i = \Omega$, and $P(B_i) > 0$ for all i. Then

$$
P(B_j \mid A) = \frac{P(A \mid B_j)P(B_j)}{\sum\limits_{i=1}^{n} P(A \mid B_i)P(B_i)}
$$

The proof of Bayes' rule follows exactly as in the preceding discussion. ∎

E X A M P L E **E** Diamond and Forrester (1979) applied Bayes' rule to the diagnosis of coronary artery disease. A procedure called cardiac fluoroscopy is used to determine whether there is calcification of coronary arteries and thereby to diagnose coronary artery disease. From the test, it can be determined if 0, 1, 2, or 3 coronary arteries are calcified. Let T_0, T_1, T_2, T_3 denote these events. Let $D+$ or $D-$ denote the event that disease is present or absent, respectively. Diamond and Forrester presented the following table, based on medical studies:

i	$P(T_i \mid D+)$	$P(T_i \mid D-)$
0	.42	.96
1	.24	.02
2	.20	.02
3	.15	.00

According to Bayes' rule,

$$P(D+ \mid T_i) = \frac{P(T_i \mid D+)P(D+)}{P(T_i \mid D+)P(D+) + P(T_i \mid D-)P(D-)}$$

Thus, if the initial probabilities $P(D+)$ and $P(D-)$ are known, the probability that a patient has coronary artery disease can be calculated.

Let us consider two specific cases. For the first, suppose that a male between the ages of 30 and 39 suffers from nonanginal chest pain. For such a patient, it is known from medical statistics that $P(D+) \approx .05$. Suppose that the test shows that no arteries are calcified. From the preceding equation,

$$P(D+ \mid T_0) = \frac{.42 \times .05}{.42 \times .05 + .96 \times .95} = .02$$

It is unlikely that the patient has coronary artery disease. On the other hand, suppose that the test shows that one artery is calcified. Then

$$P(D+ \mid T_1) = \frac{.24 \times .05}{.24 \times .05 + .02 \times .95} = .39$$

Now it is more likely that this patient has coronary artery disease, but by no means certain.

As a second case, suppose that the patient is a male between ages 50 and 59 who suffers typical angina. For such a patient, $P(D+) = .92$. For him, we find that

$$P(D+ \mid T_0) = \frac{.42 \times .92}{.42 \times .92 + .96 \times .08} = .83$$

$$P(D+ \mid T_1) = \frac{.24 \times .92}{.24 \times .92 + .02 \times .08} = .99$$

Comparing the two patients, we see the strong influence of the prior probability, $P(D+)$. ∎

E X A M P L E **F**   Polygraph tests (lie-detector tests) are often routinely administered to employees or prospective employees in sensitive positions. Let $+$ denote the event that the polygraph reading is positive, indicating that the subject is lying; let T denote the event that the subject is telling the truth; and let L denote the event that the subject is lying. According to studies of polygraph reliability (Gastwirth 1987),

$$P(+ \mid L) = .88$$

from which it follows that $P(- \mid L) = .12$ also

$$P(- \mid T) = .86$$

from which it follows that $P(+ \mid T) = .14$. In words, if a person is lying, the probability that this is detected by the polygraph is .88, whereas if he is telling the truth, the polygraph indicates that he is telling the truth with probability .86. Now suppose that polygraphs are routinely administered to screen employees for security reasons, and that on a particular question the vast majority of subjects have no reason to lie so that $P(T) = .99$, whereas $P(L) = .01$. A subject produces a positive response on the polygraph. What is the probability that the polygraph is incorrect and that she is in fact telling the truth? We can evaluate this probability with Bayes' rule:

$$
\begin{aligned}
P(T \mid +) &= \frac{P(+ \mid T)P(T)}{P(+ \mid T)P(T) + P(+ \mid L)P(L)} \\
&= \frac{(.14)(.99)}{(.14)(.99) + (.88)(.01)} \\
&= .94
\end{aligned}
$$

Thus, in screening this population of largely innocent people, 94% of the positive polygraph readings will be in error. Most of those placed under suspicion because of the polygraph result will, in fact, be innocent. This example illustrates some of the dangers in using screening procedures on large populations.   ∎

Bayes' rule is the fundamental mathematical ingredient of a subjective, or "Bayesian," approach to epistemology, theories of evidence, and theories of learning. According to this point of view, an individual's beliefs about the world can be coded in probabilities. For example, an individual's belief that it will hail tomorrow can be represented by a probability $P(H)$. This probability varies from individual to individual. In principle, each individual's probability can be ascertained, or elicited, by offering him or her a series of bets at different odds.

According to Bayesian theory, our beliefs are modified as we are confronted with evidence. If, initially, my probability for a hypothesis is $P(H)$, after seeing evidence E (e.g., a weather forecast), my probability becomes $P(H|E)$. $P(E|H)$ is often easier to evaluate than $P(H|E)$. In this case, the application of Bayes' rule gives

$$
P(H|E) = \frac{P(E|H)P(H)}{P(E|H)P(H) + P(E|\bar{H})P(\bar{H})}
$$

where $\bar{H}$ is the event that H does not hold. This point can be illustrated by the preceding polygraph example. Suppose an investigator is questioning a particular suspect and that the investigator's prior opinion that the suspect is telling the truth is $P(T)$. Then, upon observing a positive polygraph reading, his opinion becomes $P(T|+)$. Note that different investigators will have different prior probabilities $P(T)$ for different suspects, and thus different posterior probabilities.

As appealing as this formulation might be, a long line of research has demonstrated that humans are actually not very good at doing probability calculations in evaluating evidence. For example, Tversky and Kahneman (1974) presented subjects with the following question: "If Linda is a 31-year-old single woman who is outspoken on social issues such as disarmament and equal rights, which of the following statements is more likely to be true?

• Linda is bank teller.
• Linda is a bank teller and active in the feminist movement."

More than 80% of those questioned chose the second statement, despite Property C of Section 1.3.

Even highly trained professionals are not good at doing probability calculations, as illustrated by the following example of Eddy (1982), regarding interpreting the results from mammogram screening. One hundred physicians were presented with the following information:

- In the absence of any special information, the probability that a woman (of the age and health status of this patient) has breast cancer is 1%.
- If the patient has breast cancer, the probability that the radiologist will correctly diagnose it is 80%.
- If the patient has a benign lesion (no breast cancer), the probability that the radiologist will incorrectly diagnose it as cancer is 10%.

They were then asked, "What is the probability that a patient with a positive mammogram actually has breast cancer?"

Ninety-five of the 100 physicians estimated the probability to be about 75%. The correct probability, as given by Bayes' rule, is 7.5%. (You should check this.) So even experts radically overestimate the strength of the evidence provided by a positive outcome on the screening test.

Thus the Bayesian probability calculus does not describe the way people actually assimilate evidence. Advocates for Bayesian learning theory might assert that the theory describes the way people "should think." A softer point of view is that Bayesian learning theory is a model for learning, and it has the merit of being a simple model that can be programmed on computers. Probability theory in general, and Bayesian learning theory in particular, are part of the core of artificial intelligence.

1.6 Independence

Intuitively, we would say that two events, A and B, are independent if knowing that one had occurred gave us no information about whether the other had occurred; that is, $P(A \mid B) = P(A)$ and $P(B \mid A) = P(B)$. Now, if

$$P(A) = P(A \mid B) = \frac{P(A \cap B)}{P(B)}$$

then

$$P(A \cap B) = P(A)P(B)$$

We will use this last relation as the definition of independence. Note that it is symmetric in A and in B, and does not require the existence of a conditional probability, that is, $P(B)$ can be 0.

DEFINITION

A and B are said to be independent events if $P(A \cap B) = P(A)P(B)$. ■

E X A M P L E **A**   A card is selected randomly from a deck. Let A denote the event that it is an ace and D the event that it is a diamond. Knowing that the card is an ace gives no information about its suit. Checking formally that the events are independent, we have $P(A) = \frac{4}{52} = \frac{1}{13}$ and $P(D) = \frac{1}{4}$. Also, $A \cap D$ is the event that the card is the ace of diamonds and $P(A \cap D) = \frac{1}{52}$. Since $P(A)P(D) = (\frac{1}{4}) \times (\frac{1}{13}) = \frac{1}{52}$, the events are in fact independent.   ■

E X A M P L E   **B**   A system is designed so that it fails only if a unit and a backup unit both fail. Assuming that these failures are independent and that each unit fails with probability p, the system fails with probability p^2. If, for example, the probability that any unit fails during a given year is .1, then the probability that the system fails is .01, which represents a considerable improvement in reliability.   ■

Things become more complicated when we consider more than two events. For example, suppose we know that events A, B, and C are **pairwise independent** (any two are independent). We would like to be able to say that they are all independent based on the assumption that knowing something about two of the events does not tell us anything about the third, for example, $P(C \mid A \cap B) = P(C)$. But as the following example shows, pairwise independence does not guarantee mutual independence.

E X A M P L E   **C**   A fair coin is tossed twice. Let A denote the event of heads on the first toss, B the event of heads on the second toss, and C the event that exactly one head is thrown. A and B are clearly independent, and $P(A) = P(B) = P(C) = .5$. To see that A and C are independent, we observe that $P(C \mid A) = .5$. But

$$P(A \cap B \cap C) = 0 \neq P(A)P(B)P(C)$$ ■

To encompass situations such as that in Example C, we define a collection of events, $A_1, A_2, \ldots, A_n$, to be **mutually independent** if for any subcollection, $A_{i_1}, \ldots, A_{i_m}$,

$$P(A_{i_1} \cap \cdots \cap A_{i_m}) = P(A_{i_1}) \cdots P(A_{i_m})$$

E X A M P L E   **D**   We return to Example B of Section 1.3 (infectivity of AIDS). Suppose that virus transmissions in 500 acts of intercourse are mutually independent events and that the probability of transmission in any one act is 1/500. Under this model, what is the probability of infection? It is easier to first find the probability of the complement of this event. Let $C_1, C_2, \ldots, C_{500}$ denote the events that virus transmission does not occur during encounters $1, 2, \ldots, 500$. Then the probability of no infection is

$$P(C_1 \cap C_2 \cap \cdots \cap C_{500}) = \left(1 - \frac{1}{500}\right)^{500} = .37$$

so the probability of infection is $1 - .37 = .63$, not 1, which is the answer produced by incorrectly adding probabilities.   ■

EXAMPLE E   Consider a circuit with three relays (Figure 1.4). Let A_i denote the event that the ith relay works, and assume that $P(A_i) = p$ and that the relays are mutually independent. If F denotes the event that current flows through the circuit, then $F = A_3 \cup (A_1 \cap A_2)$ and, from the addition law and the assumption of independence,

$$P(F) = P(A_3) + P(A_1 \cap A_2) - P(A_1 \cap A_2 \cap A_3) = p + p^2 - p^3 \qquad \blacksquare$$

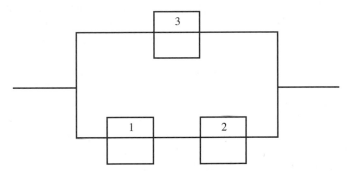

FIGURE **1.4**   Circuit with three relays.

EXAMPLE F   Suppose that a system consists of components connected in a series, so the system fails if any one component fails. If there are n mutually independent components and each fails with probability p, what is the probability that the system will fail?

It is easier to find the probability of the complement of this event; the system works if and only if all the components work, and this situation has probability $(1 - p)^n$. The probability that the system fails is then $1 - (1 - p)^n$. For example, if $n = 10$ and $p = .05$, the probability that the system works is only $.95^{10} = .60$, and the probability that the system fails is .40.

Suppose, instead, that the components are connected in parallel, so the system fails only when all components fail. In this case, the probability that the system fails is only $.05^{10} = 9.8 \times 10^{-14}$. $\blacksquare$

Calculations like those in Example F are made in reliability studies for systems consisting of quite complicated networks of components. The absolutely crucial assumption is that the components are independent of one another. Theoretical studies of the reliability of nuclear power plants have been criticized on the grounds that they incorrectly assume independence of the components.

EXAMPLE G   *Matching DNA Fragments*
Fragments of DNA are often compared for similarity, for example, across species. A simple way to make a comparison is to count the number of locations, or sites, at which these fragments agree. For example, consider these two sequences, which agree at three sites: fragment 1: AGATCAGT; and fragment 2: TGGATACT.

Many such comparisons are made, and to sort the wheat from the chaff, a probability model is often used. A comparison is deemed interesting if the number of matches is much larger than would be expected by chance alone. This requires a chance model; a simple one stipulates that the nucleotide at each site of fragment 1 occurs randomly with probabilities p_{A1}, p_{G1}, p_{C1}, p_{T1}, and that the second fragment is similarly composed with probabilities $p_{A2}, \ldots, p_{T2}$. What is the chance that the fragments match at a particular site if in fact the identity of the nucleotide on fragment 1 is independent of that on fragment 2? The match probability can be calculated using the law of total probability:

$$P(\text{match}) = P(\text{match}|A \text{ on fragment } 1)P(A \text{ on fragment } 1) +$$
$$\ldots + P(\text{match}|T \text{ on fragment } 1)P(T \text{ on fragment } 1)$$
$$= p_{A2}p_{A1} + p_{G2}p_{G1} + p_{C2}p_{C1} + p_{T2}p_{T1}$$

The problem of determining the probability that they match at k out of a total of n sites is discussed later.   ■

1.7 Concluding Remarks

This chapter provides a simple axiomatic development of the mathematical theory of probability. Some subtle issues that arise in a careful analysis of infinite sample spaces have been neglected. Such issues are typically addressed in graduate-level courses in measure theory and probability theory. Certain philosophical questions have also been avoided. One might ask what is meant by the statement "The probability that this coin will land heads up is $\frac{1}{2}$." Two commonly advocated views are the **frequentist approach** and the **Bayesian approach.** According to the frequentist approach, the statement means that if the experiment were repeated many times, the long-run average number of heads would tend to $\frac{1}{2}$. According to the Bayesian approach, the statement is a quantification of the speaker's uncertainty about the outcome of the experiment and thus is a personal or subjective notion; the probability that the coin will land heads up may be different for different speakers, depending on their experience and knowledge of the situation. There has been vigorous and occasionally acrimonious debate among proponents of various versions of these points of view.

In this and ensuing chapters, there are many examples of the use of probability as a model for various phenomena. In any such modeling endeavor, an idealized mathematical theory is hoped to provide an adequate match to characteristics of the phenomenon under study. The standard of adequacy is relative to the field of study and the modeler's goals.

1.8 Problems

1. A coin is tossed three times and the sequence of heads and tails is recorded.
   a. List the sample space.
   b. List the elements that make up the following events: (1) A = at least two heads, (2) B = the first two tosses are heads, (3) C = the last toss is a tail.
   c. List the elements of the following events: (1) A^c, (2) $A \cap B$, (3) $A \cup C$.

2. Two six-sided dice are thrown sequentially, and the face values that come up are recorded.

   a. List the sample space.

   b. List the elements that make up the following events: (1) A = the sum of the two values is at least 5, (2) B = the value of the first die is higher than the value of the second, (3) C = the first value is 4.

   c. List the elements of the following events: (1) $A \cap C$, (2) $B \cup C$, (3) $A \cap (B \cup C)$.

3. An urn contains three red balls, two green balls, and one white ball. Three balls are drawn without replacement from the urn, and the colors are noted in sequence. List the sample space. Define events A, B, and C as you wish and find their unions and intersections.

4. Draw Venn diagrams to illustrate De Morgan's laws:

$$(A \cup B)^c = A^c \cap B^c$$

$$(A \cap B)^c = A^c \cup B^c$$

5. Let A and B be arbitrary events. Let C be the event that either A occurs or B occurs, but not both. Express C in terms of A and B using any of the basic operations of union, intersection, and complement.

6. Verify the following extension of the addition rule (a) by an appropriate Venn diagram and (b) by a formal argument using the axioms of probability and the propositions in this chapter.

$$P(A \cup B \cup C) = P(A) + P(B) + P(C) - P(A \cap B)$$

$$- P(A \cap C) - P(B \cap C) + P(A \cap B \cap C)$$

7. Prove Bonferroni's inequality:

$$P(A \cap B) \geq P(A) + P(B) - 1$$

8. Prove that

$$P\left(\bigcup_{i=1}^{n} A_i\right) \leq \sum_{i=1}^{n} P(A_i)$$

9. The weather forecaster says that the probability of rain on Saturday is 25% and that the probability of rain on Sunday is 25%. Is the probability of rain during the weekend 50%? Why or why not?

10. Make up another example of Simpson's paradox by changing the numbers in Example B of Section 1.4.

11. The first three digits of a university telephone exchange are 452. If all the sequences of the remaining four digits are equally likely, what is the probability that a randomly selected university phone number contains seven distinct digits?

12. In a game of poker, five players are each dealt 5 cards from a 52-card deck. How many ways are there to deal the cards?

13. In a game of poker, what is the probability that a five-card hand will contain (a) a straight (five cards in unbroken numerical sequence), (b) four of a kind, and (c) a full house (three cards of one value and two cards of another value)?

14. The four players in a bridge game are each dealt 13 cards. How many ways are there to do this?

15. How many different meals can be made from four kinds of meat, six vegetables, and three starches if a meal consists of one selection from each group?

16. How many different letter arrangements can be obtained from the letters of the word *statistically*, using all the letters?

17. In acceptance sampling, a purchaser samples 4 items from a lot of 100 and rejects the lot if 1 or more are defective. Graph the probability that the lot is accepted as a function of the percentage of defective items in the lot.

18. A lot of n items contains k defectives, and m are selected randomly and inspected. How should the value of m be chosen so that the probability that at least one defective item turns up is .90? Apply your answer to (a) $n = 1000$, $k = 10$, and (b) $n = 10,000$, $k = 100$.

19. A committee consists of five Chicanos, two Asians, three African Americans, and two Caucasians.

   a. A subcommittee of four is chosen at random. What is the probability that all the ethnic groups are represented on the subcommittee?
   b. Answer the question for part (a) if a subcommittee of five is chosen.

20. A deck of 52 cards is shuffled thoroughly. What is the probability that the four aces are all next to each other?

21. A fair coin is tossed five times. What is the probability of getting a sequence of three heads?

22. A standard deck of 52 cards is shuffled thoroughly, and n cards are turned up. What is the probability that a face card turns up? For what value of n is this probability about .5?

23. How many ways are there to place n indistinguishable balls in n urns so that exactly one urn is empty?

24. If n balls are distributed randomly into k urns, what is the probability that the last urn contains j balls?

25. A woman getting dressed up for a night out is asked by her significant other to wear a red dress, high-heeled sneakers, and a wig. In how many orders can she put on these objects?

26. The game of Mastermind starts in the following way: One player selects four pegs, each peg having six possible colors, and places them in a line. The second player then tries to guess the sequence of colors. What is the probability of guessing correctly?

27. If a five-letter word is formed at random (meaning that all sequences of five letters are equally likely), what is the probability that no letter occurs more than once?

28. How many ways are there to encode the 26-letter English alphabet into 8-bit binary words (sequences of eight 0s and 1s)?

29. A poker player is dealt three spades and two hearts. He discards the two hearts and draws two more cards. What is the probability that he draws two more spades?

30. A group of 60 second graders is to be randomly assigned to two classes of 30 each. (The random assignment is ordered by the school district to ensure against any bias.) Five of the second graders, Marcelle, Sarah, Michelle, Katy, and Camerin, are close friends. What is the probability that they will all be in the same class? What is the probability that exactly four of them will be? What is the probability that Marcelle will be in one class and her friends in the other?

31. Six male and six female dancers perform the Virginia reel. This dance requires that they form a line consisting of six male/female pairs. How many such arrangements are there?

32. A wine taster claims that she can distinguish four vintages of a particular Cabernet. What is the probability that she can do this by merely guessing? (She is confronted with four unlabeled glasses.)

33. An elevator containing five people can stop at any of seven floors. What is the probability that no two people get off at the same floor? Assume that the occupants act independently and that all floors are equally likely for each occupant.

34. Prove the following identity:

$$\sum_{k=0}^{n} \binom{n}{k}\binom{m-n}{n-k} = \binom{m}{n}$$

(*Hint*: How can each of the summands be interpreted?)

35. Prove the following two identities both algebraically and by interpreting their meaning combinatorially.
   a. $\binom{n}{r} = \binom{n}{n-r}$
   b. $\binom{n}{r} = \binom{n-1}{r-1} + \binom{n-1}{r}$

36. What is the coefficient of $x^3 y^4$ in the expansion of $(x+y)^7$?

37. What is the coefficient of $x^2 y^2 z^3$ in the expansion of $(x+y+z)^7$?

38. A child has six blocks, three of which are red and three of which are green. How many patterns can she make by placing them all in a line? If she is given three white blocks, how many total patterns can she make by placing all nine blocks in a line?

39. A monkey at a typewriter types each of the 26 letters of the alphabet exactly once, the order being random.
   a. What is the probability that the word *Hamlet* appears somewhere in the string of letters?

b. How many independent monkey typists would you need in order that the probability that the word appears is at least .90?

40. In how many ways can two octopi shake hands? (There are a number of ways to interpret this question—choose one.)

41. A drawer of socks contains seven black socks, eight blue socks, and nine green socks. Two socks are chosen in the dark.

    a. What is the probability that they match?
    b. What is the probability that a black pair is chosen?

42. How many ways can 11 boys on a soccer team be grouped into 4 forwards, 3 midfielders, 3 defenders, and 1 goalie?

43. A software development company has three jobs to do. Two of the jobs require three programmers, and the other requires four. If the company employs ten programmers, how many different ways are there to assign them to the jobs?

44. In how many ways can 12 people be divided into three groups of 4 for an evening of bridge? In how many ways can this be done if the 12 consist of six pairs of partners?

45. Show that if the conditional probabilities exist, then

$$P(A_1 \cap A_2 \cap \cdots \cap A_n)$$
$$= P(A_1)P(A_2 \mid A_1)P(A_3 \mid A_1 \cap A_2) \cdots P(A_n \mid A_1 \cap A_2 \cap \cdots \cap A_{n-1})$$

46. Urn A has three red balls and two white balls, and urn B has two red balls and five white balls. A fair coin is tossed. If it lands heads up, a ball is drawn from urn A; otherwise, a ball is drawn from urn B.

    a. What is the probability that a red ball is drawn?
    b. If a red ball is drawn, what is the probability that the coin landed heads up?

47. Urn A has four red, three blue, and two green balls. Urn B has two red, three blue, and four green balls. A ball is drawn from urn A and put into urn B, and then a ball is drawn from urn B.

    a. What is the probability that a red ball is drawn from urn B?
    b. If a red ball is drawn from urn B, what is the probability that a red ball was drawn from urn A?

48. An urn contains three red and two white balls. A ball is drawn, and then it and another ball of the same color are placed back in the urn. Finally, a second ball is drawn.

    a. What is the probability that the second ball drawn is white?
    b. If the second ball drawn is white, what is the probability that the first ball drawn was red?

49. A fair coin is tossed three times.

    a. What is the probability of two or more heads given that there was at least one head?
    b. What is the probability given that there was at least one tail?

50. Two dice are rolled, and the sum of the face values is six. What is the probability that at least one of the dice came up a three?

51. Answer Problem 50 again given that the sum is less than six.

52. Suppose that 5 cards are dealt from a 52-card deck and the first one is a king. What is the probability of at least one more king?

53. A fire insurance company has high-risk, medium-risk, and low-risk clients, who have, respectively, probabilities .02, .01, and .0025 of filing claims within a given year. The proportions of the numbers of clients in the three categories are .10, .20, and .70, respectively. What proportion of the claims filed each year come from high-risk clients?

54. This problem introduces a simple meteorological model, more complicated versions of which have been proposed in the meteorological literature. Consider a sequence of days and let R_i denote the event that it rains on day i. Suppose that $P(R_i \mid R_{i-1}) = \alpha$ and $P(R_i^c \mid R_{i-1}^c) = \beta$. Suppose further that only today's weather is relevant to predicting tomorrow's; that is, $P(R_i \mid R_{i-1} \cap R_{i-2} \cap \cdots \cap R_0) = P(R_i \mid R_{i-1})$.

  a. If the probability of rain today is p, what is the probability of rain tomorrow?
  b. What is the probability of rain the day after tomorrow?
  c. What is the probability of rain n days from now? What happens as n approaches infinity?

55. This problem continues Example D of Section 1.5 and concerns occupational mobility.

  a. Find $P(M_1 \mid M_2)$ and $P(L_1 \mid L_2)$.
  b. Find the proportions that will be in the three occupational levels in the third generation. To do this, assume that a son's occupational status depends on his father's status, but that given his father's status, it does not depend on his grandfather's.

56. A couple has two children. What is the probability that both are girls given that the oldest is a girl? What is the probability that both are girls given that one of them is a girl?

57. There are three cabinets, A, B, and C, each of which has two drawers. Each drawer contains one coin; A has two gold coins, B has two silver coins, and C has one gold and one silver coin. A cabinet is chosen at random, one drawer is opened, and a silver coin is found. What is the probability that the other drawer in that cabinet contains a silver coin?

58. A teacher tells three boys, Drew, Chris, and Jason, that two of them will have to stay after school to help her clean erasers and that one of them will be able to leave. She further says that she has made the decision as to who will leave and who will stay at random by rolling a special three-sided Dungeons and Dragons die. Drew wants to leave to play soccer and has a clever idea about how to increase his chances of doing so. He figures that one of Jason and Chris will certainly stay and asks the teacher to tell him the name of one of the two who will stay. Drew's idea is that if, for example, Jason is named, then he and Chris are left and they each have a probability .5 of leaving; similarly, if Chris is named, Drew's probability of leaving is still .5. Thus, by merely asking the teacher a question, Drew will increase his probability of leaving from $\frac{1}{3}$ to $\frac{1}{2}$. What do you think of this scheme?

59. A box has three coins. One has two heads, one has two tails, and the other is a fair coin with one head and one tail. A coin is chosen at random, is flipped, and comes up heads.

    a. What is the probability that the coin chosen is the two-headed coin?

    b. What is the probability that if it is thrown another time it will come up heads?

    c. Answer part (a) again, supposing that the coin is thrown a second time and comes up heads again.

60. A factory runs three shifts. In a given day, 1% of the items produced by the first shift are defective, 2% of the second shift's items are defective, and 5% of the third shift's items are defective. If the shifts all have the same productivity, what percentage of the items produced in a day are defective? If an item is defective, what is the probability that it was produced by the third shift?

61. Suppose that chips for an integrated circuit are tested and that the probability that they are detected if they are defective is .95, and the probability that they are declared sound if in fact they are sound is .97. If .5% of the chips are faulty, what is the probability that a chip that is declared faulty is sound?

62. Show that if $P(A \mid E) \geq P(B \mid E)$ and $P(A \mid E^c) \geq P(B \mid E^c)$, then $P(A) \geq P(B)$.

63. Suppose that the probability of living to be older than 70 is .6 and the probability of living to be older than 80 is .2. If a person reaches her 70th birthday, what is the probability that she will celebrate her 80th?

64. If B is an event, with $P(B) > 0$, show that the set function $Q(A) = P(A \mid B)$ satisfies the axioms for a probability measure. Thus, for example,

$$P(A \cup C \mid B) = P(A \mid B) + P(C \mid B) - P(A \cap C \mid B)$$

65. Show that if A and B are independent, then A and B^c as well as A^c and B^c are independent.

66. Show that $\emptyset$ is independent of A for any A.

67. Show that if A and B are independent, then

$$P(A \cup B) = P(A) + P(B) - P(A)P(B)$$

68. If A is independent of B and B is independent of C, then A is independent of C. Prove this statement or give a counterexample if it is false.

69. If A and B are disjoint, can they be independent?

70. If $A \subset B$, can A and B be independent?

71. Show that if A, B, and C are mutually independent, then $A \cap B$ and C are independent and $A \cup B$ and C are independent.

72. Suppose that n components are connected in series. For each unit, there is a backup unit, and the system fails if and only if both a unit and its backup fail. Assuming that all the units are independent and fail with probability p, what is the probability that the system works? For $n = 10$ and $p = .05$, compare these results with those of Example F in Section 1.6.

73. A system has n independent units, each of which fails with probability p. The system fails only if k or more of the units fail. What is the probability that the system fails?

74. What is the probability that the following system works if each unit fails independently with probability p (see Figure 1.5)?

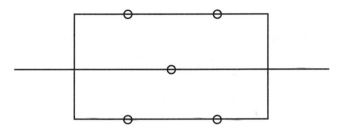

FIGURE **1.5**

75. This problem deals with an elementary aspect of a simple branching process. A population starts with one member; at time $t = 1$, it either divides with probability p or dies with probability $1 - p$. If it divides, then both of its children behave independently with the same two alternatives at time $t = 2$. What is the probability that there are no members in the third generation? For what value of p is this probability equal to .5?

76. Here is a simple model of a queue. The queue runs in discrete time ($t = 0, 1, 2, \ldots$), and at each unit of time the first person in the queue is served with probability p and, independently, a new person arrives with probability q. At time $t = 0$, there is one person in the queue. Find the probabilities that there are $0, 1, 2, 3$ people in line at time $t = 2$.

77. A player throws darts at a target. On each trial, independently of the other trials, he hits the bull's-eye with probability .05. How many times should he throw so that his probability of hitting the bull's-eye at least once is .5?

78. This problem introduces some aspects of a simple genetic model. Assume that genes in an organism occur in pairs and that each member of the pair can be either of the types a or A. The possible genotypes of an organism are then AA, Aa, and aa (Aa and aA are equivalent). When two organisms mate, each independently contributes one of its two genes; either one of the pair is transmitted with probability .5.

    a. Suppose that the genotypes of the parents are AA and Aa. Find the possible genotypes of their offspring and the corresponding probabilities.

b. Suppose that the probabilities of the genotypes AA, Aa, and aa are p, $2q$, and r, respectively, in the first generation. Find the probabilities in the second and third generations, and show that these are the same. This result is called the Hardy-Weinberg Law.

c. Compute the probabilities for the second and third generations as in part (b) but under the additional assumption that the probabilities that an individual of type AA, Aa, or aa survives to mate are u, v, and w, respectively.

79. Many human diseases are genetically transmitted (for example, hemophilia or Tay-Sachs disease). Here is a simple model for such a disease. The genotype aa is diseased and dies before it mates. The genotype Aa is a carrier but is not diseased. The genotype AA is not a carrier and is not diseased.

   a. If two carriers mate, what are the probabilities that their offspring are of each of the three genotypes?

   b. If the male offspring of two carriers is not diseased, what is the probability that he is a carrier?

   c. Suppose that the nondiseased offspring of part (b) mates with a member of the population for whom no family history is available and who is thus assumed to have probability p of being a carrier (p is a very small number). What are the probabilities that their first offspring has the genotypes AA, Aa, and aa?

   d. Suppose that the first offspring of part (c) is not diseased. What is the probability that the father is a carrier in light of this evidence?

80. If a parent has genotype Aa, he transmits either A or a to an offspring (each with a $\frac{1}{2}$ chance). The gene he transmits to one offspring is independent of the one he transmits to another. Consider a parent with three children and the following events: $A = \{$children 1 and 2 have the same gene$\}$, $B = \{$children 1 and 3 have the same gene$\}$, $C = \{$children 2 and 3 have the same gene$\}$. Show that these events are pairwise independent but not mutually independent.

CHAPTER 2

Random Variables

2.1 Discrete Random Variables

A random variable is essentially a random number. As motivation for a definition, let us consider an example. A coin is thrown three times, and the sequence of heads and tails is observed; thus,

$$\Omega = \{hhh, hht, htt, hth, ttt, tth, thh, tht\}$$

Examples of random variables defined on Ω are (1) the total number of heads, (2) the total number of tails, and (3) the number of heads minus the number of tails. Each of these is a real-valued function defined on Ω; that is, each is a rule that assigns a real number to every point $\omega \in \Omega$. Since the outcome in Ω is random, the corresponding number is random as well.

In general, a random variable is a function from Ω to the real numbers. Because the outcome of the experiment with sample space Ω is random, the number produced by the function is random as well. It is conventional to denote random variables by italic uppercase letters from the end of the alphabet. For example, we might define X to be the total number of heads in the experiment described above. A **discrete random variable** is a random variable that can take on only a finite or at most a countably infinite number of values. The random variable X just defined is a discrete random variable since it can take on only the values 0, 1, 2, and 3. For an example of a random variable that can take on a countably infinite number of values, consider an experiment that consists of tossing a coin until a head turns up and defining Y to be the total number of tosses. The possible values of Y are 0, 1, 2, 3, ... In general, a countably infinite set is one that can be put into one-to-one correspondence with the integers.

If the coin is fair, then each of the outcomes in Ω above has probability $\frac{1}{8}$, from which the probabilities that X takes on the values 0, 1, 2, and 3 can be easily computed:

$$P(X = 0) = \tfrac{1}{8}$$
$$P(X = 1) = \tfrac{3}{8}$$
$$P(X = 2) = \tfrac{3}{8}$$
$$P(X = 3) = \tfrac{1}{8}$$

Generally, the probability measure on the sample space determines the probabilities of the various values of X; if those values are denoted by $x_1, x_2, \ldots$, then there is a function p such that $p(x_i) = P(X = x_i)$ and $\sum_i p(x_i) = 1$. This function is called the **probability mass function,** or the **frequency function,** of the random variable X. Figure 2.1 shows a graph of $p(x)$ for the coin tossing experiment. The frequency function describes completely the probability properties of the random variable.

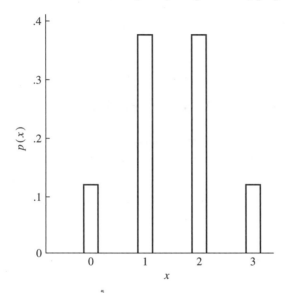

FIGURE **2.1** A probability mass function.

In addition to the frequency function, it is sometimes convenient to use the **cumulative distribution function (cdf)** of a random variable, which is defined to be

$$F(x) = P(X \leq x), \qquad -\infty < x < \infty$$

Cumulative distribution functions are usually denoted by uppercase letters and frequency functions by lowercase letters. Figure 2.2 is a graph of the cumulative distribution function of the random variable X of the preceding paragraph. Note that the cdf jumps wherever $p(x) > 0$ and that the jump at x_i is $p(x_i)$. For example, if $0 < x < 1$, $F(x) = \tfrac{1}{8}$; at $x = 1$, $F(x)$ jumps to $F(1) = \tfrac{4}{8} = \tfrac{1}{2}$. The jump at $x = 1$ is $p(1) = \tfrac{3}{8}$. The cumulative distribution function is non-decreasing and satisfies

$$\lim_{x \to -\infty} F(x) = 0 \qquad \text{and} \qquad \lim_{x \to \infty} F(x) = 1.$$

Chapter 3 will cover in detail the joint frequency functions of several random variables defined on the same sample space, but it is useful to define here the concept

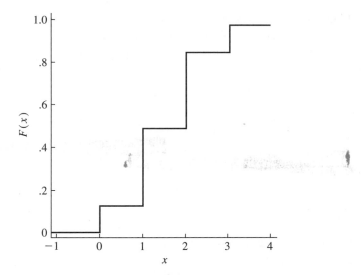

FIGURE **2.2** The cumulative distribution function corresponding to Figure 2.1.

of independence of random variables. In the case of two discrete random variables X and Y, taking on possible values $x_1, x_2, \ldots$, and $y_1, y_2, \ldots$, X and Y are said to be **independent** if, for all i and j,

$$P(X = x_i \text{ and } Y = y_j) = P(X = x_i)P(Y = y_j)$$

The definition is extended to collections of more than two discrete random variables in the obvious way; for example, X, Y, and Z are said to be mutually independent if, for all i, j, and k,

$$P(X = x_i, Y = y_j, Z = z_k) = P(X = x_i)P(Y = y_j)P(Z = z_k)$$

We next discuss some common discrete distributions that arise in applications.

2.1.1 Bernoulli Random Variables

A Bernoulli random variable takes on only two values: 0 and 1, with probabilities $1 - p$ and p, respectively. Its frequency function is thus

$$p(1) = p$$
$$p(0) = 1 - p$$
$$p(x) = 0, \qquad \text{if } x \neq 0 \text{ and } x \neq 1$$

An alternative and sometimes useful representation of this function is

$$p(x) = \begin{cases} p^x(1 - p)^{1-x}, & \text{if } x = 0 \text{ or } x = 1 \\ 0, & \text{otherwise} \end{cases}$$

If A is an event, then the **indicator random variable,** I_A, takes on the value 1 if A occurs and the value 0 if A does not occur:

$$I_A(\omega) = \begin{cases} 1, & \text{if } \omega \in A \\ 0, & \text{otherwise} \end{cases}$$

I_A is a Bernoulli random variable. In applications, Bernoulli random variables often occur as indicators. A Bernoulli random variable might take on the value 1 or 0 according to whether a guess was a success or a failure.

2.1.2 The Binomial Distribution

Suppose that n independent experiments, or trials, are performed, where n is a fixed number, and that each experiment results in a "success" with probability p and a "failure" with probability $1 - p$. The total number of successes, X, is a binomial random variable with parameters n and p. For example, a coin is tossed 10 times and the total number of heads is counted ("head" is identified with "success").

The probability that $X = k$, or $p(k)$, can be found in the following way: Any particular sequence of k successes occurs with probability $p^k(1 - p)^{n-k}$, from the multiplication principle. The total number of such sequences is $\binom{n}{k}$, since there are $\binom{n}{k}$ ways to assign k successes to n trials. $P(X = k)$ is thus the probability of any particular sequence times the number of such sequences:

$$p(k) = \binom{n}{k} p^k(1 - p)^{n-k}$$

Two binomial frequency functions are shown in Figure 2.3. Note how the shape varies as a function of p.

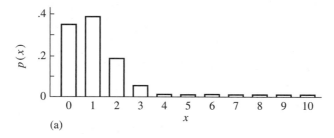

(a)

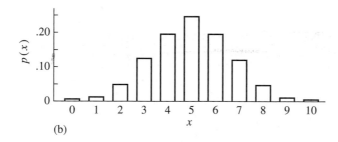

(b)

FIGURE 2.3   Binomial frequency functions, (a) $n = 10$ and $p = .1$ and (b) $n = 10$ and $p = .5$.

EXAMPLE A   Tay-Sachs disease is a rare but fatal disease of genetic origin occurring chiefly in infants and children, especially those of Jewish or eastern European extraction. If a couple are both carriers of Tay-Sachs disease, a child of theirs has probability .25 of being born with the disease. If such a couple has four children, what is the frequency function for the number of children who will have the disease?

We assume that the four outcomes are independent of each other, so, if X denotes the number of children with the disease, its frequency function is

$$p(k) = \binom{4}{k}.25^k \times .75^{4-k}, \qquad k = 0, 1, 2, 3, 4$$

These probabilities are given in the following table:

k	$p(k)$
0	.316
1	.422
2	.211
3	.047
4	.004

■

E X A M P L E **B**   If a single bit (0 or 1) is transmitted over a noisy communications channel, it has probability p of being incorrectly transmitted. To improve the reliability of the transmission, the bit is transmitted n times, where n is odd. A decoder at the receiving end, called a majority decoder, decides that the correct message is that carried by a majority of the received bits. Under a simple noise model, each bit is independently subject to being corrupted with the same probability p. The number of bits that is in error, X, is thus a binomial random variable with n trials and probability p of success on each trial (in this case, and frequently elsewhere, the word *success* is used in a generic sense; here a success is an error). Suppose, for example, that $n = 5$ and $p = .1$. The probability that the message is correctly received is the probability of two or fewer errors, which is

$$\sum_{k=0}^{2} \binom{n}{k} p^k (1 - p)^{n-k} = p^0(1 - p)^5 + 5p(1 - p)^4 + 10p^2(1 - p)^3 = .9914$$

The result is a considerable improvement in reliability.   ■

E X A M P L E **C**   *DNA Matching*

We continue Example G of Section 1.6. There we derived the probability p that two fragments agree at a particular site under the assumption that the nucleotide probabilities were the same at every site and the identities on fragment 1 were independent of those on fragment 2. To find the probability of the total number of matches, further assumptions must be made. Suppose that the fragments are each of length n and that the nucleotide identities are independent from site to site as well as between fragments. Thus, the identity of the nucleotide at site 1 of fragment 1 is independent of the identity at site 2, etc. We did not make this assumption in Example G in Section 1.6; in that case, the identity at site 2 could have depended on the identity at site 1, for example. Now, under the current assumption, the two fragments agree at each site with probability p as calculated in Example G of Section 1.6, and agreement is independent from site to site. So, the total number of agreements is a binomial random variable with n trials and probability p of success.   ■

A random variable with a binomial distribution can be expressed in terms of independent Bernoulli random variables, a fact that will be quite useful for analyzing some properties of binomial random variables in later chapters of this book. Specifically, let $X_1, X_2, \ldots, X_n$ be independent Bernoulli random variables with $p(X_i = 1) = p$. Then $Y = X_1 + X_2 + \cdots + X_n$ is a binomial random variable.

2.1.3 The Geometric and Negative Binomial Distributions

The **geometric distribution** is also constructed from independent Bernoulli trials, but from an infinite sequence. On each trial, a success occurs with probability p, and X is the total number of trials up to and including the first success. So that $X = k$, there must be $k - 1$ failures followed by a success. From the independence of the trials, this occurs with probability

$$p(k) = P(X = k) = (1 - p)^{k-1} p, \qquad k = 1, 2, 3, \ldots$$

Note that these probabilities sum to 1:

$$\sum_{k=1}^{\infty} (1 - p)^{k-1} p = p \sum_{j=0}^{\infty} (1 - p)^j = 1$$

E X A M P L E **A**   The probability of winning in a certain state lottery is said to be about $\frac{1}{9}$. If it is exactly $\frac{1}{9}$, the distribution of the number of tickets a person must purchase up to and including the first winning ticket is a geometric random variable with $p = \frac{1}{9}$. Figure 2.4 shows the frequency function.   ■

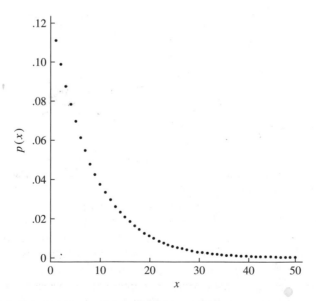

F I G U R E **2.4**   The probability mass function of a geometric random variable with $p = \frac{1}{9}$.

The **negative binomial distribution** arises as a generalization of the geometric distribution. Suppose that a sequence of independent trials, each with probability of success p, is performed until there are r successes in all; let X denote the total number of trials. To find $P(X = k)$, we can argue in the following way: Any particular such sequence has probability $p^r(1 - p)^{k-r}$, from the independence assumption. The last trial is a success, and the remaining $r - 1$ successes can be assigned to the remaining $k - 1$ trials in $\binom{k-1}{r-1}$ ways. Thus,

$$P(X = k) = \binom{k-1}{r-1} p^r(1-p)^{k-r}$$

It is sometimes helpful in analyzing properties of the negative binomial distribution to note that a negative binomial random variable can be expressed as the sum of r independent geometric random variables: the number of trials up to and including the first success plus the number of trials after the first success up to and including the second success, ... plus the number of trials from the $(r - 1)$st success up to and including the rth success.

EXAMPLE B   Continuing Example A, the distribution of the number of tickets purchased up to and including the second winning ticket is negative binomial:

$$p(k) = (k - 1)p^2(1 - p)^{k-2}$$

This frequency function is shown in Figure 2.5. ■

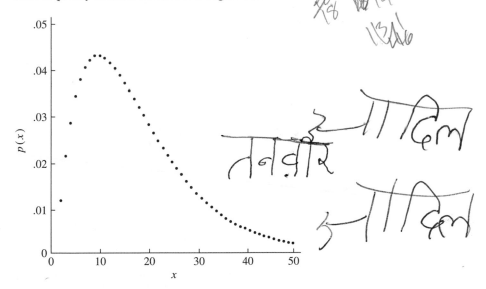

FIGURE 2.5   The probability mass function of a negative binomial random variable with $p = \frac{1}{9}$ and $r = 2$.

The definitions of the geometric and negative binomial distributions vary slightly from one textbook to another; for example, instead of X being the total number of trials in the definition of the geometric distribution, X is sometimes defined as the total number of failures.

2.1.4   The Hypergeometric Distribution

The **hypergeometric distribution** was introduced in Chapter 1 but was not named there. Suppose that an urn contains n balls, of which r are black and $n-r$ are white. Let X denote the number of black balls drawn when taking m balls without replacement. Following the line of reasoning of Examples H and I of Section 1.4.2,

$$P(X = k) = \frac{\binom{r}{k}\binom{n-r}{m-k}}{\binom{n}{m}}$$

X is a hypergeometric random variable with parameters r, n, and m.

EXAMPLE A   As explained in Example G of Section 1.4.2, a player in the California lottery chooses 6 numbers from 53 and the lottery officials later choose 6 numbers at random. Let X equal the number of matches. Then

$$P(X = k) = \frac{\binom{6}{k}\binom{47}{6-k}}{\binom{53}{6}}$$

The probability mass function of X is displayed in the following table:

k	0	1	2	3	4	5	6
$p(k)$	.468	.401	.117	.014	7.06×10^{-4}	1.22×10^{-5}	4.36×10^{-8}

∎

2.1.5   The Poisson Distribution

The **Poisson frequency function** with parameter λ ($\lambda > 0$) is

$$P(X = k) = \frac{\lambda^k}{k!} e^{-\lambda}, \qquad k = 0, 1, 2, \ldots$$

Since $e^{\lambda} = \sum_{k=0}^{\infty}(\lambda^k / k!)$, it follows that the frequency function sums to 1. Figure 2.6 shows four Poisson frequency functions. Note how the shape varies as a function of λ.

The **Poisson distribution** can be derived as the limit of a binomial distribution as the number of trials, n, approaches infinity and the probability of success on each trial, p, approaches zero in such a way that $np = \lambda$. The binomial frequency function is

$$p(k) = \frac{n!}{k!(n-k)!} p^k (1-p)^{n-k}$$

Setting $np = \lambda$, this expression becomes

$$p(k) = \frac{n!}{k!(n-k)!} \left(\frac{\lambda}{n}\right)^k \left(1 - \frac{\lambda}{n}\right)^{n-k}$$

$$= \frac{\lambda^k}{k!} \frac{n!}{(n-k)!} \frac{1}{n^k} \left(1 - \frac{\lambda}{n}\right)^n \left(1 - \frac{\lambda}{n}\right)^{-k}$$

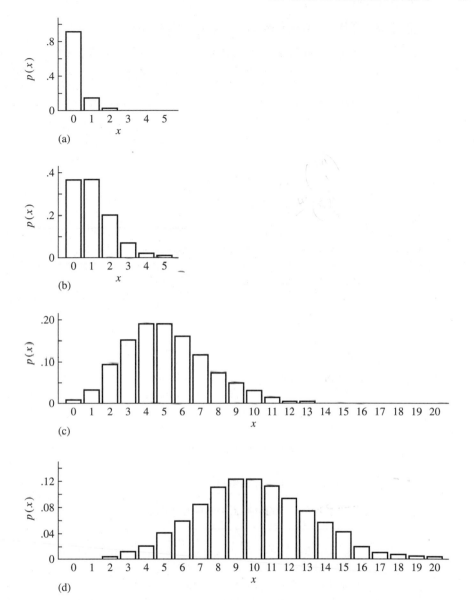

FIGURE **2.6**   Poisson frequency functions, (a) $\lambda = .1$, (b) $\lambda = 1$, (c) $\lambda = 5$, (d) $\lambda = 10$.

As $n \to \infty$,

$$\frac{\lambda}{n} \to 0$$

$$\frac{n!}{(n-k)!n^k} \to 1$$

$$\left(1 - \frac{\lambda}{n}\right)^n \to e^{-\lambda}$$

and

$$\left(1 - \frac{\lambda}{n}\right)^{-k} \rightarrow 1$$

We thus have

$$p(k) \rightarrow \frac{\lambda^k e^{-\lambda}}{k!}$$

which is the Poisson frequency function.

E X A M P L E **A**   Two dice are rolled 100 times, and the number of double sixes, X, is counted. The distribution of X is binomial with $n = 100$ and $p = \frac{1}{36} = .0278$. Since n is large and p is small, we can approximate the binomial probabilities by Poisson probabilities with $\lambda = np = 2.78$. The exact binomial probabilities and the Poisson approximations are shown in the following table:

k	Binomial Probability	Poisson Approximation
0	.0596	.0620
1	.1705	.1725
2	.2414	.2397
3	.2255	.2221
4	.1564	.1544
5	.0858	.0858
6	.0389	.0398
7	.0149	.0158
8	.0050	.0055
9	.0015	.0017
10	.0004	.0005
11	.0001	.0001

The approximation is quite good. ■

The Poisson frequency function can be used to approximate binomial probabilities for large n and small p. This suggests how Poisson distributions can arise in practice. Suppose that X is a random variable that equals the number of times some event occurs in a given interval of time. Heuristically, let us think of dividing the interval into a very large number of small subintervals of equal length, and let us assume that the subintervals are so small that the probability of more than one event in a subinterval is negligible relative to the probability of one event, which is itself very small. Let us also assume that the probability of an event is the same in each subinterval and that whether an event occurs in one subinterval is independent of what happens in the other subintervals. The random variable X is thus nearly a binomial random variable, with the subintervals consitituting the trials, and, from the limiting result above, X has nearly a Poisson distribution.

The preceding argument is not formal, of course, but merely suggestive. But, in fact, it can be made rigorous. The important assumptions underlying it are (1) what happens in one subinterval is independent of what happens in any other subinterval, (2) the probability of an event is the same in each subinterval, and (3) events do not happen simultaneously. The same kind of argument can be made if we are concerned with an area or a volume of space rather than with an interval on the real line.

The Poisson distribution is of fundamental theoretical and practical importance. It has been used in many areas, including the following:

- The Poisson distribution has been used in the analysis of telephone systems. The number of calls coming into an exchange during a unit of time might be modeled as a Poisson variable if the exchange services a large number of customers who act more or less independently.
- One of the earliest uses of the Poisson distribution was to model the number of alpha particles emitted from a radioactive source during a given period of time.
- The Poisson distribution has been used as a model by insurance companies. For example, the number of freak acidents, such as falls in the shower, for a large population of people in a given time period might be modeled as a Poisson distribution, because the accidents would presumably be rare and independent (provided there was only one person in the shower).
- The Poisson distribution has been used by traffic engineers as a model for light traffic. The number of vehicles that pass a marker on a roadway during a unit of time can be counted. If traffic is light, the individual vehicles act independently of each other. In heavy traffic, however, one vehicle's movement may influence another's, so the approximation might not be good.

E X A M P L E **B**   This amusing classical example is from von Bortkiewicz (1898). The number of fatalities that resulted from being kicked by a horse was recorded for 10 corps of Prussian cavalry over a period of 20 years, giving 200 corps-years worth of data. These data and the probabilities from a Poisson model with $\lambda = .61$ are displayed in the following table. The first column of the table gives the number of deaths per year, ranging from 0 to 4. The second column lists how many times that number of deaths was observed. Thus, for example, in 65 of the 200 corps-years, there was one death. In the third column of the table, the observed numbers are converted to relative frequencies by dividing them by 200. The fourth column of the table gives Poisson probabilities with the parameter $\lambda = .61$. In Chapters 8 and 9, we discuss how to choose a parameter value to fit a theoretical probability model to observed frequencies and methods for testing goodness of fit. For now, we will just remark that the value $\lambda = .61$ was chosen to match the average number of deaths per year.

Number of Deaths per Year	Observed	Relative Frequency	Poisson Probability
0	109	.545	.543
1	65	.325	.331
2	22	.110	.101
3	3	.015	.021
4	1	.005	.003

∎

The Poisson distribution often arises from a model called a **Poisson process** for the distribution of random events in a set S, which is typically one-, two-, or three-dimensional, corresponding to time, a plane, or a volume of space. Basically, this model states that if $S_1, S_2, \ldots, S_n$ are disjoint subsets of S, then the numbers of events in these subsets, $N_1, N_2, \ldots, N_n$, are independent random variables that follow Poisson distributions with parameters $\lambda|S_1|, \lambda|S_2|, \ldots, \lambda|S_n|$, where $|S_i|$ denotes the measure of S_i (length, area, or volume, for example). The crucial assumptions here are that events in disjoint subsets are independent of each other and that the Poisson parameter for a subset is proportional to the subset's size. Later, we will see that this latter assumption implies that the average number of events in a subset is proportional to its size.

EXAMPLE C Suppose that an office receives telephone calls as a Poisson process with $\lambda = .5$ per min. The number of calls in a 5-min. interval follows a Poisson distribution with parameter $\omega = 5\lambda = 2.5$. Thus, the probability of no calls in a 5-min. interval is $e^{-2.5} = .082$. The probability of exactly one call is $2.5e^{-2.5} = .205$. ∎

EXAMPLE D Figure 2.7 shows four realizations of a Poisson process with $\lambda = 25$ in the unit square, $0 \le x \le 1$ and $0 \le y \le 1$. It is interesting that the eye tends to perceive patterns, such

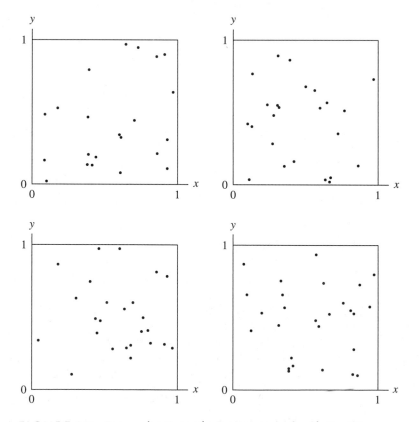

FIGURE 2.7 Four realizations of a Poisson process with $\lambda = 25$.

as clusters of points and large blank spaces. But by the nature of a Poisson process, the locations of the points have no relationship to one another, and these patterns are entirely a result of chance. ■

2.2 Continuous Random Variables

In applications, we are often interested in random variables that can take on a continuum of values rather than a finite or countably infinite number. For example, a model for the lifetime of an electronic component might be that it is random and can be any positive real number. For a continuous random variable, the role of the frequency function is taken by a **density function,** $f(x)$, which has the properties that $f(x) \geq 0$, f is piecewise continuous, and $\int_{-\infty}^{\infty} f(x) \, dx = 1$. If X is a random variable with a density function f, then for any $a < b$, the probability that X falls in the interval (a, b) is the area under the density function between a and b:

$$P(a < X < b) = \int_a^b f(x) \, dx$$

EXAMPLE A A **uniform random variable** on the interval [0, 1] is a model for what we mean when we say "choose a number at random between 0 and 1." Any real number in the interval is a possible outcome, and the probability model should have the property that the probability that X is in any subinterval of length h is equal to h. The following density function does the job:

$$f(x) = \begin{cases} 1, & 0 \leq x \leq 1 \\ 0, & x < 0 \text{ or } x > 1 \end{cases}$$

This is called the **uniform density** on [0, 1]. The uniform density on a general interval [a, b] is

$$f(x) = \begin{cases} 1/(b - a), & a \leq x \leq b \\ 0, & x < a \text{ or } x > b \end{cases}$$ ■

One consequence of this definition is that the probability that a continuous random variable X takes on any particular value is 0:

$$P(X = c) = \int_c^c f(x) \, dx = 0$$

Although this may seem strange initially, it is really quite natural. If the uniform random variable of Example A had a positive probability of being any particular number, it should have the same probability for any number in [0, 1], in which case the sum of the probabilities of any countably infinite subset of [0, 1] (for example, the rational numbers) would be infinite. If X is a continuous random variable, then

$$P(a < X < b) = P(a \leq X < b) = P(a < X \leq b)$$

Note that this is not true for a discrete random variable.

For small δ, if f is continuous at x,

$$P\left(x - \frac{\delta}{2} \le X \le x + \frac{\delta}{2}\right) = \int_{x-\delta/2}^{x+\delta/2} f(u)\, du \approx \delta f(x)$$

Therefore, the probability of a small interval around x is proportional to $f(x)$. It is sometimes useful to employ differential notation: $P(x \le X \le x + dx) = f(x)\, dx$.

The cumulative distribution function of a continuous random variable X is defined in the same way as for a discrete random variable:

$$F(x) = P(X \le x)$$

$F(x)$ can be expressed in terms of the density function:

$$F(x) = \int_{-\infty}^{x} f(u)\, du$$

From the fundamental theorem of calculus, if f is continuous at x, $f(x) = F'(x)$.

The cdf can be used to evaluate the probability that X falls in an interval:

$$P(a \le X \le b) = \int_{a}^{b} f(x)\, dx = F(b) - F(a)$$

E X A M P L E **B**   From this definition, we see that the cdf of a uniform random variable on $[0, 1]$ (Example A) is

$$F(x) = \begin{cases} 0, & x \le 0 \\ x, & 0 \le x \le 1 \\ 1, & x \ge 1 \end{cases}$$ ■

Suppose that F is the cdf of a continuous random variable and is strictly increasing on some interval I, and that $F = 0$ to the left of I and $F = 1$ to the right of I; I may be unbounded. Under this assumption, the inverse function F^{-1} is well defined; $x = F^{-1}(y)$ if $y = F(x)$. The pth **quantile** of the distribution F is defined to be that value x_p such that $F(x_p) = p$, or $P(X \le x_p) = p$. Under the preceding assumption stated, x_p is uniquely defined as $x_p = F^{-1}(p)$; see Figure 2.8. Special cases are $p = \frac{1}{2}$, which corresponds to the **median** of F; and $p = \frac{1}{4}$ and $p = \frac{3}{4}$, which correspond to the lower and upper **quartiles** of F.

E X A M P L E **C**   Suppose that $F(x) = x^2$ for $0 \le x \le 1$. This statement is shorthand for the more explicit statement

$$F(x) = \begin{cases} 0, & x \le 0 \\ x^2, & 0 \le x \le 1 \\ 1, & x \ge 1 \end{cases}$$

To find F^{-1}, we solve $y = F(x) = x^2$ for x, obtaining $x = F^{-1}(y) = \sqrt{y}$. The median is $F^{-1}(.5) = .707$, the lower quartile is $F^{-1}(.25) = .50$, and the upper quartile is $F^{-1}(.75) = .866$. ■

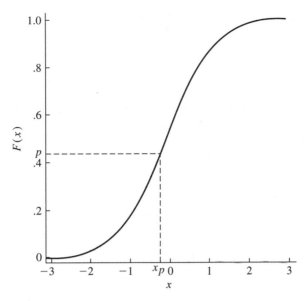

FIGURE **2.8**   A cdf F and F^{-1}.

EXAMPLE **D**   *Value at Risk*

Financial firms need to quantify and monitor the risk of their investments. **Value at Risk (VaR)** is a widely used measure of potential losses. It involves two parameters: a time horizon and a level of confidence. For example, if the VaR of an institution is $10 million with a one-day horizon and a level of confidence of 95%, the interpretation is that there is a 5% chance of losses exceeding $10 million. Such a loss should be anticipated about once in 20 days.

To see how VaR is computed, suppose the current value of the investment is V_0 and the future value is V_1. The return on the investment is $R = (V_1 - V_0)/V_0$, which is modeled as a continuous random variable with cdf $F_R(r)$. Let the desired level of confidence be denoted by $1 - \alpha$. We want to find v^*, the VaR. Then

$$\alpha = P(V_0 - V_1 \geq v^*)$$

$$= P\left(\frac{V_1 - V_0}{V_0} \leq -\frac{v^*}{V_0}\right)$$

$$= F_R\left(-\frac{v^*}{V_0}\right)$$

Thus, $-v^*/V_0$ is the α quantile, r_α; and $v^* = -V_0 r_\alpha$. The VaR is minus the current value times the α quantile of the return distribution.   ■

We next discuss some density functions that commonly arise in practice.

2.2.1 The Exponential Density

The exponential density function is

$$f(x) = \begin{cases} \lambda e^{-\lambda x}, & x \geq 0 \\ 0, & x < 0 \end{cases}$$

Like the Poisson distribution, the exponential density depends on a single parameter, $\lambda > 0$, and it would therefore be more accurate to refer to it as the family of exponential densities. Several exponential densities are shown in Figure 2.9. Note that as λ becomes larger, the density drops off more rapidly.

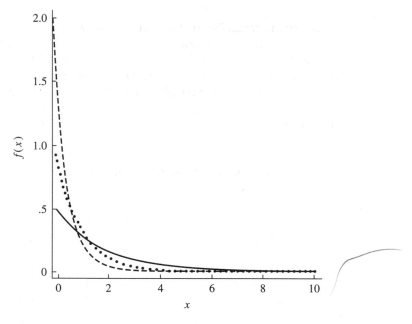

FIGURE 2.9 Exponential densities with $\lambda = .5$ (solid), $\lambda = 1$ (dotted), and $\lambda = 2$ (dashed).

The cumulative distribution function is easily found:

$$F(x) = \int_{-\infty}^{x} f(u)\, du = \begin{cases} 1 - e^{-\lambda x}, & x \geq 0 \\ 0, & x < 0 \end{cases}$$

The median of an exponential distribution, η, say, is readily found from the cdf. We solve $F(\eta) = \frac{1}{2}$:

$$1 - e^{-\lambda \eta} = \tfrac{1}{2}$$

from which we have

$$\eta = \frac{\log 2}{\lambda}$$

The exponential distribution is often used to model lifetimes or waiting times, in which context it is conventional to replace x by t. Suppose that we consider modeling the lifetime of an electronic component as an exponential random variable, that the component has lasted a length of time s, and that we wish to calculate the probability that it will last at least t more time units; that is, we wish to find $P(T > t+s \mid T > s)$:

$$P(T > t + s \mid T > s) = \frac{P(T > t + s \text{ and } T > s)}{P(T > s)}$$

$$= \frac{P(T > t + s)}{P(T > s)}$$

$$= \frac{e^{-\lambda(t+s)}}{e^{-\lambda s}}$$

$$= e^{-\lambda t}$$

We see that the probability that the unit will last t more time units does not depend on s. The exponential distribution is consequently said to be **memoryless;** it is clearly not a good model for human lifetimes, since the probability that a 16-year-old will live at least 10 more years is not the same as the probability that an 80-year-old will live at least 10 more years. It can be shown that the exponential distribution is characterized by this memoryless property—that is, the memorylessness implies that the distribution is exponential. It may be somewhat surprising that a qualitative characterization, the property of memorylessness, actually determines the form of this density function.

The memoryless character of the exponential distribution follows directly from its relation to a Poisson process. Suppose that events occur in time as a Poisson process with parameter λ and that an event occurs at time t_0. Let T denote the length of time until the next event. The density of T can be found as follows:

$$P(T > t) = P(\text{no events in } (t_0, t_0 + t))$$

Since the number of events in the interval $(t_0, t_0 + t)$, which is of length t, follows a Poisson distribution with parameter λt, this probability is $e^{-\lambda t}$, and thus T follows an exponential distribution with parameter λ. We can continue in this fashion. Suppose that the next event occurs at time t_1; the distribution of time until the third event is again exponential by the same analysis and, from the independence property of the Poisson process, is independent of the length of time between the first two events. Generally, the times between events of a Poisson process are independent, identically distributed, exponential random variables.

Proteins and other biologically important molecules are regulated in various ways. Some undergo aging and are thus more likely to degrade when they are old than when they are young. If a molecule was not subject to aging, but its chance of degradation was the same at any age, its lifetime would follow an exponential distribution.

E X A M P L E   **A**   Muscle and nerve cell membranes contain large numbers of channels through which selected ions can pass when the channels are open. Using sophisticated experimental techniques, neurophysiologists can measure the resulting current that flows through a single channel, and experimental records often indicate that a channel opens and closes at seemingly random times. In some cases, simple kinetic models predict that the duration of the open time should be exponentially distributed.

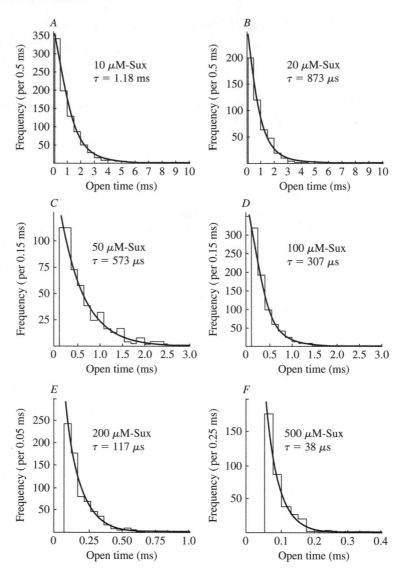

FIGURE **2.10** Histograms of open times at varying concentrations of suxamethonium and fitted exponential densities.

Marshall et al. (1990) studied the action of a channel-blocking agent (suxamethonium) on a channel (the nicotinic receptor of frog muscle). Figure 2.10 displays histograms of open times and fitted exponential distributions at a range of concentrations of suxamethonium. In this example, the exponential distribution is parametrized as $f(t) = (1/\tau)\exp(-t/\tau)$. τ is thus in units of time, whereas λ is in units of the reciprocal of time. From the figure, we see that the intervals become shorter and that the parameter τ decreases with increasing concentrations of the blocker. It can also be seen, especially at higher concentrations, that very short intervals are not recorded because of limitations of the instrumentation. ∎

2.2.2   The Gamma Density

The gamma density function depends on two parameters, α and λ:

$$g(t) = \frac{\lambda^{\alpha}}{\Gamma(\alpha)} t^{\alpha-1} e^{-\lambda t}, \qquad t \geq 0$$

For $t < 0$, $g(t) = 0$. So that the density be well defined and integrate to 1, $\alpha > 0$ and $\lambda > 0$. The gamma function, $\Gamma(x)$, is defined as

$$\Gamma(x) = \int_0^{\infty} u^{x-1} e^{-u} \, du, \qquad x > 0$$

Some properties of the gamma function are developed in the problems at the end of this chapter.

Note that if $\alpha = 1$, the gamma density coincides with the exponential density. The parameter α is called a **shape parameter** for the gamma density, and λ is called a **scale parameter.** Varying α changes the shape of the density, whereas varying λ corresponds to changing the units of measurement (say, from seconds to minutes) and does not affect the shape of the density.

Figure 2.11 shows several gamma densities. Gamma densities provide a fairly flexible class for modeling nonnegative random variables.

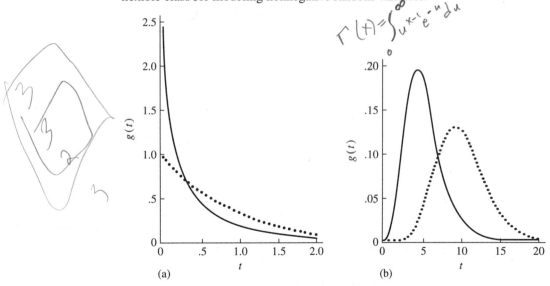

FIGURE 2.11   Gamma densities, (a) $\alpha = .5$ (solid) and $\alpha = 1$ (dotted) and (b) $\alpha = 5$ (solid) and $\alpha = 10$ (dotted); $\lambda = 1$ in all cases.

EXAMPLE A   The patterns of occurrence of earthquakes in terms of time, space, and magnitude are very erratic, and attempts are sometimes made to construct probabilistic models for these events. The models may be used in a purely descriptive manner or, more ambitiously, for purposes of predicting future occurrences and consequent damage.

Figure 2.12 shows the fit of a gamma density and an exponential density to the observed times separating a sequence of small earthquakes (Udias and Rice, 1975). The gamma density clearly gives a better fit ($\alpha = .509$ and $\lambda = .00115$). Note that an

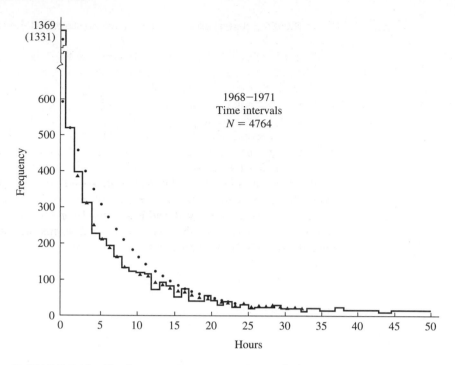

FIGURE 2.12   Fit of gamma density (triangles) and of exponential density (circles) to times between microearthquakes.

exponential model for interoccurrence times would be memoryless; that is, knowing that an earthquake had not occurred in the last t time units would tell us nothing about the probability of occurrence during the next s time units. The gamma model does not have this property. In fact, although we will not show this, the gamma model with these parameter values has the character that there is a large likelihood that the next earthquake will immediately follow any given one and this likelihood decreases monotonically with time.   ∎

2.2.3   The Normal Distribution

The normal distribution plays a central role in probability and statistics, for reasons that will become apparent in later chapters of this book. This distribution is also called the Gaussian distribution after Carl Friedrich Gauss, who proposed it as a model for measurement errors. The central limit theorem, which will be discussed in Chapter 6, justifies the use of the normal distribution in many applications. Roughly, the central limit theorem says that if a random variable is the sum of a large number of independent random variables, it is approximately normally distributed. The normal distribution has been used as a model for such diverse phenomena as a person's height, the distribution of IQ scores, and the velocity of a gas molecule. The density function of the normal distribution depends on two parameters, μ and σ (where $-\infty < \mu < \infty, \sigma > 0$):

$$f(x) = \frac{1}{\sigma\sqrt{2\pi}}e^{-(x-\mu)^2/2\sigma^2}, \qquad -\infty < x < \infty$$

The parameters μ and σ are called the **mean** and **standard deviation** of the normal density.

The cdf cannot be evaluated in closed form from this density function (the integral that defines the cdf cannot be evaluated by an explicit formula but must be found numerically). A problem at the end of this chapter asks you to show that the normal density just given integrates to one.

As shorthand for the statement "X follows a normal distribution with parameters μ and σ," it is convenient to use $X \sim N(\mu, \sigma^2)$. From the form of the density function, we see that the density is symmetric about μ, $f(\mu - x) = f(\mu + x)$, where it has a maximum, and that the rate at which it falls off is determined by σ. Figure 2.13 shows several normal densities. Normal densities are sometimes referred to as *bell-shaped curves*. The special case for which $\mu = 0$ and $\sigma = 1$ is called the **standard normal density.** Its cdf is denoted by Φ and its density by ϕ (not to be confused with the empty set). The relationship between the general normal density and the standard normal density will be developed in the next section.

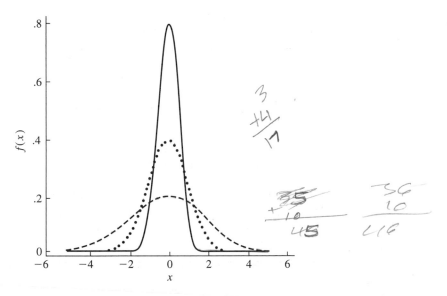

FIGURE **2.13**   Normal densities, $\mu = 0$ and $\sigma = .5$ (solid), $\mu = 0$ and $\sigma = 1$ (dotted), and $\mu = 0$ and $\sigma = 2$ (dashed).

E X A M P L E **A**   Acoustic recordings made in the ocean contain substantial background noise. To detect sonar signals of interest, it is useful to characterize this noise as accurately as possible. In the Arctic, much of the background noise is produced by the cracking and straining of ice. Veitch and Wilks (1985) studied recordings of Arctic undersea noise and characterized the noise as a mixture of a Gaussian component and occasional large-amplitude bursts. Figure 2.14 is a trace of one recording that includes a burst. Figure 2.15 shows a Gaussian distribution fit to observations from a "quiet" (nonbursty) period of this noise.   ■

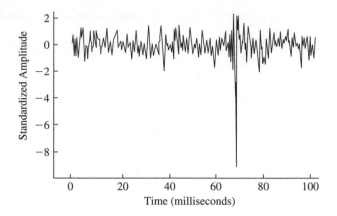

FIGURE **2.14**   A record of undersea noise containing a large burst.

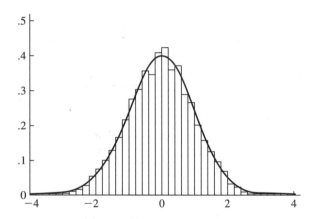

FIGURE **2.15**   A histogram from a "quiet" period of undersea noise with a fitted normal density.

EXAMPLE **B**   Turbulent air flow is sometimes modeled as a random process. Since the velocity of the flow at any point is subject to the influence of a large number of random eddies in the neighborhood of that point, one might expect from the central limit theorem that the velocity would be normally distributed. Van Atta and Chen (1968) analyzed data gathered in a wind tunnel. Figure 2.16, taken from their paper, shows a normal distribution fit to 409,600 observations of one component of the velocity; the fit is remarkably good.   ■

EXAMPLE **C**   *S&P 500*
The Standard and Poors 500 is an index of important U.S. stocks; each stock's weight in the index is proportional to its market value. Individuals can invest in mutual funds that track the index. The top panel of Figure 2.17 shows the sequential values of the

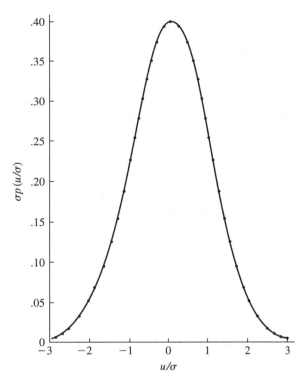

FIGURE **2.16**   A normal density (solid line) fit to 409,600 measurements of one component of the velocity of a turbulent wind flow. The dots show the values from a histogram.

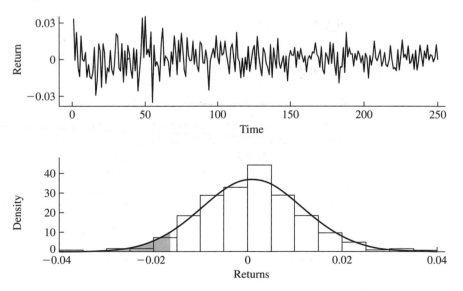

FIGURE **2.17**   Returns on the S&P 500 during 2003 (top panel) and a normal curve fitted to their histogram (bottom panel). The region area to the left of the 0.05 quantile is shaded.

returns during 2003. The average return during this period was 0.1% per day, and we can see from the figure that daily fluctuations were as large as 3% or 4%. The lower panel of the figure shows a histogram of the returns and a fitted normal density with $\mu = 0.001$ and $\sigma = 0.01$.

A financial company could use the fitted normal density in calculating its Value at Risk (see Example D of Section 2.2). Using a time horizon of one day and a confidence level of 95%, the VaR is the current investment in the index, V_0, multiplied by the negative of the 0.05 quantile of the distribution of returns. In this case, the quantile can be calculated to be -0.0165, so the VaR is $.0165V_0$. Thus, if V_0 is $10 million, the VaR is $165,000. The company can have 95% "confidence" that its losses will not exceed that amount on a given day. However, it should not be surprised if that amount is exceeded about once in every 20 trading days. ∎

2.2.4 The Beta Density

The beta density is useful for modeling random variables that are restricted to the interval [0, 1]:

$$f(u) = \frac{\Gamma(a+b)}{\Gamma(a)\Gamma(b)} u^{a-1}(1-u)^{b-1}, \qquad 0 \le u \le 1$$

Figure 2.18 shows beta densities for various values of a and b. Note that the case $a = b = 1$ is the uniform distribution. The beta distribution is important in Bayesian statistics, as you will see later.

2.3 Functions of a Random Variable

Suppose that a random variable X has a density function $f(x)$. We often need to find the density function of $Y = g(X)$ for some given function g. For example, X might be the velocity of a particle of mass m, and we might be interested in the probability density function of the particle's kinetic energy, $Y = \frac{1}{2}mX^2$. Often, the density and cdf of X are denoted by f_X and F_X; and those of Y, by f_Y and F_Y. To illustrate techniques for solving such a problem, we first develop some useful facts about the normal distribution.

Suppose that $X \sim N(\mu, \sigma^2)$ and that $Y = aX + b$, where $a > 0$. The cumulative distribution function of Y is

$$F_Y(y) = P(Y \le y)$$
$$= P(aX + b \le y)$$
$$= P\left(X \le \frac{y-b}{a}\right)$$
$$= F_X\left(\frac{y-b}{a}\right)$$

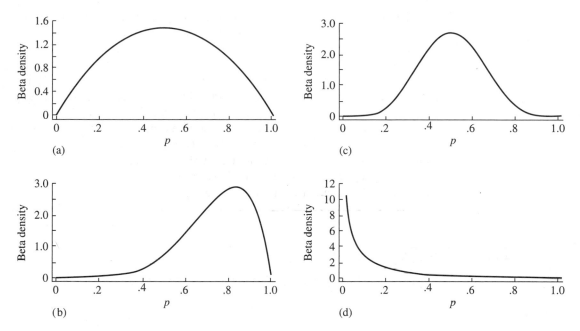

FIGURE **2.18** Beta density functions for various values of a and b: (a) $a = 2, b = 2$; (b) $a = 6, b = 2$; (c) $a = 6, b = 6$; and (d) $a = .5, b = 4$.

Thus,

$$f_Y(y) = \frac{d}{dy} F_X \left(\frac{y - b}{a} \right)$$

$$= \frac{1}{a} f_X \left(\frac{y - b}{a} \right)$$

Up to this point, we have not used the assumption of normality at all, so this result holds for a general continuous random variable, provided that F_X is appropriately differentiable. If f_X is a normal density function with parameters μ and σ, we find that, after substitution,

$$f_Y(y) = \frac{1}{a\sigma\sqrt{2\pi}} \exp \left[-\frac{1}{2} \left(\frac{y - b - a\mu}{a\sigma} \right)^2 \right]$$

From this, we see that Y follows a normal distribution with parameters $a\mu + b$ and $a\sigma$.

The case for which $a < 0$ can be analyzed similarly (see Problem 57 in the end-of-chapter problems), yielding the following proposition.

PROPOSITION A

If $X \sim N(\mu, \sigma^2)$ and $Y = aX + b$, then $Y \sim N(a\mu + b, a^2\sigma^2)$. ∎

This proposition is quite useful for finding probabilities from the normal distribution. Suppose that $X \sim N(\mu, \sigma^2)$ and we wish to find $P(x_0 < X < x_1)$ for some numbers x_0 and x_1. Consider the random variable

$$Z = \frac{X - \mu}{\sigma} = \frac{X}{\sigma} - \frac{\mu}{\sigma}$$

Applying Proposition A with $a = 1/\sigma$ and $b = -\mu/\sigma$, we see that $Z \sim N(0, 1)$, that is, Z follows a standard normal distribution. Therefore,

$$
\begin{aligned}
F_X(x) &= P(X \le x) \\
&= P\left(\frac{X - \mu}{\sigma} \le \frac{x - \mu}{\sigma}\right) \\
&= P\left(Z \le \frac{x - \mu}{\sigma}\right) \\
&= \Phi\left(\frac{x - \mu}{\sigma}\right)
\end{aligned}
$$

We thus have

$$
\begin{aligned}
P(x_0 < X < x_1) &= F_X(x_1) - F_X(x_0) \\
&= \Phi\left(\frac{x_1 - \mu}{\sigma}\right) - \Phi\left(\frac{x_0 - \mu}{\sigma}\right)
\end{aligned}
$$

Thus, probabilities for general normal random variables can be evaluated in terms of probabilities for standard normal random variables. This is quite useful, since tables need to be made up only for the standard normal distribution rather than separately for every μ and σ.

EXAMPLE A   Scores on a certain standardized test, IQ scores, are approximately normally distributed with mean $\mu = 100$ and standard deviation $\sigma = 15$. Here we are referring to the distribution of scores over a very large population, and we approximate that discrete cumulative distribution function by a normal continuous cumulative distribution function. An individual is selected at random. What is the probability that his score X satisfies $120 < X < 130$?

We can calculate this probability by using the standard normal distribution as follows:

$$
\begin{aligned}
P(120 < X < 130) &= P\left(\frac{120 - 100}{15} < \frac{X - 100}{15} < \frac{130 - 100}{15}\right) \\
&= P(1.33 < Z < 2)
\end{aligned}
$$

where Z follows a standard normal distribution. Using a table of the standard normal distribution (Table 2 of Appendix B), this probability is

$$
\begin{aligned}
P(1.33 < Z < 2) &= \Phi(2) - \Phi(1.33) \\
&= .9772 - .9082 \\
&= .069
\end{aligned}
$$

Thus, approximately 7% of the population will have scores in this range.   ∎

E X A M P L E **B**   Let $X \sim N(\mu, \sigma^2)$, and find the probability that X is less than σ away from μ; that is, find $P(|X - \mu| < \sigma)$.

This probability is

$$P(-\sigma < X - \mu < \sigma) = P\left(-1 < \frac{X - \mu}{\sigma} < 1\right)$$

$$= P(-1 < Z < 1)$$

where Z follows a standard normal distribution. From tables of the standard normal distribution, this last probability is

$$\Phi(1) - \Phi(-1) = .68$$

Thus, a normal random variable is within 1 standard deviation of its mean with probability .68. ∎

We now turn to another example involving the normal distribution.

E X A M P L E **C**   Find the density of $X = Z^2$, where $Z \sim N(0, 1)$.

Here, we have

$$F_X(x) = P(X \le x)$$

$$= P(-\sqrt{x} \le Z \le \sqrt{x})$$

$$= \Phi(\sqrt{x}) - \Phi(-\sqrt{x})$$

We find the density of X by differentiating the cdf. Since $\Phi'(x) = \phi(x)$, the chain rule gives

$$f_X(x) = \tfrac{1}{2}x^{-1/2}\phi(\sqrt{x}) + \tfrac{1}{2}x^{-1/2}\phi(-\sqrt{x})$$

$$= x^{-1/2}\phi(\sqrt{x})$$

In the last step, we used the symmetry of ϕ. Evaluating the last expression, we find

$$f_X(x) = \frac{x^{-1/2}}{\sqrt{2\pi}}e^{-x/2}, \qquad x \ge 0$$

We can recognize this as a gamma density by making use of a principle of general utility. Suppose two densities are of the forms $k_1 h(x)$ and $k_2 h(x)$; then, because they both integrate to 1, $k_1 = k_2$. Now, comparing the form of $f(x)$ given here to that of the gamma density with $\alpha = \lambda = \tfrac{1}{2}$, we recognize by this reasoning that $f(x)$ is a gamma density and that $\Gamma\left(\tfrac{1}{2}\right) = \sqrt{\pi}$. This density is also called the **chi-square density** with 1 degree of freedom. ∎

As another example, let us consider the following.

E X A M P L E **D**   Let U be a uniform random variable on $[0, 1]$, and let $V = 1/U$. To find the density of V, we first find the cdf:

$$F_V(v) = P(V \leq v)$$

$$= P\left(\frac{1}{U} \leq v\right)$$

$$= P\left(U \geq \frac{1}{v}\right)$$

$$= 1 - \frac{1}{v}$$

This expression is valid for $v \geq 1$; for $v < 1$, $F_V(v) = 0$. We can now find the density by differentiation:

$$f_V(v) = \frac{1}{v^2}, \qquad 1 \leq v < \infty$$

■

Looking back over these examples, we see that we have gone through the same basic steps in each case: first finding the cdf of the transformed variable, then differentiating to find the density, and then specifying in what region the result holds. These same steps can be used to prove the following general result.

PROPOSITION **B**

Let X be a continuous random variable with density $f(x)$ and let $Y = g(X)$ where g is a differentiable, strictly monotonic function on some interval I. Suppose that $f(x) = 0$ if x is not in I. Then Y has the density function

$$f_Y(y) = f_X(g^{-1}(y))\left|\frac{d}{dy}g^{-1}(y)\right|$$

for y such that $y = g(x)$ for some x, and $f_Y(y) = 0$ if $y \neq g(x)$ for any x in I. Here g^{-1} is the inverse function of g; that is, $g^{-1}(y) = x$ if $y = g(x)$. ■

For any specific problem, proceeding from scratch is usually easier than deciphering the notation and applying the proposition.

We conclude this section by developing some results relating the uniform distribution to other continuous distributions. Throughout, we consider a random variable X, with density f and cdf F, where F is strictly increasing on some interval I, $F = 0$ to the left of I, and $F = 1$ to the right of I. I may be a bounded interval or an unbounded interval such as the whole real line. $F^{-1}(x)$ is then well defined for $x \in I$.

PROPOSITION C

Let $Z = F(X)$; then Z has a uniform distribution on $[0, 1]$.

Proof

$$P(Z \leq z) = P(F(X) \leq z) = P(X \leq F^{-1}(z)) = F(F^{-1}(z)) = z$$

This is the uniform cdf.   ■

PROPOSITION D

Let U be uniform on $[0, 1]$, and let $X = F^{-1}(U)$. Then the cdf of X is F.

Proof

$$P(X \leq x) = P(F^{-1}(U) \leq x) = P(U \leq F(x)) = F(x)$$ ■

This last proposition is quite useful in generating pseudorandom numbers with a given cdf F. Many computer packages have routines for generating pseudorandom numbers that are uniform on $[0, 1]$. These numbers are called **pseudorandom** because they are generated according to some rule or algorithm and thus are not "really" random. Proposition D tells us that to generate random variables with cdf F, we just apply F^{-1} to uniform random numbers. This is quite practical as long as F^{-1} can be calculated easily.

EXAMPLE E   Suppose that, as part of simulation study, we want to generate random variables from an exponential distribution. For example, the performance of large queueing networks is often assessed by simulation. One aspect of such a simulation involves generating random time intervals between customer arrivals, which might be assumed to be exponentially distributed. If we have access to a uniform random number generator, we can apply Proposition D to generate exponential random numbers. The cdf is $F(t) = 1 - e^{-\lambda t}$. F^{-1} can be found by solving $x = 1 - e^{-\lambda t}$ for t:

$$e^{-\lambda t} = 1 - x$$
$$-\lambda t = \log(1 - x)$$
$$t = -\log(1 - x)/\lambda$$

Thus, if U is uniform on $[0, 1]$, then $T = -\log(1 - U)/\lambda$ is an exponential random variable with parameter λ. This can be simplified slightly by noting that $V = 1 - U$ is also uniform on $[0, 1]$ since

$$P(V \leq v) = P(1 - U \leq v) = P(U \geq 1 - v) = 1 - (1 - v) = v$$

We may thus take $T = -\log(V)/\lambda$, where V is uniform on $[0, 1]$.   ■

2.4 Concluding Remarks

This chapter introduced the concept of a random variable, one of the fundamental ideas of probability theory. A fully rigorous discussion of random variables requires a background in measure theory. The development here is sufficient for the needs of this course.

Discrete and continuous random variables have been defined, and it should be mentioned that more general random variables can also be defined and are useful on occasion. In particular, it makes sense to consider random variables that have both a discrete and a continuous component. For example, the lifetime of a transistor might be 0 with some probability $p > 0$ if it does not function at all; if it does function, the lifetime could be modeled as a continuous random variable.

2.5 Problems

1. Suppose that X is a discrete random variable with $P(X = 0) = .25$, $P(X = 1) = .125$, $P(X = 2) = .125$, and $P(X = 3) = .5$. Graph the frequency function and the cumulative distribution function of X.

2. An experiment consists of throwing a fair coin four times. Find the frequency function and the cumulative distribution function of the following random variables: (a) the number of heads before the first tail, (b) the number of heads following the first tail, (c) the number of heads minus the number of tails, and (d) the number of tails times the number of heads.

3. The following table shows the cumulative distribution function of a discrete random variable. Find the frequency function.

k	$F(k)$
0	0
1	.1
2	.3
3	.7
4	.8
5	1.0

4. If X is an integer-valued random variable, show that the frequency function is related to the cdf by $p(k) = F(k) - F(k - 1)$.

5. Show that $P(u < X \leq v) = F(v) - F(u)$ for any u and v in the cases that (a) X is a discrete random variable and (b) X is a continuous random variable.

6. Let A and B be events, and let I_A and I_B be the associated indicator random variables. Show that

$$I_{A \cap B} = I_A I_B = \min(I_A, I_B)$$

and

$$I_{A \cup B} = \max(I_A, I_B)$$

7. Find the cdf of a Bernoulli random variable.

8. Show that the binomial probabilities sum to 1.

9. For what values of p is a two-out-of-three majority decoder better than transmission of the message once?

10. Appending three extra bits to a 4-bit word in a particular way (a Hamming code) allows detection and correction of up to one error in any of the bits. If each bit has probability .05 of being changed during communication, and the bits are changed independently of each other, what is the probability that the word is correctly received (that is, 0 or 1 bit is in error)? How does this probability compare to the probability that the word will be transmitted correctly with no check bits, in which case all four bits would have to be transmitted correctly for the word to be correct?

11. Consider the binomial distribution with n trials and probability p of success on each trial. For what value of k is $P(X = k)$ maximized? This value is called the **mode** of the distribution. (*Hint:* Consider the ratio of successive terms.)

12. Which is more likely: 9 heads in 10 tosses of a fair coin or 18 heads in 20 tosses?

13. A multiple-choice test consists of 20 items, each with four choices. A student is able to eliminate one of the choices on each question as incorrect and chooses randomly from the remaining three choices. A passing grade is 12 items or more correct.
    a. What is the probability that the student passes?
    b. Answer the question in part (a) again, assuming that the student can eliminate two of the choices on each question.

14. Two boys play basketball in the following way. They take turns shooting and stop when a basket is made. Player A goes first and has probability p_1 of making a basket on any throw. Player B, who shoots second, has probability p_2 of making a basket. The outcomes of the successive trials are assumed to be independent.
    a. Find the frequency function for the total number of attempts.
    b. What is the probability that player A wins?

15. Two teams, A and B, play a series of games. If team A has probability .4 of winning each game, is it to its advantage to play the best three out of five games or the best four out of seven? Assume the outcomes of successive games are independent.

16. Show that if n approaches ∞ and r/n approaches p and m is fixed, the hypergeometric frequency function tends to the binomial frequency function. Give a heuristic argument for why this is true.

17. Suppose that in a sequence of independent Bernoulli trials, each with probability of success p, the number of failures up to the first success is counted. What is the frequency function for this random variable?

18. Continuing with Problem 17, find the frequency function for the number of failures up to the rth success.

19. Find an expression for the cumulative distribution function of a geometric random variable.

20. If X is a geometric random variable with $p = .5$, for what value of k is $P(X \le k) \approx .99$?

21. If X is a geometric random variable, show that

$$P(X > n + k - 1 | X > n - 1) = P(X > k)$$

In light of the construction of a geometric distribution from a sequence of independent Bernoulli trials, how can this be interpreted so that it is "obvious"?

22. Three identical fair coins are thrown simultaneously until all three show the same face. What is the probability that they are thrown more than three times?

23. In a sequence of independent trials with probability p of success, what is the probability that there are r successes before the kth failure?

24. (Banach Match Problem) A pipe smoker carries one box of matches in his left pocket and one box in his right. Initially, each box contains n matches. If he needs a match, the smoker is equally likely to choose either pocket. What is the frequency function for the number of matches in the other box when he first discovers that one box is empty?

25. The probability of being dealt a royal straight flush (ace, king, queen, jack, and ten of the same suit) in poker is about 1.3×10^{-8}. Suppose that an avid poker player sees 100 hands a week, 52 weeks a year, for 20 years.

  a. What is the probability that she is never dealt a royal straight flush dealt?
  b. What is the probability that she is dealt exactly two royal straight flushes?

26. The university administration assures a mathematician that he has only 1 chance in 10,000 of being trapped in a much-maligned elevator in the mathematics building. If he goes to work 5 days a week, 52 weeks a year, for 10 years, and always rides the elevator up to his office when he first arrives, what is the probability that he will never be trapped? That he will be trapped once? Twice? Assume that the outcomes on all the days are mutually independent (a dubious assumption in practice).

27. Suppose that a rare disease has an incidence of 1 in 1000. Assuming that members of the population are affected independently, find the probability of k cases in a population of 100,000 for $k = 0, 1, 2$.

28. Let $p_0, p_1, \ldots, p_n$ denote the probability mass function of the binomial distribution with parameters n and p. Let $q = 1 - p$. Show that the binomial probabilities can be computed recursively by $p_0 = q^n$ and

$$p_k = \frac{(n - k + 1)p}{kq} p_{k-1}, \qquad k = 1, 2, \ldots, n$$

Use this relation to find $P(X \le 4)$ for $n = 9000$ and $p = .0005$.

29. Show that the Poisson probabilities $p_0, p_1, \ldots$ can be computed recursively by $p_0 = \exp(-\lambda)$ and

$$p_k = \frac{\lambda}{k} p_{k-1}, \qquad k = 1, 2, \ldots$$

Use this scheme to find $P(X \le 4)$ for $\lambda = 4.5$ and compare to the results of Problem 28.

30. Suppose that in a city, the number of suicides can be approximated by a Poisson process with $\lambda = .33$ per month.

   a. Find the probability of k suicides in a year for $k = 0, 1, 2, \ldots$. What is the most probable number of suicides?
   b. What is the probability of two suicides in one week?

31. Phone calls are received at a certain residence as a Poisson process with parameter $\lambda = 2$ per hour.

   a. If Diane takes a 10-min. shower, what is the probability that the phone rings during that time?
   b. How long can her shower be if she wishes the probability of receiving no phone calls to be at most .5?

32. For what value of k is the Poisson frequency function with parameter λ maximized? (*Hint:* Consider the ratio of consecutive terms.)

33. Let $F(x) = 1 - \exp(-\alpha x^\beta)$ for $x \ge 0, \alpha > 0, \beta > 0$, and $F(x) = 0$ for $x < 0$. Show that F is a cdf, and find the corresponding density.

34. Let $f(x) = (1 + \alpha x)/2$ for $-1 \le x \le 1$ and $f(x) = 0$ otherwise, where $-1 \le \alpha \le 1$. Show that f is a density, and find the corresponding cdf. Find the quartiles and the median of the distribution in terms of α.

35. Sketch the pdf and cdf of a random variable that is uniform on $[-1, 1]$.

36. If U is a uniform random variable on $[0, 1]$, what is the distribution of the random variable $X = [nU]$, where $[t]$ denotes the greatest integer less than or equal to t?

37. A line segment of length 1 is cut once at random. What is the probability that the longer piece is more than twice the length of the shorter piece?

38. If f and g are densities, show that $\alpha f + (1 - \alpha)g$ is a density, where $0 \le \alpha \le 1$.

39. The **Cauchy** cumulative distribution function is

$$F(x) = \frac{1}{2} + \frac{1}{\pi} \tan^{-1}(x), \qquad -\infty < x < \infty$$

   a. Show that this is a cdf.
   b. Find the density function.
   c. Find x such that $P(X > x) = .1$.

40. Suppose that X has the density function $f(x) = cx^2$ for $0 \le x \le 1$ and $f(x) = 0$ otherwise.

   a. Find c.   b. Find the cdf.   c. What is $P(.1 \le X < .5)$?

41. Find the upper and lower quartiles of the exponential distribution.

42. Find the probability density for the distance from an event to its nearest neighbor for a Poisson process in the plane.

43. Find the probability density for the distance from an event to its nearest neighbor for a Poisson process in three-dimensional space.

44. Let T be an exponential random variable with parameter λ. Let X be a discrete random variable defined as $X = k$ if $k \leq T < k+1, k = 0, 1, \ldots$. Find the frequency function of X.

45. Suppose that the lifetime of an electronic component follows an exponential distribution with $\lambda = .1$.

   a. Find the probability that the lifetime is less than 10.
   b. Find the probability that the lifetime is between 5 and 15.
   c. Find t such that the probability that the lifetime is greater than t is .01.

46. T is an exponential random variable, and $P(T < 1) = .05$. What is λ?

47. If $\alpha > 1$, show that the gamma density has a maximum at $(\alpha - 1)/\lambda$.

48. Show that the gamma density integrates to 1.

49. The gamma function is a generalized factorial function.

   a. Show that $\Gamma(1) = 1$.
   b. Show that $\Gamma(x + 1) = x\Gamma(x)$. (*Hint:* Use integration by parts.)
   c. Conclude that $\Gamma(n) = (n - 1)!$, for $n = 1, 2, 3, \ldots$.
   d. Use the fact that $\Gamma(\frac{1}{2}) = \sqrt{\pi}$ to show that, if n is an odd integer,

$$\Gamma\left(\frac{n}{2}\right) = \frac{\sqrt{\pi}(n-1)!}{2^{n-1}\left(\frac{n-1}{2}\right)!}$$

50. Show by a change of variables that

$$\Gamma(x) = 2\int_0^\infty t^{2x-1}e^{-t^2}\, dt$$

$$= \int_{-\infty}^\infty e^{xt}e^{-e^t}\, dt$$

51. Show that the normal density integrates to 1. (*Hint:* First make a change of variables to reduce the integral to that for the standard normal. The problem is then to show that $\int_{-\infty}^\infty \exp(-x^2/2)\, dx = \sqrt{2\pi}$. Square both sides and reexpress the problem as that of showing

$$\left(\int_{-\infty}^\infty \exp(-x^2/2)\, dx\right)\left(\int_{-\infty}^\infty \exp(-y^2/2)\, dy\right) = 2\pi$$

Finally, write the product of integrals as a double integral and change to polar coordinates.)

52. Suppose that in a certain population, individuals' heights are approximately normally distributed with parameters $\mu = 70$ and $\sigma = 3$ in.

    a. What proportion of the population is over 6 ft. tall?
    b. What is the distribution of heights if they are expressed in centimeters? In meters?

53. Let X be a normal random variable with $\mu = 5$ and $\sigma = 10$. Find (a) $P(X > 10)$, (b) $P(-20 < X < 15)$, and (c) the value of x such that $P(X > x) = .05$.

54. If $X \sim N(\mu, \sigma^2)$, show that $P(|X - \mu| \leq .675\sigma) = .5$.

55. $X \sim N(\mu, \sigma^2)$, find the value of c in terms of σ such that $P(\mu - c \leq X \leq \mu + c) = .95$.

56. If $X \sim N(0, \sigma^2)$, find the density of $Y = |X|$.

57. $X \sim N(\mu, \sigma^2)$ and $Y = aX + b$, where $a < 0$, show that $Y \sim N(a\mu + b, a^2\sigma^2)$.

58. If U is uniform on $[0, 1]$, find the density function of $\sqrt{U}$.

59. If U is uniform on $[-1, 1]$, find the density function of U^2.

60. Find the density function of $Y = e^Z$, where $Z \sim N(\mu, \sigma^2)$. This is called the **lognormal density,** since $\log Y$ is normally distributed.

61. Find the density of cX when X follows a gamma distribution. Show that only λ is affected by such a transformation, which justifies calling λ a scale parameter.

62. Show that if X has a density function f_X and $Y = aX + b$, then

$$f_Y(y) = \frac{1}{|a|} f_X\left(\frac{y - b}{a}\right)$$

63. Suppose that Θ follows a uniform distribution on the interval $[-\pi/2, \pi/2]$. Find the cdf and density of $\tan \Theta$.

64. A particle of mass m has a random velocity, V, which is normally distributed with parameters $\mu = 0$ and σ. Find the density function of the kinetic energy, $E = \frac{1}{2}mV^2$.

65. How could random variables with the following density function be generated from a uniform random number generator?

$$f(x) = \frac{1 + \alpha x}{2}, \qquad -1 \leq x \leq 1, \qquad -1 \leq \alpha \leq 1$$

66. Let $f(x) = \alpha x^{-\alpha - 1}$ for $x \geq 1$ and $f(x) = 0$ otherwise, where α is a positive parameter. Show how to generate random variables from this density from a uniform random number generator.

67. The **Weibull** cumulative distribution function is

$$F(x) = 1 - e^{-(x/\alpha)^\beta}, \qquad x \geq 0, \qquad \alpha > 0, \qquad \beta > 0$$

    a. Find the density function.

     b. Show that if W follows a Weibull distribution, then $X = (W/\alpha)^\beta$ follows an exponential distribution.

     c. How could Weibull random variables be generated from a uniform random number generator?

68. If the radius of a circle is an exponential random variable, find the density function of the area.

69. If the radius of a sphere is an exponential random variable, find the density function of the volume.

70. Let U be a uniform random variable. Find the density function of $V = U^{-\alpha}$, $\alpha > 0$. Compare the rates of decrease of the tails of the densities as a function of α. Does the comparison make sense intuitively?

71. This problem shows one way to generate discrete random variables from a uniform random number generator. Suppose that F is the cdf of an integer-valued random variable; let U be uniform on $[0, 1]$. Define a random variable $Y = k$ if $F(k-1) < U \le F(k)$. Show that Y has cdf F. Apply this result to show how to generate geometric random variables from uniform random variables.

72. One of the most commonly used (but not one of the best) methods of generating pseudorandom numbers is the linear congruential method, which works as follows. Let x_0 be an initial number (the "seed"). The sequence is generated recursively as

$$x_n = (ax_{n-1} + c) \bmod m$$

     a. Choose values of a, c, and m, and try this out. Do the sequences "look" random?

     b. Making good choices of a, c, and m involves both art and theory. The following are some values that have been proposed: (1) $a = 69069, c = 0, m = 2^{31}$; (2) $a = 65539, c = 0, m = 2^{31}$. The latter is an infamous generator called RANDU. Try out these schemes, and examine the results.

CHAPTER 3

Joint Distributions

3.1 Introduction

This chapter is concerned with the joint probability structure of two or more random variables defined on the same sample space. Joint distributions arise naturally in many applications, of which the following are illustrative:

- In ecological studies, counts of several species, modeled as random variables, are often made. One species is often the prey of another; clearly, the number of predators will be related to the number of prey.
- The joint probability distribution of the x, y, and z components of wind velocity can be experimentally measured in studies of atmospheric turbulence.
- The joint distribution of the values of various physiological variables in a population of patients is often of interest in medical studies.
- A model for the joint distribution of age and length in a population of fish can be used to estimate the age distribution from the length distribution. The age distribution is relevant to the setting of reasonable harvesting policies.

The joint behavior of two random variables, X and Y, is determined by the cumulative distribution function

$$F(x, y) = P(X \leq x, Y \leq y)$$

regardless of whether X and Y are continuous or discrete. The cdf gives the probability that the point (X, Y) belongs to a semi-infinite rectangle in the plane, as shown in Figure 3.1. The probability that (X, Y) belongs to a given rectangle is, from Figure 3.2,

$$P(x_1 < X \leq x_2, y_1 < Y \leq y_2) = F(x_2, y_2) - F(x_2, y_1) - F(x_1, y_2)$$
$$+ F(x_1, y_1)$$

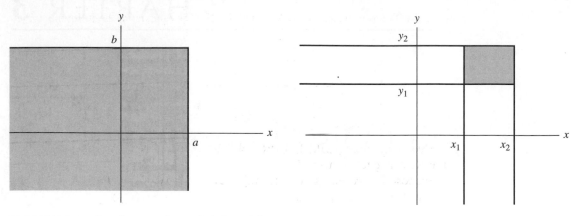

FIGURE 3.1 $F(a, b)$ gives the probability of the shaded rectangle.

FIGURE 3.2   The probability of the shaded rectangle can be found by subtracting from the probability of the (semi-infinite) rectangle having the upper-right corner (x_2, y_2) the probabilities of the (x_1, y_2) and (x_2, y_1) rectangles, and then adding back in the probability of the (x_1, y_1) rectangle.

The probability that (X, Y) belongs to a set A, for a large enough class of sets for practical purposes, can be determined by taking limits of intersections and unions of rectangles. In general, if $X_1, \ldots, X_n$ are jointly distributed random variables, their joint cdf is

$$F(x_1, x_2, \ldots, x_n) = P(X_1 \leq x_1, X_2 \leq x_2, \ldots, X_n \leq x_n)$$

Two- and higher-dimensional versions of density functions and frequency functions exist. We will start with a detailed description of such functions for the discrete case, since it is the easier one to understand.

3.2 Discrete Random Variables

Suppose that X and Y are discrete random variables defined on the same sample space and that they take on values $x_1, x_2, \ldots$, and $y_1, y_2, \ldots$, respectively. Their **joint frequency function,** or joint probability mass function, $p(x, y)$, is

$$p(x_i, y_j) = P(X = x_i, Y = y_j)$$

A simple example will illustrate this concept. A fair coin is tossed three times; let X denote the number of heads on the first toss and Y the total number of heads. From the sample space, which is

$$\Omega = \{hhh, hht, hth, htt, thh, tht, tth, ttt\}$$

we see that the joint frequency function of X and Y is as given in the following table:

x	y 0	1	2	3
0	$\frac{1}{8}$	$\frac{2}{8}$	$\frac{1}{8}$	0
1	0	$\frac{1}{8}$	$\frac{2}{8}$	$\frac{1}{8}$

Thus, for example, $p(0, 2) = P(X = 0, Y = 2) = \frac{1}{8}$. Note that the probabilities in the preceding table sum to 1.

Suppose that we wish to find the frequency function of Y from the joint frequency function. This is straightforward:

$$p_Y(0) = P(Y = 0)$$
$$= P(Y = 0, X = 0) + P(Y = 0, X = 1)$$
$$= \frac{1}{8} + 0$$
$$= \frac{1}{8}$$
$$p_Y(1) = P(Y = 1)$$
$$= P(Y = 1, X = 0) + P(Y = 1, X = 1)$$
$$= \frac{3}{8}$$

In general, to find the frequency function of Y, we simply sum down the appropriate column of the table. For this reason, p_Y is called the **marginal frequency function** of Y. Similarly, summing across the rows gives

$$p_X(x) = \sum_i p(x, y_i)$$

which is the marginal frequency function of X.

The case for several random variables is analogous. If $X_1, \ldots, X_m$ are discrete random variables defined on the same sample space, their joint frequency function is

$$p(x_1, \ldots, x_m) = P(X_1 = x_1, \ldots, X_m = x_m)$$

The marginal frequency function of X_1, for example, is

$$p_{X_1}(x_1) = \sum_{x_2 \cdots x_m} p(x_1, x_2, \ldots, x_m)$$

The two-dimensional marginal frequency function of X_1 and X_2, for example, is

$$p_{X_1 X_2}(x_1, x_2) = \sum_{x_3 \cdots x_m} p(x_1, x_2, \ldots, x_m)$$

EXAMPLE A   *Multinomial Distribution*

The multinomial distribution, an important generalization of the binomial distribution, arises in the following way. Suppose that each of n independent trials can result in one of r types of outcomes and that on each trial the probabilities of the r outcomes are $p_1, p_2, \ldots, p_r$. Let N_i be the total number of outcomes of type i in the n trials, $i = 1, \ldots, r$. To calculate the joint frequency function, we observe that any particular sequence of trials giving rise to $N_1 = n_1, N_2 = n_2, \ldots, N_r = n_r$ occurs with probability $p_1^{n_1} p_2^{n_2} \cdots p_r^{n_r}$. From Proposition C in Section 1.4.2, we know that there are $n!/(n_1! n_2! \cdots n_r!)$ such sequences, and thus the joint frequency function is

$$p(n_1, \ldots, n_r) = \binom{n}{n_1 \cdots n_r} p_1^{n_1} p_2^{n_2} \cdots p_r^{n_r}$$

The marginal distribution of any particular N_i can be obtained by summing the joint frequency function over the other n_j. This formidable algebraic task can be avoided, however, by noting that N_i can be interpreted as the number of successes in n trials, each of which has probability p_i of success and $1 - p_i$ of failure. Therefore, N_i is a binomial random variable, and

$$p_{N_i}(n_i) = \binom{n}{n_i} p_i^{n_i} (1 - p_i)^{n - n_i}$$

The multinomial distribution is applicable in considering the probabilistic properties of a **histogram.** As a concrete example, suppose that 100 independent observations are taken from a uniform distribution on [0, 1], that the interval [0, 1] is partitioned into 10 equal bins, and that the counts $n_1, \ldots, n_{10}$ in each of the 10 bins are recorded and graphed as the heights of vertical bars above the respective bins. The joint distribution of the heights is multinomial with $n = 100$ and $p_i = .1, i = 1, \ldots, 10$. Figure 3.3 shows four histograms constructed in this manner from a pseudorandom

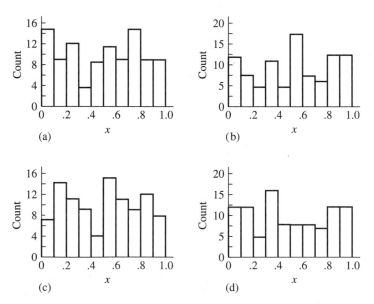

FIGURE 3.3   Four histograms, each formed from 100 independent uniform random numbers.

number generator; the figure illustrates the sort of random fluctuations that can be expected in histograms. ∎

3.3 Continuous Random Variables

Suppose that X and Y are continuous random variables with a joint cdf, $F(x, y)$. Their **joint density function** is a piecewise continuous function of two variables, $f(x, y)$. The density function $f(x, y)$ is nonnegative and $\int_{-\infty}^{\infty} \int_{-\infty}^{\infty} f(x, y)\, dy\, dx = 1$. For any "reasonable" two-dimensional set A

$$P((X, Y) \in A) = \iint_A f(x, y)\, dy\, dx$$

In particular, if $A = \{(X, Y)|X \leq x \text{ and } Y \leq y\}$,

$$F(x, y) = \int_{-\infty}^{x} \int_{-\infty}^{y} f(u, v)\, dv\, du$$

From the fundamental theorem of multivariable calculus, it follows that

$$f(x, y) = \frac{\partial^2}{\partial x \partial y} F(x, y)$$

wherever the derivative is defined.

For small δ_x and δ_y, if f is continuous at (x, y),

$$P(x \leq X \leq x + \delta_x, y \leq Y \leq y + \delta_y) = \int_{x}^{x+\delta_x} \int_{y}^{y+\delta_y} f(u, v)\, dv\, du$$
$$\approx f(x, y)\delta_x \delta_y$$

Thus, the probability that (X, Y) is in a small neighborhood of (x, y) is proportional to $f(x, y)$. Differential notation is sometimes useful:

$$P(x \leq X \leq x + dx, y \leq Y \leq y + dy) = f(x, y)\, dx\, dy$$

EXAMPLE A   Consider the bivariate density function

$$f(x, y) = \frac{12}{7}(x^2 + xy), \qquad 0 \leq x \leq 1, \qquad 0 \leq y \leq 1$$

which is plotted in Figure 3.4. $P(X > Y)$ can be found by integrating f over the set $\{(x, y)|0 \leq y \leq x \leq 1\}$:

$$P(X > Y) = \frac{12}{7} \int_{0}^{1} \int_{0}^{x} (x^2 + xy)\, dy\, dx$$

$$= \frac{9}{14}$$

∎

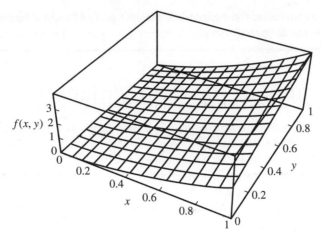

FIGURE **3.4**   The density function $f(x, y) = \frac{12}{7}(x^2 + xy)$, $0 \le x \le 1$, $0 \le y \le 1$.

The **marginal cdf** of X, or F_X, is

$$F_X(x) = P(X \le x)$$
$$= \lim_{y \to \infty} F(x, y)$$
$$= \int_{-\infty}^{x} \int_{-\infty}^{\infty} f(u, y) \, dy \, du$$

From this, it follows that the density function of X alone, known as the **marginal density** of X, is

$$f_X(x) = F_X'(x) = \int_{-\infty}^{\infty} f(x, y) \, dy$$

In the discrete case, the marginal frequency function was found by summing the joint frequency function over the other variable; in the continuous case, it is found by integration.

EXAMPLE **B**   Continuing Example A, the marginal density of X is

$$f_X(x) = \frac{12}{7} \int_{0}^{1} (x^2 + xy) \, dy$$
$$= \frac{12}{7} \left(x^2 + \frac{x}{2}\right)$$

A similar calculation shows that the marginal density of Y is $f_Y(y) = \frac{12}{7}(\frac{1}{3} + y/2)$. ■

For several jointly continuous random variables, we can make the obvious generalizations. The joint density function is a function of several variables, and the marginal density functions are found by integration. There are marginal density functions of various dimensions. Suppose that X, Y, and Z are jointly continuous random variables with density function $f(x, y, z)$. The one-dimensional marginal distribution of X is

$$f_X(x) = \int_{-\infty}^{\infty} \int_{-\infty}^{\infty} f(x, y, z) \, dy \, dz$$

and the two-dimensional marginal distribution of X and Y is

$$f_{XY}(x, y) = \int_{-\infty}^{\infty} f(x, y, z) \, dz$$

E X A M P L E **C**   *Farlie-Morgenstern Family*
If $F(x)$ and $G(y)$ are one-dimensional cdfs, it can be shown that, for any α for which $|\alpha| \leq 1$,

$$H(x, y) = F(x)G(y)\{1 + \alpha[1 - F(x)][1 - G(y)]\}$$

is a bivariate cumulative distribution function. Because $\lim_{x \to \infty} F(x) = \lim_{y \to \infty} F(y) = 1$, the marginal distributions are

$$H(x, \infty) = F(x)$$
$$H(\infty, y) = G(y)$$

In this way, an infinite number of different bivariate distributions with given marginals can be constructed.

As an example, we will construct bivariate distributions with marginals that are uniform on $[0, 1]$ $[F(x) = x, 0 \leq x \leq 1$, and $G(y) = y, 0 \leq y \leq 1]$. First, with $\alpha = -1$, we have

$$H(x, y) = xy[1 - (1 - x)(1 - y)]$$
$$= x^2 y + y^2 x - x^2 y^2, \qquad 0 \leq x \leq 1, \qquad 0 \leq y \leq 1$$

The bivariate density is

$$h(x, y) = \frac{\partial^2}{\partial x \partial y} H(x, y)$$
$$= 2x + 2y - 4xy, \qquad 0 \leq x \leq 1, \qquad 0 \leq y \leq 1$$

The density is shown in Figure 3.5. Perhaps you can imagine integrating over y (pushing all the mass onto the x axis) to produce a marginal uniform density for x.

Next, if $\alpha = 1$,

$$H(x, y) = xy[1 + (1 - x)(1 - y)]$$
$$= 2xy - x^2 y - y^2 x + x^2 y^2, \qquad 0 \leq x \leq 1, \qquad 0 \leq y \leq 1$$

The density is

$$h(x, y) = 2 - 2x - 2y + 4xy, \qquad 0 \leq x \leq 1, \qquad 0 \leq y \leq 1$$

This density is shown in Figure 3.6.

We just constructed two different bivariate distributions, both of which have uniform marginals. ∎

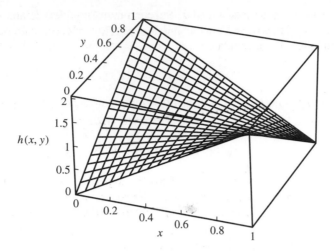

FIGURE **3.5**   The joint density $h(x, y) = 2x + 2y - 4xy$, where $0 \leq x \leq 1$ and $0 \leq y \leq 1$, which has uniform marginal densities.

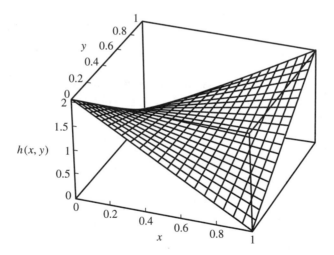

FIGURE **3.6**   The joint density $h(x, y) = 2 - 2x - 2y + 4xy$, where $0 \leq x \leq 1$ and $0 \leq y \leq 1$, which has uniform marginal densities.

A **copula** is a joint cumulative distribution function of random variables that have uniform marginal distributions. The functions $H(x, y)$ in the preceding example are copulas. Note that a copula $C(u, v)$ is nondecreasing in each variable, because it is a cdf. Also, $P(U \leq u) = C(u, 1) = u$ and $C(1, v) = v$, since the marginal distributions are uniform. We will restrict ourselves to copulas that have densities, in which case the density is

$$c(u, v) = \frac{\partial^2}{\partial u \partial v} C(u, v) \geq 0$$

Now, suppose that X and Y are continuous random variables with cdfs $F_X(x)$ and $F_Y(y)$. Then $U = F_X(x)$ and $V = F_Y(y)$ are uniform random variables (Proposition 2.3C). For a copula $C(u, v)$, consider the joint distribution defined by

$$F_{XY}(x, y) = C(F_X(x), F_Y(y))$$

Since $C(F_X(x), 1) = F_X(x)$, the marginal cdfs corresponding to F_{XY} are $F_X(x)$ and $F_Y(y)$. Using the chain rule, the corresponding density is

$$f_{XY}(x, y) = c(F_X(x), F_Y(y)) f_X(x) f_Y(y)$$

This construction points out that from the ingredients of two marginal distributions and *any* copula, a joint distribution with those marginals can be constructed. It is thus clear that the marginal distributions do not determine the joint distribution. The dependence between the random variables is captured in the copula. Copulas are not just academic curiousities—they have been extensively used in financial statistics in recent years to model dependencies in the returns of financial instruments.

E X A M P L E **D**   Consider the following joint density:

$$f(x, y) = \begin{cases} \lambda^2 e^{-\lambda y}, & 0 \le x \le y, \lambda > 0 \\ 0, & \text{elsewhere} \end{cases}$$

This joint density is plotted in Figure 3.7. To find the marginal densities, it is helpful to draw a picture showing where the density is nonzero to aid in determining the limits of integration (see Figure 3.8).

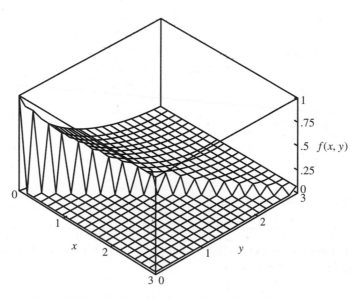

FIGURE **3.7** The joint density of Example D.

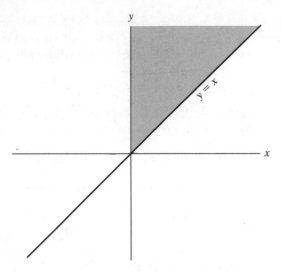

FIGURE **3.8**   The joint density of Example D is nonzero over the shaded region of the plane.

First consider the marginal density $f_X(x) = \int_{-\infty}^{\infty} f_{XY}(x, y)dy$. Since $f(x, y) = 0$ for $x \geq y$,

$$f_X(x) = \int_x^{\infty} \lambda^2 e^{-\lambda y} dy = \lambda e^{-\lambda x}, \qquad x \geq 0$$

and we see that the marginal distribution of X is exponential. Next, because $f_{XY}(x, y) = 0$ for $x \leq 0$ and $x > y$,

$$f_Y(y) = \int_0^y \lambda^2 e^{-\lambda y} dx = \lambda^2 y e^{-\lambda y}, \qquad y \geq 0$$

The marginal distribution of Y is a gamma distribution.   ∎

In some applications, it is useful to analyze distributions that are uniform over some region of space. For example, in the plane, the random point (X, Y) is uniform over a region, R, if for any $A \subset R$,

$$P((X, Y) \in A) = \frac{|A|}{|R|}$$

where $|\ |$ denotes area.

EXAMPLE E   A point is chosen randomly in a disk of radius 1. Since the area of the disk is π,

$$f(x, y) = \begin{cases} \frac{1}{\pi}, & \text{if } x^2 + y^2 \leq 1 \\ 0, & \text{otherwise} \end{cases}$$

We can calculate the distribution of R, the distance of the point from the origin. $R \leq r$ if the point lies in a disk of radius r. Since this disk has area πr^2,

$$F_R(r) = P(R \leq r) = \frac{\pi r^2}{\pi} = r^2$$

The density function of R is thus $f_R(r) = 2r, 0 \leq r \leq 1$.

Let us now find the marginal density of the x coordinate of the random point:

$$f_X(x) = \int_{-\infty}^{\infty} f(x, y)\, dy$$

$$= \frac{1}{\pi} \int_{-\sqrt{1-x^2}}^{\sqrt{1-x^2}} dy$$

$$= \frac{2}{\pi}\sqrt{1 - x^2}, \qquad -1 \leq x \leq 1$$

Note that we chose the limits of integration carefully; outside these limits the joint density is zero. (Draw a picture of the region over which $f(x, y) > 0$ and indicate the preceding limits of integration.) By symmetry, the marginal density of Y is

$$f_Y(y) = \frac{2}{\pi}\sqrt{1 - y^2}, \qquad -1 \leq y \leq 1 \qquad \blacksquare$$

E X A M P L E   **F**   *Bivariate Normal Density*

The bivariate normal density is given by the complicated expression

$$f(x, y) = \frac{1}{2\pi \sigma_X \sigma_Y \sqrt{1 - \rho^2}} \exp\left(-\frac{1}{2(1 - \rho^2)}\left[\frac{(x - \mu_X)^2}{\sigma_X^2} + \frac{(y - \mu_Y)^2}{\sigma_Y^2}\right.\right.$$

$$\left.\left. - \frac{2\rho(x - \mu_X)(y - \mu_Y)}{\sigma_X \sigma_Y}\right]\right)$$

One of the earliest uses of this bivariate density was as a model for the joint distribution of the heights of fathers and sons. The density depends on five parameters:

$$-\infty < \mu_X < \infty \qquad -\infty < \mu_Y < \infty$$

$$\sigma_X > 0 \qquad\qquad \sigma_Y > 0$$

$$-1 < \rho < 1$$

The contour lines of the density are the lines in the xy plane on which the joint density is constant. From the preceding equation, we see that $f(x, y)$ is constant if

$$\frac{(x - \mu_X)^2}{\sigma_X^2} + \frac{(y - \mu_Y)^2}{\sigma_Y^2} - \frac{2\rho(x - \mu_X)(y - \mu_Y)}{\sigma_X \sigma_Y} = \text{constant}$$

The locus of such points is an ellipse centered at (μ_X, μ_Y). If $\rho = 0$, the axes of the ellipse are parallel to the x and y axes, and if $\rho \neq 0$, they are tilted. Figure 3.9 shows several bivariate normal densities, and Figure 3.10 shows the corresponding elliptical contours.

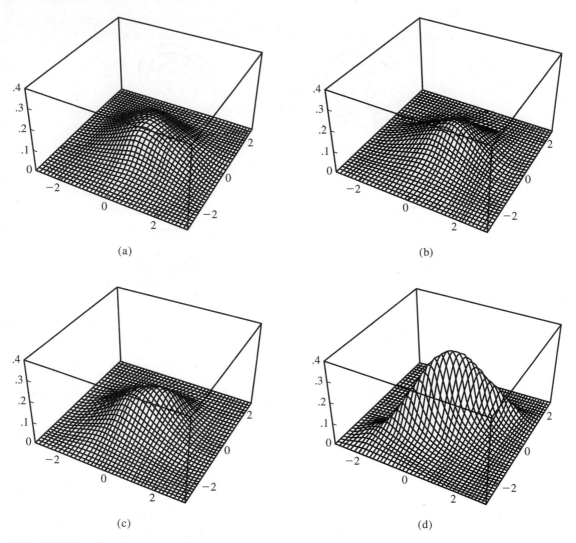

FIGURE 3.9   Bivariate normal densities with $\mu_X = \mu_Y = 0$ and $\sigma_X = \sigma_Y = 1$ and (a) $\rho = 0$, (b) $\rho = .3$, (c) $\rho = .6$, (d) $\rho = .9$.

The marginal distributions of X and Y are $N(\mu_X, \sigma_X^2)$ and $N(\mu_Y, \sigma_Y^2)$, respectively, as we will now demonstrate. The marginal density of X is

$$f_X(x) = \int_{-\infty}^{\infty} f_{XY}(x, y)\, dy$$

Making the changes of variables $u = (x - \mu_X)/\sigma_X$ and $v = (y - \mu_Y)/\sigma_Y$ gives us

$$f_X(x) = \frac{1}{2\pi \sigma_X \sqrt{1 - \rho^2}} \int_{-\infty}^{\infty} \exp\left[-\frac{1}{2(1 - \rho^2)}(u^2 + v^2 - 2\rho uv)\right] dv$$

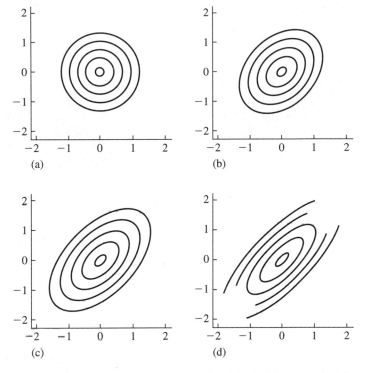

FIGURE **3.10** The elliptical contours of the bivariate normal densities of Figure 3.9.

To evaluate this integral, we use the technique of completing the square. Using the identity

$$u^2 + v^2 - 2\rho uv = (v - \rho u)^2 + u^2(1 - \rho^2)$$

we have

$$f_X(x) = \frac{1}{2\pi\sigma_X\sqrt{1-\rho^2}} e^{-u^2/2} \int_{-\infty}^{\infty} \exp\left[-\frac{1}{2(1-\rho^2)}(v-\rho u)^2\right] dv$$

Finally, recognizing the integral as that of a normal density with mean ρu and variance $(1 - \rho^2)$, we obtain

$$f_X(x) = \frac{1}{\sigma_X\sqrt{2\pi}} e^{-(1/2)\left[(x-\mu_X)^2/\sigma_X^2\right]}$$

which is a normal density, as was to be shown. Thus, for example, the marginal distributions of x and y in Figure 3.9 are all standard normal, even though the joint distributions of (a)–(d) are quite different from each other. ∎

We saw in our discussion of copulas earlier in this section that marginal densities do not determine joint densities. For example, we can take both marginal densities to be normal with parameters $\mu = 0$ and $\sigma = 1$ and use the Farlie-Morgenstern

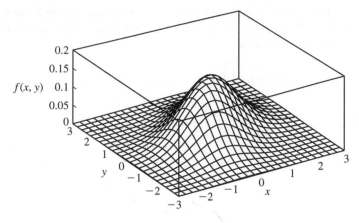

FIGURE **3.11**   A bivariate density that has normal marginals but is not bivariate normal. The contours of the density shown in the xy plane are not elliptical.

copula with density $c(u, v) = 2 - 2u - 2v + 4uv$. Denoting the normal density and cumulative distribution functions by $\phi(x)$ and $\Phi(x)$, the bivariate density is

$$f(x, y) = (2 - 2\Phi(x) - 2\Phi(y) + 4\Phi(x)\Phi(y))\phi(x)\phi(y)$$

This density and its contours are shown in Figure 3.11. Note that the contours are not elliptical. This bivariate density has normal marginals, but it is not a bivariate normal density.

3.4 Independent Random Variables

DEFINITION

Random variables $X_1, X_2, \ldots, X_n$ are said to be *independent* if their joint cdf factors into the product of their marginal cdf's:

$$F(x_1, x_2, \ldots, x_n) = F_{X_1}(x_1)F_{X_2}(x_2) \cdots F_{X_n}(x_n)$$

for all $x_1, x_2, \ldots, x_n$. ∎

The definition holds for both continuous and discrete random variables. For discrete random variables, it is equivalent to state that their joint frequency function factors; for continuous random variables, it is equivalent to state that their joint density function factors. To see why this is true, consider the case of two jointly continuous random variables, X and Y. If they are independent, then

$$F(x, y) = F_X(x)F_Y(y)$$

and taking the second mixed partial derivative makes it clear that the density function factors. On the other hand, if the density function factors, then the joint cdf can be expressed as a product:

$$F(x, y) = \int_{-\infty}^{x} \int_{-\infty}^{y} f_X(u) f_Y(v) \, dv \, du$$

$$= \left[\int_{-\infty}^{x} f_X(u) \, du \right] \left[\int_{-\infty}^{y} f_Y(v) \, dv \right] = F_X(x) F_Y(y)$$

It can be shown that the definition implies that if X and Y are independent, then

$$P(X \in A, Y \in B) = P(X \in A) P(Y \in B)$$

It can also be shown that if g and h are functions, then $Z = g(X)$ and $W = h(Y)$ are independent as well. A sketch of an argument goes like this (the details are beyond the level of this course): We wish to find $P(Z \leq z, W \leq w)$. Let $A(z)$ be the set of x such that $g(x) \leq z$, and let $B(w)$ be the set of y such that $h(y) \leq w$. Then

$$P(Z \leq z, W \leq w) = P(X \in A(z), Y \in B(w))$$

$$= P(X \in A(z)) P(Y \in B(w))$$

$$= P(Z \leq z) P(W \leq w)$$

E X A M P L E   **A**   Suppose that the point (X, Y) is uniformly distributed on the square $S = \{(x, y) \mid -1/2 \leq x \leq 1/2, -1/2 \leq y \leq 1/2\}$: $f_{XY}(x, y) = 1$ for (x, y) in S and 0 elsewhere. Make a sketch of this square. You can visualize that the marginal distributions of X and Y are uniform on $[-1/2, 1/2]$. For example, the marginal density at a point x, $-1/2 \leq x \leq 1/2$ is found by integrating (*summing*) the joint density over the vertical line that meets the horizontal axis at x. Thus, $f_X(x) = 1, -1/2 \leq x \leq 1/2$ and $f_Y(y) = 1$, and $-1/2 \leq y \leq 1/2$. The joint density is equal to the product of the marginal densities, so X and Y are independent. You should be able to see from our sketch that knowing the value of X gives no information about the possible values of Y. ∎

E X A M P L E   **B**   Now consider rotating the square of the previous example by $90°$ to form a diamond. Sketch this diamond. From the sketch, you can see that the marginal density of X is nonnegative for $-1/2 \leq x \leq 1/2$ as before, but it is not uniform, and similarly for the marginal density of Y. Thus, for example, $f_X(.9) > 0$ and $f_Y(.9) > 0$. But from the sketch you can also see that $f_{XY}(.9, .9) = 0$. Thus, X and Y are not independent. Finally, the sketch shows you that knowing the value of X— for example, $X = .9$— constrains the possible values of Y. ∎

EXAMPLE C   *Farlie-Morgenstern Family*
From Example C in Section 3.3, we see that X and Y are independent only if $\alpha = 0$, since only in this case does the joint cdf H factor into the product of the marginals F and G.   ∎

EXAMPLE D   If X and Y follow a bivariate normal distribution (Example F from Section 3.3) and $\rho = 0$, their joint density factors into the product of two normal densities, and therefore X and Y are independent.   ∎

EXAMPLE E   Suppose that a node in a communications network has the property that if two packets of information arrive within time τ of each other, they "collide" and then have to be retransmitted. If the times of arrival of the two packets are independent and uniform on $[0, T]$, what is the probability that they collide?

The times of arrival of two packets, T_1 and T_2, are independent and uniform on $[0, T]$, so their joint density is the product of the marginals, or

$$f(t_1, t_2) = \frac{1}{T^2}$$

for t_1 and t_2 in the square with sides $[0, T]$. Therefore, (T_1, T_2) is uniformly distributed over the square. The probability that the two packets collide is proportional to the area of the shaded strip in Figure 3.12. Each of the unshaded triangles of the figure has area $(T - \tau)^2/2$, and thus the area of the shaded area is $T^2 - (T - \tau)^2$. Integrating $f(t_1, t_2)$ over this area gives the desired probability: $1 - (1 - \tau/T)^2$.   ∎

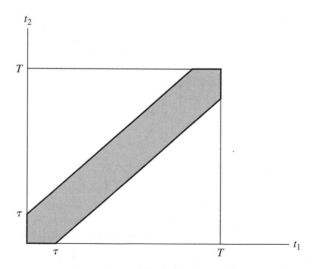

FIGURE **3.12**   The probability that the two packets collide is proportional to the area of the shaded region $|t_1 - t_2| < \tau$

3.5 Conditional Distributions

3.5.1 The Discrete Case

If X and Y are jointly distributed discrete random variables, the conditional probability that $X = x_i$ given that $Y = y_j$ is, if $p_Y(y_j) > 0$,

$$P(X = x_i | Y = y_j) = \frac{P(X = x_i, Y = y_j)}{P(Y = y_j)}$$

$$= \frac{p_{XY}(x_i, y_j)}{p_Y(y_j)}$$

This probability is defined to be zero if $p_Y(y_j) = 0$. We will denote this conditional probability by $p_{X|Y}(x|y)$. Note that this function of x is a genuine frequency function since it is nonnegative and sums to 1 and that $p_{Y|X}(y|x) = p_Y(y)$ if X and Y are independent.

E X A M P L E **A**   We return to the simple discrete distribution considered in Section 3.2, reproducing the table of values for convenience here:

	y			
x	0	1	2	3
0	$\frac{1}{8}$	$\frac{2}{8}$	$\frac{1}{8}$	0
1	0	$\frac{1}{8}$	$\frac{2}{8}$	$\frac{1}{8}$

The conditional frequency function of X given $Y = 1$ is

$$p_{X|Y}(0|1) = \frac{\frac{2}{8}}{\frac{3}{8}} = \frac{2}{3}$$

$$p_{X|Y}(1|1) = \frac{\frac{1}{8}}{\frac{3}{8}} = \frac{1}{3}$$ ■

The definition of the conditional frequency function just given can be reexpressed as

$$p_{XY}(x, y) = p_{X|Y}(x|y)p_Y(y)$$

(the multiplication law of Chapter 1). This useful equation gives a relationship between the joint and conditional frequency functions. Summing both sides over all values of y, we have an extremely useful application of the law of total probability:

$$p_X(x) = \sum_y p_{X|Y}(x|y)p_Y(y)$$

EXAMPLE B   Suppose that a particle counter is imperfect and independently detects each incoming particle with probability p. If the distribution of the number of incoming particles in a unit of time is a Poisson distribution with parameter λ, what is the distribution of the number of counted particles?

Let N denote the true number of particles and X the counted number. From the statement of the problem, the conditional distribution of X given $N = n$ is binomial, with n trials and probability p of success. By the law of total probability,

$$P(X = k) = \sum_{n=0}^{\infty} P(N = n)P(X = k|N = n)$$

$$= \sum_{n=k}^{\infty} \frac{\lambda^n e^{-\lambda}}{n!} \binom{n}{k} p^k (1 - p)^{n-k}$$

$$= \frac{(\lambda p)^k}{k!} e^{-\lambda} \sum_{n=k}^{\infty} \lambda^{n-k} \frac{(1 - p)^{n-k}}{(n - k)!}$$

$$= \frac{(\lambda p)^k}{k!} e^{-\lambda} \sum_{j=0}^{\infty} \frac{\lambda^j (1 - p)^j}{j!}$$

$$= \frac{(\lambda p)^k}{k!} e^{-\lambda} e^{\lambda(1-p)}$$

$$= \frac{(\lambda p)^k}{k!} e^{-\lambda p}$$

We see that the distribution of X is a Poisson distribution with parameter λp. This model arises in other applications as well. For example, N might denote the number of traffic accidents in a given time period, with each accident being fatal or nonfatal; X would then be the number of fatal accidents.   ∎

3.5.2 The Continuous Case

In analogy with the definition in the preceding section, if X and Y are jointly continuous random variables, the **conditional density** of Y given X is defined to be

$$f_{Y|X}(y|x) = \frac{f_{XY}(x, y)}{f_X(x)}$$

if $0 < f_X(x) < \infty$, and 0 otherwise. This definition is in accord with the result to which a differential argument would lead. We would define $f_{Y|X}(y|x)\, dy$ as $P(y \leq Y \leq y + dy | x \leq X \leq x + dx)$ and calculate

$$P(y \leq Y \leq y + dy | x \leq X \leq x + dx) = \frac{f_{XY}(x, y)\, dx\, dy}{f_X(x)\, dx} = \frac{f_{XY}(x, y)}{f_X(x)} dy$$

Note that the rightmost expression is interpreted as a function of y, x being fixed. The numerator is the joint density $f_{XY}(x, y)$, viewed as a function of y for fixed x: you can visualize it as the curve formed by slicing through the joint density function perpendicular to the x axis. The denominator normalizes that curve to have unit area.

The joint density can be expressed in terms of the marginal and conditional densities as follows:

$$f_{XY}(x, y) = f_{Y|X}(y|x) f_X(x)$$

Integrating both sides over x allows the marginal density of Y to be expressed as

$$f_Y(y) = \int_{-\infty}^{\infty} f_{Y|X}(y|x) f_X(x) \, dx$$

which is the law of total probability for the continuous case.

EXAMPLE A   In Example D in Section 3.3, we saw that

$$f_{XY}(x, y) = \lambda^2 e^{-\lambda y}, \qquad 0 \le x \le y$$
$$f_X(x) = \lambda e^{-\lambda x}, \qquad x \ge 0$$
$$f_Y(y) = \lambda^2 y e^{-\lambda y}, \qquad y \ge 0$$

Let us find the conditional densities. Before doing the formal calculations, it is informative to examine the joint density for x and y, respectively, held constant. If x is constant, the joint density decays exponentially in y for $y \ge x$; if y is constant, the joint density is constant for $0 \le x \le y$. (See Figure 3.7.) Now let us find the conditional densities according to the preceding definition. First,

$$f_{Y|X}(y|x) = \frac{\lambda^2 e^{-\lambda y}}{\lambda e^{-\lambda x}} = \lambda e^{-\lambda(y-x)}, \qquad y \ge x$$

The conditional density of Y given $X = x$ is exponential on the interval $[x, \infty)$. Expressing the joint density as

$$f_{XY}(x, y) = f_{Y|X}(y|x) f_X(x)$$

we see that we could generate X and Y according to f_{XY} in the following way: First, generate X as an exponential random variable (f_X), and then generate Y as another exponential random variable ($f_{Y|X}$) on the interval $[x, \infty)$. From this representation, we see that Y may be interpreted as the sum of two independent exponential random variables and that the distribution of this sum is gamma, a fact that we will derive later by a different method.

Now,

$$f_{X|Y}(x|y) = \frac{\lambda^2 e^{-\lambda y}}{\lambda^2 y e^{-\lambda y}} = \frac{1}{y}, \qquad 0 \le x \le y$$

The conditional density of X given $Y = y$ is uniform on the interval $[0, y]$. Finally, expressing the joint density as

$$f_{XY}(x, y) = f_{X|Y}(x|y) f_Y(y)$$

we see that alternatively we could generate X and Y according to the density f_{XY} by first generating Y from a gamma density and then generating X uniformly on $[0, y]$. Another interpretation of this result is that, conditional on the sum of two independent exponential random variables, the first is uniformly distributed.   ∎

EXAMPLE B   *Stereology*

In metallography and other applications of quantitative microscopy, aspects of a three-dimensional structure are deduced from studying two-dimensional cross sections. Concepts of probability and statistics play an important role (DeHoff and Rhines 1968). In particular, the following problem arises. Spherical particles are dispersed in a medium (grains in a metal, for example); the density function of the radii of the spheres can be denoted as $f_R(r)$. When the medium is sliced, two-dimensional, circular cross sections of the spheres are observed; let the density function of the radii of these circles be denoted by $f_X(x)$. How are these density functions related?

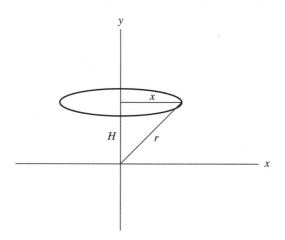

FIGURE **3.13**   A plane slices a sphere of radius r at a distance H from its center, producing a circle of radius x.

To derive the relationship, we assume that the cross-sectioning plane is chosen at random, fix $R = r$, and find the conditional density $f_{X|R}(x|r)$. As shown in Figure 3.13, let H denote the distance from the center of the sphere to the planar cross section. By our assumption, H is uniformly distributed on $[0, r]$, and $X = \sqrt{r^2 - H^2}$. We can thus find the conditional distribution of X given $R = r$:

$$F_{X|R}(x|r) = P(X \leq x)$$
$$= P(\sqrt{r^2 - H^2} \leq x)$$
$$= P(H \geq \sqrt{r^2 - x^2})$$
$$= 1 - \frac{\sqrt{r^2 - x^2}}{r}, \qquad 0 \leq x \leq r$$

Differentiating, we find

$$f_{X|R}(x|r) = \frac{x}{r\sqrt{r^2 - x^2}}, \qquad 0 \le x \le r$$

The marginal density of X is, from the law of total probability,

$$f_X(x) = \int_{-\infty}^{\infty} f_{X|R}(x|r) f_R(r) \, dr$$

$$= \int_x^{\infty} \frac{x}{r\sqrt{r^2 - x^2}} f_R(r) \, dr$$

[The limits of integration are x and ∞ since for $r \le x$, $f_{X|R}(x|r) = 0$.] This equation is called *Abel's equation*. In practice, the marginal density f_X can be approximated by making measurements of the radii of cross-sectional circles. Then the problem becomes that of trying to solve for an approximation to f_R, since it is the distribution of spherical radii that is of real interest. ∎

EXAMPLE C   *Bivariate Normal Density*
The conditional density of Y given X is the ratio of the bivariate normal density to a univariate normal density. After some messy algebra, this ratio simplifies to

$$f_{Y|X}(y|x) = \frac{1}{\sigma_Y \sqrt{2\pi(1 - \rho^2)}} \exp\left(-\frac{1}{2} \frac{\left[y - \mu_Y - \rho \frac{\sigma_Y}{\sigma_X}(x - \mu_X) \right]^2}{\sigma_Y^2(1 - \rho^2)} \right)$$

This is a normal density with mean $\mu_Y + \rho(x - \mu_X)\sigma_Y/\sigma_X$ and variance $\sigma_Y^2(1 - \rho^2)$. The conditional distribution of Y given X is a univariate normal distribution.

In Example B in Section 2.2.3, the distribution of the velocity of a turbulent wind flow was shown to be approximately normally distributed. Van Atta and Chen (1968) also measured the joint distribution of the velocity at a point at two different times, t and $t + \tau$. Figure 3.14 shows the measured conditional density of the velocity, v_2, at time $t + \tau$, given various values of v_1. There is a systematic departure from the normal distribution. Therefore, it appears that, even though the velocity is normally distributed, the joint distribution of v_1 and v_2 is not bivariate normal. This should not be totally unexpected, since the relation of v_1 and v_2 must conform to equations of motion and continuity, which may not permit a joint normal distribution. ∎

Example C illustrates that even when two random variables are marginally normally distributed, they need not be jointly normally distributed.

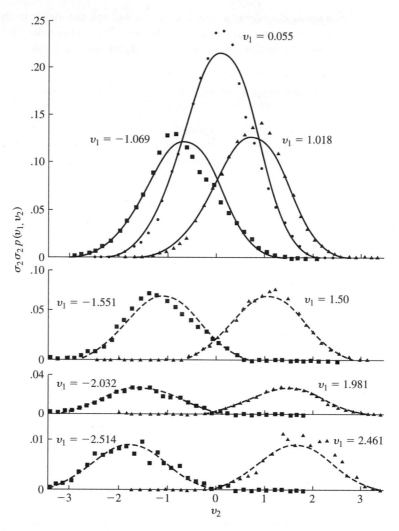

F I G U R E **3.14**   The conditional densities of v_2 given v_1 for selected values of v_1, where v_1 and v_2 are components of the velocity of a turbulent wind flow at different times. The solid lines are the conditional densities according to a normal fit, and the triangles and squares are empirical values determined from 409,600 observations.

E X A M P L E **D**   *Rejection Method*

The **rejection method** is commonly used to generate random variables from a density function, especially when the inverse of the cdf cannot be found in closed form and therefore the inverse cdf method, Proposition D in Section 2.3, cannot be used. Suppose that f is a density function that is nonzero on an interval $[a, b]$ and zero outside the interval (a and b may be infinite). Let $M(x)$ be a function such that $M(x) \geq f(x)$ on $[a, b]$, and let

$$m(x) = \frac{M(x)}{\int_a^b M(x)\,dx}$$

be a probability density function. As we will see, the idea is to choose M so that it is easy to generate random variables from m. If $[a, b]$ is finite, m can be chosen to be the uniform distribution on $[a, b]$. The algorithm is as follows:

Step 1:   Generate T with the density m.

Step 2:   Generate U, uniform on $[0, 1]$ and independent of T. If $M(T) \times U \leq f(T)$, then let $X = T$ (accept T). Otherwise, go to Step 1 (reject T).

See Figure 3.15. From the figure, we can see that a geometrical interpretation of this algorithm is as follows: Throw a dart that lands uniformly in the rectangular region of the figure. If the dart lands below the curve $f(x)$, record its x coordinate; otherwise, reject it.

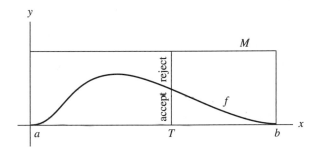

FIGURE 3.15   Illustration of the rejection method.

We must check that the density function of the random variable X thus obtained is in fact f:

$$P(x \leq X \leq x + dx) = P(x \leq T \leq x + dx \mid \text{accept})$$

$$= \frac{P(x \leq T \leq x + dx \ \text{and accept})}{P(\text{accept})}$$

$$= \frac{P(\text{accept}|x \leq T \leq x + dx)P(x \leq T \leq x + dx)}{P(\text{accept})}$$

First consider the numerator of this expression. We have

$$P(\text{accept}|x \leq T \leq x + dx) = P(U \leq f(x)/M(x)) = \frac{f(x)}{M(x)}$$

so that the numerator is

$$\frac{m(x) \, dx \ f(x)}{M(x)} = \frac{f(x) \, dx}{\int_a^b M(x) \, dx}$$

From the law of total probability, the denominator is

$$P(\text{accept}) = P(U \leq f(T)/M(T))$$

$$= \int_a^b \frac{f(t)}{M(t)} m(t) \, dt = \frac{1}{\int_a^b M(t) \, dt}$$

where the last two steps follow from the definition of m and since f integrates to 1. Finally, we see that the numerator over the denominator is $f(x) \, dx$.   ∎

In order for the rejection method to be computationally efficient, the algorithm should lead to acceptance with high probability; otherwise, many rejection steps may have to be looped through for each acceptance.

E X A M P L E **E**   *Bayesian Inference*

A freshly minted coin has a certain probability of coming up heads if it is spun on its edge, but that probability is not necessarily equal to $\frac{1}{2}$. Now suppose it is spun n times and comes up heads X times. What has been learned about the chance the coin comes up heads? We will go through a Bayesian treatment of this problem. Let Θ denote the probability that the coin will come up heads. We represent our knowledge about Θ before gathering any data by a probability density on [0, 1], called the **prior density.** If we are totally ignorant about Θ, we might represent our state of knowledge by a uniform density on [0, 1]:

$$f_\Theta(\theta) = 1, \quad 0 \le \theta \le 1.$$

We will see how observing X changes our knowledge about Θ, transforming the prior distribution into a "posterior" distribution.

Given a value θ, X follows a binomial distribution with n trials and probability of success θ:

$$f_{X|\Theta}(x|\theta) = \binom{n}{x}\theta^x(1-\theta)^{n-x}, \qquad x = 0, 1, \ldots, n$$

Now Θ is continuous and X is discrete, and they have a joint probability distribution:

$$f_{\Theta,X}(\theta, x) = f_{X|\Theta}(x|\theta)f_\Theta(\theta)$$

$$= \binom{n}{x}\theta^x(1-\theta)^{n-x}, \qquad x = 0, 1, \ldots, n, \;\; 0 \le \theta \le 1$$

This is a density function in θ and a probability mass function in x, an object of a kind we have not seen before. We can calculate the marginal density X by integrating the joint over θ:

$$f_X(x) = \int_0^1 \binom{n}{x}\theta^x(1-\theta)^{n-x}d\theta$$

We can calculate this formidable looking integral by a trick. First write

$$\binom{n}{x} = \frac{n!}{x!(n-x)!} = \frac{\Gamma(n+1)}{\Gamma(x+1)\Gamma(n-x+1)}$$

(If k is an integer, $\Gamma(k) = (k-1)!$; see Problem 49 in Chapter 2). Recall the beta density (Section 2.2.4)

$$g(u) = \frac{\Gamma(a+b)}{\Gamma(a)\Gamma(b)}u^{a-1}(1-u)^{b-1}, \qquad 0 \le u \le 1$$

The fact that this density integrates to 1 tells us that

$$\int_0^1 u^{a-1}(1-u)^{b-1}du = \frac{\Gamma(a)\Gamma(b)}{\Gamma(a+b)}$$

Thus, identifying u with θ, $a - 1$ with x, and $b - 1$ with $n - x$,

$$f_X(x) = \frac{\Gamma(n+1)}{\Gamma(x+1)\Gamma(n-x+1)} \int_0^1 \theta^x(1-\theta)^{n-x}d\theta$$

$$= \frac{\Gamma(n+1)}{\Gamma(x+1)\Gamma(n-x+1)} \frac{\Gamma(x+1)\Gamma(n-x+1)}{\Gamma(n+2)}$$

$$= \frac{1}{n+1}, \qquad x = 0, 1, \ldots, n$$

Thus, if our prior on θ is uniform, each outcome of X is *a priori* equally likely.

Our knowledge about Θ having observed $X = x$ is quantified in the conditional density of Θ given $X = x$:

$$f_{\Theta|X}(\theta|x) = \frac{f_{\Theta,X}(\theta, x)}{f_X(x)}$$

$$= (n+1)\binom{n}{x}\theta^x(1-\theta)^{n-x}$$

$$= (n+1)\frac{\Gamma(n+1)}{\Gamma(x+1)\Gamma(n-x+1)}\theta^x(1-\theta)^{n-x}$$

$$= \frac{\Gamma(n+2)}{\Gamma(x+1)\Gamma(n-x+1)}\theta^x(1-\theta)^{n-x}$$

The relationship $x\Gamma(x) = \Gamma(x+1)$ has been used in the second step (see Problem 49, Chapter 2). Bear in mind that for each fixed x, this is a function of θ—the posterior density of θ given x—which quantifies our opinion about Θ having observed x heads in n spins. The posterior density is a beta density with parameters $a = x + 1$, $b = n - x + 1$.

A one-Euro coin has the number 1 on one face and a bird on the other face. I spun such a coin 20 times: the 1 came up 13 of the 20 times. Using the prior, $\Theta \sim U[0, 1]$, the posterior is beta with $a = x + 1 = 14$ and $b = n - x + 1 = 8$. Figure 3.16 shows this posterior, which represents my opinion if I was initially totally ignorant of θ and then observed thirteen 1s in 20 spins. From the figure, it is extremely unlikely that $\theta < 0.25$, for example. My probability, or belief, that θ is greater than $\frac{1}{2}$ is the area under the density to the right of $\frac{1}{2}$, which can be calculated to be 0.91. I can be 91% certain that θ is greater than $\frac{1}{2}$.

We need to distinguish between the steps of the preceding probability calculations, which are are mathematically straightforward; and the interpretation of the results, which goes beyond the mathematics and requires a model that belief can be expressed in terms of probability and revised using the laws of probability. See Figure 3.16. ∎

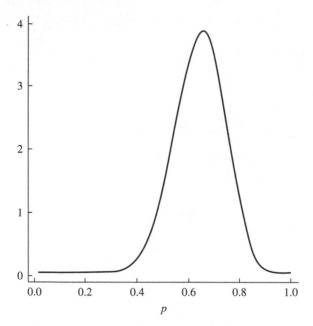

FIGURE **3.16**   Beta density with parameters $a = 14$ and $b = 8$.

3.6 Functions of Jointly Distributed Random Variables

The distribution of a function of a single random variable was developed in Section 2.3. In this section, that development is extended to several random variables, but first some important special cases are considered.

3.6.1 Sums and Quotients

Suppose that X and Y are discrete random variables taking values on the integers and having the joint frequency function $p(x, y)$, and let $Z = X + Y$. To find the frequency function of Z, we note that $Z = z$ whenever $X = x$ and $Y = z - x$, where x is an integer. The probability that $Z = z$ is thus the sum over all x of these joint probabilities, or

$$p_Z(z) = \sum_{x=-\infty}^{\infty} p(x, z - x)$$

If X and Y are independent so that $p(x, y) = p_X(x)p_Y(y)$, then

$$p_Z(z) = \sum_{x=-\infty}^{\infty} p_X(x)p_Y(z - x)$$

This sum is called the **convolution** of the sequences p_X and p_Y.

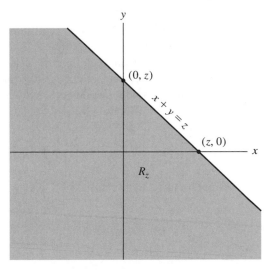

FIGURE **3.17** $X + Y \leq z$ whenever (X, Y) is in the shaded region R_z.

The continuous case is very similar. Supposing that X and Y are continuous random variables, we first find the cdf of Z and then differentiate to find the density. Since $Z \leq z$ whenever the point (X, Y) is in the shaded region R_z shown in Figure 3.17, we have

$$F_Z(z) = \iint\limits_{R_z} f(x, y) \, dx \, dy$$

$$= \int_{-\infty}^{\infty} \int_{-\infty}^{z-x} f(x, y) \, dy \, dx$$

In the inner integral, we make the change of variables $y = v - x$ to obtain

$$F_Z(z) = \int_{-\infty}^{\infty} \int_{-\infty}^{z} f(x, v - x) \, dv \, dx$$

$$= \int_{-\infty}^{z} \int_{-\infty}^{\infty} f(x, v - x) \, dx \, dv$$

Differentiating, we have, if $\int_{-\infty}^{\infty} f(x, z - x) \, dx$ is continuous at z,

$$f_Z(z) = \int_{-\infty}^{\infty} f(x, z - x) \, dx$$

which is the obvious analogue of the result for the discrete case.

If X and Y are independent,

$$f_Z(z) = \int_{-\infty}^{\infty} f_X(x) f_Y(z - x) \, dx$$

This integral is called the **convolution** of the functions f_X and f_Y.

EXAMPLE A   Suppose that the lifetime of a component is exponentially distributed and that an identical and independent backup component is available. The system operates as long as one of the components is functional; therefore, the distribution of the life of the system is that of the sum of two independent exponential random variables. Let T_1 and T_2 be independent exponentials with parameter λ, and let $S = T_1 + T_2$.

$$f_S(s) = \int_0^s \lambda e^{-\lambda t} \lambda e^{-\lambda(s-t)} dt$$

It is important to note the limits of integration. Beyond these limits, one of the two component densities is zero. When dealing with densities that are nonzero only on some subset of the real line, we must always be careful. Continuing, we have

$$f_S(s) = \lambda^2 \int_0^s e^{-\lambda s} dt$$
$$= \lambda^2 s e^{-\lambda s}$$

This is a gamma distribution with parameters 2 and λ (compare with Example A in Section 3.5.2).   ∎

Let us next consider the quotient of two continuous random variables. The derivation is very similar to that for the sum of such variables, given previously: We first find the cdf and then differentiate to find the density. Suppose that X and Y are continuous with joint density function f and that $Z = Y/X$. Then $F_Z(z) = P(Z \le z)$ is the probability of the set of (x, y) such that $y/x \le z$. If $x > 0$, this is the set $y \le xz$; if $x < 0$, it is the set $y \ge xz$. Thus,

$$F_Z(z) = \int_{-\infty}^0 \int_{xz}^{\infty} f(x, y) \, dy \, dx + \int_0^{\infty} \int_{-\infty}^{xz} f(x, y) \, dy \, dx$$

To remove the dependence of the inner integrals on x, we make the change of variables $y = xv$ in the inner integrals and obtain

$$F_Z(z) = \int_{-\infty}^0 \int_z^{-\infty} xf(x, xv) \, dv \, dx + \int_0^{\infty} \int_{-\infty}^z xf(x, xv) \, dv \, dx$$
$$= \int_{-\infty}^0 \int_{-\infty}^z (-x)f(x, xv) \, dv \, dx + \int_0^{\infty} \int_{-\infty}^z xf(x, xv) \, dv \, dx$$
$$= \int_{-\infty}^z \int_{-\infty}^{\infty} |x| f(x, xv) \, dx \, dv$$

Finally, differentiating (again under an assumption of continuity), we find

$$f_Z(z) = \int_{-\infty}^{\infty} |x| f(x, xz) \, dx$$

In particular, if X and Y are independent,

$$f_Z(z) = \int_{-\infty}^{\infty} |x| f_X(x) f_Y(xz) \, dx$$

EXAMPLE **B**   Suppose that X and Y are independent standard normal random variables and that $Z = Y/X$. We then have

$$f_Z(z) = \int_{-\infty}^{\infty} \frac{|x|}{2\pi} e^{-x^2/2} e^{-x^2 z^2/2} \, dx$$

From the symmetry of the integrand about zero,

$$f_Z(z) = \frac{1}{\pi} \int_{0}^{\infty} x e^{-x^2((z^2+1)/2)} \, dx$$

Cauchy Density

To simplify this, we make the change of variables $u = x^2$ to obtain

$$f_Z(z) = \frac{1}{2\pi} \int_{0}^{\infty} e^{-u((z^2+1)/2)} \, du$$

Next, using the fact that $\int_0^\infty \lambda \exp(-\lambda x) \, dx = 1$ with $\lambda = (z^2 + 1)/2$, we get

$$f_Z(z) = \frac{1}{\pi(z^2 + 1)}, \qquad -\infty < z < \infty$$

This density is called the **Cauchy density.** Like the standard normal density, the Cauchy density is symmetric about zero and bell-shaped, but the tails of the Cauchy tend to zero very slowly compared to the tails of the normal. This can be interpreted as being because of a substantial probability that X in the quotient Y/X is near zero.   ∎

Example B indicates one method of generating Cauchy random variables—we can generate independent standard normal random variables and form their quotient. The next section shows how to generate standard normals.

3.6.2 The General Case

The following example illustrates the concepts that are important to the general case of functions of several random variables and is also interesting in its own right.

EXAMPLE **A**   Suppose that X and Y are independent standard normal random variables, which means that their joint distribution is the standard bivariate normal distribution, or

$$f_{XY}(x, y) = \frac{1}{2\pi} e^{-(x^2/2)-(y^2/2)}$$

We change to polar coordinates and then reexpress the density in this new coordinate system ($R \geq 0, 0 \leq \Theta \leq 2\pi$):

$$R = \sqrt{X^2 + Y^2}$$

$$\Theta = \begin{cases} \tan^{-1}\left(\frac{Y}{X}\right), & \text{if } X > 0 \\[1.5em] \tan^{-1}\left(\frac{Y}{X}\right) + \pi, & \text{if } X < 0 \\[1.5em] \frac{\pi}{2}\operatorname{sgn}(Y), & \text{if } X = 0, Y \neq 0 \\[1.5em] 0, & \text{if } X = 0, Y = 0 \end{cases}$$

(The range of the inverse tangent function is taken to be $-\frac{\pi}{2} < \Theta < \frac{\pi}{2}$.) The inverse transformation is

$$X = R \cos \Theta$$
$$Y = R \sin \Theta$$

The joint density of R and Θ is

$$f_{R\Theta}(r, \theta)\, dr\, d\theta = P(r \leq R \leq r + dr, \theta \leq \Theta \leq \theta + d\theta)$$

This probability is equal to the area of the shaded patch in Figure 3.18 times $f_{XY}[x(r, \theta), y(r, \theta)]$. The area in question is clearly $r\, dr\, d\theta$, so

$$P(r \leq R \leq r + dr, \theta \leq \Theta \leq \theta + d\theta) = f_{XY}(r \cos \theta, r \sin \theta)r\, dr\, d\theta$$

and

$$f_{R\Theta}(r, \theta) = r f_{XY}(r \cos \theta, r \sin \theta)$$

Thus,

$$f_{R\Theta}(r, \theta) = \frac{r}{2\pi} e^{[-(r^2 \cos^2 \theta)/2 - (r^2 \sin^2 \theta)/2]}$$
$$= \frac{1}{2\pi} r e^{-r^2/2}$$

From this, we see that the joint density factors implying that R and Θ are independent random variables, that Θ is uniform on $[0, 2\pi]$, and that R has the density

$$f_R(r) = r e^{-r^2/2}, \qquad r \geq 0$$

which is called the **Rayleigh density.**

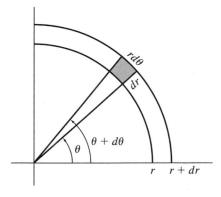

FIGURE **3.18** The area of the shaded patch is $r\, dr\, d\theta$.

An interesting relationship can be found by changing variables again, letting $T = R^2$. Using the standard techniques for finding the density of a function of a single random variable, we obtain

$$f_T(t) = \frac{1}{2}e^{-t/2}, \qquad t \geq 0$$

This is an exponential distribution with parameter $\frac{1}{2}$. Because R and Θ are independent, so are T and Θ, and the joint density of the latter pair is

$$f_{T\Theta}(t, \theta) = \frac{1}{2\pi}\left(\frac{1}{2}\right)e^{-t/2}$$

We have thus arrived at a characterization of the standard bivariate normal distribution: Θ is uniform on $[0, 2\pi]$, and R^2 is exponential with parameter $\frac{1}{2}$. (Also, from Example B in Section 3.6.1, $\tan \Theta$ follows a Cauchy distribution.)

These relationships can be used to construct an algorithm for generating standard normal random variables, which is quite useful since Φ, the cdf, and Φ^{-1} cannot be expressed in closed form. First, generate U_1 and U_2, which are independent and uniform on $[0, 1]$. Then $-2 \log U_1$ is exponential with parameter $\frac{1}{2}$, and $2\pi U_2$ is uniform on $[0, 2\pi]$. It follows that

$$X = \sqrt{-2 \log U_1} \cos(2\pi U_2)$$

and

$$Y = \sqrt{-2 \log U_1} \sin(2\pi U_2)$$

are independent standard normal random variables. This method of generating normally distributed random variables is sometimes called the **polar method.** ∎

For the general case, suppose that X and Y are jointly distributed continuous random variables, that X and Y are mapped onto U and V by the transformation

$$u = g_1(x, y)$$
$$v = g_2(x, y)$$

and that the transformation can be inverted to obtain

$$x = h_1(u, v)$$
$$y = h_2(u, v)$$

Assume that g_1 and g_2 have continuous partial derivatives and that the Jacobian

$$J(x, y) = \det \begin{bmatrix} \dfrac{\partial g_1}{\partial x} & \dfrac{\partial g_1}{\partial y} \\[2mm] \dfrac{\partial g_2}{\partial x} & \dfrac{\partial g_2}{\partial y} \end{bmatrix} = \left(\frac{\partial g_1}{\partial x}\right)\left(\frac{\partial g_2}{\partial y}\right) - \left(\frac{\partial g_2}{\partial x}\right)\left(\frac{\partial g_1}{\partial y}\right) \neq 0$$

for all x and y. This leads directly to the following result.

PROPOSITION A

Under the assumptions just stated, the joint density of U and V is

$$f_{UV}(u, v) = f_{XY}(h_1(u, v), h_2(u, v))|J^{-1}(h_1(u, v), h_2(u, v))|$$

for (u, v) such that $u = g_1(x, y)$ and $v = g_2(x, y)$ for some (x, y) and 0 elsewhere.   ■

We will not prove Proposition A here. It follows from the formula established in advanced calculus for a change of variables in multiple integrals. The essential elements of the proof follow the discussion in Example A.

EXAMPLE B   To illustrate the formalism, let us redo Example A. The roles of u and v are played by r and θ:

$$r = \sqrt{x^2 + y^2}$$

$$\theta = \tan^{-1}\left(\frac{y}{x}\right)$$

The inverse transformation is

$$x = r\cos\theta$$

$$y = r\sin\theta$$

After some algebra, we obtain the partial derivatives:

$$\frac{\partial r}{\partial x} = \frac{x}{\sqrt{x^2 + y^2}} \qquad \frac{\partial r}{\partial y} = \frac{y}{\sqrt{x^2 + y^2}}$$

$$\frac{\partial \theta}{\partial x} = \frac{-y}{x^2 + y^2} \qquad \frac{\partial \theta}{\partial y} = \frac{x}{x^2 + y^2}$$

The Jacobian is the determinant of the matrix of these expressions, or

$$J(x, y) = \frac{1}{\sqrt{x^2 + y^2}} = \frac{1}{r}$$

Proposition A therefore says that

$$f_{R\Theta}(r, \theta) = r f_{XY}(r\cos\theta, r\sin\theta)$$

for $r \geq 0, 0 \leq \theta \leq 2\pi$, and 0 elsewhere, which is the same as the result we obtained by a direct argument in Example A.   ■

Proposition A extends readily to transformations of more than two random variables. If $X_1, \ldots, X_n$ have the joint density function $f_{X_1 \cdots X_n}$ and

$$Y_i = g_i(X_1, \ldots, X_n), \qquad i = 1, \ldots, n$$

$$X_i = h_i(Y_1, \ldots, Y_n), \qquad i = 1, \ldots, n$$

and if $J(x_1, \ldots, x_n)$ is the determinant of the matrix with the ij entry $\partial g_i / \partial x_j$, then the joint density of $Y_1, \ldots, Y_n$ is

$$f_{Y_1 \cdots Y_n}(y_1, \ldots, y_n) = f_{X_1 \cdots X_n}(x_1, \ldots, x_n)|J^{-1}(x_1, \ldots, x_n)|$$

wherein each x_i is expressed in terms of the y's; $x_i = h_i(y_1, \ldots, y_n)$.

E X A M P L E C   Suppose that X_1 and X_2 are independent standard normal random variables and that

$$Y_1 = X_1$$
$$Y_2 = X_1 + X_2$$

We will show that the joint distribution of Y_1 and Y_2 is bivariate normal. The Jacobian of the transformation is simply

$$J(x, y) = \det \begin{bmatrix} 1 & 0 \\ 1 & 1 \end{bmatrix} = 1$$

Since the inverse transformation is $x_1 = y_1$ and $x_2 = y_2 - y_1$, from Proposition A the joint density of Y_1 and Y_2 is

$$f_{Y_1 Y_2}(y_1, y_2) = \frac{1}{2\pi} \exp\left[-\frac{1}{2}\left[y_1^2 + (y_2 - y_1)^2\right]\right]$$
$$= \frac{1}{2\pi} \exp\left[-\frac{1}{2}\left(2y_1^2 + y_2^2 - 2y_1 y_2\right)\right]$$

This can be recognized to be a bivariate normal density, the parameters of which can be identified by comparing the constants in this expression with the general form of the bivariate normal (see Example F of Section 3.3). First, since the exponential contains only quadratic terms in y_1 and y_2, we have $\mu_{Y_1} = \mu_{Y_2} = 0$. (If μ_{Y_1} were nonzero, for example, examination of the equation for the bivariate density in Example F of Section 3.3 shows that there would be a term $y_1 \mu_{Y_1}$.) Next, from the constant that occurs in front of the exponential, we have

$$\sigma_{Y_1} \sigma_{Y_2} \sqrt{1 - \rho^2} = 1$$

From the coefficient of y_1 we have

$$\sigma_{Y_1}^2 (1 - \rho^2) = \frac{1}{2}$$

Dividing the second relationship into the square of the first gives $\sigma_{Y_2}^2 = 2$. From the coefficient of y_2, we have

$$\sigma_{Y_2}^2 (1 - \rho^2) = 1$$

from which it follows that $\rho^2 = \frac{1}{2}$.

From the sign of the cross product, we see that $\rho = 1/\sqrt{2}$. Finally, we have $\sigma_{Y_1}^2 = 1$. We thus see that this linear transformation of two independent standard normal random variables follows a bivariate normal distribution. This is a special case of a more general result: A nonsingular linear transformation of two random variables whose joint distribution is bivariate normal yields two random variables whose joint distribution is still bivariate normal, although with different parameters. (See Problem 58.) ∎

3.7 Extrema and Order Statistics

This section is concerned with ordering a collection of independent continuous random variables. In particular, let us assume that $X_1, X_2, \ldots, X_n$ are independent random variables with the common cdf F and density f. Let U denote the maximum of the X_i and V the minimum. The cdfs of U and V, and therefore their densities, can be found by a simple trick.

First, we note that $U \leq u$ if and only if $X_i \leq u$ for all i. Thus,

$$F_U(u) = P(U \leq u)$$
$$= P(X_1 \leq u)P(X_2 \leq u) \cdots P(X_n \leq u)$$
$$= [F(u)]^n$$

Differentiating, we find the density,

$$f_U(u) = nf(u)[F(u)]^{n-1}$$

Similarly, $V \geq v$ if and only if $X_i \geq v$ for all i. Thus,

$$1 - F_V(v) = [1 - F(v)]^n$$

and

$$F_V(v) = 1 - [1 - F(v)]^n$$

The density function of V is therefore

$$f_V(v) = nf(v)[1 - F(v)]^{n-1}$$

EXAMPLE A   Suppose that n system components are connected in series, which means that the system fails if any one of them fails, and that the lifetimes of the components, $T_1, \ldots, T_n$, are independent random variables that are exponentially distributed with parameter λ: $F(t) = 1 - e^{-\lambda t}$. The random variable that represents the length of time the system operates is V, which is the minimum of the T_i and by the preceding result has the density

$$f_V(v) = n\lambda e^{-\lambda v}(e^{-\lambda v})^{n-1}$$
$$= n\lambda e^{-n\lambda v}$$

We see that V is exponentially distributed with parameter $n\lambda$. ∎

EXAMPLE B   Suppose that a system has components as described in Example A but connected in parallel, which means that the system fails only when they all fail. The system's lifetime is thus the maximum of n exponential random variables and has the density

$$f_U(u) = n\lambda e^{-\lambda u}(1 - e^{-\lambda u})^{n-1}$$

By expanding the last term using the binomial theorem, we see that this density is a weighted sum of exponential terms rather than a simple exponential density.   ■

We will now derive the preceding results once more, by the differential technique, and generalize them. To find $f_U(u)$, we observe that $u \leq U \leq u + du$ if one of the nX_i falls in the interval $(u, u + du)$ and the other $(n - 1)X_i$ fall to the left of u. The probability of any particular such arrangement is $[F(u)]^{n-1}f(u)du$, and because there are n such arrangements,

$$f_U(u) = n[F(u)]^{n-1}f(u)$$

Now we again assume that $X_1, \ldots, X_n$ are independent continuous random variables with density $f(x)$. We sort the X_i and denote by $X_{(1)} < X_{(2)} < \cdots < X_{(n)}$ the **order statistics.** Note that X_1 is not necessarily equal to $X_{(1)}$. (In fact, this equality holds with probability n^{-1}.) Thus, $X_{(n)}$ is the maximum, and $X_{(1)}$ is the minimum. If n is odd, say, $n = 2m + 1$, then $X_{(m+1)}$ is called the **median** of the X_i.

THEOREM A

The density of $X_{(k)}$, the kth-order statistic, is

$$f_k(x) = \frac{n!}{(k-1)!(n-k)!}f(x)F^{k-1}(x)[1 - F(x)]^{n-k}$$

Proof

We will use a differential argument to derive this result heuristically. (The alternative approach of first deriving the cdf and then differentiating is developed in Problem 66 at the end of this chapter.) The event $x \leq X_{(k)} \leq x + dx$ occurs if $k - 1$ observations are less than x, one observation is in the interval $[x, x + dx]$, and $n - k$ observations are greater than $x + dx$. The probability of any particular arrangement of this type is $f(x)F^{k-1}(x)[1 - F(x)]^{n-k}dx$, and, by the multinomial theorem, there are $n!/[(k - 1)!1!(n - k)!]$ such arrangements, which completes the argument.   ■

E X A M P L E C   For the case where the X_i are uniform on $[0, 1]$, the density of the kth-order statistic reduces to

$$\frac{n!}{(k-1)!(n-k)!}x^{k-1}(1 - x)^{n-k}, \qquad 0 \leq x \leq 1$$

This is the beta density. An interesting by-product of this result is that since the density integrates to 1,

$$\int_0^1 x^{k-1}(1-x)^{n-k}dx = \frac{(k-1)!(n-k)!}{n!}$$ ∎

Joint distributions of order statistics can also be worked out. For example, to find the joint density of the minimum and maximum, we note that $x \le X_{(1)} \le x + dx$ and $y \le X_{(n)} \le y + dy$ if one X_i falls in $[x, x + dx]$, one falls in $[y, y + dy]$, and $n - 2$ fall in $[x, y]$. There are $n(n-1)$ ways to choose the minimum and maximum, and thus $V = X_{(1)}$ and $U = X_{(n)}$ have the joint density

$$f(u, v) = n(n-1)f(v)f(u)[F(u) - F(v)]^{n-2}, \qquad u \ge v$$

For example, for the uniform case,

$$f(u, v) = n(n-1)(u - v)^{n-2}, \qquad 1 \ge u \ge v \ge 0$$

The range of $X_{(1)}, \ldots, X_{(n)}$ is $R = X_{(n)} - X_{(1)}$. Using the same kind of analysis we used in Section 3.6.1 to derive the distribution of a sum, we find

$$f_R(r) = \int_{-\infty}^{\infty} f(v + r, v)\, dv$$

E X A M P L E **D**   Find the distribution of the range, $U - V$, for the uniform $[0, 1]$ case. The integrand is $f(v + r, v) = n(n-1)r^{n-2}$ for $0 \le v \le v + r \le 1$ or, equivalently, $0 \le v \le 1 - r$. Thus,

$$f_R(r) = \int_0^{1-r} n(n-1)r^{n-2}\, dv = n(n-1)r^{n-2}(1 - r), \qquad 0 \le r \le 1$$

The corresponding cdf is

$$F_R(r) = nr^{n-1} - (n-1)r^n, \qquad 0 \le r \le 1$$ ∎

E X A M P L E **E**   *Tolerance Interval*
If a large number of independent random variables having the common density function f are observed, it seems intuitively likely that most of the probability mass of the density $f(x)$ is contained in the interval $(X_{(1)}, X_{(n)})$ and unlikely that a future observation will lie outside this interval. In fact, very precise statements can be made. For example, the amount of the probability mass in the interval is $F(X_{(n)}) - F(X_{(1)})$, a random variable that we will denote by Q. From Proposition C of Section 2.3, the distribution of $F(X_i)$ is uniform; therefore, the distribution of Q is the distribution of $U_{(n)} - U_{(1)}$, which is the range of n independent uniform random variables. Thus, $P(Q > \alpha)$, the probability that more than $100\alpha\%$ of the probability mass is contained in the range is from Example D,

$$P(Q > \alpha) = 1 - n\alpha^{n-1} + (n-1)\alpha^n$$

For example, if $n = 100$ and $\alpha = .95$, this probability is .96. In words, this means that the probability is .96 that the range of 100 independent random variables covers 95% or more of the probability mass, or, with probability .96, 95% of all further observations from the same distribution will fall between the minimum and maximum. This statement does not depend on the actual form of the distribution.   ■

3.8 Problems

1. The joint frequency function of two discrete random variables, X and Y, is given in the following table:

	x			
y	1	2	3	4
1	.10	.05	.02	.02
2	.05	.20	.05	.02
3	.02	.05	.20	.04
4	.02	.02	.04	.10

   a. Find the marginal frequency functions of X and Y.
   b. Find the conditional frequency function of X given $Y = 1$ and of Y given $X = 1$.

2. An urn contains p black balls, q white balls, and r red balls; and n balls are chosen without replacement.

   a. Find the joint distribution of the numbers of black, white, and red balls in the sample.
   b. Find the joint distribution of the numbers of black and white balls in the sample.
   c. Find the marginal distribution of the number of white balls in the sample.

3. Three players play 10 independent rounds of a game, and each player has probability $\frac{1}{3}$ of winning each round. Find the joint distribution of the numbers of games won by each of the three players.

4. A sieve is made of a square mesh of wires. Each wire has diameter d, and the holes in the mesh are squares whose side length is w. A spherical particle of radius r is dropped on the mesh. What is the probability that it passes through? What is the probability that it fails to pass through if it is dropped n times? (Calculations such as these are relevant to the theory of sieving for analyzing the size distribution of particulate matter.)

5. (Buffon's Needle Problem) A needle of length L is dropped randomly on a plane ruled with parallel lines that are a distance D apart, where $D \geq L$. Show that the probability that the needle comes to rest crossing a line is $2L/(\pi D)$. Explain how this gives a mechanical means of estimating the value of π.

6. A point is chosen randomly in the interior of an ellipse:

$$\frac{x^2}{a^2} + \frac{y^2}{b^2} = 1$$

Find the marginal densities of the x and y coordinates of the point.

7. Find the joint and marginal densities corresponding to the cdf

$$F(x, y) = (1 - e^{-\alpha x})(1 - e^{-\beta y}), \qquad x \geq 0, \qquad y \geq 0, \qquad \alpha > 0, \qquad \beta > 0$$

8. Let X and Y have the joint density

$$f(x, y) = \frac{6}{7}(x + y)^2, \qquad 0 \leq x \leq 1, \qquad 0 \leq y \leq 1$$

a. By integrating over the appropriate regions, find (i) $P(X > Y)$, (ii) $P(X + Y \leq 1)$, (iii) $P(X \leq \frac{1}{2})$.
b. Find the marginal densities of X and Y.
c. Find the two conditional densities.

9. Suppose that (X, Y) is uniformly distributed over the region defined by $0 \leq y \leq 1 - x^2$ and $-1 \leq x \leq 1$.

a. Find the marginal densities of X and Y.
b. Find the two conditional densities.

10. A point is uniformly distributed in a unit sphere in three dimensions.

a. Find the marginal densities of the x, y, and z coordinates.
b. Find the joint density of the x and y coordinates.
c. Find the density of the xy coordinates conditional on $Z = 0$.

11. Let U_1, U_2, and U_3 be independent random variables uniform on $[0, 1]$. Find the probability that the roots of the quadratic $U_1 x^2 + U_2 x + U_3$ are real.

12. Let

$$f(x, y) = c(x^2 - y^2)e^{-x}, \qquad 0 \leq x < \infty, \qquad -x \leq y < x$$

a. Find c.
b. Find the marginal densities.
c. Find the conditional densities.

13. A fair coin is thrown once; if it lands heads up, it is thrown a second time. Find the frequency function of the total number of heads.

14. Suppose that

$$f(x, y) = xe^{-x(y+1)}, \qquad 0 \leq x < \infty, \qquad 0 \leq y < \infty$$

a. Find the marginal densities of X and Y. Are X and Y independent?
b. Find the conditional densities of X and Y.

15. Suppose that X and Y have the joint density function

$$f(x, y) = c\sqrt{1 - x^2 - y^2}, \qquad x^2 + y^2 \leq 1$$

a. Find c.
$(x+1)(x+1),$
$x^2 + 2x + 1 (x+1) = x^3 + 2x^2 + x + x^2 + 2x + 1$
$x^3 + 3x^2 + 3x + 1$

b. Sketch the joint density.

c. Find $P(X^2 + Y^2 \leq \frac{1}{2})$.

d. Find the marginal densities of X and Y. Are X and Y independent random variables?

e. Find the conditional densities.

16. What is the probability density of the time between the arrival of the two packets of Example E in Section 3.4?

17. Let (X, Y) be a random point chosen uniformly on the region $R = \{(x, y) : |x| + |y| \leq 1\}$.

a. Sketch R.

b. Find the marginal densities of X and Y using your sketch. Be careful of the range of integration.

c. Find the conditional density of Y given X.

18. Let X and Y have the joint density function

$$f(x, y) = k(x - y), \qquad 0 \leq y \leq x \leq 1$$

and 0 elsewhere.

a. Sketch the region over which the density is positive and use it in determining limits of integration to answer the following questions.

b. Find k.

c. Find the marginal densities of X and Y.

d. Find the conditional densities of Y given X and X given Y.

19. Suppose that two components have independent exponentially distributed lifetimes, T_1 and T_2, with parameters α and β, respectively. Find (a) $P(T_1 > T_2)$ and (b) $P(T_1 > 2T_2)$.

20. If X_1 is uniform on $[0, 1]$, and, conditional on X_1, X_2, is uniform on $[0, X_1]$, find the joint and marginal distributions of X_1 and X_2.

21. An instrument is used to measure very small concentrations, X, of a certain chemical in soil samples. Suppose that the values of X in those soils in which the chemical is present is modeled as a random variable with density function $f(x)$. The assay of a soil reports a concentration only if the chemical is first determined to be present. At very low concentrations, however, the chemical may fail to be detected even if it is present. This phenomenon is modeled by assuming that if the concentration is x, the chemical is detected with probability $R(x)$. Let Y denote the concentration of a chemical in a soil in which it has been determined to be present. Show that the density function of Y is

$$g(y) = \frac{R(y)f(y)}{\int_0^{\infty} R(x)f(x)\, dx}$$

22. Consider a Poisson process on the real line, and denote by $N(t_1, t_2)$ the number of events in the interval (t_1, t_2). If $t_0 < t_1 < t_2$, find the conditional distribution of $N(t_0, t_1)$ given that $N(t_0, t_2) = n$. (*Hint:* Use the fact that the numbers of events in disjoint subsets are independent.)

23. Suppose that, conditional on N, X has a binomial distribution with N trials and probability p of success, and that N is a binomial random variable with m trials and probability r of success. Find the unconditional distribution of X.

24. Let P have a uniform distribution on $[0, 1]$, and, conditional on $P = p$, let X have a Bernoulli distribution with parameter p. Find the conditional distribution of P given X.

25. Let X have the density function f, and let $Y = X$ with probability $\frac{1}{2}$ and $Y = -X$ with probability $\frac{1}{2}$. Show that the density of Y is symmetric about zero—that is, $f_Y(y) = f_Y(-y)$.

26. Spherical particles whose radii have the density function $f_R(r)$ are dropped on a mesh as in Problem 4. Find an expression for the density function of the particles that pass through.

27. Prove that X and Y are independent if and only if $f_{X|Y}(x|y) = f_X(x)$ for all x and y.

28. Show that $C(u, v) = uv$ is a copula. Why is it called "the independence copula"?

29. Use the Farlie-Morgenstern copula to construct a bivariate density whose marginal densities are exponential. Find an expression for the joint density.

30. For $0 \le \alpha \le 1$ and $0 \le \beta \le 1$, show that $C(u, v) = \min(u^{1-\alpha}v, uv^{1-\beta})$ is a copula (the Marshall-Olkin copula). What is the joint density?

31. Suppose that (X, Y) is uniform on the disk of radius 1 as in Example E of Section 3.3. Without doing any calculations, argue that X and Y are not independent.

32. Continuing Example E of Section 3.5.2, suppose you had to guess a value of θ. One plausible guess would be the value of θ that maximizes the posterior density. Find that value. Does the result make intuitive sense?

33. Suppose that, as in Example E of Section 3.5.2, your prior opinion that the coin will land with heads up is represented by a uniform density on $[0, 1]$. You now spin the coin repeatedly and record the number of times, N, until a heads comes up. So if heads comes up on the first spin, $N = 1$, etc.

  a. Find the posterior density of Θ given N.
  b. Do this with a newly minted penny and graph the posterior density.

34. This problem continues Example E of Section 3.5.2. In that example, the prior opinion for the value of Θ was represented by the uniform density. Suppose that the prior density had been a beta density with parameters $a = b = 3$, reflecting a stronger prior belief that the chance of a 1 was near $\frac{1}{2}$. Graph this prior density. Following the reasoning of the example, find the posterior density, plot it, and compare it to the posterior density shown in the example.

35. Find a newly minted penny. Place it on its edge and spin it 20 times. Following Example E of Section 3.5.2, calculate and graph the posterior distribution. Spin another 20 times, and calculate and graph the posterior based on all 40 spins. What happens as you increase the number of spins?

36. Let $f(x) = (1 + \alpha x)/2$, for $-1 \le x \le 1$ and $-1 \le \alpha \le 1$.

  a. Describe an algorithm to generate random variables from this density using the rejection method.

  b. Write a computer program to do so, and test it out.

37. Let $f(x) = 6x^2(1 - x)^2$, for $-1 \le x \le 1$.

  a. Describe an algorithm to generate random variables from this density using the rejection method. In what proportion of the trials will the acceptance step be taken?

  b. Write a computer program to do so, and test it out.

38. Show that the number of iterations necessary to generate a random variable using the rejection method is a geometric random variable, and evaluate the parameter of the geometric frequency function. Show that in order to keep the number of iterations small, $M(x)$ should be chosen to be close to $f(x)$.

39. Show that the following method of generating discrete random variables works (D. R. Fredkin). Suppose, for concreteness, that X takes on values $0, 1, 2, \ldots$ with probabilities $p_0, p_1, p_2, \ldots$. Let U be a uniform random variable. If $U < p_0$, return $X = 0$. If not, replace U by $U - p_0$, and if the new U is less than p_1, return $X = 1$. If not, decrement U by p_1, compare U to p_2, etc.

40. Suppose that X and Y are discrete random variables with a joint probability mass function $p_{XY}(x, y)$. Show that the following procedure generates a random variable $X \sim p_{X|Y}(x|y)$.

  a. Generate $X \sim p_X(x)$.

  b. Accept X with probability $p(y|X)$.

  c. If X is accepted, terminate and return X. Otherwise go to Step a.

  Now suppose that X is uniformly distributed on the integers $1, 2, \ldots, 100$ and that given $X = x$, Y is uniform on the integers $1, 2, \ldots, x$. You observe $Y = 44$. What does this tell you about X? Simulate the distribution of X, given $Y = 44$, 1000 times and make a histogram of the value obtained. How would you estimate $E(X|Y = 44)$?

41. How could you extend the procedure of the previous problem in the case that X and Y are continuous random variables?

42. a. Let T be an exponential random variable with parameter λ; let W be a random variable independent of T, which is ± 1 with probability $\frac{1}{2}$ each; and let $X = WT$. Show that the density of X is

$$f_X(x) = \frac{\lambda}{2} e^{-\lambda|x|}$$

which is called the **double exponential density**.

  b. Show that for some constant c,

$$\frac{1}{\sqrt{2\pi}} e^{-x^2/2} \le c e^{-|x|}$$

Use this result and that of part (a) to show how to use the rejection method to generate random variables from a standard normal density.

43. Let U_1 and U_2 be independent and uniform on $[0, 1]$. Find and sketch the density function of $S = U_1 + U_2$.

44. Let N_1 and N_2 be independent random variables following Poisson distributions with parameters λ_1 and λ_2. Show that the distribution of $N = N_1 + N_2$ is Poisson with parameter $\lambda_1 + \lambda_2$.

45. For a Poisson distribution, suppose that events are independently labeled A and B with probabilities $p_A + p_B = 1$. If the parameter of the Poisson distribution is λ, show that the number of events labeled A follows a Poisson distribution with parameter $p_A \lambda$.

46. Let X and Y be jointly continuous random variables. Find an expression for the density of $Z = X - Y$.

47. Let X and Y be independent standard normal random variables. Find the density of $Z = X + Y$, and show that Z is normally distributed as well. (*Hint:* Use the technique of completing the square to help in evaluating the integral.)

48. Let T_1 and T_2 be independent exponentials with parameters λ_1 and λ_2. Find the density function of $T_1 + T_2$.

49. Find the density function of $X + Y$, where X and Y have a joint density as given in Example D in Section 3.3.

50. Suppose that X and Y are independent discrete random variables and each assumes the values 0, 1, and 2 with probability $\frac{1}{3}$ each. Find the frequency function of $X + Y$.

51. Let X and Y have the joint density function $f(x, y)$, and let $Z = XY$. Show that the density function of Z is

$$f_Z(z) = \int_{-\infty}^{\infty} f\left(y, \frac{z}{y}\right) \frac{1}{|y|} \, dy$$

52. Find the density of the quotient of two independent uniform random variables.

53. Consider forming a random rectangle in two ways. Let $U_1, U_2,$ and U_3 be independent random variables uniform on $[0, 1]$. One rectangle has sides U_1 and U_2, and the other is a square with sides U_3. Find the probability that the area of the square is greater than the area of the other rectangle.

54. Let $X, Y,$ and Z be independent $N(0, \sigma^2)$. Let $\Theta, \Phi,$ and R be the corresponding random variables that are the spherical coordinates of (X, Y, Z):

$$x = r \sin \phi \cos \theta$$
$$y = r \sin \phi \sin \theta$$
$$z = r \cos \phi$$
$$0 \le \phi \le \pi, \qquad 0 \le \theta \le 2\pi$$

Find the joint and marginal densities of Θ, Φ, and R. (*Hint: $dx\,dy\,dz = r^2 \sin \phi$ $dr\,d\theta\,d\phi$.*)

55. A point is generated on a unit disk in the following way: The radius, R, is uniform on $[0, 1]$, and the angle Θ is uniform on $[0, 2\pi]$ and is independent of R.
    a. Find the joint density of $X = R \cos \Theta$ and $Y = R \sin \Theta$.
    b. Find the marginal densities of X and Y.
    c. Is the density uniform over the disk? If not, modify the method to produce a uniform density.

56. If X and Y are independent exponential random variables, find the joint density of the polar coordinates R and Θ of the point (X, Y). Are R and Θ independent?

57. Suppose that Y_1 and Y_2 follow a bivariate normal distribution with parameters $\mu_{Y_1} = \mu_{Y_2} = 0$, $\sigma_{Y_1}^2 = 1$, $\sigma_{Y_2}^2 = 2$, and $\rho = 1/\sqrt{2}$. Find a linear transformation $x_1 = a_{11}y_1 + a_{12}y_2$, $x_2 = a_{21}y_1 + a_{22}y_2$ such that x_1 and x_2 are independent standard normal random variables. (*Hint:* See Example C of Section 3.6.2.)

58. Show that if the joint distribution of X_1 and X_2 is bivariate normal, then the joint distribution of $Y_1 = a_1 X_1 + b_1$ and $Y_2 = a_2 X_2 + b_2$ is bivariate normal.

59. Let X_1 and X_2 be independent standard normal random variables. Show that the joint distribution of

$$Y_1 = a_{11} X_1 + a_{12} X_2 + b_1$$
$$Y_2 = a_{21} X_1 + a_{22} X_2 + b_2$$

is bivariate normal.

60. Using the results of the previous problem, describe a method for generating pseudorandom variables that have a bivariate normal distribution from independent pseudorandom uniform variables.

61. Let X and Y be jointly continuous random variables. Find an expression for the joint density of $U = a + bX$ and $V = c + dY$.

62. If X and Y are independent standard normal random variables, find $P(X^2 + Y^2 \le 1)$.

63. Let X and Y be jointly continuous random variables.
    a. Develop an expression for the joint density of $X + Y$ and $X - Y$.
    b. Develop an expression for the joint density of XY and Y/X.
    c. Specialize the expressions from parts (a) and (b) to the case where X and Y are independent.

64. Find the joint density of $X + Y$ and X/Y, where X and Y are independent exponential random variables with parameter λ. Show that $X + Y$ and X/Y are independent.

65. Suppose that a system's components are connected in series and have lifetimes that are independent exponential random variables with parameters λ_i. Show that the lifetime of the system is exponential with parameter $\sum \lambda_i$.

66. Each component of the following system (Figure 3.19) has an independent exponentially distributed lifetime with parameter λ. Find the cdf and the density of the system's lifetime.

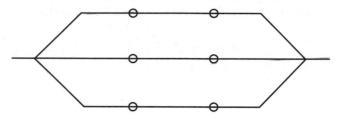

FIGURE **3.19**

67. A card contains n chips and has an error-correcting mechanism such that the card still functions if a single chip fails but does not function if two or more chips fail. If each chip has a lifetime that is an independent exponential with parameter λ, find the density function of the card's lifetime.

68. Suppose that a queue has n servers and that the length of time to complete a job is an exponential random variable. If a job is at the top of the queue and will be handled by the next available server, what is the distribution of the waiting time until service? What is the distribution of the waiting time until service of the next job in the queue?

69. Find the density of the minimum of n independent Weibull random variables, each of which has the density

$$f(t) = \beta\alpha^{-\beta}t^{\beta-1}e^{-(t/\alpha)^{\beta}}, \qquad t \geq 0$$

70. If five numbers are chosen at random in the interval $[0, 1]$, what is the probability that they all lie in the middle half of the interval?

71. Let $X_1, \ldots, X_n$ be independent random variables, each with the density function f. Find an expression for the probability that the interval $(-\infty, X_{(n)}]$ encompasses at least $100\nu\%$ of the probability mass of f.

72. Let $X_1, X_2, \ldots, X_n$ be independent continuous random variables each with cumulative distribution function F. Show that the joint cdf of $X_{(1)}$ and $X_{(n)}$ is

$$F(x, y) = F^n(y) - [F(y) - F(x)]^n, \qquad x \leq y$$

73. If $X_1, \ldots, X_n$ are independent random variables, each with the density function f, show that the joint density of $X_{(1)}, \ldots, X_{(n)}$ is

$$n!f(x_1)f(x_2)\cdots f(x_n), \qquad x_1 < x_2 < \cdots < x_n$$

74. Let U_1, U_2, and U_3 be independent uniform random variables.
    a. Find the joint density of $U_{(1)}$, $U_{(2)}$, and $U_{(3)}$.
    b. The locations of three gas stations are independently and randomly placed along a mile of highway. What is the probability that no two gas stations are less than $\frac{1}{3}$ mile apart?

75. Use the differential method to find the joint density of $X_{(i)}$ and $X_{(j)}$, where $i < j$.

76. Prove Theorem A of Section 3.7 by finding the cdf of $X_{(k)}$ and differentiating. (*Hint:* $X_{(k)} \leq x$ if and only if k or more of the X_i are less than or equal to x. The number of X_i less than or equal to x is a binomial random variable.)

77. Find the density of $U_{(k)} - U_{(k-1)}$ if the $U_i, i = 1, \ldots, n$ are independent uniform random variables. This is the density of the spacing between adjacent points chosen uniformly in the interval $[0, 1]$.

78. Show that

$$\int_0^1 \int_0^y (y - x)^n \, dx \, dy = \frac{1}{(n + 1)(n + 2)}$$

79. If T_1 and T_2 are independent exponential random variables, find the density function of $R = T_{(2)} - T_{(1)}$.

80. Let $U_1, \ldots, U_n$ be independent uniform random variables, and let V be uniform and independent of the U_i.
   a. Find $P(V \leq U_{(n)})$.
   b. Find $P(U_{(1)} < V < U_{(n)})$.

81. Do both parts of Problem 80 again, assuming that the U_i and V have the density function f and the cdf F, with F^{-1} uniquely defined. *Hint:* $F(U_i)$ has a uniform distribution.

CHAPTER 4

Expected Values

4.1 The Expected Value of a Random Variable

The concept of the expected value of a random variable parallels the notion of a weighted average. The possible values of the random variable are weighted by their probabilities, as specified in the following definition.

> **DEFINITION**
>
> If X is a discrete random variable with frequency function $p(x)$, the expected value of X, denoted by $E(X)$, is
>
> $$E(X) = \sum_i x_i\, p(x_i)$$
>
> provided that $\sum_i |x_i| p(x_i) < \infty$. If the sum diverges, the expectation is undefined. ∎

$E(X)$ is also referred to as the **mean** of X and is often denoted by μ or μ_X. It might be helpful to think of the expected value of X as the center of mass of the frequency function. Imagine placing the masses $p(x_i)$ at the points x_i on a beam; the balance point of the beam is the expected value of X.

EXAMPLE A *Roulette*

A roulette wheel has the numbers 1 through 36, as well as 0 and 00. If you bet $1 that an odd number comes up, you win or lose $1 according to whether that event occurs. If X denotes your net gain, $X = 1$ with probability $\frac{18}{38}$ and $X = -1$ with probability $\frac{20}{38}$. The expected value of X is

$$E(X) = 1 \times \frac{18}{38} + (-1) \times \frac{20}{38} = -\frac{1}{19}$$

Thus, your expected loss is about $.05. In Chapter 5, it will be shown that this coincides in the limit with the actual average loss per game if you play a long sequence of independent games. ∎

E X A M P L E **B**   *Expectation of a Geometric Random Variable*

Suppose that items produced in a plant are independently defective with probability p. Items are inspected one by one until a defective item is found. On the average, how many items must be inspected?

The number of items inspected, X, is a geometric random variable, with $P(X = k) = q^{k-1}p$, where $q = 1 - p$. Therefore,

$$E(X) = \sum_{k=1}^{\infty} kpq^{k-1} = p \sum_{k=1}^{\infty} kq^{k-1}$$

We use a trick to calculate the sum. Since $kq^{k-1} = \dfrac{d}{dq}q^k$, we interchange the operations of summation and differentiation to obtain

$$E(X) = p\frac{d}{dq} \sum_{k=1}^{\infty} q^k = p\frac{d}{dq} \frac{q}{1-q}$$

$$= \frac{p}{(1-q)^2} = \frac{1}{p}$$

It can be shown that the interchange of differentiation and summation is justified. Thus, for example, if 10% of the items are defective, an average of 10 items must be examined to find one that is defective, as might have been guessed. ∎

E X A M P L E **C**   *Poisson Distribution*

The expected value of a Poisson random variable is

$$E(X) = \sum_{k=0}^{\infty} \frac{k\lambda^k}{k!} e^{-\lambda}$$

$$= \lambda e^{-\lambda} \sum_{k=1}^{\infty} \frac{\lambda^{k-1}}{(k-1)!}$$

$$= \lambda e^{-\lambda} \sum_{j=0}^{\infty} \frac{\lambda^j}{j!}$$

Since $\sum_{j=0}^{\infty}(\lambda^j/j!) = e^{\lambda}$, we have $E(X) = \lambda$. The parameter λ of the Poisson distribution can thus be interpreted as the average count. ∎

EXAMPLE D *St. Petersburg Paradox*

A gambler has the following strategy for playing a sequence of games: He starts off betting $1; if he loses, he doubles his bet; and he continues to double his bet until he finally wins. To analyze this scheme, suppose that the game is fair and that he wins or loses the amount he bets. At trial 0, he bets $1; if he loses, he bets $2 at trial 1; and if he has not won by the kth trial, he bets 2^k. When he finally wins, he will be $1 ahead, which can be checked by going through the scheme for the first few values of k. This seems like a foolproof way to win $1. What could be wrong with it?

Let X denote the amount of money bet on the very last game (the game he wins). Because the probability that k losses are followed by one win is $2^{-(k+1)}$,

$$P(X = 2^k) = \frac{1}{2^{k+1}}$$

and

$$E(X) = \sum_{n=0}^{\infty} n P(X = n)$$

$$= \sum_{k=0}^{\infty} 2^k \frac{1}{2^{k+1}} = \infty$$

Formally, $E(X)$ is not defined. Practically, the analysis shows a flaw in this scheme, which is that it does not take into account the enormous amount of capital required. ∎

The definition of expectation for a continuous random variable is a fairly obvious extension of the discrete case—summation is replaced by integration.

> **DEFINITION**
>
> If X is a continuous random variable with density $f(x)$, then
>
> $$E(X) = \int_{-\infty}^{\infty} x f(x) \, dx$$
>
> provided that $\int |x| f(x) dx < \infty$. If the integral diverges, the expectation is undefined. ∎

Again $E(X)$ can be regarded as the center of mass of the density. We next consider some examples.

EXAMPLE E *Gamma Density*

If X follows a gamma density with parameters α and λ,

$$E(X) = \int_0^{\infty} \frac{\lambda^\alpha}{\Gamma(\alpha)} x^\alpha e^{-\lambda x} \, dx$$

This integral is easy to evaluate once we realize that $\lambda^{\alpha+1} x^{\alpha} e^{-\lambda x} / \Gamma(\alpha+1)$ is a gamma density and therefore integrates to 1. We thus have

$$\int_0^{\infty} x^{\alpha} e^{-\lambda x} dx = \frac{\Gamma(\alpha+1)}{\lambda^{\alpha+1}}$$

from which it follows that

$$E(X) = \frac{\lambda^{\alpha}}{\Gamma(\alpha)} \left[\frac{\Gamma(\alpha+1)}{\lambda^{\alpha+1}} \right]$$

Finally, using the relation $\Gamma(\alpha+1) = \alpha\Gamma(\alpha)$, we find

$$E(X) = \frac{\alpha}{\lambda}$$

For the exponential density, $\alpha = 1$, so $E(X) = 1/\lambda$. This may be contrasted to the median of the exponential density, which was found in Section 2.2.1 to be $\log 2/\lambda$. The mean and the median can both be interpreted as "typical" values of X, but they measure different attributes of the probability distribution. ∎

E X A M P L E **F**   *Normal Distribution*
From the definition of the expectation, we have

$$E(X) = \frac{1}{\sigma\sqrt{2\pi}} \int_{-\infty}^{\infty} x e^{-\frac{1}{2}\frac{(x-\mu)^2}{\sigma^2}} dx$$

Making the change of variables $z = x - \mu$ changes this equation to

$$E(X) = \frac{1}{\sigma\sqrt{2\pi}} \int_{-\infty}^{\infty} z e^{-z^2/2\sigma^2} dz + \frac{\mu}{\sigma\sqrt{2\pi}} \int_{-\infty}^{\infty} e^{-z^2/2\sigma^2} dz$$

The first integral is 0 since the contributions from $z < 0$ cancel those from $z > 0$, and the second integral is μ because the normal density integrates to 1. Thus,

$$E(X) = \mu$$

The parameter μ of the normal density is the expectation, or mean value. We could have made the derivation much shorter by claiming that it was "obvious" that since the center of symmetry of the density is μ, the expectation must be μ. ∎

E X A M P L E **G**   *Cauchy Density*
Recall that the Cauchy density is

$$E(X) = \int_{-\infty}^{\infty} x f(x) dx$$

$$f(x) = \frac{1}{\pi} \left(\frac{1}{1+x^2} \right), \qquad -\infty < x < \infty$$

The density is symmetric about zero, so it would seem that $E(X) = 0$. However,

$$\int_{-\infty}^{\infty} \frac{|x|}{1+x^2} = \infty$$

Therefore, the expectation does not exist. The reason that it fails to exist is, that the density decreases so slowly that very large values of X can occur with substantial probability.   ∎

The expected value can be interpreted as a long-run average. In Chapter 5, it will be shown that if $E(X)$ exists and if X_1, X_2, ... is a sequence of independent random variables with the same distribution as X, and if $S_n = \sum_{i=1}^{n} X_i$, then, as $n \to \infty$,

$$\frac{S_n}{n} \to E(X)$$

This statement will be made more precise in Chapter 5. For now, a simple empirical demonstration will be sufficient.

EXAMPLE **H**   Using a pseudorandom number generator, a sequence X_1, X_2, ... of independent standard normal random variables was generated, as well as a sequence Y_1, Y_2, ... of independent Cauchy random variables. Figure 4.1 shows the graphs of

$$G(n) = \frac{1}{n} \sum_{i=1}^{n} X_i \quad \text{and} \quad C(n) = \frac{1}{n} \sum_{i=1}^{n} Y_i \qquad n = 1, 2, \ldots, 500$$

Note how $G(n)$ appears to be tending to a limit, whereas $C(n)$ does not.   ∎

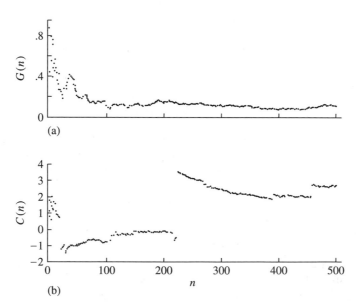

FIGURE **4.1**   The average of n independent random variables as a function of n for (a) normal random variables and (b) Cauchy random variables.

We conclude this section with a simple result that is of great utility in probability theory.

THEOREM A *Markov's Inequality*

If X is a random variable with $P(X \geq 0) = 1$ and for which $E(X)$ exists, then $P(X \geq t) \leq E(X)/t$.

Proof

We will prove this for the discrete case; the continuous case is entirely analogous.

$$E(X) = \sum_x xp(x)$$

$$= \sum_{x < t} xp(x) + \sum_{x \geq t} xp(x)$$

All the terms in the sums are nonnegative because X takes on only nonnegative values. Thus

$$E(X) \geq \sum_{x \geq t} xp(x)$$

$$\geq \sum_{x \geq t} tp(x) = tP(X \geq t) \qquad \blacksquare$$

This result says that the probability that X is much bigger than $E(X)$ is small. Suppose that in the theorem, we let $t = kE(X)$; then according to the result, $P(X > kE(X)) \leq k^{-1}$. This holds for any nonnegative random variable, regardless of its probability distribution.

4.1.1 Expectations of Functions of Random Variables

We often need to find $E[g(X)]$, where X is a random variable and g is a fixed function. For example, according to the kinetic theory of gases, the magnitude of the velocity of a gas molecule is random and its probability density is given by

$$f_X(x) = \frac{\sqrt{2/\pi}}{\sigma^3} x^2 e^{-\frac{1}{2}\frac{x^2}{\sigma^2}}$$

(This is Maxwell's distribution: the parameter σ depends on the temperature of the gas.) From this density, we can find the average velocity, but suppose that we are interested in finding the average kinetic energy $Y = \frac{1}{2}mX^2$, where m is the mass of the molecule. The straightforward way to do this would seem to be the following: Let $Y = g(X)$; find the density or frequency function of Y, say, f_Y; and then compute $E(Y)$ from the definition. It turns out, fortunately, that the process is much simpler than that.

THEOREM A

Suppose that $Y = g(X)$.

a. If X is discrete with frequency function $p(x)$, then

$$E(Y) = \sum_x g(x)p(x)$$

provided that $\sum |g(x)|p(x) < \infty$.

b. If X is continuous with density function $f(x)$, then

$$E(Y) = \int_{-\infty}^{\infty} g(x)f(x)\,dx$$

provided that $\int |g(x)|f(x)\,dx < \infty$.

Proof

We will prove this result for the discrete case. The basic argument is the same for the continuous case, but making that proof rigorous requires some advanced theory of integration. By definition,

$$E(Y) = \sum_i y_i p_Y(y_i)$$

Let A_i denote the set of x's mapped to y_i by g; that is, $x \in A_i$ if $g(x) = y_i$. Then

$$p_Y(y_i) = \sum_{x \in A_i} p(x)$$

and

$$E(Y) = \sum_i y_i \sum_{x \in A_i} p(x)$$

$$= \sum_i \sum_{x \in A_i} y_i p(x)$$

$$= \sum_i \sum_{x \in A_i} g(x)p(x)$$

$$= \sum_x g(x)p(x)$$

This last step follows because the A_i are disjoint and every x belongs to some A_i. ∎

It is worth pointing out explicitly that $E[g(X)] \neq g[E(X)]$; that is, the average value of the function is not equal to the function of the average value. Suppose, for example, that X takes on values 1 and 2, each with probability $\frac{1}{2}$, so $E(X) = \frac{3}{2}$. Let $Y = 1/X$. Then $E(Y)$ is clearly $1 \times .5 + .5 \times .5 = .75$, but $1/E(X) = \frac{2}{3}$.

EXAMPLE A   Let us now apply Theorem A to find the average kinetic energy of a gas molecule.

$$E(Y) = \int_0^\infty \frac{1}{2} mx^2 f_X(x)\, dx$$

$$= \frac{m}{2} \frac{\sqrt{2/\pi}}{\sigma^3} \int_0^\infty x^4 e^{-\frac{1}{2}\frac{x^2}{\sigma^2}}\, dx$$

To evaluate the integral, we make the change of variables $u = x^2/2\sigma^2$ to reduce it to

$$\frac{2m\sigma^2}{\sqrt{\pi}} \int_0^\infty u^{3/2} e^{-u}\, du = \frac{2m\sigma^2}{\sqrt{\pi}} \Gamma\left(\frac{5}{2}\right)$$

Finally, using the facts $\Gamma(\frac{1}{2}) = \sqrt{\pi}$ and $\Gamma(\alpha + 1) = \alpha\Gamma(\alpha)$, we have

$$E(Y) = \tfrac{3}{2}m\sigma^2$$

■

Now suppose that $Y = g(X_1, \ldots, X_n)$, where X_i have a joint distribution, and that we want to find $E(Y)$. We do not have to find the density or frequency function of Y, which again could be a formidable task.

THEOREM B

Suppose that $X_1, \ldots, X_n$ are jointly distributed random variables and $Y = g(X_1, \ldots, X_n)$.

a. If the X_i are discrete with frequency function $p(x_1, \ldots, x_n)$, then

$$E(Y) = \sum_{x_1, \ldots, x_n} g(x_1, \ldots, x_n) p(x_1, \ldots, x_n)$$

provided that $\sum_{x_1, \ldots, x_n} |g(x_1, \ldots, x_n)| p(x_1, \ldots, x_n) < \infty$.
b. If the X_i are continuous with joint density function $f(x_1, \ldots, x_n)$, then

$$E(Y) = \int \int \cdots \int g(x_1, \ldots, x_n) f(x_1, \ldots, x_n) dx_1 \cdots dx_n$$

provided that the integral with $|g|$ in place of g converges.

Proof

The proof is similar to that of Theorem A.   ■

EXAMPLE B   A stick of unit length is broken randomly in two places. What is the average length of the middle piece?

We interpret this question to mean that the locations of the two break points are independent uniform random variables U_1 and U_2. Therefore, we need to compute $E|U_1 - U_2|$. Theorem B tells us that we do not need to find the density function of $|U_1 - U_2|$ and that we can just integrate $|u_1 - u_2|$ against the joint density of U_1 and U_2, $f(u_1, u_2) = 1, 0 \le u_1 \le 1, 0 \le u_2 \le 1$. Thus,

$$E|U_1 - U_2| = \int_0^1 \int_0^1 |u_1 - u_2| \, du_1 \, du_2$$

$$= \int_0^1 \int_0^{u_1} (u_1 - u_2) \, du_2 \, du_1 + \int_0^1 \int_{u_1}^1 (u_2 - u_1) \, du_2 \, du_1$$

With some care, we find the expectation to be $\frac{1}{3}$. This is in accord with the intuitive argument that the smaller of U_1 and U_2 should be $\frac{1}{3}$ on the average and the larger should be $\frac{2}{3}$ on the average, which means that the average difference should be $\frac{1}{3}$.   ■

We note the following immediate consequence of Theorem B.

COROLLARY A

If X and Y are independent random variables and g and h are fixed functions, then $E[g(X)h(Y)] = \{E[g(X)]\}\{E[h(Y)]\}$, provided that the expectations on the right-hand side exist.   ■

In particular, if X and Y are independent, $E(XY) = E(X)E(Y)$. The proof of this corollary is left to Problem 29 of the end-of-chapter problems.

4.1.2 Expectations of Linear Combinations of Random Variables

One of the most useful properties of the expectation is that it is a linear operation. Suppose that you were told that the average temperature on July 1 in a certain location was 70°F, and you were asked what the average temperature in degrees Celsius was. You can simply convert to degrees Celsius and obtain $\frac{5}{9} \times 70 - 17.7 = 21.2$°C. The notion of the average value of a random variable, which we have defined as the expected value of a random variable, behaves in the same fashion. If $Y = aX + b$, then $E(Y) = aE(X) + b$. More generally, this property extends to linear combinations of random variables.

THEOREM A

If $X_1, \ldots, X_n$ are jointly distributed random variables with expectations $E(X_i)$ and Y is a linear function of the X_i, $Y = a + \sum_{i=1}^{n} b_i X_i$, then

$$E(Y) = a + \sum_{i=1}^{n} b_i E(X_i)$$

Proof

We will prove this for the continuous case. The proof in the discrete case is parallel and is left to Problem 24 at the end of this chapter. For notational simplicity, we take $n = 2$. From Theorem B of Section 4.1.1, we have

$$E(Y) = \int \int (a + b_1 x_1 + b_2 x_2) f(x_1, x_2)\, dx_1\, dx_2$$

$$= a \int \int f(x_1, x_2)\, dx_1\, dx_2 + b_1 \int \int x_1 f(x_1, x_2)\, dx_1\, dx_2$$

$$+ b_2 \int \int x_2 f(x_1, x_2)\, dx_1\, dx_2$$

The first double integral of the last expression is merely the integral of the bivariate density, which is equal to 1. The second double integral can be evaluated as follows:

$$\int \int x_1 f(x_1, x_2)\, dx_1\, dx_2 = \int x_1 \left[\int f(x_1, x_2)\, dx_2 \right] dx_1$$

$$= \int x_1 f_{x_1}(x_1)\, dx_1$$

$$= E(X_1)$$

A similar evaluation for the third double integral brings us to

$$E(Y) = a + b_1 E(X_1) + b_2 E(X_2)$$

This proves the theorem once we check that the expectation is well defined, or that

$$\int \int |a + b_1 x_1 + b_2 x_2|\, f(x_1, x_2)\, dx_1\, dx_2 < \infty$$

This can be verified using the inequality

$$|a + b_1 x_1 + b_2 x_2| \le |a| + |b_1||x_1| + |b_2||x_2|$$

and the assumption that the $E(X_i)$ exist. ∎

Theorem A is extremely useful. We will illustrate its utility with several examples.

E X A M P L E   **A**   Suppose that we wish to find the expectation of a binomial random variable, Y. From the binomial frequency function,

$$E(Y) = \sum_{k=0}^{n} \binom{n}{k} k p^k (1-p)^{n-k}$$

It is not immediately obvious how to evaluate this sum. We can, however, represent Y as the sum of Bernoulli random variables, X_i, which equal 1 or 0 depending on whether there is success or failure on the ith trial,

$$Y = \sum_{i=1}^{n} X_i$$

Because $E(X_i) = 0 \times (1-p) + 1 \times p = p$, it follows immediately that $E(Y) = np$.

An application of the binomial distribution and its expectation occurs in "shotgun sequencing" in genomics, a method of trying to figure out the sequence of letters that make up a long segment of DNA. It is technically too difficult to sequence the entire segment at once if it is very long. The basic idea of shotgun sequencing is to chop the DNA randomly into many small fragments, sequence each fragment, and then somehow assemble the fragments into one long "contig." The hope is that if there are many fragments, their overlaps can be used to assemble the contig.

Suppose, then, that the length of the DNA sequence is G and that there are N fragments each of length L. G might be at least 100,000 and L about 500. Assume that the left end of each fragment is equally likely to be at positions $1, 2, \ldots, G - L + 1$. What is the probability that a particular location $x \in \{L, L+1, \ldots, G\}$ is covered by at least one fragment? How many fragments are expected to cover a particular location? (The positions $\{1, 2, \ldots, L-1\}$ are not included in this discussion because the boundary effect makes them a little different; for example, the only fragment that covers position 1 has its left end at position 1.) To answer these questions, first consider a single fragment. The chance that it covers x equals the chance that its left end is in one of the L locations $\{x - L + 1, x - L + 2, \ldots, x\}$, and because the location of the left end is uniform, this probability is

$$p = \frac{L}{G - L + 1} \approx \frac{L}{G}$$

where the approximation holds because $G \gg L$. Thus, the distribution of W, the number of fragments that cover a particular location, is binomial with parameters N and p.

From the binomial probability formula, the chance of coverage is

$$P(W > 0) = 1 - P(W = 0) = 1 - \left(1 - \frac{L}{G}\right)^N$$

Since N is large and p is small, the distribution of W is nearly Poisson with parameter $\lambda = Np = NL/G$. From the Poisson probability formula, $P(W = 0) \approx e^{-NL/G}$, so the probability that a particular location is covered is approximately $1 - e^{-NL/G}$. Observe that NL is the total length of all the fragments; the ratio NL/G is called the *coverage*. Calculations of this kind are thus useful in deciding how many fragments to use. If the coverage is 8, for example, the chance that a site is covered is .9997.

Overlap of fragments is important when trying to assemble them. Since W is a binomial random variable, the expected number of fragments that cover a given site is $Np = NL/G$, precisely the coverage.

We can also now answer this closely related question: How many sites do we expect to be entirely missed? We will calculate this using indicator random variables: let I_x equal 1 if site x is missed and 0 elsewhere. Then

$$E(I_x) = 1 \times P(I_x = 1) + 0 \times P(I_x = 0) = e^{-NL/G}.$$

The number of sites that are not covered is

$$V = \sum_{x=1}^{G} I_x$$

and from the linearity of expectation

$$E(V) = \sum_{x=1}^{G} E(I_x) \approx Ge^{-NL/G}.$$

The length of the human genome is approximately $G = 3 \times 10^9$, so with eight times coverage, we would expect about a million sites to be missed. ∎

EXAMPLE **B**   *Coupon Collection*
Suppose that you collect coupons, that there are n distinct types of coupons, and that on each trial you are equally likely to get a coupon of any of the types. How many trials would you expect to go through until you had a complete set of coupons? (This might be a model for collecting baseball cards or for certain grocery store promotions.)

The solution of this problem is greatly simplified by representing the number of trials as a sum. Let X_1 be the number of trials up to and including the trial on which the first coupon is collected: $X_1 = 1$. Let X_2 be the number of trials from that point up to and including the trial on which the next coupon different from the first is obtained; let X_3 be the number of trials from that point up to and including the trial on which the third distinct coupon is collected; and so on, up to X_n. Then the total number of trials, X, is the sum of the X_i, $i = 1, 2, \ldots, n$.

We now find the distribution of X_r. At this point, $r - 1$ of n coupons have been collected, so on each trial the probability of success is $(n - r + 1)/n$. Therefore, X_r is a geometric random variable, with $E(X_r) = n/(n - r + 1)$. (See Example B of Section 4.1.) Thus,

$$E(X) = \sum_{r=1}^{n} E(X_r)$$

$$= \frac{n}{n} + \frac{n}{n-1} + \frac{n}{n-2} + \cdots + \frac{n}{1}$$

$$= n \sum_{r=1}^{n} \frac{1}{r}$$

For example, if there are 10 types of coupons, the expected number of trials necessary to obtain at least one of each kind is 29.3.

Finally, we note the following famous approximation:

$$\sum_{r=1}^{n} \frac{1}{r} = \log n + \gamma + \varepsilon_n$$

where log is the natural log or $\log_e$ (unless otherwise specified, log means natural log throughout this text), γ is Euler's constant, $\gamma = .57\ldots$, and ε_n approaches zero as n approaches infinity. Using this approximation for $n = 10$, we find that the approximate expected number of trials is 28.8. Generally, we see that the expected number of trials grows at the rate $n \log n$, or slightly faster than n. ∎

E X A M P L E **C**   *Group Testing*
Suppose that a large number, n, of blood samples are to be screened for a relatively rare disease. If each sample is assayed individually, n tests will be required. On the other hand, if each sample is divided in half and one of the halves is put into a pool with all the other halves, the pooled lot can be tested. Then, provided that the test method is sensitive enough, if this test is negative, no further assays are necessary and only one test has to be performed. If the test on the pooled blood is positive, each reserved half-sample can be tested individually. In this case, a total of $n + 1$ tests will be required. It is therefore plausible, assuming that the disease is rare, that some savings can be achieved through this pooling procedure.

To analyze this more quantitatively, let us first generalize the scheme and suppose that the n samples are first grouped into m groups of k samples each, or $n = mk$. Each group is then tested; if a group tests positively, each individual in the group is tested. If X_i is the number of tests run on the ith group, the total number of tests run is $N = \sum_{i=1}^{m} X_i$, and the expected total number of tests is

$$E(N) = \sum_{i=1}^{m} E(X_i)$$

Let us find $E(X_i)$. If the probability of a negative on any individual sample is p, then the X_i take on the value 1 with probability p^k or the value $k + 1$ with probability $1 - p^k$. Thus,

$$E(X_i) = p^k + (k + 1)(1 - p^k)$$
$$= k + 1 - kp^k$$

We now have

$$E(N) = m(k + 1) - mkp^k = n\left(1 + \frac{1}{k} - p^k\right)$$

Recalling that n tests are necessary with no pooling, we see that the factor $(1 + 1/k - p^k)$ is the average number of samples used in group testing as a proportion of n. Figure 4.2 shows this proportion as a function of k for $p = .99$. From the figure, we

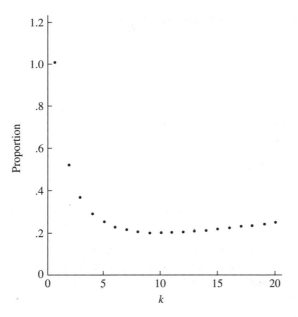

F I G U R E **4.2**   The proportion of n in the average number of samples tested using group testing as a function of k.

see that for group testing with a group size of about 10, only 20% of the number of tests used with the straightforward method are needed on the average.   ■

E X A M P L E **D**   *Counting Word Occurrences in DNA Sequences*
Here we consider another example from genomics, and one that again illustrates the power of using indicator random variables. In searching for patterns in DNA sequences, there might be reason to expect that a "word" such as TATA would occur more frequently than expected in a random sequence. Or suppose we want to identify regions of a DNA sequence in which the occurrence of the word is unusually large. To quantify these ideas, we need to specify the meaning of *random*. In this example, we will take it to mean that the sequence is randomly composed of the letters A,C,G, and T in the sense that the letters at sites are independent and, at every site, each letter has probability $\frac{1}{4}$.

We also need to be careful to specify how we count. Consider the following sequence

ACTATATAGATATA

We will count overlaps, so in the preceding sequence, TATA occurs three times. Now suppose that the sequence is of length N and that the word is of length q. Let I_n be an indicator random variable taking on the value 1 if the word begins at position n and 0 otherwise: $P(I_n = 1) = \left(\frac{1}{4}\right)^q$ from the assumption of independence and $E(I_n) = P(I_n = 1)$. Now the total number of times the word occurs is

$$W = \sum_{n=1}^{N-q+1} I_n$$

and

$$E(W) = \sum_{n=1}^{N-q+1} E(I_n) = (N - q + 1)\left(\frac{1}{4}\right)^q$$

Note that the I_n are not independent—for example, in the case of the word TATA, if $I_1 = 1$, then $I_2 = 0$. Thus W is not a binomial random variable. But despite the lack of independence, we can find $E(W)$ by expressing W as a linear combination of indicator variables. ∎

E X A M P L E **E**   *Investment Portfolios*

An investor plans to apportion an amount of capital, C_0, between two investments, placing a fraction π, $0 \le \pi \le 1$, in one investment and a fraction $1 - \pi$ in the other for a fixed period of time. Denoting the returns (final value divided by initial value) on the investments by R_1 and R_2, her capital at the end of the period will be $C_1 = \pi C_0 R_1 + (1 - \pi)C_0 R_2$. Her return will then be

$$R = \frac{C_1}{C_0}$$
$$= \pi R_1 + (1 - \pi)R_2$$

Suppose that the returns are unknown, as would be the case if they were stocks, for example, and that they are hence modeled as random variables, with expected values $E(R_1)$ and $E(R_2)$. Then her expected return is

$$E(R) = \pi E(R_1) + (1 - \pi)E(R_2)$$

How should she choose π? A simple solution would apparently be to choose $\pi = 1$ if $E(R_1) > E(R_2)$ and $\pi = 0$ otherwise. But there is more to the story as we will see later. ∎

4.2 Variance and Standard Deviation

The expected value of a random variable is its average value and can be viewed as an indication of the central value of the density or frequency function. The expected value is therefore sometimes referred to as a **location parameter.** The median of a distribution is also a location parameter, one that does not necessarily equal the mean. This section introduces another parameter, the **standard deviation** of a random variable, which is an indication of how dispersed the probability distribution is about its center, of how spread out on the average are the values of the random variable about its expectation. We first define the **variance** of a random variable and then define the standard deviation in terms of the variance.

DEFINITION

If X is a random variable with expected value $E(X)$, the variance of X is

$$\text{Var}(X) = E\{[X - E(X)]^2\}$$

provided that the expectation exists. The standard deviation of X is the square root of the variance. ∎

If X is a discrete random variable with frequency function $p(x)$ and expected value $\mu = E(X)$, then according to the definition and Theorem A of Section 4.1.1,

$$\text{Var}(X) = \sum_i (x_i - \mu)^2 p(x_i)$$

whereas if X is a continuous random variable with density function $f(x)$ and $E(X) = \mu$

$$\text{Var}(X) = \int_{-\infty}^{\infty} (x - \mu)^2 f(x)\, dx$$

The variance is often denoted by σ^2 and the standard deviation by σ. From the preceding definition, the variance of X is the average value of the squared deviation of X from its mean. If X has units of meters, for example, the variance has units of meters squared, and the standard deviation has units of meters. Although we are often interested ultimately in the standard deviation rather than the variance, it is usually easier to find the variance first and then take the square root.

The variance of a random variable changes in a simple way under linear transformations.

THEOREM **A**

If $\text{Var}(X)$ exists and $Y = a + bX$, then $\text{Var}(Y) = b^2 \text{Var}(X)$.

Proof

Since $E(Y) = a + bE(X)$,

$$E[(Y - E(Y))^2] = E\{[a + bX - a - bE(X)]^2\}$$
$$= E\{b^2[X - E(X)]^2\}$$
$$= b^2 E\{[X - E(X)]^2\}$$
$$= b^2 \text{Var}(X)$$

∎

This result seems reasonable once you realize that the addition of a constant does not affect the variance, since the variance is a measure of the spread around a center and the center has merely been shifted.

The standard deviation transforms in a natural way: $\sigma_Y = |b|\sigma_X$. Thus, if the units of measurement are changed from meters to centimeters, for example, the standard deviation is simply multiplied by 100.

EXAMPLE A   *Bernoulli Distribution*

If X has a Bernoulli distribution—that is, X takes on values 0 and 1 with probability $1 - p$ and p, respectively—then we have seen (Example A of Section 4.1.2) that $E(X) = p$. By the definition of variance,

$$\text{Var}(X) = (0 - p)^2 \times (1 - p) + (1 - p)^2 \times p$$
$$= p^2 - p^3 + p - 2p^2 + p^3$$
$$= p(1 - p)$$

Note that the expression $p(1 - p)$ is a quadratic with a maximum at $p = \frac{1}{2}$. If p is 0 or 1, the variance is 0, which makes sense since the probability distribution is concentrated at a single point and the random variable is not variable at all. The distribution is most dispersed when $p = \frac{1}{2}$.   ∎

EXAMPLE B   *Normal Distribution*

We have seen that $E(X) = \mu$. Then

$$\text{Var}(X) = E[(X - \mu)^2] = \frac{1}{\sigma\sqrt{2\pi}} \int_{-\infty}^{\infty} (x - \mu)^2 e^{-\frac{1}{2}\frac{(x-\mu)^2}{\sigma^2}} \, dx$$

Making the change of variables $z = (x - \mu)/\sigma$ changes the right-hand side to

$$\frac{\sigma^2}{\sqrt{2\pi}} \int_{-\infty}^{\infty} z^2 e^{-z^2/2} \, dz$$

Finally, making the change of variables $u = z^2/2$ reduces the integral to a gamma function, and we find that $\text{Var}(X) = \sigma^2$.   ∎

The following theorem gives an alternative way of calculating the variance.

THEOREM B

The variance of X, if it exists, may also be calculated as follows:

$$\text{Var}(X) = E(X^2) - [E(X)]^2$$

Proof

Denote $E(X)$ by μ.

$$Var(X) = E[(X - \mu)^2]$$
$$= E(X^2 - 2\mu X + \mu^2)$$

By the linearity of the expectation, this becomes

$$Var(X) = E(X^2) - 2\mu E(X) + \mu^2$$
$$= E(X^2) - 2\mu^2 + \mu^2$$
$$= E(X^2) - \mu^2$$

as was to be shown.   ■

According to Theorem B, the variance of X can be found in two steps: First find $E(X)$, and then find $E(X^2)$.

EXAMPLE C   *Uniform Distribution*

Let us apply Theorem B to find the variance of a random variable that is uniform on $[0, 1]$. We know that $E(X) = \frac{1}{2}$; next we need to find $E(X^2)$:

$$E(X^2) = \int_0^1 x^2 \, dx = \frac{1}{3}$$

We thus have

$$Var(X) = \frac{1}{3} - \left(\frac{1}{2}\right)^2 = \frac{1}{12}$$ ■

It was stated earlier that the variance or standard deviation of a random variable gives an indication as to how spread out its possible values are. A famous inequality, **Chebyshev's inequality,** lends a quantitative aspect to this indication.

THEOREM C   *Chebyshev's Inequality*

Let X be a random variable with mean μ and variance σ^2. Then, for any $t > 0$,

$$P(|X - \mu| > t) \leq \frac{\sigma^2}{t^2}$$

Proof

Let $Y = (X - \mu)^2$. Then $E(Y) = \sigma^2$, and the result follows by applying Markov's inequality to Y.   ■

Theorem C says that if σ^2 is very small, there is a high probability that X will not deviate much from μ. For another interpretation, we can set $t = k\sigma$ so that the inequality becomes

$$P(|X - \mu| \geq k\sigma) \leq \frac{1}{k^2}$$

For example, the probability that X is more than 4σ away from μ is less than or equal to $\frac{1}{16}$. These results hold for any random variable with any distribution provided the variance exists. In particular cases, the bounds are often much narrower. For example, if X is normally distributed, we find from tables of the normal distribution that $P(|X - \mu| > 2\sigma) = .05$ (compared to $\frac{1}{4}$ obtained from Chebyshev's inequality).

Chebyshev's inequality has the following consequence.

COROLLARY A

If $\text{Var}(X) = 0$, then $P(X = \mu) = 1$.

Proof

We will give a proof by contradiction. Suppose that $P(X = \mu) < 1$. Then, for some $\varepsilon > 0$, $P(|X - \mu| \geq \varepsilon) > 0$. However, by Chebyshev's inequality, for any $\varepsilon > 0$,

$$P(|X - \mu| \geq \varepsilon) = 0 \qquad \blacksquare$$

EXAMPLE **D** *Investment Portfolios*

We continue Example E in Section 4.1.2. Suppose that one of the two investments is risky and the other is risk free. The first might be a stock and the other an insured savings account. The stock has a return R_1, which is modeled as a random variable with expectation $\mu_1 = 0.10$ and standard deviation $\sigma_1 = 0.075$. The standard deviation is a measure of risk—a large standard deviation means that the returns fluctuate a lot so that the investor might be lucky and get a large return, but might also be unlucky and lose a lot. Suppose that the savings account has a certain return $R_2 = 0.03$. The expected value of this return is $\mu_2 = 0.03$ and its standard deviation is 0—it is risk free. If the investor places a fraction π_1 in the stock and a fraction $\pi_2 = 1 - \pi_1$ in the savings account, her return is

$$R = \pi_1 R_1 + (1 - \pi_1) R_2$$

and her expected return is

$$E(R) = \pi_1 \mu_1 + (1 - \pi_1) \mu_2$$

Since $\mu_1 > \mu_2$, her expected return is maximized by $\pi_1 = 1$, putting all her money in the stock. However, this point of view is too narrow; it does not take into account the risk of the stock. By Theorem A

$$\text{Var}(R) = \pi_1^2 \sigma_1^2$$

and the standard deviation of the return is $\sigma_R = \pi_1 \sigma_1$. The larger π_1, the larger the expected return, but also the larger the risk. In choosing π_1, the investor has to balance the risk she is willing to take against the expected gain; the desired balance will be different for different investors. If she is risk averse, she will choose a small value of π_1, being leery of volatile investments. By tracing out the expected return and the standard deviation as functions of π_1, she can strike a balance with which she is comfortable. ∎

4.2.1 A Model for Measurement Error

Values of physical constants are not precisely known but must be determined by experimental procedures. Such seemingly simple operations as weighting an object, determining a voltage, or measuring an interval of time are actually quite complicated when all the details and possible sources of error are taken into account. The National Institute of Standards and Technology (NIST) in the United States and similar agencies in other countries are charged with developing and maintaining measurement standards. Such agencies employ probabilists and statisticians as well as physical scientists in this endeavor.

A distinction is usually made between random and systematic measurement errors. A sequence of repeated independent measurements made with no deliberate change in the apparatus or experimental procedure may not yield identical values, and the uncontrollable fluctuations are often modeled as random. At the same time, there may be errors that have the same effect on every measurement; equipment may be slightly out of calibration, for example, or there may be errors associated with the theory underlying the method of measurement. If the true value of the quantity being measured is denoted by x_0, the measurement, X, is modeled as

$$X = x_0 + \beta + \varepsilon$$

where β is the constant, or systematic, error and ε is the random component of the error; ε is a random variable with $E(\varepsilon) = 0$ and $\text{Var}(\varepsilon) = \sigma^2$. We then have

$$E(X) = x_0 + \beta$$

and

$$\text{Var}(X) = \sigma^2$$

β is often called the **bias** of the measurement procedure. The two factors affecting the size of the error are the bias and the size of the variance, σ^2. A perfect measurement would have $\beta = 0$ and $\sigma^2 = 0$.

E X A M P L E **A**   *Measurement of the Gravity Constant*
This and the next example are taken from an interesting and readable paper by Youden (1972), a statistician at NIST. Measurement of the acceleration due to gravity at Ottawa was done 32 times with each of two different methods (Preston-Thomas et al. 1960). The results are displayed as histograms in Figure 4.3. There is clearly some systematic difference between the two methods as well as some variation within each method. It

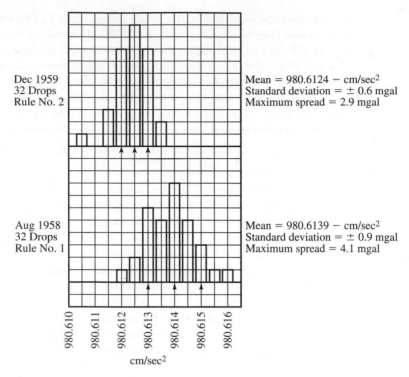

FIGURE 4.3   Histograms of two sets of measurements of the acceleration due to gravity.

appears that the two biases are unequal. The results from Rule 1 are more scattered than those of Rule 2, and their standard deviation is larger.   ■

An overall measure of the size of the measurement error that is often used is the **mean squared error,** which is defined as

$$\text{MSE} = E[(X - x_0)^2]$$

The mean squared error, which is the expected squared deviation of X from x_0, can be decomposed into contributions from the bias and the variance.

THEOREM A

$\text{MSE} = \beta^2 + \sigma^2.$

Proof

From Theorem B of Section 4.2,

$$E[(X - x_0)^2] = \text{Var}(X - x_0) + [E(X - x_0)]^2$$
$$= \text{Var}(X) + \beta^2$$
$$= \sigma^2 + \beta^2$$

    ■

Measurements are often reported in the form 102 ± 1.6, for example. Although it is not always clear what precisely is meant by such notation, 102 is the experimentally determined value and 1.6 is some measure of the error. It is often claimed or hoped that β is negligible relative to σ, and in that case 1.6 represents σ or some multiple of σ. In the graphical presentation of experimentally obtained data, error bars, usually of width σ or some multiple of σ, are placed around measured values. In some cases, efforts are made to bound the magnitude of β, and the bound is incorporated into the error bars in some fashion.

EXAMPLE **B**   *Measurement of the Speed of Light*

Figure 4.4, taken from McNish (1962) and discussed by Youden (1972), shows 24 independent determinations of the speed of light, c, with error bars. The right column of the figure contains codes for the experimental methods used to obtain the measurements; for example, G denotes a method called the geodimeter method. The range of values for c is about 3.5 km/sec, and many of the errors are less than .5 km/sec. Examination of the figure makes it clear that the error bars are too small and that the spread of values cannot be accounted for by different experimental techniques alone—the geodimeter method produced both the smallest and the next to largest value for c. Youden remarks, "Surely the evidence suggests that individual investigators are unable to set realistic limits of error to their reported values." He goes on to suggest that the differences are largely a result of calibration errors for equipment.   ■

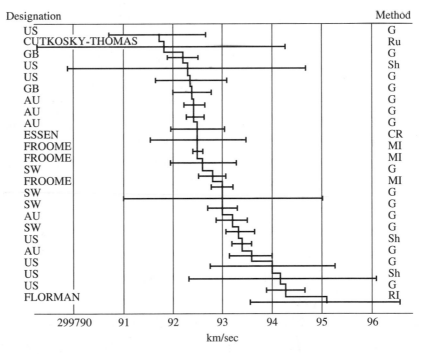

FIGURE **4.4**   A plot of 24 independent determinations of the speed of light with the reported error bars. The investigator or country is listed in the left column, and the experimental method is coded in the right column.

4.3 Covariance and Correlation

The variance of a random variable is a measure of its variability, and the **covariance** of two random variables is a measure of their joint variability, or their degree of association. After defining covariance, we will develop some of its properties and discuss a measure of association called correlation, which is defined in terms of covariance. You may find this material somewhat formal and abstract at first, but as you use them, covariance, correlation, and their properties will begin to seem natural and familiar.

DEFINITION

If X and Y are jointly distributed random variables with expectations μ_X and μ_Y, respectively, the covariance of X and Y is

$$\text{Cov}(X,\ Y) = E[(X - \mu_X)(Y - \mu_Y)]$$

provided that the expectation exists.   ■

The covariance is the average value of the product of the deviation of X from its mean and the deviation of Y from its mean. If the random variables are positively associated—that is, when X is larger than its mean, Y tends to be larger than its mean as well—the covariance will be positive. If the association is negative—that is, when X is larger than its mean, Y tends to be smaller than its mean—the covariance is negative. These statements will be expanded in the discussion of correlation.

By expanding the product and using the linearity of the expectation, we obtain an alternative expression for the covariance, paralleling Theorem B of Section 4.2:

$$\begin{aligned}
\text{Cov}(X,\ Y) &= E(XY - X\mu_Y - Y\mu_X + \mu_X\mu_Y) \\
&= E(XY) - E(X)\mu_Y - E(Y)\mu_X + \mu_X\mu_Y \\
&= E(XY) - E(X)E(Y)
\end{aligned}$$

In particular, if X and Y are independent, then $E(XY) = E(X)E(Y)$ and $\text{Cov}(X,\ Y) = 0$ (but the converse is not true). See Problems 59 and 60 at the end of this chapter for examples.

EXAMPLE A   Let us return to the bivariate uniform distributions of Example C in Section 3.3. First, note that since the marginal distributions are uniform, $E(X) = E(Y) = \frac{1}{2}$. For the case $\alpha = -1$, the joint density of X and Y is $f(x, y) = (2x + 2y - 4xy)$, $0 \le x \le 1$, $0 \le y \le 1$.

$$\begin{aligned}
E(XY) &= \int \int xyf(x,\ y)\,dx\,dy \\
&= \int_0^1 \int_0^1 xy(2x + 2y - 4xy)\,dx\,dy \\
&= \tfrac{2}{9}
\end{aligned}$$

Thus,

$$\text{Cov}(X, Y) = \tfrac{2}{9} - \left(\tfrac{1}{2}\right)\left(\tfrac{1}{2}\right) = -\tfrac{1}{36}$$

The covariance is negative, indicating a negative relationship between X and Y. In fact, from Figure 3.5, we see that if X is less than its mean, $\tfrac{1}{2}$, then Y tends to be larger than its mean, and vice versa. A similar analysis shows that when $\alpha = 1$, $\text{Cov}(X, Y) = \tfrac{1}{36}$. ∎

We will now develop an expression for the covariance of linear combinations of random variables, proceeding in a number of small steps. First, since $E(a + X) = a + E(X)$,

$$\begin{aligned}
\text{Cov}(a + X, Y) &= E\{[a + X - E(a + X)][Y - E(Y)]\} \\
&= E\{[X - E(X)][Y - E(Y)]\} \\
&= \text{Cov}(X, Y)
\end{aligned}$$

Next, since $E(aX) = aE(X)$,

$$\begin{aligned}
\text{Cov}(aX, bY) &= E\{[aX - aE(X)][bY - bE(Y)]\} \\
&= E\{ab[X - E(X)][Y - E(Y)]\} \\
&= abE\{[X - E(X)][Y - E(Y)]\} \\
&= ab\,\text{Cov}(X, Y)
\end{aligned}$$

Next, we consider $\text{Cov}(X, Y + Z)$:

$$\begin{aligned}
\text{Cov}(X, Y + Z) &= E([X - E(X)]\{[Y - E(Y)] + [Z - E(Z)]\}) \\
&= E\{[X - E(X)][Y - E(Y)] + [X - E(X)][Z - E(Z)]\} \\
&= E\{[X - E(X)][Y - E(Y)]\} \\
&\quad + E\{[X - E(X)][Z - E(Z)]\} \\
&= \text{Cov}(X, Y) + \text{Cov}(X, Z)
\end{aligned}$$

We can now put these results together to find $\text{Cov}(aW + bX, cY + dZ)$:

$$\begin{aligned}
\text{Cov}(aW + bX, cY + dZ) &= \text{Cov}(aW + bX, cY) + \text{Cov}(aW + bX, dZ) \\
&= \text{Cov}(aW, cY) + \text{Cov}(bX, cY) + \text{Cov}(aW, dZ) \\
&\quad + \text{Cov}(bX, dZ) \\
&= ac\,\text{Cov}(W, Y) + bc\,\text{Cov}(X, Y) + ad\,\text{Cov}(W, Z) \\
&\quad + bd\,\text{Cov}(X, Z)
\end{aligned}$$

In general, the same kind of argument gives the following important bilinear property of covariance.

THEOREM A

Suppose that $U = a + \sum_{i=1}^{n} b_i X_i$ and $V = c + \sum_{j=1}^{m} d_j Y_j$. Then

$$\text{Cov}(U, V) = \sum_{i=1}^{n} \sum_{j=1}^{m} b_i d_j \text{Cov}(X_i, Y_j) \qquad \blacksquare$$

This theorem has many applications. In particular, since $\text{Var}(X) = \text{Cov}(X, X)$,

$$\text{Var}(X + Y) = \text{Cov}(X + Y, X + Y)$$
$$= \text{Var}(X) + \text{Var}(Y) + 2\text{Cov}(X, Y)$$

More generally, we have the following result for the variance of a linear combination of random variables.

COROLLARY A

$\text{Var}(a + \sum_{i=1}^{n} b_i X_i) = \sum_{i=1}^{n} \sum_{j=1}^{n} b_i b_j \text{Cov}(X_i, X_j).$ $\qquad \blacksquare$

If the X_i are independent, then $\text{Cov}(X_i, X_j) = 0$ for $i \neq j$, and we have another corollary.

COROLLARY B

$\text{Var}(\sum_{i=1}^{n} X_i) = \sum_{i=1}^{n} \text{Var}(X_i)$, if the X_i are independent. $\qquad \blacksquare$

Corollary B is very useful. Note that $E(\sum X_i) = \sum E(X_i)$ whether or not the X_i are independent, but it is generally *not* the case that $\text{Var}(\sum X_i) = \sum \text{Var}(X_i)$.

E X A M P L E   **B**   Finding the variance of a binomial random variable from the definition of variance and the frequency function of the binomial distribution is not easy (try it). But expressing a binomial random variable as a sum of independent Bernoulli random variables makes the computation of the variance trivial. Specifically, if Y is a binomial random variable, it can be expressed as $Y = X_1 + X_2 + \cdots + X_n$, where the X_i are independent Bernoulli random variables with $P(X_i = 1) = p$. We saw earlier (Example A in Section 4.2) that $\text{Var}(X_i) = p(1 - p)$, from which it follows from Corollary B that $\text{Var}(Y) = np(1 - p)$. $\qquad \blacksquare$

E X A M P L E   **C**   *Random Walk*
A drunken walker starts out at a point x_0 on the real line. He takes a step of length X_1, which is a random variable with expected value μ and variance σ, and his position

at that time is $S(1) = x_0 + X_1$. He then takes another step of length X_2, which is independent of X_1 with the same mean and standard deviation. His position after n such steps is $S(n) = x_0 + \sum_{i=1}^{n} X_i$. Then

$$E(S(n)) = x_0 + E\left(\sum_{i=1}^{n} X_i\right) = x_0 + n\mu$$

$$\mathrm{Var}(S(n)) = \mathrm{Var}\left(\sum_{i=1}^{n} X_i\right) = n\sigma^2$$

He thus expects to be at the position $x_0 + n\mu$ with an uncertainty as measured by the standard deviation of $\sqrt{n}\sigma$. Note that if $\mu > 0$, for example, for large values of n he will be to the right of the point x_0 with very high probability (using Chebyshev's inequality).

Random walks have found applications in many areas of science. Brownian motion is a continuous time version of a random walk with the steps being normally distributed random variables. The name derives from observations of the biologist Robert Brown in 1827 of the apparently spontaneous motion of pollen grains suspended in water. This was later explained by Einstein to be due to the collisions of the grains with randomly moving water molecules.

The theory of Brownian motion was developed by Louis Bachelier in 1900 in his PhD thesis "The theory of speculation," which related random walks to the evolution of stock prices. If the value of a stock evolves through time as a random walk, its short-term behavior is unpredictable. The efficient market hypothesis states that stock prices already reflect all known information so that the future price is random and unknowable. The solid line in Figure 4.5 shows the value of the S&P 500 during

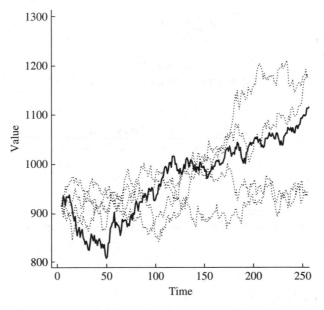

FIGURE **4.5** The solid line is the value of the S&P 500 during 2003. The dashed lines are simulations of random walks.

2003. The average of the increments (steps) was 0.81 and the standard deviation was 9.82. The dashed lines are simulations of random walks with the same intial value and increments that were normally distributed random variables with $\mu = 0.81$ and $\sigma = 9.82$. Notice the long stretches of upturns and downturns that occurred in the random walks as the markets reacted in ways that would have been explained *ex post facto* by analysts. See Malkiel (2004) for a popular exposition of the implications of random walk theory for stock market investors.   ■

The **correlation coefficient** is defined in terms of the covariance.

DEFINITION

If X and Y are jointly distributed random variables and the variances and covariances of both X and Y exist and the variances are nonzero, then the correlation of X and Y, denoted by ρ, is

$$\rho = \frac{\text{Cov}(X, Y)}{\sqrt{\text{Var}(X)\text{Var}(Y)}}$$ ■

Note that because of the way the ratio is formed, the correlation is a dimensionless quantity (it has no units, such as inches, since the units in the numerator and denominator cancel). From the properties of the variance and covariance that we have established, it follows easily that if X and Y are both subjected to linear transformations (such as changing their units from inches to meters), the correlation coefficient does not change. Since it does not depend on the units of measurement, ρ is in many cases a more useful measure of association than is the covariance.

EXAMPLE D   Let us return to the bivariate uniform distribution of Example A. Because X and Y are marginally uniform, $\text{Var}(X) = \text{Var}(Y) = \frac{1}{12}$. In the one case ($\alpha = -1$), we found $\text{Cov}(X, Y) = -\frac{1}{36}$, so

$$\rho = -\tfrac{1}{36} \times 12 = -\tfrac{1}{3}$$

In the other case ($\alpha = 1$), the covariance was $\frac{1}{36}$, so the correlation is $\frac{1}{3}$.   ■

The following notation and relationship are often useful. The standard deviations of X and Y are denoted by σ_X and σ_Y and their covariance by σ_{XY}. We thus have

$$\rho = \frac{\sigma_{XY}}{\sigma_X \sigma_Y}$$

and

$$\sigma_{XY} = \rho \sigma_X \sigma_Y$$

The following theorem states some further properties of ρ.

THEOREM **B**

$-1 \leq \rho \leq 1$. Furthermore, $\rho = \pm 1$ if and only if $P(Y = a + bX) = 1$ for some constants a and b.

Proof

Since the variance of a random variable is nonnegative,

$$0 \leq \text{Var}\left(\frac{X}{\sigma_X} + \frac{Y}{\sigma_Y}\right)$$

$$= \text{Var}\left(\frac{X}{\sigma_X}\right) + \text{Var}\left(\frac{Y}{\sigma_Y}\right) + 2\text{Cov}\left(\frac{X}{\sigma_X}, \frac{Y}{\sigma_Y}\right)$$

$$= \frac{\text{Var}(X)}{\sigma_X^2} + \frac{\text{Var}(Y)}{\sigma_Y^2} + \frac{2\text{Cov}(X, Y)}{\sigma_X \sigma_Y}$$

$$= 2(1 + \rho) \qquad 0 \leq \text{Var}\left(\frac{X}{\sigma_X} + \right.$$

From this, we see that $\rho \geq -1$. Similarly,

$$0 \leq \text{Var}\left(\frac{X}{\sigma_X} - \frac{Y}{\sigma_Y}\right) = 2(1 - \rho)$$

implies that $\rho \leq 1$. Suppose that $\rho = 1$. Then

$$\text{Var}\left(\frac{X}{\sigma_X} - \frac{Y}{\sigma_Y}\right) = 0$$

which by Corollary A of Section 4.2 implies that

$$P\left(\frac{X}{\sigma_X} - \frac{Y}{\sigma_Y} = c\right) = 1$$

for some constant, c. This is equivalent to $P(Y = a + bX) = 1$ for some a and b. A similar argument holds for $\rho = -1$. ∎

EXAMPLE **E** *Investment Portfolio*

We are now in a position to further develop the investment theory discussed in Section 4.1.2, Example E, and Section 4.2, Example D. Please review those examples before continuing. We first consider the simple example of two securities, assuming that they have the same expected returns $\mu_1 = \mu_2 = \mu$ and their returns are uncorrelated: $\sigma_{ij} = \text{Cov}(R_i, R_j) = 0$. For a portfolio $(\pi, 1 - \pi)$, the expected return is

$$E(R(\pi)) = \pi\mu + (1 - \pi)\mu = \mu$$

so that when considering expected return only, the choice of π makes no difference. However, taking risk into account,

$$\text{Var}(R(\pi)) = \pi^2\sigma_1^2 + (1 - \pi)^2\sigma_2^2.$$

Minimizing this with respect to π gives the optimal portfolio

$$\pi_{\text{opt}} = \frac{\sigma_2^2}{\sigma_1^2 + \sigma_2^2}$$

For example, if the investments are equally risky, $\sigma_1 = \sigma_2 = \sigma$, then $\pi = 1/2$, so the best strategy is to split her total investment equally between the two securities. If she does so, the variance of her return is,

$$\text{Var}\left(R\left(\tfrac{1}{2}\right)\right) = \frac{\sigma^2}{2}$$

whereas if she put all her money in one security, the variance of her return would be σ^2. The expected return is the same in both cases. This is a particularly simple example of the value of diversification of investments.

Suppose now that the two securities do not have the same expected returns, $\mu_1 < \mu_2$. Let the standard deviations of the returns be σ_1 and σ_2; usually less risky investments have lower expected returns, $\sigma_1 < \sigma_2$. Furthermore, the two returns may be correlated: $\text{Cov}(R_1, R_2) = \rho\sigma_1\sigma_2$. Corresponding to the portfolio $(\pi, 1 - \pi)$, we have expected return

$$E(R(\pi)) = \pi\mu_1 + (1 - \pi)\mu_2$$

and the variance of the return is

$$\text{Var}(R(\pi)) = \pi^2\sigma_1^2 + 2\pi(1 - \pi)\rho\sigma_1\sigma_2 + (1 - \pi)^2\sigma_2^2$$

Comparing this to the result when the returns were independent, we see the risk is lower when the returns are independent than when they are positively correlated. It would thus be better to invest in two unrelated or weakly related market sectors than to make two investments in the same sector. In deciding the choice of the portfolio vector, the investor can study how the risk (the standard deviation of $R(\pi)$) changes as the expected return increases, and balance expected return versus risk.

In actual investment decisions, many more than two possible investments are involved, but the basic idea remains the same. Suppose there are n possible investments. Let the portfolio weights be denoted by the vector $\pi = (\pi_1, \pi_2, \ldots, \pi_n)$. Let $E(R_i) = \mu_i$, $\text{Cov}(R_i, R_j) = \sigma_{ij}$ (so, in particular, $\text{Var}(R_i)$ is denoted by σ_{ii}), then

$$E(R(\pi)) = \sum \pi_i\mu_i$$

and

$$\text{Var}(R(\pi)) = \sum_{i=1}^{n}\sum_{j=1}^{n} \pi_i\pi_j\sigma_{ij}.$$

The investment decision, the choice of the portfolio vector π, is often couched as that of maximizing expected return subject to the risk being less than some value the individual investor is willing to tolerate. Some investors are more risk averse than others, so the portfolio vectors will differ from investor to investor. Equivalently, the decision may be phrased as that of finding the portfolio vector with the minimum risk subject to a desired return; there may well be many portfolio choices that give the

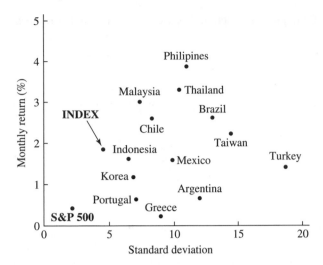

F I G U R E **4.6** The benefit of diversification. The monthly average return from January 1992 to June 1994 of 13 stock markets, plotted against their standard deviations. The performance of the Standard and Poor's 500 index of U.S. stocks is plotted for comparison.

same expected return, and the wise investor would choose the one among them that had the lowest risk.

As a general rule, risk is reduced by diversification and can be decreased with only a small sacrifice of returns. Figure 4.6 from Bernstein (1996, p. 254) illustrates this point empirically. The point labeled "Index" shows the monthly average versus standard deviation for an investment that was equally weighted across all the markets. A reasonably high return with relatively little risk would thus have been obtained by spreading investments equally over the 13 stock markets. In fact, the risk is less than that of any of the individual markets. Note that these emerging markets were riskier than the U.S. market, but that they were more profitable. ■

E X A M P L E **F**   *Bivariate Normal Distribution*

We will show that the covariance of X and Y when they follow a bivariate normal distribution is $\rho \sigma_X \sigma_Y$, which means that ρ is the correlation coefficient. The covariance is

$$\mathrm{Cov}(X, Y) = \int_{-\infty}^{\infty} \int_{-\infty}^{\infty} (x - \mu_X)(y - \mu_Y) f(x, y)\, dx\, dy$$

Making the changes of variables $u = (x - \mu_X)/\sigma_X$ and $v = (y - \mu_Y)/\sigma_Y$ changes the right-hand side to

$$\frac{\sigma_X \sigma_Y}{2\pi \sqrt{1 - \rho^2}} \int_{-\infty}^{\infty} \int_{-\infty}^{\infty} uv \exp\left[-\frac{1}{2(1 - \rho^2)}(u^2 + v^2 - 2\rho uv)\right] du\, dv$$

As in Example F in Section 3.3, we use the technique of completing the square to rewrite this expression as

$$\frac{\sigma_X \sigma_Y}{2\pi \sqrt{1 - \rho^2}} \int_{-\infty}^{\infty} v \exp(-v^2/2) \left(\int_{-\infty}^{\infty} u \exp\left[-\frac{1}{2(1 - \rho^2)}(u - \rho v)^2 \right] du \right) dv$$

The inner integral is the mean of an $N[\rho v, (1 - \rho^2)]$ random variable, lacking only the normalizing constant $[2\pi(1 - \rho^2)]^{-1/2}$, and we thus have

$$\mathrm{Cov}(X, Y) = \frac{\rho \sigma_X \sigma_Y}{\sqrt{2\pi}} \int_{-\infty}^{\infty} v^2 e^{-v^2/2} dv = \rho \sigma_X \sigma_Y$$

as was to be shown. ∎

The correlation coefficient ρ measures the strength of the linear relationship between X and Y (compare with Figure 3.9). Correlation also affects the appearance of a scatterplot, which is constructed by generating n independent pairs (X_i, Y_i), where $i = 1, \ldots, n$, and plotting the points. Figure 4.7 shows scatterplots of 100 pairs of pseudorandom bivariate normal random variables for various values of ρ. Note that the clouds of points are roughly elliptical in shape.

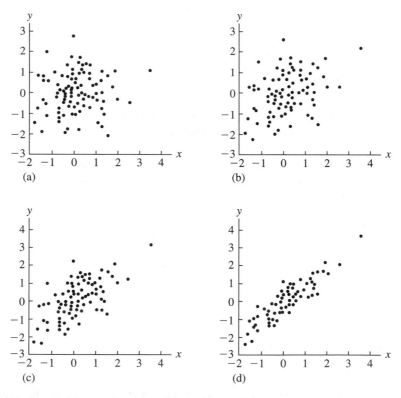

FIGURE 4.7   Scatterplots of 100 independent pairs of bivariate normal random variables, (a) $\rho = 0$, (b) $\rho = .3$, (c) $\rho = .6$, (d) $\rho = .9$.

4.4 Conditional Expectation and Prediction

4.4.1 Definitions and Examples

In Section 3.5, conditional frequency functions and density functions were defined. We noted that these had the properties of ordinary frequency and density functions. In particular, associated with a conditional distribution is a conditional mean. Suppose that Y and X are discrete random variables and that the conditional frequency function of Y given x is $p_{Y|X}(y|x)$. The **conditional expectation** of Y given $X = x$ is

$$E(Y|X = x) = \sum_y y p_{Y|X}(y|x)$$

For the continuous case, we have

$$E(Y|X = x) = \int y f_{Y|X}(y|x)\,dy$$

More generally, the conditional expectation of a function $h(Y)$ is

$$E[h(Y)|X = x] = \int h(y) f_{Y|X}(y|x)\,dy$$

in the continuous case. A similar equation holds in the discrete case.

EXAMPLE A   Consider a Poisson process on $[0, 1]$ with mean λ, and let N be the number of points in $[0, 1]$. For $p < 1$, let X be the number of points in $[0, p]$. Find the conditional distribution and conditional mean of X given $N = n$.

We first find the joint distribution: $P(X = x, N = n)$, which is the probability of x events in $[0, p]$ and $n - x$ events in $[p, 1]$. From the assumption of a Poisson process, the counts in the two intervals are independent Poisson random variables with parameters $p\lambda$ and $(1 - p)\lambda$, so

$$p_{XN}(x, n) = \frac{(p\lambda)^x e^{-p\lambda}}{x!} \frac{[(1 - p)\lambda]^{n-x} e^{-(1-p)\lambda}}{(n - x)!}$$

The marginal distribution of N is Poisson, so the conditional frequency function of X is, after some algebra,

$$p_{X|N}(x|n) = \frac{p_{XN}(x, n)}{p_N(n)}$$

$$= \frac{n!}{x!(n - x)!} p^x (1 - p)^{n-x}$$

This is the binomial distribution with parameters n and p. The conditional expectation is thus by Example A of Section 4.1.2, np.   ■

EXAMPLE B   *Bivariate Normal Distribution*

From Example C in Section 3.5.2, if Y and X follow a bivariate normal distribution, the conditional density of Y given X is

$$f_{Y|X}(y|x) = \frac{1}{\sigma_Y\sqrt{2\pi(1-\rho^2)}} \exp\left(-\frac{1}{2}\frac{\left[y-\mu_Y-\rho\dfrac{\sigma_Y}{\sigma_X}(x-\mu_X)\right]^2}{\sigma_Y^2(1-\rho^2)}\right)$$

This is a normal density with mean $\mu_Y + \rho(x-\mu_X)\sigma_Y/\sigma_X$ and variance $\sigma_Y^2(1-\rho^2)$. The former is the conditional mean and the latter the conditional variance of Y given $X = x$.

Note that the conditional mean is a linear function of X and that as $|\rho|$ increases, the conditional variance decreases; both of these facts are suggested by the elliptical contours of the joint density. To see this more exactly, consider the case in which $\sigma_X = \sigma_Y = 1$ and $\mu_X = \mu_Y = 0$. The contours then are ellipses satisfying

$$\rho^2 x^2 - 2\rho xy + y^2 = \text{constant}$$

The major and minor axes of such an ellipse are at $45°$ and $135°$. The conditional expectation of Y given $X = x$ is the line $y = \rho x$; note that this line does not lie along the major axis of the ellipse. Figure 4.8 shows such a bivariate normal distribution with $\rho = 0.5$. The curved lines of the bivariate density correspond to the conditional density of Y given various values of x, but they are not normalized to integrate to 1. The contours of the bivariate normal are the ellipses shown in the xy plane as dashed curves, with the major axis shown by the straight dashed line. The conditional expectation of Y given $X = x$ is shown as a function of x by the solid line in the plane. Note that it is not the major axis of the ellipse.

This phenomenon was noted by Sir Francis Galton (1822–1911) who studied the relationship of the heights of sons to that of their fathers. He observed that

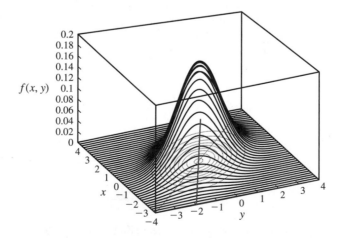

FIGURE 4.8   Bivariate normal density with correlation, $\rho = 0.5$. The conditional expectation of Y given $X = x$ is shown as the solid line in the xy plane.

sons of very tall fathers were shorter on average than their fathers and that sons of very short fathers were on average taller. The empirical relationship is shown in Figure 14.19. ■

Assuming that the conditional expectation of Y given $X = x$ exists for every x in the range of X, it is a well-defined function of X and hence is a random variable, which we write as $E(Y|X)$. For instance, in Example A we found that $E(X|N = n) = np$; thus, $E(X|N) = Np$ is a random variable that is a function of N. Provided that the appropriate sums or integrals converge, this random variable has an expectation and a variance. Its expectation is $E[E(Y|X)]$; for this expression, note that since $E(Y|X)$ is a random variable that is a function of X, the outer expectation can be taken with respect to the distribution of X (Theorem A of Section 4.1.1). The following theorem says that the average (expected) value of Y can be found by first conditioning on X, finding $E(Y|X)$, and then averaging this quantity with respect to X.

> **THEOREM A**
>
> $E(Y) = E[E(Y|X)]$.
>
> **Proof**
>
> We will prove this for the discrete case. The continuous case is proved similarly. Using Theorem 4.1.1A we need to show that
>
> $$E(Y) = \sum_x E(Y|X = x)p_X(x)$$
>
> where
>
> $$E(Y|X = x) = \sum_y y p_{Y|X}(y|x)$$
>
> Interchanging the order of summation gives us
>
> $$\sum_x E(Y|X = x)p_X(x) = \sum_y y \sum_x p_{Y|X}(y|x)p_X(x)$$
>
> (It can be shown that this interchange can be made.) From the law of total probability, we have
>
> $$p_Y(y) = \sum_x p_{Y|X}(y|x)p_X(x)$$
>
> Therefore,
>
> $$\sum_y y \sum_x p_{Y|X}(y|x)p_X(x) = \sum_y y p_Y(y) = E(Y) \qquad ■$$

Theorem A gives what might be called a **law of total expectation:** The expectation of a random variable Y can be calculated by weighting the conditional expectations appropriately and summing or integrating.

EXAMPLE C   Suppose that in a system, a component and a backup unit both have mean lifetimes equal to μ. If the component fails, the system automatically substitutes the backup unit, but there is probability p that something will go wrong and it will fail to do so. Let T be the total lifetime, and let $X = 1$ if the substitution of the backup takes place successfully, and $X = 0$ if it does not. Thus the total lifetime is the lifetime of the first component if the backup fails and is the sum of the lifetimes of the original and the backup units if the backup is successfully made. Then

$$E(T|X = 1) = 2\mu$$
$$E(T|X = 0) = \mu$$

Thus,

$$E(T) = E(T|X = 1)P(X = 1) + E(T|X = 0)P(X = 0) = \mu(2 - p) \qquad \blacksquare$$

EXAMPLE D   *Random Sums*
This example introduces sums of the type

$$T = \sum_{i=1}^{N} X_i$$

where N is a random variable with a finite expectation and the X_i are random variables that are independent of N and have the common mean $E(X)$. Such sums arise in a variety of applications. An insurance company might receive N claims in a given period of time, and the amounts of the individual claims might be modeled as random variables $X_1, X_2, \ldots$. The random variable N could denote the number of customers entering a store and X_i the expenditure of the ith customer, or N could denote the number of jobs in a single-server queue and X_i the service time for the ith job. For this last case, T is the time to serve all the jobs in the queue. According to Theorem A,

$$E(T) = E[E(T|N)]$$

Since $E(T|N = n) = nE(X)$, $E(T|N) = NE(X)$ and thus

$$E(T) = E[NE(X)] = E(N)E(X)$$

This agrees with the intuitive guess that the average time to complete N jobs, where N is random, is the average value of N times the average amount of time to complete a job. $\qquad \blacksquare$

We have seen that the expectation of the random variable $E(Y|X)$ is $E(Y)$. We now find its variance.

THEOREM **B**

$\text{Var}(Y) = \text{Var}[E(Y|X)] + E[\text{Var}(Y|X)]$.

Proof

We will explain what is meant by the notation in the course of the proof. First,

$$\text{Var}(Y|X = x) = E(Y^2|X = x) - [E(Y)|X = x)]^2$$

which is defined for all values of x. Thus, just as we defined $E(Y|X)$ to be a random variable by letting X be random, we can define $\text{Var}(Y|X)$ as a random variable. In particular, $\text{Var}(Y|X)$ has the expectation $E[\text{Var}(Y|X)]$. Since

$$\text{Var}(Y|X) = E(Y^2|X) - [E(Y|X)]^2$$
$$E[\text{Var}(Y|X)] = E[E(Y^2|X)] - E\{[E(Y|X)]^2\}$$

Also,

$$\text{Var}[E(Y|X)] = E\{[E(Y|X)]^2\} - \{E[E(Y|X)]\}^2$$

The final piece that we need is

$$\text{Var}(Y) = E(Y^2) - [E(Y)]^2$$
$$= E[E(Y^2|X)] - \{E[E(Y|X)]\}^2$$

by the law of total expectation. Now we can put all the pieces together:

$$\text{Var}(Y) = E[E(Y^2|X)] - \{E[E(Y|X)]\}^2$$
$$= E[E(Y^2|X)] - E\{[E(Y|X)]^2\} + E\{[E(Y|X)]^2\} - \{E[E(Y|X)]\}^2$$
$$= E[\text{Var}(Y|X)] + \text{Var}[E(Y|X)] \qquad ■$$

EXAMPLE **E**   *Random Sums*

Let us continue Example D but with the additional assumptions that the X_i are independent random variables with the same mean, $E(X)$, and the same variance, $\text{Var}(X)$, and that $\text{Var}(N) < \infty$. According to Theorem B,

$$\text{Var}(T) = E[\text{Var}(T|N)] + \text{Var}[E(T|N)]$$

Because $E(T|N) = NE(X)$,

$$\text{Var}[E(T|N)] = [E(X)]^2 \text{Var}(N)$$

Also, since $\text{Var}(T|N = n) = \text{Var}(\sum_{i=1}^{n} X_i) = n\,\text{Var}(X)$,

$$\text{Var}(T|N) = N\,\text{Var}(X)$$

and

$$E[\text{Var}(T|N)] = E(N)\text{Var}(X)$$

We thus have

$$\text{Var}(T) = [E(X)]^2 \text{Var}(N) + E(N)\text{Var}(X)$$

If N is fixed, say, $N = n$, then $\text{Var}(T) = n\,\text{Var}(X)$. Thus, we see from the preceding equation that extra variability occurs in T because N is random.

As a concrete example, suppose that the number of insurance claims in a certain time period has expected value equal to 900 and standard deviation equal to 30, as would be the case if the number were a Poisson random variable with expected value 900. Suppose that the average claim value is $1000 and the standard deviation is $500. Then the expected value of the total, T, of the claims is $E(T) = \$900,000$ and the variance of T is

$$\begin{aligned} \text{Var}\,(T) &= 1000^2 \times 900 + 900 \times 500^2 \\ &= 1.125 \times 10^9 \end{aligned}$$

The standard deviation of T is the square root of the variance, $33,541. The insurance company could then plan on total claims of $900,000 plus or minus a few standard deviations (by Chebyshev's inequality). Observe that if the total number of claims were not variable but were fixed at $N = 900$, the variance of the total claims would be given by $E(N)\text{Var}(X)$ in the preceding expression. The result would be a standard deviation equal to $15,000. The variability in the number of claims thus contributes substantially to the uncertainty in the total.   ∎

4.4.2   Prediction

This section treats the problem of predicting the value of one random variable from another. We might wish, for example, to measure the value of some physical quantity, such as pressure, using an instrument. The actual pressures to be measured are unknown and variable, so we might model them as values of a random variable, Y. Assume that measurements are to be taken by some instrument that produces a response, X, related to Y in some fashion but corrupted by random noise as well; X might represent current flow, for example. Y and X have some joint distribution, and we wish to predict the actual pressure, Y, from the instrument response, X.

As another example, in forestry, the volume of a tree is sometimes estimated from its diameter, which is easily measured. For a whole forest, it is reasonable to model diameter (X) and volume (Y) as random variables with some joint distribution, and then attempt to predict Y from X.

Let us first consider a relatively trivial situation: the problem of predicting Y by means of a constant value, c. If we wish to choose the "best" value of c, we need some measure of the effectiveness of a prediction. One that is amenable to mathematical analysis and that is widely used is the mean squared error:

$$\text{MSE} = E[(Y - c)^2]$$

This is the average squared error of prediction, the averaging being done with respect to the distribution of Y. The problem then becomes finding the value of c that minimizes the mean squared error. To solve this problem, we denote $E(Y)$ by μ and observe that (see Theorem A of Section 4.2.1)

$$E[(Y - c)^2] = \text{Var}(Y - c) + [E(Y - c)]^2$$
$$= \text{Var}(Y) + (\mu - c)^2$$

The first term of the last expression does not depend on c, and the second term is minimized for $c = \mu$, which is the optimal choice of c.

Now let us consider predicting Y by some function $h(X)$ in order to minimize MSE $= E\{[Y - h(X)]^2\}$. From Theorem A of Section 4.4.1, the right-hand side can be expressed as

$$E\{[Y - h(X)]^2\} = E(E\{[Y - h(X)]^2|X\})$$

The outer expectation is with respect to X. For every x, the inner expectation is minimized by setting $h(x)$ equal to the constant $E(Y|X = x)$, from the result of the preceding paragraph. We thus have that the minimizing function $h(X)$ is

$$h(X) = E(Y|X)$$

E X A M P L E **A**   For the bivariate normal distribution, we found that

$$E(Y|X) = \mu_Y + \rho \frac{\sigma_Y}{\sigma_X}(X - \mu_X)$$

This linear function of X is thus the minimum mean squared error predictor of Y from X.   ■

A practical limitation of the optimal prediction scheme is that its implementation depends on knowing the joint distribution of Y and X in order to find $E(Y|X)$, and often this information is not available, not even approximately. For this reason, we can try to attain the more modest goal of finding the optimal *linear* predictor of Y. (In Example A, it turned out that the best predictor was linear, but this is not generally the case.) That is, rather than finding the best function h among all functions, we try to find the best function of the form $h(x) = \alpha + \beta x$. This merely requires optimizing over the two parameters α and β. Now

$$E[(Y - \alpha - \beta X)^2] = \text{Var}(Y - \alpha - \beta X) + [E(Y - \alpha - \beta X)]^2$$
$$= \text{Var}(Y - \beta X) + [E(Y - \alpha - \beta X)]^2$$

The first term of the last expression does not depend on α, so α can be chosen so as to minimize the second term. To do this, note that

$$E(Y - \alpha - \beta X) = \mu_Y - \alpha - \beta \mu_X$$

and that the right-hand side is zero, and hence its square is minimized, if

$$\alpha = \mu_Y - \beta \mu_X$$

As for the first term,

$$\text{Var}(Y - \beta X) = \sigma_Y^2 + \beta^2 \sigma_X^2 - 2\beta \sigma_{XY}$$

where $\sigma_{XY} = \mathrm{Cov}(X, Y)$. This is a quadratic function of β, and the minimum is found by setting the derivative with respect to β equal to zero, which yields

$$\beta = \frac{\sigma_{XY}}{\sigma_X^2} = \rho \frac{\sigma_Y}{\sigma_X}$$

ρ is the correlation coefficient. Substituting in these values of α and β, we find that the minimum mean squared error predictor, which we denote by $\hat{Y}$, is

$$\hat{Y} = \alpha + \beta X$$
$$= \mu_Y + \frac{\sigma_{XY}}{\sigma_X^2}(X - \mu_X)$$

The mean squared prediction error is then

$$\mathrm{Var}(Y - \beta X) = \sigma_Y^2 + \frac{\sigma_{XY}^2}{\sigma_X^4}\sigma_X^2 - 2\frac{\sigma_{XY}}{\sigma_X^2}\sigma_{XY}$$
$$= \sigma_Y^2 - \frac{\sigma_{XY}^2}{\sigma_X^2}$$
$$= \sigma_Y^2 - \rho^2\sigma_Y^2$$
$$= \sigma_Y^2(1 - \rho^2)$$

Note that the optimal linear predictor depends on the joint distribution of X and Y only through their means, variances, and covariance. Thus, in practice, it is generally easier to construct the optimal linear predictor or an approximation to it than to construct the general optimal predictor $E(Y|X)$. Second, note that the form of the optimal linear predictor is the same as that of $E(Y|X)$ for the bivariate normal distribution. Third, note that the mean squared prediction error depends only on σ_Y and ρ and that it is small if ρ is close to $+1$ or -1. Here we see again, from a different point of view, that the correlation coefficient is a measure of the strength of the linear relationship between X and Y.

EXAMPLE B   Suppose that two examinations are given in a course. As a probability model, we regard the scores of a student on the first and second examinations as jointly distributed random variables X and Y. Suppose for simplicity that the exams are scaled to have the same means $\mu = \mu_X = \mu_Y$ and standard deviations $\sigma = \sigma_X = \sigma_Y$. Then, the correlation between X and Y is $\rho = \sigma_{XY}/\sigma^2$ and the best linear predictor is $\hat{Y} = \mu + \rho(X - \mu)$, so

$$\hat{Y} - \mu = \rho(X - \mu)$$

Notice that by this equation we predict the student's score on the second examination to differ from the overall mean μ by less than did the score on the first examination. If the correlation ρ is positive, this is encouraging for a student who scores below the mean on the first exam, since our best prediction is that his score on the next exam will be closer to the mean. On the other hand, it's bad news for the student who scored above the mean on the first exam, since our best prediction is that she will score closer to the mean on the next exam. This phenomenon is often referred to as *regression to the mean*. ∎

4.5 The Moment-Generating Function

This section develops and applies some of the properties of the moment-generating function. It turns out, despite its unlikely appearance, to be a very useful tool that can dramatically simplify certain calculations.

The **moment-generating function (mgf)** of a random variable X is $M(t) = E(e^{tX})$ if the expectation is defined. In the discrete case,

$$M(t) = \sum_x e^{tx} p(x)$$

and in the continuous case,

$$M(t) = \int_{-\infty}^{\infty} e^{tx} f(x) \, dx$$

The expectation, and hence the moment-generating function, may or may not exist for any particular value of t. In the continuous case, the existence of the expectation depends on how rapidly the tails of the density decrease; for example, because the tails of the Cauchy density die down at the rate x^{-2}, the expectation does not exist for any t and the moment-generating function is undefined. The tails of the normal density die down at the rate e^{-x^2}, so the integral converges for all t.

PROPERTY A

If the moment-generating function exists for t in an open interval containing zero, it uniquely determines the probability distribution. ∎

We cannot prove this important property here—its proof depends on properties of the Laplace transform. Note that Property A says that if two random variables have the same mgf in an open interval containing zero, they have the same distribution. For some problems, we can find the mgf and then deduce the unique probability distribution corresponding to it.

The rth moment of a random variable is $E(X^r)$ if the expectation exists. We have already encountered the first and second moments earlier in this chapter, that is, $E(X)$ and $E(X^2)$. Central moments rather than ordinary moments are often used: The rth central moment is $E\{[X - E(X)]^r\}$. The variance is the second central moment and is a measure of dispersion about the mean. The third central moment, called the **skewness,** is used as a measure of the asymmetry of a density or a frequency function about its mean; if a density is symmetric about its mean, the skewness is zero (see Problem 78 at the end of this chapter). As its name implies, the moment-generating function has something to do with moments. To see this, consider the continuous case:

$$M(t) = \int_{-\infty}^{\infty} e^{tx} f(x) \, dx$$

The derivative of $M(t)$ is

$$M'(t) = \frac{d}{dt} \int_{-\infty}^{\infty} e^{tx} f(x)\, dx$$

It can be shown that differentiation and integration can be interchanged, so that

$$M'(t) = \int_{-\infty}^{\infty} x e^{tx} f(x)\, dx$$

and

$$M'(0) = \int_{-\infty}^{\infty} x f(x)\, dx = E(X)$$

Differentiating r times, we find

$$M^{(r)}(0) = E(X^r)$$

It can further be argued that if the moment-generating function exists in an interval containing zero, then so do all the moments. We thus have the following property.

PROPERTY B

If the moment-generating function exists in an open interval containing zero, then $M^{(r)}(0) = E(X^r)$. ∎

To find the moments of a random variable from the definition of expectation, we must sum a series or carry out an integration. The utility of Property B is that, if the mgf can be found, the process of integration or summation, which may be difficult, can be replaced by the process of differentiation, which is mechanical. We now illustrate these concepts using some familiar distributions.

EXAMPLE A   *Poisson Distribution*
By definition,

$$M(t) = \sum_{k=0}^{\infty} e^{tk} \frac{\lambda^k}{k!} e^{-\lambda}$$

$$= \sum_{k=0}^{\infty} \frac{(\lambda e^t)^k}{k!} e^{-\lambda}$$

$$= e^{-\lambda} e^{\lambda e^t}$$

$$= e^{\lambda(e^t - 1)}$$

The sum converges for all t. Differentiating, we have

$$M'(t) = \lambda e^t e^{\lambda(e^t - 1)}$$

$$M''(t) = \lambda e^t e^{\lambda(e^t - 1)} + \lambda^2 e^{2t} e^{\lambda(e^t - 1)}$$

Evaluating these derivatives at $t = 0$, we find

$$E(X) = \lambda$$
$$E(X^2) = \lambda^2 + \lambda$$

from which it follows that

$$\text{Var}(X) = E(X^2) - [E(X)]^2 = \lambda$$

We have found that the mean and the variance of a Poisson distribution are equal. ∎

E X A M P L E **B**   *Gamma Distribution*

The mgf of a gamma distribution is

$$M(t) = \int_0^\infty e^{tx} \frac{\lambda^\alpha}{\Gamma(\alpha)} x^{\alpha-1} e^{-\lambda x} \, dx$$

$$= \frac{\lambda^\alpha}{\Gamma(\alpha)} \int_0^\infty x^{\alpha-1} e^{x(t-\lambda)} \, dx$$

The latter integral converges for $t < \lambda$ and can be evaluated by relating it to the gamma density having parameters α and $\lambda - t$. We thus obtain

$$M(t) = \frac{\lambda^\alpha}{\Gamma(\alpha)} \left(\frac{\Gamma(\alpha)}{(\lambda - t)^\alpha} \right) = \left(\frac{\lambda}{\lambda - t} \right)^\alpha$$

Differentiating, we find

$$M'(0) = E(X) = \frac{\alpha}{\lambda}$$

$$M''(0) = E(X^2) = \frac{\alpha(\alpha + 1)}{\lambda^2}$$

From these equations, we find that

$$\text{Var}(X) = E(X^2) - [E(X)]^2$$

$$= \frac{\alpha(\alpha + 1)}{\lambda^2} - \frac{\alpha^2}{\lambda^2}$$

$$= \frac{\alpha}{\lambda^2}$$ ∎

E X A M P L E **C**   *Standard Normal Distribution*

For the standard normal distribution, we have

$$M(t) = \frac{1}{\sqrt{2\pi}} \int_{-\infty}^\infty e^{tx} e^{-x^2/2} \, dx$$

The integral converges for all t and can be evaluated using the technique of completing the square. Since

$$\frac{x^2}{2} - tx = \frac{1}{2}(x^2 - 2tx + t^2) - \frac{t^2}{2}$$

$$= \frac{1}{2}(x - t)^2 - \frac{t^2}{2}$$

therefore,

$$M(t) = \frac{e^{t^2/2}}{\sqrt{2\pi}} \int_{-\infty}^{\infty} e^{-(x-t)^2/2} \, dx$$

Making the change of variables $u = x - t$ and using the fact that the standard normal density integrates to 1, we find that

$$M(t) = e^{t^2/2}$$

From this result, we easily see that $E(X) = 0$ and $\text{Var}(X) = 1$. ∎

Let us continue with the development of the properties of the moment-generating function.

PROPERTY C

If X has the mgf $M_X(t)$ and $Y = a + bX$, then Y has the mgf $M_Y(t) = e^{at} M_X(bt)$.

Proof

$$M_Y(t) = E(e^{tY})$$

$$= E(e^{at + btX})$$

$$= E(e^{at} e^{btX})$$

$$= e^{at} E(e^{btX})$$

$$= e^{at} M_X(bt)$$ ∎

EXAMPLE D   *General Normal Distribution*
If Y follows a general normal distribution with parameters μ and σ, the distribution of Y is the same as that of $\mu + \sigma X$, where X follows a standard normal distribution. Thus, from Example C and Property C,

$$M_Y(t) = e^{\mu t} M_X(\sigma t) = e^{\mu t} e^{\sigma^2 t^2/2}$$ ∎

PROPERTY **D**

If X and Y are independent random variables with mgf's M_X and M_Y and $Z = X + Y$, then $M_Z(t) = M_X(t)M_Y(t)$ on the common interval where both mgf's exist.

Proof

$$M_Z(t) = E(e^{tZ})$$
$$= E(e^{tX + tY})$$
$$= E(e^{tX}e^{tY})$$

From the assumption of independence,

$$M_Z(t) = E(e^{tX})E(e^{tY})$$
$$= M_X(t)M_Y(t) \qquad \blacksquare$$

By induction, Property D can be extended to sums of several independent random variables. This is one of the most useful properties of the moment-generating function. The next three examples show how it can be used to easily derive results that would take a lot more work to achieve without recourse to the mgf.

E X A M P L E **E**   The sum of independent Poisson random variables is a Poisson random variable: If X is Poisson with parameter λ and Y is Poisson with parameter μ, then $X + Y$ is Poisson with parameter $\lambda + \mu$, since

$$e^{\lambda(e^t - 1)}e^{\mu(e^t - 1)} = e^{(\lambda + \mu)(e^t - 1)} \qquad \blacksquare$$

E X A M P L E **F**   If X follows a gamma distribution with parameters α_1 and λ and Y follows a gamma distribution with parameters α_2 and λ, the mgf of $X + Y$ is

$$\left(\frac{\lambda}{\lambda - t}\right)^{\alpha_1} \left(\frac{\lambda}{\lambda - t}\right)^{\alpha_2} = \left(\frac{\lambda}{\lambda - t}\right)^{\alpha_1 + \alpha_2}$$

where $t < \lambda$. The right-hand expression is the mgf of a gamma distribution with parameters λ and $\alpha_1 + \alpha_2$. It follows from this that the sum of n independent exponential random variables with parameter λ follows a gamma distribution with parameters n and λ. Thus, the time between n consecutive events of a Poisson process in time follows a gamma distribution. Assuming that the service times in a queue are independent exponential random variables, the length of time to serve n customers follows a gamma distribution. $\qquad \blacksquare$

EXAMPLE G   If $X \sim N(\mu, \sigma^2)$ and, independent of X, $Y \sim N(v, \tau^2)$, then the mgf of $X + Y$ is

$$e^{\mu t} e^{t^2 \sigma^2 / 2} e^{vt} e^{t^2 \tau^2 / 2} = e^{(\mu + v)t} e^{t^2 (\sigma^2 + \tau^2)/2}$$

which is the mgf of a normal distribution with mean $\mu + v$ and variance $\sigma^2 + \tau^2$. The sum of independent normal random variables is thus normal.   ∎

The preceding three examples are atypical. In general, if two independent random variables follow some type of distribution, it is not necessarily true that their sum follows the same type of distribution. For example, the sum of two gamma random variables having different values for the parameter λ does not follow a gamma distribution, as can be easily seen from the mgf.

We now apply moment-generating functions to random sums of the type introduced in Section 4.4.1. Suppose that

$$S = \sum_{i=1}^{N} X_i$$

where the X_i are independent and have the same mgf, M_X, and where N has the mgf M_N and is independent of the X_i. By conditioning, we have

$$M_S(t) = E[E(e^{tS}|N)]$$

Given $N = n$, $M_S(t) = [M_X(t)]^n$ from Property D. We thus have

$$M_S(t) = E[M_X(t)^N]$$
$$= E(e^{N \log M_X(t)})$$
$$= M_N[\log M_X(t)]$$

(We must carefully note the values of t for which this is defined.)

EXAMPLE H   *Compound Poisson Distribution*
This example presents a model that occurs for certain chain reactions, or "cascade" processes. When a single primary electron, having been accelerated in an electrical field, hits a plate, several secondary electrons are produced. In a multistage multiplying tube, each of these secondary electrons hits another plate and thereby produces a number of tertiary electrons. The process can continue through several stages in this manner. Woodward (1948) considered models of this type in which the number of electrons produced by the impact of a single electron on the plate is random and, in particular, in which the number of secondary electrons follows a Poisson distribution. The number of electrons produced at the third stage is described by a random sum of the type just described, where N is the number of secondary electrons and X_i is the number of electrons produced by the ith secondary electron. Suppose that the X_i are independent Poisson random variables with parameter λ and that N is a Poisson random variable with parameter μ. According to the preceding result, the mgf of S,

the total number of particles, is

$$M_S(t) = \exp[\mu(e^{\lambda(e^t - 1)} - 1)]$$ ∎

Example H illustrates the utility of the mgf. It would have been more difficult to find the probability mass function of the number of particles at the third stage. By differentiating the mgf, we can find the moments of the probability mass function (see Problem 98 at the end of this chapter).

If X and Y have a joint distribution, their joint moment-generating function is defined as

$$M_{XY}(s, t) = E(e^{sX + tY})$$

which is a function of two variables, s and t. If the joint mgf is defined on an open set containing the origin, it uniquely determines the joint distribution. The mgf of the marginal distribution of X alone is

$$M_X(s) = M_{XY}(s, 0)$$

and similarly for Y. It can be shown that X and Y are independent if and only if their joint mgf factors into the product of the mgf's of the marginal distributions. $E(XY)$ and other higher-order joint moments can be obtained from the joint mgf by differentiation. Analogous properties hold for the joint mgf of several random variables.

The major limitation of the mgf is that it may not exist. The **characteristic function** of a random variable X is defined to be

$$\phi(t) = E(e^{itX})$$

where $i = \sqrt{-1}$. In the continuous case,

$$\phi(t) = \int_{-\infty}^{\infty} e^{itx} f(x) \, dx$$

This integral converges for all values of t, since $|e^{itx}| \leq 1$. The characteristic function is thus defined for all distributions. Its properties are similar to those of the mgf: Moments can be obtained by differentiation, the characteristic function changes simply under linear transformations, and the characteristic function of a sum of independent random variables is the product of their characteristic functions. But using the characteristic function requires some familiarity with the techniques of complex variables.

4.6 Approximate Methods

In many applications, only the first two moments of a random variable, and not the entire probability distribution, are known, and even these may be known only approximately. We will see in Chapter 5 that repeated independent observations of a random variable allow reliable estimates to be made of its mean and variance. Suppose that we know the expectation and the variance of a random variable X but not the entire distribution, and that we are interested in the mean and variance of $Y = g(X)$ for some fixed function g. For example, we might be able to measure X and determine its mean and variance, but really be interested in Y, which is related to X in a known way. We might want to know $\text{Var}(Y)$, at least approximately, in order to assess the accuracy of the indirect measurement process. From the results given in this chapter, we cannot in general find $E(Y) = \mu_Y$ and $\text{Var}(Y) = \sigma_Y^2$ from $E(X) = \mu_X$ and $\text{Var}(X) = \sigma_X^2$, unless the function g is linear. However, if g is nearly linear in a range in which X has high probability, it can be approximated by a linear function and approximate moments of Y can be found.

In proceeding as just described, we follow a tack often taken in applied mathematics: When confronted with a nonlinear problem that we cannot solve, we linearize. In probability and statistics, this method is called **propagation of error,** or the δ **method.** Linearization is carried out through a Taylor series expansion of g about μ_X. To the first order,

$$Y = g(X) \approx g(\mu_X) + (X - \mu_X)g'(\mu_X)$$

We have expressed Y as approximately equal to a linear function of X. Recalling that if $U = a + bV$, then $E(U) = a + bE(V)$ and $\text{Var}(U) = b^2\text{Var}(V)$, we find

$$\mu_Y \approx g(\mu_X)$$
$$\sigma_Y^2 \approx \sigma_X^2[g'(\mu_X)]^2$$

We know that in general $E(Y) \neq g(E(X))$, as given by the approximation. In fact, we can carry out the Taylor series expansion to the second order to get an improved approximation of μ_Y:

$$Y = g(X) \approx g(\mu_X) + (X - \mu_X)g'(\mu_X) + \tfrac{1}{2}(X - \mu_X)^2 g''(\mu_X)$$

Taking the expectation of the right-hand side, we have, since $E(X - \mu_X) = 0$,

$$E(Y) \approx g(\mu_X) + \tfrac{1}{2}\sigma_X^2 g''(\mu_X)$$

How good such approximations are depends on how nonlinear g is in a neighborhood of μ_X and on the size of σ_X. From Chebyshev's inequality, we know that X is unlikely to be many standard deviations away from μ_X; if g can be reasonably well approximated in this range by a linear function, the approximations for the moments will be reasonable as well.

E X A M P L E A   The relation of voltage, current, and resistance is $V = IR$. Suppose that the voltage is held constant at a value V_0 across a medium whose resistance fluctuates randomly as a result, say, of random fluctuations at the molecular level. The current therefore also varies randomly. Suppose that it can be determined experimentally to have mean $\mu_I \neq 0$ and variance σ_I^2. We wish to find the mean and variance of the resistance, R, and since we do not know the distribution of I, we must resort to an approximation.

We have

$$R = g(I) = \frac{V_0}{I}$$

$$g'(\mu_I) = -\frac{V_0}{\mu_I^2} \qquad g''(\mu_I) = \frac{2V_0}{\mu_I^3}$$

Thus,

$$\mu_R \approx \frac{V_0}{\mu_I} + \frac{V_0}{\mu_I^3}\sigma_I^2$$

$$\sigma_R^2 \approx \frac{V_0^2}{\mu_I^4}\sigma_I^2$$

We see that the variability of R depends on both the mean level of I and the variance of I. This makes sense, since if I is quite small, small variations in I will result in large variations in $R = V_0/I$, whereas if I is large, small variations will not affect R as much. The second-order correction factor for μ_R also depends on μ_I and is large if μ_I is small. In fact, when I is near zero, the function $g(I) = V_0/I$ is quite nonlinear, and the linearization is not a good approximation. ∎

EXAMPLE B   This example examines the accuracy of the approximations using a simple test case. We choose the function $g(x) = \sqrt{x}$ and consider two cases: X uniform on $[0, 1]$, and X uniform on $[1, 2]$. The graph of $g(x)$ in Figure 4.9 shows that g is more nearly linear in the latter case, so we would expect the approximations to work better there.

Let $Y = \sqrt{X}$; because X is uniform on $[0, 1]$,

$$E(Y) = \int_0^1 \sqrt{x}\, dx = \tfrac{2}{3}$$

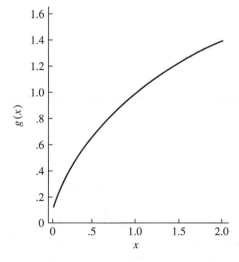

FIGURE **4.9**   The function $g(x) = \sqrt{x}$ is more nearly linear over the interval $[1, 2]$ than over the interval $[0, 1]$.

and

$$E(Y^2) = \int_0^1 x \, dx = \tfrac{1}{2}$$

so $\text{Var}(Y) = \tfrac{1}{2} - \left(\tfrac{2}{3}\right)^2 = \tfrac{1}{18}$ and $\sigma_Y = .236$. These results are exact.

Using the approximation method, we first calculate

$$g'(x) = \tfrac{1}{2}x^{-1/2}$$
$$g''(x) = -\tfrac{1}{4}x^{-3/2}$$

Since X is uniform on $[0, 1]$, $\mu_X = \tfrac{1}{2}$, and evaluating the derivatives at this value gives us

$$g'(\mu_X) = \frac{\sqrt{2}}{2}$$

$$g''(\mu_X) = -\frac{\sqrt{2}}{2}$$

We know that $\text{Var}(X) = \tfrac{1}{12}$ for a random variable uniform on $[0, 1]$, so the approximations are

$$E(Y) \approx \sqrt{\frac{1}{2}} - \frac{1}{2}\left(\frac{\sqrt{2}}{12 \times 2}\right) = .678$$

$$\text{Var}(Y) \approx \tfrac{1}{2} \times \tfrac{1}{12} = .042$$

$$\sigma_Y \approx .204$$

The approximation to the mean is .678, and compared to the actual value of .667, it is off by about 1.6%. The approximation to the standard deviation is .204, and compared to the actual value of .236, it is off by 13%.

Now let us consider the case in which X is uniform on $[1, 2]$. Proceeding as before, we find that $y = \sqrt{x}$ has mean 1.219. The variance and standard deviation are .0142 and .119, respectively. To compare these to the approximations, we note that $\mu_X = \tfrac{3}{2}$ and $\text{Var}(X) = \tfrac{1}{12}$ (the random variable uniform on $[1, 2]$ can be obtained by adding the constant 1 to a random variable uniform on $[0, 1]$; compare with Theorem A in Section 4.2). We find

$$g'(\mu_X) = .408$$
$$g''(\mu_X) = -.136$$

so the approximations are

$$E(Y) \approx \sqrt{\frac{3}{2}} - \frac{1}{2}\left(\frac{.136}{12}\right) = 1.219$$

$$\text{Var}(Y) \approx \frac{.408^2}{12} = .0138$$

$$\sigma_Y \approx .118$$

These values are much closer to the actual values than are the approximations for the first case. ∎

Suppose that we have $Z = g(X, Y)$, a function of two variables. We can again carry out Taylor series expansions to approximate the mean and variance of Z. To the first order, letting μ denote the point (μ_X, μ_Y),

$$Z = g(X, Y) \approx g(\mu) + (X - \mu_X)\frac{\partial g(\mu)}{\partial x} + (Y - \mu_Y)\frac{\partial g(\mu)}{\partial y}$$

The notation $\partial g(\mu)/\partial x$ means that the derivative is evaluated at the point μ. Here Z is expressed as approximately equal to a linear function of X and Y, and the mean and variance of this linear function are easily calculated to be

$$E(Z) \approx g(\mu)$$

and

$$\text{Var}(Z) \approx \sigma_X^2 \left(\frac{\partial g(\mu)}{\partial x}\right)^2 + \sigma_Y^2 \left(\frac{\partial g(\mu)}{\partial y}\right)^2 + 2\sigma_{XY}\left(\frac{\partial g(\mu)}{\partial x}\right)\left(\frac{\partial g(\mu)}{\partial y}\right)$$

(For the latter calculation, see Corollary A in Section 4.3.) As is the case with a single variable, a second-order expansion can be used to obtain an improved estimate of $E(Z)$:

$$Z = g(X, Y) \approx g(\mu) + (X - \mu_X)\frac{\partial g(\mu)}{\partial x} + (Y - \mu_Y)\frac{\partial g(\mu)}{\partial y}$$

$$+ \frac{1}{2}(X - \mu_X)^2\frac{\partial^2 g(\mu)}{\partial x^2} + \frac{1}{2}(Y - \mu_Y)^2\frac{\partial^2 g(\mu)}{\partial y^2}$$

$$+ (X - \mu_X)(Y - \mu_Y)\frac{\partial^2 g(\mu)}{\partial x \partial y}$$

Taking expectations term by term on the right-hand side yields

$$E(Z) \approx g(\mu) + \frac{1}{2}\sigma_X^2\frac{\partial^2 g(\mu)}{\partial x^2} + \frac{1}{2}\sigma_Y^2\frac{\partial^2 g(\mu)}{\partial y^2} + \sigma_{XY}\frac{\partial^2 g(\mu)}{\partial x \partial y}$$

The general case of a function of n variables can be worked out similarly; the basic concepts are illustrated by the two-variable case.

E X A M P L E **C**   *Expectation and Variance of a Ratio*

Let us consider the case where $Z = Y/X$, which arises frequently in practice. For example, a chemist might measure the concentrations of two substances, both with some measurement error that is indicated by their standard deviations, and then report the relative concentrations in the form of a ratio. What is the approximate standard deviation of the ratio, Z?

Using the method of propagation of error derived above, for $g(x, y) = y/x$, we have

$$\frac{\partial g}{\partial x} = \frac{-y}{x^2} \qquad \frac{\partial g}{\partial y} = \frac{1}{x}$$

$$\frac{\partial^2 g}{\partial x^2} = \frac{2y}{x^3} \qquad \frac{\partial^2 g}{\partial y^2} = 0$$

$$\frac{\partial^2 g}{\partial x \partial y} = \frac{-1}{x^2}$$

Evaluating these derivatives at (μ_X, μ_Y) and using the preceding result, we find, if $\mu_X \neq 0$,

$$E(Z) \approx \frac{\mu_Y}{\mu_X} + \sigma_X^2 \frac{\mu_Y}{\mu_X^3} - \frac{\sigma_{XY}}{\mu_X^2}$$

$$= \frac{\mu_Y}{\mu_X} + \frac{1}{\mu_X^2}\left(\sigma_X^2 \frac{\mu_Y}{\mu_X} - \rho\sigma_X\sigma_Y\right)$$

From this equation, we see that the difference between $E(Z)$ and μ_Y/μ_X depends on several factors. If σ_X and σ_Y are small—that is, if the two concentrations are measured quite accurately—the difference is small. If μ_X is small, the difference is relatively large. Finally, correlation between X and Y affects the difference.

We now consider the variance. Again using the preceding result and evaluating the partial derivatives at (μ_X, μ_Y), we find

$$\text{Var}(Z) \approx \sigma_X^2 \frac{\mu_Y^2}{\mu_X^4} + \frac{\sigma_Y^2}{\mu_X^2} - 2\sigma_{XY}\frac{\mu_Y}{\mu_X^3}$$

$$= \frac{1}{\mu_X^2}\left(\sigma_X^2 \frac{\mu_Y^2}{\mu_X^2} + \sigma_Y^2 - 2\rho\sigma_X\sigma_Y\frac{\mu_Y}{\mu_X}\right)$$

From this equation, we see that the ratio is quite variable when μ_X is small, paralleling the results in Example A, and that correlation between X and Y, if of the same sign as μ_Y/μ_X, decreases $\text{Var}(Z)$.   ∎

4.7 Problems

1. Show that if a random variable is bounded—that is, $|X| < M < \infty$—then $E(X)$ exists.

2. If X is a discrete uniform random variable—that is, $P(X = k) = 1/n$ for $k = 1, 2, \ldots, n$—find $E(X)$ and $\text{Var}(X)$.

3. Find $E(X)$ and $\text{Var}(X)$ for Problem 3 in Chapter 2.

4. Let X have the cdf $F(x) = 1 - x^{-\alpha}, x \geq 1$.
   a. Find $E(X)$ for those values of α for which $E(X)$ exists.
   b. Find $\text{Var}(X)$ for those values of α for which it exists.

5. Let X have the density

$$f(x) = \frac{1 + \alpha x}{2}, \qquad -1 \leq x \leq 1, \qquad -1 \leq \alpha \leq 1$$

Find $E(X)$ and $\text{Var}(X)$.

6. Let X be a continuous random variable with probability density function $f(x) = 2x, 0 \le x \le 1$.

   a. Find $E(X)$.

   b. Let $Y = X^2$. Find the probability mass function of Y and use it to find $E(Y)$.

   c. Use Theorem A in Section 4.1.1 to find $E(X^2)$ and compare to your answer in part (b).

   d. Find $\text{Var}(X)$ according to the definition of variance given in Section 4.2. Also find $\text{Var}(X)$ by using Theorem B of Section 4.2.

7. Let X be a discrete random variable that takes on values $0, 1, 2$ with probabilities $\frac{1}{2}, \frac{3}{8}, \frac{1}{8}$, respectively.

   a. Find $E(X)$.

   b. Let $Y = X^2$. Find the probability mass function of Y and use it to find $E(Y)$.

   c. Use Theorem A of Section 4.1.1 to find $E(X^2)$ and compare to your answer in part (b).

   d. Find $\text{Var}(X)$ according to the definition of variance given in Section 4.2. Also find $\text{Var}(X)$ by using Theorem B in Section 4.2.

8. Show that if X is a discrete random variable, taking values on the positive integers, then $E(X) = \sum_{k=1}^{\infty} P(X \ge k)$. Apply this result to find the expected value of a geometric random variable.

9. This is a simplified inventory problem. Suppose that it costs c dollars to stock an item and that the item sells for s dollars. Suppose that the number of items that will be asked for by customers is a random variable with the frequency function $p(k)$. Find a rule for the number of items that should be stocked in order to maximize the expected income. (*Hint:* Consider the difference of successive terms.)

10. A list of n items is arranged in random order; to find a requested item, they are searched sequentially until the desired item is found. What is the expected number of items that must be searched through, assuming that each item is equally likely to be the one requested? (Questions of this nature arise in the design of computer algorithms.)

11. Referring to Problem 10, suppose that the items are not equally likely to be requested but have known probabilities $p_1, p_2, \ldots, p_n$. Suggest an alternative searching procedure that will decrease the average number of items that must be searched through, and show that in fact it does so.

12. If X is a continuous random variable with a density that is symmetric about some point, ξ, show that $E(X) = \xi$, provided that $E(X)$ exists.

13. If X is a nonnegative continuous random variable, show that

$$E(X) = \int_0^{\infty} [1 - F(x)]\,dx$$

Apply this result to find the mean of the exponential distribution.

14. Let X be a continuous random variable with the density function

$$f(x) = 2x, \qquad 0 \le x \le 1$$

   a. Find $E(X)$.
   b. Find $E(X^2)$ and Var(X).

15. Suppose that two lotteries each have n possible numbers and the same payoff. In terms of expected gain, is it better to buy two tickets from one of the lotteries or one from each?

16. Suppose that $E(X) = \mu$ and Var$(X) = \sigma^2$. Let $Z = (X - \mu)/\sigma$. Show that $E(Z) = 0$ and Var$(Z) = 1$.

17. Find (a) the expectation and (b) the variance of the kth-order statistic of a sample of n independent random variables uniform on $[0, 1]$. The density function is given in Example C in Section 3.7.

18. If $U_1, \ldots, U_n$ are independent uniform random variables, find $E(U_{(n)} - U_{(1)})$.

19. Find $E(U_{(k)} - U_{(k-1)})$, where the $U_{(i)}$ are as in Problem 18.

20. A stick of unit length is broken into two pieces. Find the expected ratio of the length of the longer piece to the length of the shorter piece.

21. A random square has a side length that is a uniform $[0, 1]$ random variable. Find the expected area of the square.

22. A random rectangle has sides the lengths of which are independent uniform random variables. Find the expected area of the rectangle, and compare this result to that of Problem 21.

23. Repeat Problems 21 and 22 assuming that the distribution of the lengths is exponential.

24. Prove Theorem A of Section 4.1.2 for the discrete case.

25. If X_1 and X_2 are independent random variables following a gamma distribution with parameters α and λ, find $E(R^2)$, where $R^2 = X_1^2 + X_2^2$.

26. Referring to Example B in Section 4.1.2, what is the expected number of coupons needed to collect r different types, where $r < n$?

27. If n men throw their hats into a pile and each man takes a hat at random, what is the expected number of matches? (*Hint:* Express the number as a sum.)

28. Suppose that n enemy aircraft are shot at simultaneously by m gunners, that each gunner selects an aircraft to shoot at independently of the other gunners, and that each gunner hits the selected aircraft with probability p. Find the expected number of aircraft hit by the gunners.

29. Prove Corollary A of Section 4.1.1.

30. Find $E[1/(X + 1)]$, where X is a Poisson random variable.

31. Let X be uniformly distributed on the interval $[1, 2]$. Find $E(1/X)$. Is $E(1/X) = 1/E(X)$?

32. Let X have a gamma distribution with parameters α and λ. For those values of α and λ for which it is defined, find $E(1/X)$.

33. Prove Chebyshev's inequality in the discrete case.

34. Let X be uniform on $[0, 1]$, and let $Y = \sqrt{X}$. Find $E(Y)$ by (a) finding the density of Y and then finding the expectation and (b) using Theorem A of Section 4.1.1.

35. Find the mean of a negative binomial random variable. (*Hint:* Express the random variable as a sum.)

36. Consider the following scheme for group testing. The original lot of samples is divided into two groups, and each of the subgroups is tested as a whole. If either subgroup tests positive, it is divided in two, and the procedure is repeated. If any of the groups thus obtained tests positive, test every member of that group. Find the expected number of tests performed, and compare it to the number performed with no grouping and with the scheme described in Example C in Section 4.1.2.

37. For what values of p is the group testing of Example C in Section 4.1.2 inferior to testing every individual?

38. This problem continues Example A of Section 4.1.2.

    a. What is the probability that a fragment is the leftmost member of a contig?

    b. What is the expected number of fragments that are leftmost members of contigs?

    c. What is the expected number of contigs?

39. Suppose that a segment of DNA of length 1,000,000 is to be shotgun sequenced with fragments of length 1000.

    a. How many fragment would be needed so that the chance of an individual site being covered is greater than 0.99?

    b. With this choice, how many sites would you expect to be missed?

40. A child types the letters Q, W, E, R, T, Y, randomly producing 1000 letters in all. What is the expected number of times that the sequence QQQQ appears, counting overlaps?

41. Continuing with the previous problem, how many times would we expect the word "TRY" to appear? Would we be surprised if it occurred 100 times? (*Hint:* Consider Markov's inequality.)

42. Let X be an exponential random variable with standard deviation σ. Find $P(|X - E(X)| > k\sigma)$ for $k = 2, 3, 4$, and compare the results to the bounds from Chebyshev's inequality.

43. Show that $\mathrm{Var}(X - Y) = \mathrm{Var}(X) + \mathrm{Var}(Y) - 2\mathrm{Cov}(X, Y)$.

44. If X and Y are independent random variables with equal variances, find $\text{Cov}(X + Y, X - Y)$.

45. Find the covariance and the correlation of N_i and N_j, where $N_1, N_2, \ldots, N_r$ are multinomial random variables. (*Hint:* Express them as sums.)

46. If $U = a + bX$ and $V = c + dY$, show that $|\rho_{UV}| = |\rho_{XY}|$.

47. If X and Y are independent random variables and $Z = Y - X$, find expressions for the covariance and the correlation of X and Z in terms of the variances of X and Y.

48. Let U and V be independent random variables with means μ and variances σ^2. Let $Z = \alpha U + V \sqrt{1 - \alpha^2}$. Find $E(Z)$ and ρ_{UZ}.

49. Two independent measurements, X and Y, are taken of a quantity μ. $E(X) = E(Y) = \mu$, but σ_X and σ_Y are unequal. The two measurements are combined by means of a weighted average to give

$$Z = \alpha X + (1 - \alpha)Y$$

where α is a scalar and $0 \leq \alpha \leq 1$.

    a. Show that $E(Z) = \mu$.
    b. Find α in terms of σ_X and σ_Y to minimize $\text{Var}(Z)$.
    c. Under what circumstances is it better to use the average $(X + Y)/2$ than either X or Y alone?

50. Suppose that X_i, where $i = 1, \ldots, n$, are independent random variables with $E(X_i) = \mu$ and $\text{Var}(X_i) = \sigma^2$. Let $\overline{X} = n^{-1} \sum_{i=1}^{n} X_i$. Show that $E(\overline{X}) = \mu$ and $\text{Var}(\overline{X}) = \sigma^2/n$.

51. Continuing Example E in Section 4.3, suppose there are n securities, each with the same expected return, that all the returns have the same standard deviations, and that the returns are uncorrelated. What is the optimal portfolio vector? Plot the risk of the optimal portfolio versus n. How does this risk compare to that incurred by putting all your money in one security?

52. Consider two securities, the first having $\mu_1 = 1$ and $\sigma_1 = 0.1$, and the second having $\mu_2 = 0.8$ and $\sigma_2 = 0.12$. Suppose that they are negatively correlated, with $\rho = -0.8$.

    a. If you could only invest in one security, which one would you choose, and why?
    b. Suppose you invest 50% of your money in each of the two. What is your expected return and what is your risk?
    c. If you invest 80% of your money in security 1 and 20% in security 2, what is your expected return and your risk?
    d. Denote the expected return and its standard deviation as functions of π by $\mu(\pi)$ and $\sigma(\pi)$. The pair $(\mu(\pi), \sigma(\pi))$ trace out a curve in the plane as π varies from 0 to 1. Plot this curve.
    e. Repeat **b–d** if the correlation is $\rho = 0.1$.

53. Show that $\text{Cov}(X, Y) \leq \sqrt{\text{Var}(X)\text{Var}(Y)}$.

54. Let X, Y, and Z be uncorrelated random variables with variances σ_X^2, σ_Y^2, and σ_Z^2, respectively. Let

$$U = Z + X$$
$$V = Z + Y$$

Find $\text{Cov}(U, V)$ and ρ_{UV}.

55. Let $T = \sum_{k=1}^{n} k X_k$, where the X_k are independent random variables with means μ and variances σ^2. Find $E(T)$ and $\text{Var}(T)$.

56. Let $S = \sum_{k=1}^{n} X_k$, where the X_k are as in Problem 55. Find the covariance and the correlation of S and T.

57. If X and Y are independent random variables, find $\text{Var}(XY)$ in terms of the means and variances of X and Y.

58. A function is measured at two points with some error (for example, the position of an object is measured at two times). Let

$$X_1 = f(x) + \varepsilon_1$$
$$X_2 = f(x + h) + \varepsilon_2$$

where ε_1 and ε_2 are independent random variables with mean zero and variance σ^2. Since the derivative of f is

$$\lim_{h \to 0} \frac{f(x + h) - f(x)}{h}$$

it is estimated by

$$Z = \frac{X_2 - X_1}{h}$$

  a. Find $E(Z)$ and $\text{Var}(Z)$. What is the effect of choosing a value of h that is very small, as is suggested by the definition of the derivative?
  b. Find an approximation to the mean squared error of Z as an estimate of $f'(x)$ using a Taylor series expansion. Can you find the value of h that minimizes the mean squared error?
  c. Suppose that f is measured at three points with some error. How could you construct an estimate of the second derivative of f, and what are the mean and the variance of your estimate?

59. Let (X, Y) be a random point uniformly distributed on a unit disk. Show that $\text{Cov}(X, Y) = 0$, but that X and Y are not independent.

60. Let Y have a density that is symmetric about zero, and let $X = SY$, where S is an independent random variable taking on the values $+1$ and -1 with probability $\frac{1}{2}$ each. Show that $\text{Cov}(X, Y) = 0$, but that X and Y are not independent.

61. In Section 3.7, the joint density of the minimum and maximum of n independent uniform random variables was found. In the case $n = 2$, this amounts to X and Y, the minimum and maximum, respectively, of two independent random variables uniform on [0, 1], having the joint density

$$f(x, y) = 2, \qquad 0 \le x \le y$$

a. Find the covariance and the correlation of X and Y. Does the sign of the correlation make sense intuitively?

b. Find $E(X|Y = y)$ and $E(Y|X = x)$. Do these results make sense intuitively?

c. Find the probability density functions of the random variables $E(X|Y)$ and $E(Y|X)$.

d. What is the linear predictor of Y in terms of X (denoted by $\hat{Y} = a + bX$) that has minimal mean squared error? What is the mean square prediction error?

e. What is the predictor of Y in terms of X $[\hat{Y} = h(X)]$ that has minimal mean squared error? What is the mean square prediction error?

62. Let X and Y have the joint distribution given in Problem 1 of Chapter 3.

a. Find the covariance and correlation of X and Y.

b. Find $E(Y|X = x)$ for $x = 1, 2, 3, 4$. Find the probability mass function of the random variable $E(Y|X)$.

63. Let X and Y have the joint distribution given in Problem 8 of Chapter 3.

a. Find the covariance and correlation of X and Y.

b. Find $E(Y|X = x)$ for $0 \le x \le 1$.

64. Let X and Y be jointly distributed random variables with correlation ρ_{XY}; define the *standardized* random variables $\tilde{X}$ and $\tilde{Y}$ as $\tilde{X} = (X - E(X))/\sqrt{\text{Var}(X)}$ and $\tilde{Y} = (Y - E(Y))/\sqrt{\text{Var}(Y)}$. Show that $\text{Cov}(\tilde{X}, \tilde{Y}) = \rho_{XY}$.

65. How has the assumption that N and the X_i are independent been used in Example D of Section 4.4.1?

66. A building contains two elevators, one fast and one slow. The average waiting time for the slow elevator is 3 min. and the average waiting time of the fast elevator is 1 min. If a passenger chooses the fast elevator with probability $\frac{2}{3}$ and the slow elevator with probability $\frac{1}{3}$, what is the expected waiting time? (Use the law of total expectation, Theorem A of Section 4.4.1, defining appropriate random variables X and Y.)

67. A random rectangle is formed in the following way: The base, X, is chosen to be a uniform [0, 1] random variable and after having generated the base, the height is chosen to be uniform on [0, X]. Use the law of total expectation, Theorem A of Section 4.4.1, to find the expected circumference and area of the rectangle.

68. Show that $E[\text{Var}(Y|X)] \le \text{Var}(Y)$.

69. Suppose that a bivariate normal distribution has $\mu_X = \mu_Y = 0$ and $\sigma_X = \sigma_Y = 1$. Sketch the contours of the density and the lines $E(Y|X = x)$ and $E(X|Y = y)$ for $\rho = 0, .5,$ and $.9$.

70. If X and Y are independent, show that $E(X|Y = y) = E(X)$.

71. Let X be a binomial random variable representing the number of successes in n independent Bernoulli trials. Let Y be the number of successes in the first m trials, where $m < n$. Find the conditional frequency function of Y given $X = x$ and the conditional mean.

72. An item is present in a list of n items with probability p; if it is present, its position in the list is uniformly distributed. A computer program searches through the list sequentially. Find the expected number of items searched through before the program terminates.

73. A fair coin is tossed n times, and the number of heads, N, is counted. The coin is then tossed N more times. Find the expected total number of heads generated by this process.

74. The number of offspring of an organism is a discrete random variable with mean μ and variance σ^2. Each of its offspring reproduces in the same manner. Find the expected number of offspring in the third generation and its variance.

75. Let T be an exponential random variable, and conditional on T, let U be uniform on $[0, T]$. Find the unconditional mean and variance of U.

76. Let the point (X, Y) be uniformly distributed over the half disk $x^2 + y^2 \leq 1$, where $y \geq 0$. If you observe X, what is the best prediction for Y? If you observe Y, what is the best prediction for X? For both questions, "best" means having the minimum mean squared error.

77. Let X and Y have the joint density

$$f(x, y) = e^{-y}, \qquad 0 \leq x \leq y$$

a. Find $\text{Cov}(X, Y)$ and the correlation of X and Y.
b. Find $E(X|Y = y)$ and $E(Y|X = x)$.
c. Find the density functions of the random variables $E(X|Y)$ and $E(Y|X)$.

78. Show that if a density is symmetric about zero, its skewness is zero.

79. Let X be a discrete random variable that takes on values $0, 1, 2$ with probabilities $\frac{1}{2}, \frac{3}{8}, \frac{1}{8}$, respectively. Find the moment-generating function of X, $M(t)$, and verify that $E(X) = M'(0)$ and that $E(X^2) = M''(0)$.

80. Let X be a continuous random variable with density function $f(x) = 2x$, $0 \leq x \leq 1$. Find the moment-generating function of X, $M(t)$, and verify that $E(X) = M'(0)$ and that $E(X^2) = M''(0)$.

81. Find the moment-generating function of a Bernoulli random variable, and use it to find the mean, variance, and third moment.

82. Use the result of Problem 81 to find the mgf of a binomial random variable and its mean and variance.

83. Show that if X_i follows a binomial distribution with n_i trials and probability of success $p_i = p$, where $i = 1, \ldots, n$ and the X_i are independent, then $\sum_{i=1}^{n} X_i$ follows a binomial distribution.

84. Referring to Problem 83, show that if the p_i are unequal, the sum does not follow a binomial distribution.

85. Find the mgf of a geometric random variable, and use it to find the mean and the variance.

86. Use the result of Problem 85 to find the mgf of a negative binomial random variable and its mean and variance.

87. Under what conditions is the sum of independent negative binomial random variables also negative binomial?

88. Let X and Y be independent random variables, and let α and β be scalars. Find an expression for the mgf of $Z = \alpha X + \beta Y$ in terms of the mgf's of X and Y.

89. Let $X_1, X_2, \ldots, X_n$ be independent normal random variables with means μ_i and variances σ_i^2. Show that $Y = \sum_{i=1}^{n} \alpha_i X_i$, where the α_i are scalars, is normally distributed, and find its mean and variance. (*Hint:* Use moment-generating functions.)

90. Assuming that $X \sim N(0, \sigma^2)$, use the mgf to show that the odd moments are zero and the even moments are

$$\mu_{2n} = \frac{(2n)!\sigma^{2n}}{2^n(n!)}$$

91. Use the mgf to show that if X follows an exponential distribution, cX ($c > 0$) does also.

92. Suppose that Θ is a random variable that follows a gamma distribution with parameters λ and α, where α is an integer, and suppose that, conditional on Θ, X follows a Poisson distribution with parameter Θ. Find the unconditional distribution of $\alpha + X$. (*Hint:* Find the mgf by using iterated conditional expectations.)

93. Find the distribution of a geometric sum of exponential random variables by using moment-generating functions.

94. If X is a nonnegative integer-valued random variable, the **probability-generating function** of X is defined to be

$$G(s) = \sum_{k=0}^{\infty} s^k p_k$$

where $p_k = P(X = k)$.

   a. Show that

$$p_k = \frac{1}{k!}\frac{d^k}{ds^k}G(s)\bigg|_{s=0}$$

b. Show that

$$\frac{dG}{ds}\bigg|_{s=1} = E(X)$$

$$\frac{d^2G}{ds^2}\bigg|_{s=1} = E[X(X-1)]$$

c. Express the probability-generating function in terms of moment-generating function.
d. Find the probability-generating function of the Poisson distribution.

95. Show that if X and Y are independent, their joint moment-generating function factors.

96. Show how to find $E(XY)$ from the joint moment-generating function of X and Y.

97. Use moment-generating functions to show that if X and Y are independent, then

$$\text{Var}(aX + bY) = a^2\text{Var}(X) + b^2\text{Var}(Y)$$

98. Find the mean and variance of the compound Poisson distribution (Example H in Section 4.5).

99. Find expressions for the approximate mean and variance of $Y = g(X)$ for (a) $g(x) = \sqrt{x}$, (b) $g(x) = \log x$, and (c) $g(x) = \sin^{-1} x$.

100. If X is uniform on $[10, 20]$, find the approximate and exact mean and variance of $Y = 1/X$, and compare them.

101. Find the approximate mean and variance of $Y = \sqrt{X}$, where X is a random variable following a Poisson distribution.

102. Two sides, x_0 and y_0, of a right triangle are independently measured as X and Y, where $E(X) = x_0$ and $E(Y) = y_0$ and $\text{Var}(X) = \text{Var}(Y) = \sigma^2$. The angle between the two sides is then determined as

$$\Theta = \tan^{-1}\left(\frac{Y}{X}\right)$$

Find the approximate mean and variance of Θ.

103. The volume of a bubble is estimated by measuring its diameter and using the relationship

$$V = \frac{\pi}{6}D^3$$

Suppose that the true diameter is 2 mm and that the standard deviation of the measurement of the diameter is .01 mm. What is the approximate standard deviation of the estimated volume?

104. The position of an aircraft relative to an observer on the ground is estimated by measuring its distance r from the observer and the angle θ that the line of sight from the observer to the aircraft makes with the horizontal. Suppose that the measurements, denoted by R and Θ, are subject to random errors and are independent of each other. The altitude of the aircraft is then estimated to be $Y = R \sin \Theta$.

a. Find an approximate expression for the variance of Y.

b. For given r, at what value of θ is the estimated altitude most variable?

CHAPTER 5

Limit Theorems

5.1 Introduction

This chapter is principally concerned with the limiting behavior of the sum of independent random variables as the number of summands becomes large. The results presented here are both intrinsically interesting and useful in statistics, since many commonly computed statistical quantities, such as averages, can be represented as sums.

5.2 The Law of Large Numbers

It is commonly believed that if a fair coin is tossed many times and the proportion of heads is calculated, that proportion will be close to $\frac{1}{2}$. John Kerrich, a South African mathematician, tested this belief empirically while detained as a prisoner during World War II. He tossed a coin 10,000 times and observed 5067 heads. The law of large numbers is a mathematical formulation of this belief. The successive tosses of the coin are modeled as independent random trials. The random variable X_i takes on the value 0 or 1 according to whether the ith trial results in a tail or a head, and the proportion of heads in n trials is

$$\overline{X}_n = \frac{1}{n} \sum_{i=1}^{n} X_i$$

The law of large numbers states that $\overline{X}_n$ approaches $\frac{1}{2}$ in a sense that is specified by the following theorem.

THEOREM A   *Law of Large Numbers*

Let $X_1, X_2, \ldots, X_i \ldots$ be a sequence of independent random variables with $E(X_i) = \mu$ and $\mathrm{Var}(X_i) = \sigma^2$. Let $\overline{X}_n = n^{-1} \sum_{i=1}^{n} X_i$. Then, for any $\varepsilon > 0$,

$$P(|\overline{X}_n - \mu| > \varepsilon) \to 0 \qquad \text{as } n \to \infty$$

Proof

We first find $E(\overline{X}_n)$ and $\mathrm{Var}(\overline{X}_n)$:

$$E(\overline{X}_n) = \frac{1}{n} \sum_{i=1}^{n} E(X_i) = \mu$$

Since the X_i are independent,

$$\mathrm{Var}(\overline{X}_n) = \frac{1}{n^2} \sum_{i=1}^{n} \mathrm{Var}(X_i) = \frac{\sigma^2}{n}$$

The desired result now follows immediately from Chebyshev's inequality, which states that

$$P(|\overline{X}_n - \mu| > \varepsilon) \leq \frac{\mathrm{Var}(\overline{X}_n)}{\varepsilon^2} = \frac{\sigma^2}{n\varepsilon^2} \to 0, \qquad \text{as } n \to \infty \qquad \blacksquare$$

In the case of a fair coin toss, the X_i are Bernoulli random variables with $p = 1/2$, $E(X_i) = 1/2$ and $\mathrm{Var}(X_i) = 1/4$. If tossed 10,000 times

$$\mathrm{Var}(\overline{X}_{10,000}) = 2.5 \times 10^{-5}$$

and the standard deviation of the average is the square root of the variance, 0.005. The proportion observed by Kerrich, 0.5067, is thus a little more than one standard deviation away from its expected value of 0.5, consistent with Chebyshev's inequality. (Recall from Section 4.2 that Chebyshev's inequality can be written in the form $P(|\overline{X}_n - \mu| > k\sigma) \leq 1/k^2$.)

If a sequence of random variables, $\{Z_n\}$, is such that $P(|Z_n - \alpha| > \varepsilon)$ approaches zero as n approaches infinity, for any $\varepsilon > 0$ and where α is some scalar, then Z_n is said to **converge in probability** to α. There is another mode of convergence, called *strong convergence* or *almost sure convergence,* which asserts more. Z_n is said to **converge almost surely to** α if for every $\varepsilon > 0$, $|Z_n - \alpha| > \varepsilon$ only a finite number of times with probability 1; that is, beyond some point in the sequence, the difference is always less than ε, but where that point is random. The version of the law of large numbers stated and proved earlier asserts that $\overline{X}_n$ converges to μ in probability. This version is usually called the *weak law of large numbers.* Under the same assumptions, a strong law of large numbers, which asserts that $\overline{X}_n$ converges almost surely to μ, can also be proved, but we will not do so.

We now consider some examples that illustrate the utility of the law of large numbers.

EXAMPLE A   *Monte Carlo Integration*
Suppose that we wish to calculate

$$I(f) = \int_0^1 f(x)\, dx$$

where the integration cannot be done by elementary means or evaluated using tables of integrals. The most common approach is to use a numerical method in which the integral is approximated by a sum; various schemes and computer packages exist for doing this. Another method, called the **Monte Carlo method,** works in the following way. Generate independent uniform random variables on [0, 1]—that is, $X_1, X_2, \ldots, X_n$—and compute

$$\hat{I}(f) = \frac{1}{n} \sum_{i=1}^n f(X_i)$$

By the law of large numbers, this should be close to $E[f(X)]$, which is simply

$$E[f(X)] = \int_0^1 f(x)\, dx = I(f)$$

This simple scheme can be easily modified in order to change the range of integration and in other ways. Compared to the standard numerical methods, it is not especially efficient in one dimension, but becomes increasingly efficient as the dimensionality of the integral grows.

As a concrete example, let us consider the evaluation of

$$I(f) = \frac{1}{\sqrt{2\pi}} \int_0^1 e^{-x^2/2}\, dx$$

The integral is that of the standard normal density, which cannot be evaluated in closed form. From the table of the normal distribution (Table 2 in Appendix B), an accurate numerical approximation is $I(f) = .3413$. If 1000 points, $X_1, \ldots, X_{1000}$, uniformly distributed over the interval $0 \le x \le 1$, are generated using a pseudorandom number generator, the integral is then approximated by

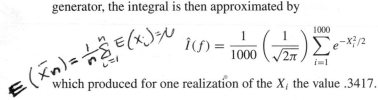

$$\hat{I}(f) = \frac{1}{1000} \left(\frac{1}{\sqrt{2\pi}} \right) \sum_{i=1}^{1000} e^{-X_i^2/2}$$

which produced for one realization of the X_i the value .3417.   ∎

EXAMPLE B   *Repeated Measurements*
Suppose that repeated independent unbiased measurements, $X_1, \ldots, X_n$, of a quantity are made. If n is large, the law of large numbers says that $\overline{X}$ will be close to the true value, μ, of the quantity, but how close $\overline{X}$ is depends not only on n but on the variance of the measurement error, σ^2, as can be seen in the proof of Theorem A.

Fortunately, σ^2 can be estimated and therefore

$$\text{Var}(\overline{X}) = \frac{\sigma^2}{n}$$

can be estimated from the data to assess the precision of $\overline{X}$. First, note that $n^{-1} \sum_{i=1}^{n} X_i^2$ converges to $E(X^2)$, from the law of large numbers. Second, it can be shown that if Z_n converges to α in probability and g is a continuous function, then

$$g(Z_n) \to g(\alpha)$$

which implies that

$$\overline{X}^2 \to [E(X)]^2$$

Finally, since $n^{-1} \sum_{i=1}^{n} X_i^2$ converges to $E(X^2)$ and $\overline{X}^2$ converges to $[E(X)]^2$, with a little additional argument it can be shown that

$$\frac{1}{n} \sum_{i=1}^{n} X_i^2 - \overline{X}^2 \to E(X^2) - [E(X)]^2 = \text{Var}(X)$$

More generally, it follows from the law of large numbers that the sample moments, $n^{-1} \sum_{i=1}^{n} X_i^r$, converge in probability to the moments of X, $E(X^r)$.   ■

E X A M P L E **C**   A muscle or nerve cell membrane contains a very large number of channels; when open, these channels allow ions to pass through. Individual channels seem to open and close randomly, and it is often assumed that in an equilibrium situation the channels open and close independently of each other and that only a very small fraction are open at any one time. Suppose then that the probability that a channel is open is p, a very small number, that there are m channels in all, and that the amount of current flowing through an individual channel is c. The number of channels open at any particular time is N, a binomial random variable with m trials and probability p of success on each trial. The total amount of current is $S = cN$ and can be measured. We then have

$$E(S) = cE(N) = cmp$$

$$\text{Var}(S) = c^2mp(1 - p)$$

and

$$\frac{\text{Var}(S)}{E(S)} = c(1 - p) \approx c$$

since p is small. Thus, through independent measurements, $S_1, \ldots, S_n$, we can estimate $E(S)$ and $\text{Var}(S)$ and therefore c, the amount of current flowing through a single channel, without knowing how many channels there are.   ■

5.3 Convergence in Distribution and the Central Limit Theorem

In applications, we often want to find $P(a < X < b)$ when we do not know the cdf of X precisely; it is sometimes possible to do this by approximating F_X. The approximation is often arrived at by some sort of limiting argument. The most famous limit theorem in probability theory is the central limit theorem, which is the main topic of this section. Before discussing the central limit theorem, we develop some introductory terminology, theory, and examples.

DEFINITION

Let $X_1, X_2, \ldots$ be a sequence of random variables with cumulative distribution functions $F_1, F_2, \ldots$, and let X be a random variable with distribution function F. We say that X_n converges in distribution to X if

$$\lim_{n \to \infty} F_n(x) = F(x)$$

at every point at which F is continuous.   ∎

Moment-generating functions are often useful for establishing the convergence of distribution functions. We know from Property A of Section 4.5 that a distribution function F_n is uniquely determined by its mgf, M_n. The following theorem, which we give without proof, states that this unique determination holds for limits as well.

THEOREM A   *Continuity Theorem*

Let F_n be a sequence of cumulative distribution functions with the corresponding moment-generating function M_n. Let F be a cumulative distribution function with the moment-generating function M. If $M_n(t) \to M(t)$ for all t in an open interval containing zero, then $F_n(x) \to F(x)$ at all continuity points of F.   ∎

EXAMPLE A   We will show that the Poisson distribution can be approximated by the normal distribution for large values of λ. This is suggested by examining Figure 2.6, which shows that as λ increases, the probability mass function of the Poisson distribution becomes more symmetric and bell-shaped.

Let $\lambda_1, \lambda_2, \ldots$ be an increasing sequence with $\lambda_n \to \infty$, and let $\{X_n\}$ be a sequence of Poisson random variables with the corresponding parameters. We know that $E(X_n) = \text{Var}(X_n) = \lambda_n$. If we wish to approximate the Poisson distribution function by a normal distribution function, the normal must have the same mean and variance as the Poisson does. In addition, if we wish to prove a limiting result, we run into the difficulty that the mean and variance are tending to infinity. This difficulty is dealt with by **standardizing** the random variables—that is, by letting

$$Z_n = \frac{X_n - E(X_n)}{\sqrt{\text{Var}(X_n)}}$$

$$= \frac{X_n - \lambda_n}{\sqrt{\lambda_n}}$$

We then have $E(Z_n) = 0$ and $\text{Var}(Z_n) = 1$, and we will show that the mgf of Z_n converges to the mgf of the standard normal distribution.

The mgf of X_n is

$$M_{X_n}(t) = e^{\lambda_n(e^t - 1)}$$

By Property C of Section 4.5, the mgf of Z_n is

$$M_{Z_n}(t) = e^{-t\sqrt{\lambda_n}} M_{X_n}\left(\frac{t}{\sqrt{\lambda_n}}\right)$$

$$= e^{-t\sqrt{\lambda_n}} e^{\lambda_n(e^{t/\sqrt{\lambda_n}} - 1)}$$

It will be easier to work with the log of this expression.

$$\log M_{Z_n}(t) = -t\sqrt{\lambda_n} + \lambda_n(e^{t/\sqrt{\lambda_n}} - 1)$$

Using the power series expansion $e^x = \sum_{k=0}^{\infty} \frac{x^k}{k!}$, we see that

$$\lim_{n \to \infty} \log M_{Z_n}(t) = \frac{t^2}{2}$$

or

$$\lim_{n \to \infty} M_{Z_n}(t) = e^{t^2/2}$$

The last expression is the mgf of the standard normal distribution.

We have shown that a standardized Poisson random variable converges in distribution to a standard normal variable as λ approaches infinity. Practically, we wish to use this limiting result as a basis for an approximation for large but finite values of λ_n. How adequate the approximation is for $\lambda = 100$, say, is a matter for theoretical and/or empirical investigation. It turns out that the approximation is increasingly good for large values of λ and that λ does not have to be all that large. (See Problem 8 at the end of this chapter.)   ■

The next example shows how the approximation of the Poisson distribution can be applied in a specific case.

EXAMPLE B   A certain type of particle is emitted at a rate of 900 per hour. What is the probability that more than 950 particles will be emitted in a given hour if the counts form a Poisson process?

Let X be a Poisson random variable with mean 900. We find $P(X > 950)$ by standardizing:

$$P(X > 950) = P\left(\frac{X - 900}{\sqrt{900}} > \frac{950 - 900}{\sqrt{900}}\right)$$

$$\approx 1 - \Phi\left(\tfrac{5}{3}\right)$$

$$= .04779$$

where Φ is the standard normal cdf. For comparison, the exact probability is .04712. ∎

We now turn to the central limit theorem, which is concerned with a limiting property of sums of random variables. If $X_1, X_2, \ldots$ is a sequence of independent random variables with mean μ and variance σ^2, and if

$$S_n = \sum_{i=1}^{n} X_i$$

we know from the law of large numbers that S_n/n converges to μ in probability. This followed from the fact that

$$\text{Var}\left(\frac{S_n}{n}\right) = \frac{1}{n^2}\text{Var}(S_n) = \frac{\sigma^2}{n} \to 0$$

The central limit theorem is concerned not with the fact that the ratio S_n/n converges to μ but with how it fluctuates around μ. To analyze these fluctuations, we standardize:

$$Z_n = \frac{S_n - n\mu}{\sigma\sqrt{n}}$$

You should verify that Z_n has mean 0 and variance 1. The central limit theorem states that the distribution of Z_n converges to the standard normal distribution.

THEOREM B   *Central Limit Theorem*

Let $X_1, X_2, \ldots$ be a sequence of independent random variables having mean 0 and variance σ^2 and the common distribution function F and moment-generating function M defined in a neighborhood of zero. Let

$$S_n = \sum_{i=1}^{n} X_i$$

Then

$$\lim_{n \to \infty} P\left(\frac{S_n}{\sigma\sqrt{n}} \leq x\right) = \Phi(x), \qquad -\infty < x < \infty$$

Proof

Let $Z_n = S_n/(\sigma\sqrt{n})$. We will show that the mgf of Z_n tends to the mgf of the standard normal distribution. Since S_n is a sum of independent random variables,

$$M_{S_n}(t) = [M(t)]^n$$

and

$$M_{Z_n}(t) = \left[M\left(\frac{t}{\sigma\sqrt{n}}\right)\right]^n$$

$M(s)$ has a Taylor series expansion about zero:

$$M(s) = M(0) + sM'(0) + \tfrac{1}{2}s^2 M''(0) + \varepsilon_s$$

where $\varepsilon_s/s^2 \to 0$ as $s \to 0$. Since $E(X) = 0$, $M'(0) = 0$, and $M''(0) = \sigma^2$. As $n \to \infty$, $t/(\sigma\sqrt{n}) \to 0$, and

$$M\left(\frac{t}{\sigma\sqrt{n}}\right) = 1 + \frac{1}{2}\sigma^2 \left(\frac{t}{\sigma\sqrt{n}}\right)^2 + \varepsilon_n$$

where $\varepsilon_n/(t^2/(n\sigma^2)) \to 0$ as $n \to \infty$. We thus have

$$M_{Z_n}(t) = \left(1 + \frac{t^2}{2n} + \varepsilon_n\right)^n$$

It can be shown that if $a_n \to a$, then

$$\lim_{n \to \infty} \left(1 + \frac{a_n}{n}\right)^n = e^a$$

From this result, it follows that

$$M_{Z_n}(t) \to e^{t^2/2} \quad \text{as } n \to \infty$$

where $\exp(t^2/2)$ is the mgf of the standard normal distribution, as was to be shown.   ■

Theorem B is one of the simplest versions of the central limit theorem; there are many central limit theorems of various degrees of abstraction and generality. We have proved Theorem B under the assumption that the moment-generating functions exist, which is a rather strong assumption. By using characteristic functions instead, we could modify the proof so that it would only be necessary that first and second moments exist. Further generalizations weaken the assumption that the X_i have the same distribution and apply to linear combinations of independent random variables. Central limit theorems have also been proved that weaken the independence assumption and allow the X_i to be dependent but not "too" dependent.

For practical purposes, especially for statistics, the limiting result in itself is not of primary interest. Statisticians are more interested in its use as an approximation with finite values of n. It is impossible to give a concise and definitive statement of how good the approximation is, but some general guidelines are available, and examining special cases can give insight. How fast the approximation becomes good depends on the distribution of the summands, the X_i. If the distribution is fairly symmetric and has tails that die off rapidly, the approximation becomes good for relatively small values of n. If the distribution is very skewed or if the tails die down very slowly, a larger value of n is needed for a good approximation. The following examples deal with two special cases.

E X A M P L E   **C**   Because the uniform distribution on [0, 1] has mean $\frac{1}{2}$ and variance $\frac{1}{12}$, the sum of 12 uniform random variables, minus 6, has mean 0 and variance 1. The distribution of this sum is quite close to normal; in fact, before better algorithms were developed, it was commonly used in computers for generating normal random variables from uniform ones. It is possible to compare the real and approximate distributions analytically, but we will content ourselves with a simple demonstration. Figure 5.1 shows a histogram of 1000 such sums with a superimposed normal density function. The fit is surprisingly good, especially considering that 12 is not usually regarded as a large value of n.   ■

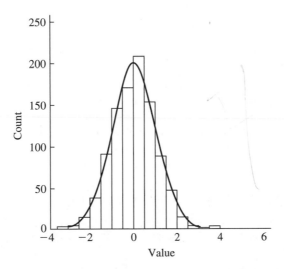

FIGURE 5.1   A histogram of 1000 values, each of which is the sum of 12 uniform $[-\frac{1}{2}, \frac{1}{2}]$ pseudorandom variables, with an approximating standard normal density.

EXAMPLE **D**   The sum of n independent exponential random variables with parameter $\lambda = 1$ follows a gamma distribution with $\lambda = 1$ and $\alpha = n$ (Example F in Section 4.5). The exponential density is quite skewed; therefore, a good approximation of a standardized gamma by a standardized normal would not be expected for small n. Figure 5.2 shows the cdf's of the standard normal and standardized gamma distributions for increasing values of n. Note how the approximation improves as n increases.   ∎

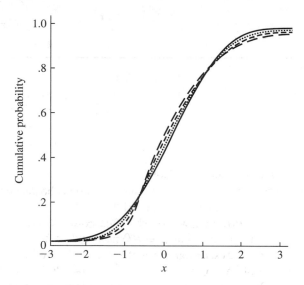

FIGURE **5.2**   The standard normal cdf (solid line) and the cdf's of standardized gamma distributions with $\alpha = 5$ (long dashes), $\alpha = 10$ (short dashes), and $\alpha = 30$ (dots).

Let us now consider some applications of the central limit theorem.

EXAMPLE **E**   *Measurement Error*

Suppose that $X_1, \ldots, X_n$ are repeated, independent measurements of a quantity, μ, and that $E(X_i) = \mu$ and $\text{Var}(X_i) = \sigma^2$. The average of the measurements, $\overline{X}$, is used as an estimate of μ. The law of large numbers tells us that $\overline{X}$ converges to μ in probability, so we can hope that $\overline{X}$ is close to μ if n is large. Chebyshev's inequality allows us to bound the probability of an error of a given size, but the central limit theorem gives a much sharper approximation to the actual error. Suppose that we wish to find $P(|\overline{X} - \mu| < c)$ for some constant c. To use the central limit theorem to approximate this probability, we first standardize, using $E(\overline{X}) = \mu$ and $\text{Var}(\overline{X}) = \sigma^2/n$:

$$P(|\overline{X} - \mu| < c) = P(-c < \overline{X} - \mu < c)$$

$$= P\left(\frac{-c}{\sigma/\sqrt{n}} < \frac{\overline{X} - \mu}{\sigma/\sqrt{n}} < \frac{c}{\sigma/\sqrt{n}} \right)$$

$$\approx \Phi\left(\frac{c\sqrt{n}}{\sigma} \right) - \Phi\left(-\frac{c\sqrt{n}}{\sigma} \right)$$

For example, suppose that 16 measurements are taken with $\sigma = 1$. The probability that the average deviates from μ by less than .5 is approximately

$$P(|\overline{X} - \mu| < .5) = \Phi(.5 \times 4) - \Phi(-.5 \times 4) = .954$$

This sort of reasoning can be turned around. That is, given c and γ, n can be found such that

$$P(|\overline{X} - \mu| < c) \geq \gamma$$

∎

E X A M P L E F *Normal Approximation to the Binomial Distribution*
Since a binomial random variable is the sum of independent Bernoulli random variables, its distribution can be approximated by a normal distribution. The approximation is best when the binomial distribution is symmetric—that is, when $p = \frac{1}{2}$. A frequently used rule of thumb is that the approximation is reasonable when $np > 5$ and $n(1 - p) > 5$. The approximation is especially useful for large values of n, for which tables are not readily available.

Suppose that a coin is tossed 100 times and lands heads up 60 times. Should we be surprised and doubt that the coin is fair?

To answer this question, we note that if the coin is fair, the number of heads, X, is a binomial random variable with $n = 100$ trials and probability of success $p = \frac{1}{2}$, so that $E(X) = np = 50$ (see Example A of Section 4.1) and $\text{Var}(X) = np(1 - p) = 25$ (see Example B of Section 4.3). We could calculate $P(X = 60)$, which would be a small number. But because there are so many possible outcomes, $P(X = 50)$ is also a small number, so this calculation would not really answer the question. Instead, we calculate the probability of a deviation as extreme as or more extreme than 60 if the coin is fair; that is, we calculate $P(X \geq 60)$. To approximate this probability from the normal distribution, we standardize:

$$P(X \geq 60) = P\left(\frac{X - \overset{E(X)}{50}}{\underset{\sqrt{\text{Var}(X)}}{5}} \geq \frac{60 - 50}{5}\right)$$

$$\approx 1 - \Phi(2)$$

$$= .0228$$

The probability is rather small, so the fairness of the coin is called into question. ∎

E X A M P L E G *Particle Size Distribution*
The distribution of the sizes of grains of particulate matter is often found to be quite skewed, with a slowly decreasing right tail. A distribution called the **lognormal** is sometimes fit to such a distribution, and X is said to follow a lognormal distribution if $\log X$ has a normal distribution. The central limit theorem gives a theoretical rationale for the use of the lognormal distribution in some situations.

Suppose that a particle of initial size y_0 is subjected to repeated impacts, that on each impact a proportion, X_i, of the particle remains, and that the X_i are modeled as independent random variables having the same distribution. After the first impact, the size of the particle is $Y_1 = X_1 y_0$; after the second impact, the size is $Y_2 = X_2 X_1 y_0$; and after the nth impact, the size is

$$Y_n = X_n X_{n-1} \cdots X_2 X_1 y_0$$

Then

$$\log Y_n = \log y_0 + \sum_{i=1}^{n} \log X_i$$

and the central limit theorem applies to $\log Y_n$. ∎

A similar construction is relevant to the theory of finance. Suppose that an initial investment of value v_0 is made and that returns occur in discrete time, for example, daily. If the return on the first day is R_1, then the value becomes $V_1 = R_1 v_0$. After day two the value is $V_2 = R_2 R_1 v_0$, and after day n the value is

$$V_n = R_n R_{n-1} \cdots R_1 v_0$$

The log value is thus

$$\log V_n = \log v_0 + \sum_{i=1}^{n} \log R_i$$

If the returns are independent random variables with the same distribution, then the distribution of $\log V_n$ is approximately normally distributed.

5.4 Problems

1. Let $X_1, X_2, \ldots$ be a sequence of independent random variables with $E(X_i) = \mu$ and $\mathrm{Var}(X_i) = \sigma_i^2$. Show that if $n^{-2} \sum_{i=1}^{n} \sigma_i^2 \to 0$, then $\overline{X} \to \mu$ in probability.

2. Let X_i be as in Problem 1 but with $E(X_i) = \mu_i$ and $n^{-1} \sum_{i=1}^{n} \mu_i \to \mu$. Show that $\overline{X} \to \mu$ in probability.

3. Suppose that the number of insurance claims, N, filed in a year is Poisson distributed with $E(N) = 10,000$. Use the normal approximation to the Poisson to approximate $P(N > 10,200)$.

4. Suppose that the number of traffic accidents, N, in a given period of time is distributed as a Poisson random variable with $E(N) = 100$. Use the normal approximation to the Poisson to find Δ such that $P(100 - \Delta < N < 100 + \Delta) \approx .9$.

5. Using moment-generating functions, show that as $n \to \infty$, $p \to 0$, and $np \to \lambda$, the binomial distribution with parameters n and p tends to the Poisson distribution.

6. Using moment-generating functions, show that as $\alpha \to \infty$ the gamma distribution with parameters α and λ, properly standardized, tends to the standard normal distribution.

7. Show that if $X_n \to c$ in probability and if g is a continuous function, then $g(X_n) \to g(c)$ in probability.

8. Compare the Poisson cdf and the normal approximation for (a) $\lambda = 10$, (b) $\lambda = 20$, and (c) $\lambda = 40$.

9. Compare the binomial cdf and the normal approximation for (a) $n = 20$ and $p = .2$, and (b) $n = 40$ and $p = .5$.

10. A six-sided die is rolled 100 times. Using the normal approximation, find the probability that the face showing a six turns up between 15 and 20 times. Find the probability that the sum of the face values of the 100 trials is less than 300.

11. A skeptic gives the following argument to show that there must be a flaw in the central limit theorem: "We know that the sum of independent Poisson random variables follows a Poisson distribution with a parameter that is the sum of the parameters of the summands. In particular, if n independent Poisson random variables, each with parameter n^{-1}, are summed, the sum has a Poisson distribution with parameter 1. The central limit theorem says that as n approaches infinity, the distribution of the sum tends to a normal distribution, but the Poisson with parameter 1 is not the normal." What do you think of this argument?

12. The central limit theorem can be used to analyze round-off error. Suppose that the round-off error is represented as a uniform random variable on $[-\frac{1}{2}, \frac{1}{2}]$. If 100 numbers are added, approximate the probability that the round-off error exceeds (a) 1, (b) 2, and (c) 5.

13. A drunkard executes a "random walk" in the following way: Each minute he takes a step north or south, with probability $\frac{1}{2}$ each, and his successive step directions are independent. His step length is 50 cm. Use the central limit theorem to approximate the probability distribution of his location after 1 h. Where is he most likely to be?

14. Answer Problem 13 under the assumption that the drunkard has some idea of where he wants to go so that he steps north with probability $\frac{2}{3}$ and south with probability $\frac{1}{3}$.

15. Suppose that you bet $5 on each of a sequence of 50 independent fair games. Use the central limit theorem to approximate the probability that you will lose more than $75.

16. Suppose that $X_1, \ldots, X_{20}$ are independent random variables with density functions

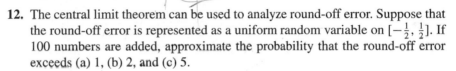

$$f(x) = 2x, \qquad 0 \le x \le 1$$

Let $S = X_1 + \cdots + X_{20}$. Use the central limit theorem to approximate $P(S \le 10)$.

17. Suppose that a measurement has mean μ and variance $\sigma^2 = 25$. Let $\overline{X}$ be the average of n such independent measurements. How large should n be so that $P(|\overline{X} - \mu| < 1) = .95$?

18. Suppose that a company ships packages that are variable in weight, with an average weight of 15 lb and a standard deviation of 10. Assuming that the packages come from a large number of different customers so that it is reasonable to model their weights as independent random variables, find the probability that 100 packages will have a total weight exceeding 1700 lb.

19. a. Use the Monte Carlo method with $n = 100$ and $n = 1000$ to estimate $\int_0^1 \cos(2\pi x)\, dx$. Compare the estimates to the exact answer.
   b. Use Monte Carlo to evaluate $\int_0^1 \cos(2\pi x^2)\, dx$. Can you find the exact answer?

20. What is the variance of the estimate of an integral by the Monte Carlo method (Example A of Section 5.2)? [*Hint:* Find $E(\hat{I}^2(f))$.] Compare the standard deviations of the estimates of part (a) of previous problem to the actual errors you made.

21. This problem introduces a variation on the Monte Carlo integration technique of Example A of Section 5.2. Suppose that we wish to evaluate

$$I(f) = \int_a^b f(x)\, dx$$

Let g be a density function on $[a, b]$. Generate $X_1, \cdots, X_n$ from g and estimate I by

$$\hat{I}(f) = \frac{1}{n} \sum_{i=1}^n \frac{f(X_i)}{g(X_i)}$$

   a. Show that $E(\hat{I}(f)) = I(f)$.
   b. Find an expression for $\text{Var}(\hat{I}(f))$. Give an example for which it is finite and an example for which it is infinite. Note that if it is finite, the law of large numbers implies that $\hat{I}(f) \to I(f)$ as $n \to \infty$.
   c. Show that if $a = 0, b = 1$, and g is uniform, this is the same Monte Carlo estimate as that of Example A of Section 5.2. Can this estimate be improved by choosing g to be other than uniform? (*Hint:* Compare variances.)

22. Use the central limit theorem to find Δ such that $P(|\hat{I}(f) - I(f)| \le \Delta) = .05$, where $\hat{I}(f)$ is the Monte Carlo estimate of $\int_0^1 \cos(2\pi x)\, dx$ based on 1000 points.

23. An irregularly shaped object of unknown area A is located in the unit square $0 \le x \le 1, 0 \le y \le 1$. Consider a random point distributed uniformly over the square; let $Z = 1$ if the point lies inside the object and $Z = 0$ otherwise. Show that $E(Z) = A$. How could A be estimated from a sequence of n independent points uniformly distributed on the square?

24. How could the central limit theorem be used to gauge the probable size of the error of the estimate of the previous problem? Denoting the estimate by $\hat{A}$, if $A = .2$, how large should n be so that $P(|\hat{A} - A| < .01) \approx .99$?

25. Let X be a continuous random variable with density function $f(x) = \frac{3}{2}x^2$, $-1 \le x \le 1$. Sketch this density function. Use the central limit theorem to sketch the approximate density function of $S = X_1 + \cdots + X_{50}$, where the X_i are independent random variables with density f. Similarly, sketch the approximate density functions of $S/50$ and $S/\sqrt{50}$. For each sketch, label at least three points on the horizontal axis.

26. Suppose that a basketball player can score on a particular shot with probability .3. Use the central limit theorem to find the approximate distribution of S, the number of successes out of 25 independent shots. Find the approximate probabilities that S is less than or equal to 5, 7, 9, and 11 and compare these to the exact probabilities.

27. Prove that if $a_n \to a$, then $(1 + a_n/n)^n \to e^a$.

28. Let f_n be a sequence of frequency functions with $f_n(x) = \frac{1}{2}$ if $x = \pm(\frac{1}{2})^n$ and $f_n(x) = 0$ otherwise. Show that $\lim f_n(x) = 0$ for all x, which means that the frequency functions do not converge to a frequency function, but that there exists a cdf F such that $\lim F_n(x) = F(x)$.

29. In addition to limit theorems that deal with sums, there are limit theorems that deal with extreme values such as maxima or minima. Here is an example. Let $U_1, \ldots, U_n$ be independent uniform random variables on $[0, 1]$, and let $U_{(n)}$ be the maximum. Find the cdf of $U_{(n)}$ and a standardized $U_{(n)}$, and show that the cdf of the standardized variable tends to a limiting value.

30. Generate a sequence $U_1, U_2, \ldots, U_{1000}$ of independent uniform random variables on a computer. Let $S_n = \sum_{i=1}^n U_i$ for $n = 1, 2, \ldots, 1000$. Plot each of the following versus n:
    a. S_n
    b. S_n/n
    c. $S_n - n/2$
    d. $(S_n - n/2)/n$
    e. $(S_n - n/2)/\sqrt{n}$

    Explain the shapes of the resulting graphs using the concepts of this chapter.

CHAPTER 6

Distributions Derived from the Normal Distribution

6.1 Introduction

This chapter assembles some results concerning three probability distributions derived from the normal distribution—the χ^2, t, and F distributions. These distributions occur in many statistical problems and will be used in later chapters.

6.2 $\chi^2, t,$ and F Distributions

DEFINITION

If Z is a standard normal random variable, the distribution of $U = Z^2$ is called the chi-square distribution with 1 degree of freedom. ∎

We have already encountered the chi-square distribution in Section 2.3, where we saw that it is a special case of the gamma distribution with parameters $\frac{1}{2}$ and $\frac{1}{2}$. The chi-square distribution with 1 degree of freedom is denoted χ_1^2. It is useful to note that if $X \sim N(\mu, \sigma^2)$, then $(X - \mu)/\sigma \sim N(0, 1)$, and therefore $[(X - \mu)/\sigma]^2 \sim \chi_1^2$.

DEFINITION

If $U_1, U_2, \ldots, U_n$ are independent chi-square random variables with 1 degree of freedom, the distribution of $V = U_1 + U_2 + \cdots + U_n$ is called the *chi-square distribution with n degrees of freedom* and is denoted by χ_n^2. ∎

From Example F in Section 4.5, we know that the sum of independent gamma random variables that have the same value of λ follows a gamma distribution, and therefore the chi-square distribution with n degrees of freedom is a gamma distribution with $\alpha = n/2$ and $\lambda = \frac{1}{2}$. Its density is

$$f(v) = \frac{1}{2^{n/2}\Gamma(n/2)} v^{(n/2)-1} e^{-v/2}, \qquad v \geq 0$$

Its moment-generating function is

$$M(t) = (1 - 2t)^{-n/2}$$

Also, $E(V) = n$ and $\text{Var}(V) = 2n$. To indicate that V follows a chi-square distribution with n degrees of freedom, we write $V \sim \chi_n^2$. A notable consequence of the definition of the chi-square distribution is that if U and V are independent and $U \sim \chi_n^2$ and $V \sim \chi_m^2$, then $U + V \sim \chi_{m+n}^2$.

We now turn to the t distribution.

DEFINITION

If $Z \sim N(0, 1)$ and $U \sim \chi_n^2$ and Z and U are independent, then the distribution of $Z/\sqrt{U/n}$ is called the **t distribution** with n degrees of freedom. ∎

PROPOSITION A

The density function of the t distribution with n degrees of freedom is

$$f(t) = \frac{\Gamma[(n+1)/2]}{\sqrt{n\pi}\,\Gamma(n/2)} \left(1 + \frac{t^2}{n}\right)^{-(n+1)/2}$$

Proof

This is proved by a standard method. The density function of $\sqrt{U/n}$ is straightforward to obtain, and the density function of the quotient of two independent random variables was derived in Section 3.6.1. The details of the proof are left as an end-of-chapter problem. ∎

From the density function of Proposition A, $f(t) = f(-t)$, so the t distribution is symmetric about zero. As the number of degrees of freedom approaches infinity, the t distribution tends to the standard normal distribution; in fact, for more than 20 or 30 degrees of freedom, the distributions are very close. Figure 6.1 shows several t densities. Note that the tails become lighter as the degrees of freedom increase.

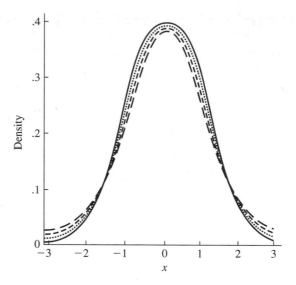

F I G U R E **6.1**   Three t densities with 5 (long dashes), 10 (short dashes), and 30 (dots) degrees of freedom and the standard normal density (solid line).

DEFINITION

Let U and V be independent chi-square random variables with m and n degrees of freedom, respectively. The distribution of

$$W = \frac{U/m}{V/n}$$

is called the F distribution with m and n degrees of freedom and is denoted by $F_{m,n}$.   ∎

PROPOSITION **B**

The density function of W is given by

$$f(w) = \frac{\Gamma[(m+n)/2]}{\Gamma(m/2)\Gamma(n/2)} \left(\frac{m}{n}\right)^{m/2} w^{m/2-1} \left(1 + \frac{m}{n}w\right)^{-(m+n)/2}, \qquad w \geq 0$$

Proof

W is the ratio of two independent random variables, and its density follows from the results given in Section 3.6.1.   ∎

It can be shown that, for $n > 2$, $E(W)$ exists and equals $n/(n-2)$. From the definitions of the t and F distributions, it follows that the square of a t_n random variable follows an $F_{1,n}$ distribution (see Problem 6 at the end of this chapter).

6.3 The Sample Mean and the Sample Variance

Let $X_1, \ldots, X_n$ be independent $N(\mu, \sigma^2)$ random variables; we sometimes refer to them as a **sample** from a normal distribution. In this section, we will find the joint and marginal distributions of

$$\overline{X} = \frac{1}{n} \sum_{i=1}^{n} X_i$$

$$S^2 = \frac{1}{n-1} \sum_{i=1}^{n} (X_i - \overline{X})^2$$

These are called the **sample mean** and the **sample variance,** respectively. First note that because $\overline{X}$ is a linear combination of independent normal random variables, it is normally distributed with

$$E(\overline{X}) = \mu$$

$$\text{Var}(\overline{X}) = \frac{\sigma^2}{n}$$

As a preliminary to showing that $\overline{X}$ and S^2 are independently distributed, we establish the following theorem.

THEOREM A

The random variable $\overline{X}$ and the vector of random variables $(X_1 - \overline{X}, X_2 - \overline{X}, \ldots, X_n - \overline{X})$ are independent.

Proof

At the level of this course, it is difficult to give a proof that provides sufficient insight into why this result is true; a rigorous proof essentially depends on geometric properties of the multivariate normal distribution, which this book does not cover. We present a proof based on moment-generating functions; in particular, we will show that the joint moment-generating function

$$M(s, t_1, \ldots, t_n) = E\{\exp[s\overline{X} + t_1(X_1 - \overline{X}) + \cdots + t_n(X_n - \overline{X})]\}$$

factors into the product of two moment-generating functions—one of $\overline{X}$ and the other of $(X_1 - \overline{X}), \ldots, (X_n - \overline{X})$. The factoring implies (Section 4.5) that the random variables are independent of each other and is accomplished through some algebraic trickery. First we observe that since

$$\sum_{i=1}^{n} t_i (X_i - \overline{X}) = \sum_{i=1}^{n} t_i X_i - n \overline{X} \overline{t}$$

then

$$s\overline{X} + \sum_{i=1}^{n} t_i(X_i - \overline{X}) = \sum_{i=1}^{n}\left[\frac{s}{n} + (t_i - \bar{t})\right]X_i$$

$$= \sum_{i=1}^{n} a_i X_i$$

where

$$a_i = \frac{s}{n} + (t_i - \bar{t})$$

Furthermore, we observe that

$$\sum_{i=1}^{n} a_i = s$$

$$\sum_{i=1}^{n} a_i^2 = \frac{s^2}{n} + \sum_{i=1}^{n}(t_i - \bar{t})^2$$

Now we have

$$M(s, t_1, \ldots, t_n) = M_{X_1 \cdots X_n}(a_1, \ldots, a_n)$$

and since the X_i are independent normal random variables, we have

$$M(s, t_1, \ldots, t_n) = \prod_{i=1}^{n} M_{X_i}(a_i)$$

$$= \prod_{i=1}^{n} \exp\left(\mu a_i + \frac{\sigma^2}{2} a_i^2\right)$$

$$= \exp\left(\mu \sum_{i=1}^{n} a_i + \frac{\sigma^2}{2} \sum_{i=1}^{n} a_i^2\right)$$

$$= \exp\left[\mu s + \frac{\sigma^2}{2}\left(\frac{s^2}{n}\right) + \frac{\sigma^2}{2} \sum_{i=1}^{n}(t_i - \bar{t})^2\right]$$

$$= \exp\left(\mu s + \frac{\sigma^2}{2n} s^2\right) \exp\left[\frac{\sigma^2}{2} \sum_{i=1}^{n}(t_i - \bar{t})^2\right]$$

The first factor is the mgf of $\overline{X}$. Since the mgf of the vector $(X_1 - \overline{X}, \ldots, X_n - \overline{X})$ can be obtained by setting $s = 0$ in M, the second factor is this mgf. ∎

COROLLARY A

$\overline{X}$ and S^2 are independently distributed.

Proof

This follows immediately since S^2 is a function of the vector $(X_1 - \overline{X}, \ldots, X_n - \overline{X})$, which is independent of $\overline{X}$. ∎

The next theorem gives the marginal distribution of S^2.

THEOREM B

The distribution of $(n-1)S^2/\sigma^2$ is the chi-square distribution with $n-1$ degrees of freedom.

Proof

We first note that

$$\frac{1}{\sigma^2} \sum_{i=1}^{n} (X_i - \mu)^2 = \sum_{i=1}^{n} \left(\frac{X_i - \mu}{\sigma} \right)^2 \sim \chi_n^2$$

Also,

$$\frac{1}{\sigma^2} \sum_{i=1}^{n} (X_i - \mu)^2 = \frac{1}{\sigma^2} \sum_{i=1}^{n} [(X_i - \overline{X}) + (\overline{X} - \mu)]^2$$

Expanding the square and using the fact that $\sum_{i=1}^{n}(X_i - \overline{X}) = 0$, we obtain

$$\frac{1}{\sigma^2} \sum_{i=1}^{n} (X_i - \mu)^2 = \frac{1}{\sigma^2} \sum_{i=1}^{n} (X_i - \overline{X})^2 + \left(\frac{\overline{X} - \mu}{\sigma/\sqrt{n}} \right)^2$$

This is a relation of the form $W = U + V$. Since U and V are independent by Corollary A, $M_W(t) = M_U(t)M_V(t)$. W and V both follow chi-square distributions, so

$$M_U(t) = \frac{M_W(t)}{M_V(t)}$$

$$= \frac{(1 - 2t)^{-n/2}}{(1 - 2t)^{-1/2}}$$

$$= (1 - 2t)^{-(n-1)/2}$$

The last expression is the mgf of a random variable with a χ_{n-1}^2 distribution. ∎

One final result concludes this chapter's collection.

COROLLARY **B**

Let $\overline{X}$ and S^2 be as given at the beginning of this section. Then

$$\frac{\overline{X} - \mu}{S/\sqrt{n}} \sim t_{n-1}$$

Proof

We simply express the given ratio in a different form:

$$\frac{\overline{X} - \mu}{S/\sqrt{n}} = \frac{\left(\dfrac{\overline{X} - \mu}{\sigma/\sqrt{n}}\right)}{\sqrt{S^2/\sigma^2}}$$

The latter is the ratio of an $N(0, 1)$ random variable to the square root of an independent random variable with a χ^2_{n-1} distribution divided by its degrees of freedom. Thus, from the definition in Section 6.2, the ratio follows a t distribution with $n - 1$ degrees of freedom. ∎

6.4 Problems

1. Prove Proposition A of Section 6.2.

2. Prove Proposition B of Section 6.2.

3. Let $\overline{X}$ be the average of a sample of 16 independent normal random variables with mean 0 and variance 1. Determine c such that

$$P(|\overline{X}| < c) = .5$$

4. If T follows a t_7 distribution, find t_0 such that (a) $P(|T| < t_0) = .9$ and (b) $P(T > t_0) = .05$.

5. Show that if $X \sim F_{n,m}$, then $X^{-1} \sim F_{m,n}$.

6. Show that if $T \sim t_n$, then $T^2 \sim F_{1,n}$.

7. Show that the Cauchy distribution and the t distribution with 1 degree of freedom are the same.

8. Show that if X and Y are independent exponential random variables with $\lambda = 1$, then X/Y follows an F distribution. Also, identify the degrees of freedom.

9. Find the mean and variance of S^2, where S^2 is as in Section 6.3.

10. Show how to use the chi-square distribution to calculate $P(a < S^2/\sigma^2 < b)$.

11. Let $X_1, \ldots, X_n$ be a sample from an $N(\mu_X, \sigma^2)$ distribution and $Y_1, \ldots, Y_m$ be an independent sample from an $N(\mu_Y, \sigma^2)$ distribution. Show how to use the F distribution to find $P(S_X^2/S_Y^2 > c)$.

CHAPTER 7

Survey Sampling

7.1 Introduction

Resting on the probabilistic foundations of the preceding chapters, this chapter marks the beginning of our study of statistics by introducing the subject of survey sampling. As well as being of considerable intrinsic interest and practical utility, the development of the elementary theory of survey sampling serves to introduce several concepts and techniques that will recur and be amplified in later chapters.

Sample surveys are used to obtain information about a large population by examining only a small fraction of that population. Sampling techniques have been used in many fields, such as the following:

- Governments survey human populations; for example, the U.S. government conducts health surveys and census surveys.
- Sampling techniques have been extensively employed in agriculture to estimate such quantities as the total acreage of wheat in a state by surveying a sample of farms.
- The Interstate Commerce Commission has carried out sampling studies of rail and highway traffic. In one such study, records of shipments of household goods by motor carriers were sampled to evaluate the accuracy of preshipment estimates of charges, claims for damages, and other variables.
- In the practice of quality control, the output of a manufacturing process may be sampled in order to examine the items for defects.
- During audits of the financial records of large companies, sampling techniques may be used when examination of the entire set of records is impractical.

The sampling techniques discussed here are probabilistic in nature—each member of the population has a specified probability of being included in the sample, and the actual composition of the sample is random. Such techniques differ markedly from the type of sampling scheme in which particular population members are included in the sample because the investigator thinks they are typical in some way. Such a scheme may be effective in some situations, but there is no way mathematically to guarantee its unbiasedness (a term that will be precisely defined later) or to estimate the magnitude of any error committed, such as that arising from estimating the population mean by the sample mean. We will see that using a random sampling technique has a consequence that estimates can be guaranteed to be unbiased and probabilistic bounds on errors can be calculated. Among the advantages of using random sampling are the following:

- The selection of sample units at random is a guard against investigator biases, even unconscious ones.
- A small sample costs far less and is much faster to survey than a complete enumeration.
- The results from a small sample may actually be more accurate than those from a complete enumeration. The quality of the data in a small sample can be more easily monitored and controlled, and a complete enumeration may require a much larger, and therefore perhaps more poorly trained, staff.
- Random sampling techniques make possible the calculation of an estimate of the error due to sampling.
- In designing a sample, it is frequently possible to determine the sample size necessary to obtain a prescribed error level.

Peck et al. (2005) contains several interesting papers about applications of sampling.

7.2 Population Parameters

This section defines those numerical characteristics, or parameters, of the population that we will estimate from a sample. We will assume that the population is of size N and that associated with each member of the population is a numerical value of interest. These numerical values will be denoted by $x_1, x_2, \cdots, x_N$. The variable x_i may be a numerical variable such as age or weight, or it may take on the value 1 or 0 to denote the presence or absence of some characteristic such as gender. We will refer to the latter situation as the dichotomous case.

EXAMPLE **A**   This is the first of many examples in this chapter in which we will illustrate ideas by using a study by Herkson (1976). The population consists of $N = 393$ short-stay hospitals. We will let x_i denote the number of patients discharged from the ith hospital during January 1968. A histogram of the population values is shown in Figure 7.1. The histogram was constructed in the following way: The number of hospitals that discharged 0–200, 201–400, ..., 2801–3000 patients were graphed as horizontal lines above the respective intervals. For example, the figure indicates that about

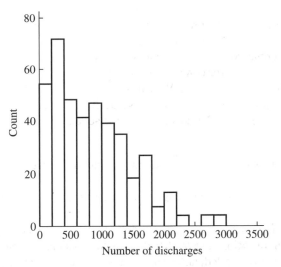

FIGURE **7.1** Histogram of the numbers of patients discharged during January 1968 from 393 short-stay hospitals.

40 hospitals discharged from 601 to 800 patients. The histogram is a convenient graphical representation of the distribution of the values in the population, being more quickly assimilated than would a list of 393 values. ∎

We will be particularly interested in the **population mean,** or average,

$$\mu = \frac{1}{N} \sum_{i=1}^{N} x_i$$

For the population of 393 hospitals, the mean number of discharges is 814.6. Note the location of this value in Figure 7.1. In the dichotomous case, where the presence or absence of a characteristic is to be determined, μ equals the proportion, p, of individuals in the population having the particular characteristic, because in the sum above, each x_i is either 0 or 1. The sum thus reduces to the number of 1s and when divided by N, gives the proportion, p.

The **population total** is

$$\tau = \sum_{i=1}^{N} x_i = N\mu$$

The total number of people discharged from the population of hospitals is $\tau = 320,138$. In the dichotomous case, the population total is the total number of members of the population possessing the characteristic of interest.

We will also need to consider the **population variance,**

$$\sigma^2 = \frac{1}{N} \sum_{i=1}^{N} (x_i - \mu)^2$$

A useful identity can be obtained by expanding the square in this equation:

$$\sigma^2 = \frac{1}{N}\left(\sum_{i=1}^{N} x_i^2 - 2\mu \sum_{i=1}^{N} x_i + N\mu^2\right)$$

$$= \frac{1}{N}\left(\sum_{i=1}^{N} x_i^2 - 2N\mu^2 + N\mu^2\right)$$

$$= \frac{1}{N}\sum_{i=1}^{N} x_i^2 - \mu^2$$

In the dichotomous case, the population variance reduces to $p(1-p)$:

$$\sigma^2 = \frac{1}{N}\sum_{i=1}^{N} x_i^2 - \mu^2$$

$$= p - p^2$$

$$= p(1-p)$$

Here we used the fact that because each x_i is 0 or 1, each x_i^2 is also 0 or 1.

The **population standard deviation** is the square root of the population variance and is used as a measure of how spread out, dispersed, or scattered the individual values are. The standard deviation is given in the same units (for example, inches) as are the population values, whereas the variance is given in those units squared. The variance of the discharges is 347,766, and the standard deviation is 589.7; examination of the histogram in Figure 7.1 makes it clear that the latter number is the more reasonable description of the spread of the population values.

7.3 Simple Random Sampling

The most elementary form of sampling is **simple random sampling** (s.r.s.): Each particular sample of size n has the same probability of occurrence; that is, each of the $\binom{N}{n}$ possible samples of size n taken without replacement has the same probability. We assume that sampling is done without replacement so that each member of the population will appear in the sample at most once. The actual composition of the sample is usually determined by using a table of random numbers or a random number generator on a computer. Conceptually, we can regard the population members as balls in an urn, a specified number of which are selected for inclusion in the sample at random and without replacement.

Because the composition of the sample is random, the sample mean is random. An analysis of the accuracy with which the sample mean approximates the population mean must therefore be probabilistic in nature. In this section, we will derive some statistical properties of the sample mean.

7.3.1   The Expectation and Variance of the Sample Mean

We will denote the sample size by n (n is less than N) and the values of the sample members by $X_1, X_2, \ldots, X_n$. It is important to realize that each X_i is a random variable. In particular, X_i is not the same as x_i: X_i is the value of the ith member of the sample, which is random and x_i is that of the ith member of the population, which is fixed.

We will consider the **sample mean,**

$$\overline{X} = \frac{1}{n} \sum_{i=1}^{n} X_i$$

as an estimate of the population mean. As an estimate of the population total, we will consider

$$T = N\overline{X}$$

Properties of T will follow readily from those of $\overline{X}$. Since each X_i is a random variable, so is the sample mean; its probability distribution is called its **sampling distribution.** In general, any numerical value, or statistic, computed from a random sample is a random variable and has an associated sampling distribution. The sampling distribution of $\overline{X}$ determines how accurately $\overline{X}$ estimates μ; roughly speaking, the more tightly the sampling distribution is centered on μ, the better the estimate.

EXAMPLE A   To illustrate the concept of a sampling distribution, let us look again at the population of 393 hospitals. In practice, of course, the population would not be known, and only one sample would be drawn. For pedagogical purposes here, we can consider the sampling distribution of the sample mean from this known population. Suppose, for example, that we want to find the sampling distribution of the mean of a sample of size 16. In principle, we could form all $\binom{393}{16}$ samples and compute the mean of each one— this would give the sampling distribution. But because the number of such samples is of the order 10^{28}, this is clearly not practical. We will thus employ a technique known as **simulation.** We can estimate the sampling distribution of the mean of a sample of size n by drawing many samples of size n, computing the mean of each sample, and then forming a histogram of the collection of sample means. Figure 7.2 shows the results of such a simulation for sample sizes of 8, 16, 32, and 64 with 500 replications for each sample size. Three features of Figure 7.2 are noteworthy:

1. All the histograms are centered about the population mean, 814.6.
2. As the sample size increases, the histograms become less spread out.
3. Although the shape of the histogram of population values (Figure 7.1) is not symmetric about the mean, the histograms in Figure 7.2 are more nearly so.

These features will be explained quantitatively.   ■

As we have said, $\overline{X}$ is a random variable whose distribution is determined by that of the X_i. We thus examine the distribution of a single sample element, X_i. It should be noted that the following lemma holds whether sampling is with or without replacement.

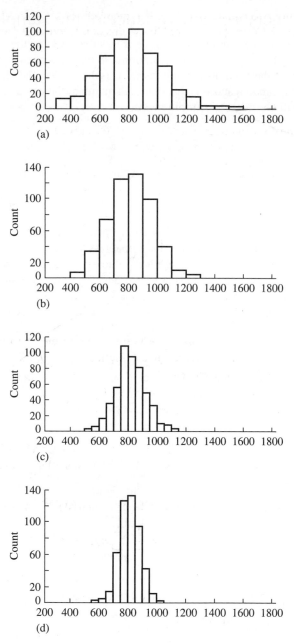

FIGURE **7.2** Histograms of the values of the mean number of discharges in 500 simple random samples from the population of 393 hospitals. Sample sizes: (a) $n = 8$, (b) $n = 16$, (c) $n = 32$, (d) $n = 64$.

We need to be careful about the values that the random variable X_i can assume. The ith sample member is equally likely to be any of the N population members. If all the population values were distinct, we would then have $P(X_1 = x_j) = 1/N$. But the population values may not be distinct (for example, in the dichotomous case there are only two values, 0 and 1). If k members of the population have the same value ζ, then $P(X_i = \zeta) = k/N$. We use this construction in proving the following lemma.

LEMMA A

Denote the distinct values assumed by the population members by $\zeta_1, \zeta_2, \ldots, \zeta_m$, and denote the number of population members that have the value ζ_j by n_j, $j = 1, 2, \ldots, m$. Then X_i is a discrete random variable with probability mass function

$$P(X_i = \zeta_j) = \frac{n_j}{N}$$

Also,

$$E(X_i) = \mu$$
$$\text{Var}(X_i) = \sigma^2$$

Proof

The only possible values that X_i can assume are $\zeta_1, \zeta_2, \ldots, \zeta_m$. Since each member of the population is equally likely to be the ith member of the sample, the probability that X_i assumes the value ζ_j is thus n_j/N. The expected value of the random variable X_i is then

$$E(X_i) = \sum_{j=1}^{m} \zeta_j P(X_i = \zeta_j) = \frac{1}{N} \sum_{j=1}^{m} n_j \zeta_j = \mu$$

The last equation follows because n_j population members have the value ζ_j and the sum is thus equal to the sum of the values of all the population members. Finally,

$$\text{Var}(X_i) = E\left(X_i^2\right) - [E(X_i)]^2$$

$$= \frac{1}{N} \sum_{j=1}^{m} n_j \zeta_j^2 - \mu^2$$

$$= \sigma^2$$

Here we have used the fact that $\sum_{i=1}^{N} x_i^2 = \sum_{j=1}^{m} n_j \zeta_j^2$ and the identity for the population variance derived in Section 7.2. ∎

As a measure of the center of the sampling distribution, we will use $E(\overline{X})$. As a measure of the dispersion of the sampling distribution about this center, we will use the standard deviation of $\overline{X}$. The key results that will be obtained shortly are that the sampling distribution is centered at μ and that its spread is inversely proportional to the square root of the sample size, n. We first show that the sampling distribution is centered at μ.

THEOREM A

With simple random sampling, $E(\overline{X}) = \mu$.

Proof

Since, from Lemma A, $E(X_i) = \mu$, it follows from Theorem A in Section 4.1.2 that

$$E(\overline{X}) = \frac{1}{n} \sum_{i=1}^{n} E(X_i) = \mu \qquad \blacksquare$$

From Theorem A, we have the following corollary.

COROLLARY A

With simple random sampling, $E(T) = \tau$.

Proof

$$\begin{aligned} E(T) &= E(N\overline{X}) \\ &= N E(\overline{X}) \\ &= N\mu \\ &= \tau \end{aligned} \qquad \blacksquare$$

In the dichotomous case, $\mu = p$, and $\overline{X}$ is the proportion of the sample that possesses the characteristic of interest. In this case, $\overline{X}$ will be denoted by $\hat{p}$. We have shown that $E(\hat{p}) = p$.

It is important to keep in mind that $\overline{X}$ is random. The result $E(\overline{X}) = \mu$ can be interpreted to mean that "on the average" $\overline{X} = \mu$. In general, if we wish to estimate a population parameter, θ say, by a function $\hat{\theta}$ of the sample, $X_1, X_2, \ldots, X_n$, and $E(\hat{\theta}) = \theta$, whatever the value of θ may be, we say that $\hat{\theta}$ is **unbiased.** Thus, $\overline{X}$ and T are unbiased estimates of μ and τ. On average they are correct. We next investigate how variable they are, by deriving their variances and standard deviations. Section 4.2.1 introduced the concepts of bias and variance in the context of a model of measurement error, and these concepts are also relevant in this new context. In Chapter 4, it was shown that

$$\text{Mean squared error} = \text{variance} + \text{bias}^2$$

Since $\overline{X}$ and T are unbiased, their mean squared errors are equal to their variances.

We next find $\text{Var}(\overline{X})$. Since $\overline{X} = n^{-1} \sum_{i=1}^{n} X_i$, it follows from Corollary A of Section 4.3 that

$$\text{Var}(\overline{X}) = \frac{1}{n^2} \sum_{i=1}^{n} \sum_{j=1}^{n} \text{Cov}(X_i, X_j)$$

Suppose that sampling were done with replacement. Then the X_i would be independent, and for $i \neq j$ we would have $\text{Cov}(X_i, X_j) = 0$, whereas $\text{Cov}(X_i, X_i) = \text{Var}(X_i) = \sigma^2$. It would then follow that

$$\text{Var } \overline{X} = \frac{1}{n^2} \sum_{i=1}^{n} \text{Var}(X_i)$$

$$= \frac{\sigma^2}{n}$$

and that the standard deviation of $\overline{X}$, also called its **standard error,** would be

$$\sigma_{\overline{X}} = \frac{\sigma}{\sqrt{n}}$$

Sampling without replacement induces dependence among the X_i, which complicates this simple result. However, we will see that if the sample size n is small relative to the population size N, the dependence is weak and this simple result holds to a good approximation.

To find the variance of the sample mean in sampling without replacement we need to find $\text{Cov}(X_i, X_j)$ for $i \neq j$.

LEMMA **B**

For simple random sampling without replacement,

$$\text{Cov}(X_i, X_j) = -\sigma^2/(N-1) \qquad \text{if } i \neq j$$

Using the identity for covariance established at the beginning of Section 4.3,

$$\text{Cov}(X_i, X_j) = E(X_i X_j) - E(X_i)E(X_j)$$

and

$$E(X_i X_j) = \sum_{k=1}^{m} \sum_{l=1}^{m} \zeta_k \zeta_l P(X_i = \zeta_k \text{ and } X_j = \zeta_l)$$

$$= \sum_{k=1}^{m} \zeta_k P(X_i = \zeta_k) \sum_{l=1}^{m} \zeta_l P(X_j = \zeta_l | X_i = \zeta_k)$$

from the multiplication law for conditional probability. Now,

$$P(X_j = \zeta_l | X_i = \zeta_k) = \begin{cases} n_l/(N-1), & \text{if } k \neq l \\ (n_l - 1)/(N-1), & \text{if } k = l \end{cases}$$

Now if we express

$$\sum_{l=1}^{m} \zeta_l P(X_j = \zeta_l | X_i = \zeta_k) = \sum_{l \neq k} \zeta_l \frac{n_l}{N-1} + \zeta_k \frac{n_k - 1}{N-1}$$

$$= \sum_{l=1}^{m} \zeta_l \frac{n_l}{N-1} - \zeta_k \frac{1}{N-1}$$

the expression for $E(X_i X_j)$ becomes

$$\sum_{k=1}^{m} \zeta_k \frac{n_k}{N} \left(\sum_{l=1}^{m} \zeta_l \frac{n_l}{N-1} - \frac{\zeta_k}{N-1} \right) = \frac{1}{N(N-1)} \left(\tau^2 - \sum_{k=1}^{m} \zeta_k^2 n_k \right)$$

$$= \frac{\tau^2}{N(N-1)} - \frac{1}{N(N-1)} \sum_{k=1}^{m} \zeta_k^2 n_k$$

$$= \frac{N\mu^2}{N-1} - \frac{1}{N-1}(\mu^2 + \sigma^2)$$

$$= \mu^2 - \frac{\sigma^2}{N-1}$$

Finally, subtracting $E(X_i)E(X_j) = \mu^2$ from the last equation, we have

$$\mathrm{Cov}(X_i, X_j) = -\frac{\sigma^2}{N-1}$$

for $i \neq j$. ∎

(Alternative proofs of Lemma B are outlined in Problems 25 and 26 at the end of this chapter.) This lemma shows that X_i and X_j are not independent of each other for $i \neq j$, but that the covariance is very small for large values of N. We are now able to derive the following theorem.

THEOREM **B**

With simple random sampling,

$$\mathrm{Var}(\overline{X}) = \frac{\sigma^2}{n} \left(\frac{N-n}{N-1} \right)$$

$$= \frac{\sigma^2}{n} \left(1 - \frac{n-1}{N-1} \right)$$

Proof

From Corollary A of Section 4.3,

$$\mathrm{Var}(\overline{X}) = \frac{1}{n^2} \sum_{i=1}^{n} \sum_{j=1}^{n} \mathrm{Cov}(X_i, X_j)$$

$$= \frac{1}{n^2} \sum_{i=1}^{n} \mathrm{Var}(X_i) + \frac{1}{n^2} \sum_{i=1}^{n} \sum_{j \neq i} \mathrm{Cov}(X_i, X_j)$$

$$= \frac{\sigma^2}{n} - \frac{1}{n^2} n(n-1) \frac{\sigma^2}{N-1}$$

After some algebra, this gives the desired result. ∎

Notice that the variance of the sample mean in sampling without replacement differs from that in sampling with replacement by the factor

$$\left(1 - \frac{n-1}{N-1}\right)$$

which is called the **finite population correction.** The ratio n/N is called the **sampling fraction.** Frequently, the sampling fraction is very small, in which case the **standard error** (standard deviation) of $\overline{X}$ is

$$\sigma_{\overline{X}} \approx \frac{\sigma}{\sqrt{n}}$$

We see that, apart from the usually small finite population correction, the spread of the sampling distribution and therefore the precision of $\overline{X}$ are determined by the sample size (n) and not by the population size (N). As will be made more explicit later, the appropriate measure of the precision of the sample mean is its standard error, which is inversely proportional to the square root of the sample size. Thus, in order to double the accuracy, the sample size must be quadrupled. (You might examine Figure 7.2 with this in mind.) The other factor that determines the accuracy of the sample mean is the population standard deviation, σ. If σ is small, the population values are not very dispersed and a small sample will be fairly accurate. But if the values are widely dispersed, a much larger sample will be required in order to attain the same accuracy.

EXAMPLE B   If the population of hospitals is sampled without replacement and the sample size is $n = 32$,

$$\sigma_{\overline{X}} = \frac{\sigma}{\sqrt{n}} \sqrt{1 - \frac{n-1}{N-1}}$$

$$= \frac{589.7}{\sqrt{32}} \sqrt{1 - \frac{31}{392}}$$

$$= 104.2 \times .96$$

$$= 100.0$$

Notice that because the sampling fraction is small, the finite population correction makes little difference. To see that $\sigma_{\overline{X}} = 100.0$ is a reasonable measure of accuracy, examine part (b) of Figure 7.2 and observe that the vast majority of sample means differed from the population mean (814) by less than two standard errors; i.e., the vast majority of sample means were in the interval (614, 1014). ∎

EXAMPLE C   Let us apply this result to the problem of estimating a proportion. In the population of hospitals, a proportion $p = .654$ had fewer than 1000 discharges. If this proportion were estimated from a sample as the sample proportion $\hat{p}$, the standard error of $\hat{p}$

could be found by applying Theorem B to this dichotomous case:

$$\sigma_{\hat{p}} = \sqrt{\frac{p(1-p)}{n}} \sqrt{1 - \frac{n-1}{N-1}}$$

For example, for $n = 32$, the standard error of $\hat{p}$ is

$$\sigma_{\hat{p}} = \sqrt{\frac{.654 \times .346}{32}} \sqrt{1 - \frac{31}{392}}$$
$$= .08 \qquad \blacksquare$$

The precision of the estimate of the population total does depend on the population size, N.

COROLLARY B

With simple random sampling,

$$\mathrm{Var}(T) = N^2 \left(\frac{\sigma^2}{n}\right) \frac{N-n}{N-1}$$

Proof

Since $T = N\overline{X}$,

$$\mathrm{Var}(T) = N^2 \, \mathrm{Var}(\overline{X}) \qquad \blacksquare$$

7.3.2 Estimation of the Population Variance

A sample survey is used to estimate population parameters, and it is desirable also to assess and quantify the variability of the estimates. In the previous section, we saw how the standard error of an estimate may be determined from the sample size and the population variance. In practice, however, the population variance will not be known, but as we will show in this section, it can be estimated from the sample. Since the population variance is the average squared deviation from the population mean, estimating it by the average squared deviation from the sample mean seems natural:

$$\hat{\sigma}^2 = \frac{1}{n} \sum_{i=1}^{n} (X_i - \overline{X})^2$$

The following theorem shows that this estimate is biased.

THEOREM **A**

With simple random sampling,

$$E(\hat{\sigma}^2) = \sigma^2 \left(\frac{n-1}{n}\right) \frac{N}{N-1}$$

Proof

Expanding the square and proceeding as in the identity for the population variance in Section 7.2, we find

$$\hat{\sigma}^2 = \frac{1}{n} \sum_{i=1}^{n} X_i^2 - \overline{X}^2$$

Thus,

$$E(\hat{\sigma}^2) = \frac{1}{n} \sum_{i=1}^{n} E\left(X_i^2\right) - E(\overline{X}^2)$$

Now, we know that

$$E\left(X_i^2\right) = \text{Var}(X_i) + [E(X_i)]^2$$
$$= \sigma^2 + \mu^2$$

Similarly, from Theorems A and B of Section 7.3.1,

$$E(\overline{X}^2) = \text{Var}(\overline{X}) + [E(\overline{X})]^2$$
$$= \frac{\sigma^2}{n}\left(1 - \frac{n-1}{N-1}\right) + \mu^2$$

Substituting these expressions for $E(X_i^2)$ and $E(\overline{X}^2)$ in the preceding equation for $E(\hat{\sigma}^2)$ gives the desired result. ∎

Because $N > n$, it follows with a little algebra that

$$\frac{n-1}{n} \frac{N}{N-1} < 1$$

so that $E(\hat{\sigma}^2) < \sigma^2$; $\hat{\sigma}^2$ thus tends to underestimate σ^2. From Theorem A, we see that an unbiased estimate of σ^2 may be obtained by multiplying $\hat{\sigma}^2$ by the factor $n(N-1)/[(n-1)N]$. Thus, an unbiased estimate of σ^2 is $\frac{1}{n-1}(1-\frac{1}{N})\sum_{i=1}^{n}(X_i - \overline{X})^2$. We also have the following corollary.

COROLLARY A

An unbiased estimate of $\text{Var}(\overline{X})$ is

$$s_{\overline{X}}^2 = \frac{\hat{\sigma}^2}{n} \left(\frac{n}{n-1}\right) \left(\frac{N-1}{N}\right) \left(\frac{N-n}{N-1}\right)$$

$$= \frac{s^2}{n} \left(1 - \frac{n}{N}\right)$$

where

$$s^2 = \frac{1}{n-1} \sum_{i=1}^{n} (X_i - \overline{X})^2$$

Proof

Since

$$\text{Var}(\overline{X}) = \frac{\sigma^2}{n} \cdot \left(\frac{N-n}{N-1}\right)$$

an unbiased estimate of $\text{Var}(\overline{X})$ may be obtained by substituting in an unbiased estimate of σ^2. Algebra then yields the desired result. ∎

Similarly, an unbiased estimate of the variance of T, the estimator of the population total, is

$$s_T^2 = N^2 s_{\overline{X}}^2$$

For the dichotomous case, in which each X_i is 0 or 1, note that

$$\frac{1}{n} \sum_{i=1}^{n} (X_i - \overline{X})^2 = \frac{1}{n} \sum_{i=1}^{n} X_i^2 - \overline{X}^2$$

$$= \hat{p}(1 - \hat{p})$$

Therefore,

$$s^2 = \frac{n}{n-1} \hat{p}(1 - \hat{p})$$

Thus, as a special case of Corollary A, we have the following corollary.

COROLLARY B

An unbiased estimate of $\text{Var}(\hat{p})$ is

$$s_{\hat{p}}^2 = \frac{\hat{p}(1 - \hat{p})}{n-1} \left(1 - \frac{n}{N}\right)$$

∎

In many cases, the sampling fraction, n/N, is small and may be neglected. Furthermore, it often makes little difference whether $n - 1$ or n is used as the divisor.

The quantities $s_{\overline{X}}$, s_T, and $s_{\hat{p}}$ are called **estimated standard errors.** If we knew them, the actual standard errors, $\sigma_{\overline{X}}$, σ_T and $\sigma_{\hat{p}}$, would be used to gauge the accuracy of the estimates $\overline{X}$, T and $\hat{p}$. If they are not known, which is the typical case, the estimated standard errors are used in their place.

E X A M P L E A

A simple random sample of 50 of the 393 hospitals was taken. From this sample, $\overline{X} = 938.5$ (recall that, in fact, $\mu = 814.6$) and $s = 614.53$ ($\sigma = 590$). An estimate of the variance of $\overline{X}$ is

$$s_{\overline{X}}^2 = \frac{s^2}{n}\left(1 - \frac{n}{N}\right) = 6592$$

The estimated standard error of $\overline{X}$ is

$$s_{\overline{X}} = 81.19$$

(Note that the true value is $\sigma_{\overline{X}} = \frac{\sigma}{\sqrt{50}}\sqrt{1 - \frac{49}{392}} = 78$.) This estimated standard error gives a rough idea of how accurate the value of $\overline{X}$ is; in this case, we see that the magnitude of the error is of the order 80, as opposed to 8 or 800, say. In fact, the error was 123.9, or about $1.5\ s_{\overline{X}}$. ∎

E X A M P L E B

From the same sample, the estimate of the total number of discharges in the population of hospitals is

$$T = N\overline{X} = 368{,}831$$

Recall that the true value of the population total is 320,139. The estimated standard error of T is

$$s_T = Ns_{\overline{X}} = 31{,}908$$

Again, this estimated standard error can be used as a rough gauge of the estimation error. ∎

E X A M P L E C

Let p be the proportion of hospitals that had fewer than 1000 discharges—that is, $p = .654$. In the sample of Example A, 26 of 50 hospitals had fewer than 1000 discharges, so

$$\hat{p} = \frac{26}{50} = .52$$

The variance of $\hat{p}$ is estimated by

$$s_{\hat{p}}^2 = \frac{\hat{p}(1 - \hat{p})}{n - 1}\left(1 - \frac{n}{N}\right) = .0045$$

Thus, the estimated standard error of $\hat{p}$ is

$$s_{\hat{p}} = .067$$

Crudely, this tells us that the error of $\hat{p}$ is in the second or first decimal place—that we are probably not so fortunate as to have an error only in the third decimal place. In fact, the error was .134 or about $2 \times s_{\hat{p}}$.   ■

These examples show how, in simple random sampling, we can not only form estimates of unknown population parameters, but can also gauge the likely size of the errors of the estimates, by estimating their standard errors from the data in the sample.

We have covered a lot of ground, and the presence of the finite population correction complicates the expressions we have derived. It is thus useful to summarize our results in the following table:

Population Parameter	Estimate	Variance of Estimate	Estimated Variance
μ	$\overline{X} = \frac{1}{n} \sum_{i=1}^{n} X_i$	$\sigma_{\overline{X}}^2 = \frac{\sigma^2}{n} \left(\frac{N-n}{N-1} \right)$	$s_{\overline{X}}^2 = \frac{s^2}{n} \left(1 - \frac{n}{N} \right)$
p	$\hat{p} = $ sample proportion	$\sigma_{\hat{p}}^2 = \frac{p(1-p)}{n} \left(\frac{N-n}{N-1} \right)$	$s_{\hat{p}}^2 = \frac{\hat{p}(1-\hat{p})}{n-1} \left(1 - \frac{n}{N} \right)$
τ	$T = N\overline{X}$	$\sigma_T^2 = N^2 \sigma_{\overline{X}}^2$	$s_T^2 = N^2 s_{\overline{X}}^2$
σ^2	$\left(1 - \frac{1}{N} \right) s^2$		

where $s^2 = \frac{1}{n-1} \sum_{i=1}^{n} (X_i - \overline{X})^2$.

The square roots of the entries in the third column are called *standard errors,* and the square roots of the entries in the fourth column are called *estimated standard errors.* The former depend on unknown population parameters, so the latter are used to gauge the accuracy of the parameter estimates. When the population is large relative to the sample size, the finite population correction can be ignored, simplifying the preceding expressions.

7.3.3   The Normal Approximation to the Sampling Distribution of $\overline{X}$

We have found the mean and the standard deviation of the sampling distribution of $\overline{X}$. Ideally, we would like to know the sampling distribution, since it would tell us everything we could hope to know about the accuracy of the estimate. Without knowledge of the population itself, however, we cannot determine the sampling distribution. In this section, we will use the central limit theorem to deduce an approximation to the sampling distribution—the normal, or Gaussian, distribution. This approximation will be used to find probabilistic bounds for the estimation error.

In Section 5.3, we considered a sequence of independent and identically distributed (i.i.d.) random variables, $X_1, X_2, \ldots$ having the common mean and variance μ and σ^2. The sample mean of $X_1, X_2, \ldots, X_n$ is

$$\overline{X}_n = \frac{1}{n} \sum_{i=1}^{n} X_i$$

This sample mean has the properties

$$E(\overline{X}_n) = \mu$$

and

$$\text{Var}(\overline{X}_n) = \frac{\sigma^2}{n}$$

The central limit theorem says that, for a fixed number z,

$$P\left(\frac{\overline{X}_n - \mu}{\sigma/\sqrt{n}} \leq z\right) \to \Phi(z) \qquad \text{as } n \to \infty$$

where Φ is the cumulative distribution function of the standard normal distribution. Using a more compact and suggestive notation, we have

$$P\left(\frac{\overline{X}_n - \mu}{\sigma_{\overline{X}_n}} \leq z\right) \to \Phi(z)$$

The context of survey sampling is not exactly like that of the central limit theorem as stated above—as we have seen, in sampling without replacement, the X_i are not independent of each other, and it makes no sense to have n tend to infinity while N remains fixed. But other central limit theorems have been proved that are appropriate to the sampling context. These show that if n is large, but still small relative to N, then $\overline{X}_n$, the mean of a simple random sample, is approximately normally distributed.

To demonstrate the use of the central limit theorem, we will apply it to approximate $P(|\overline{X} - \mu| \leq \delta)$, the probability that the error made in estimating μ by $\overline{X}$ is less than some constant δ

$$P(|\overline{X} - \mu| \leq \delta) = P(-\delta \leq \overline{X} - \mu \leq \delta)$$

$$= P\left(-\frac{\delta}{\sigma_{\overline{X}}} \leq \frac{\overline{X} - \mu}{\sigma_{\overline{X}}} \leq \frac{\delta}{\sigma_{\overline{X}}}\right)$$

$$\approx \Phi\left(\frac{\delta}{\sigma_{\overline{X}}}\right) - \Phi\left(-\frac{\delta}{\sigma_{\overline{X}}}\right)$$

$$= 2\Phi\left(\frac{\delta}{\sigma_{\overline{X}}}\right) - 1$$

since $\Phi(-z) = 1 - \Phi(z)$, from the symmetry of the standard normal distribution about zero.

E X A M P L E **A**   Let us again consider the population of 393 hospitals. The standard deviation of the mean of a sample of size $n = 64$ is, using the finite population correction,

$$\sigma_{\overline{X}} = \frac{\sigma}{\sqrt{n}}\sqrt{1 - \frac{n-1}{N-1}}$$

$$= \frac{589.7}{8}\sqrt{1 - \frac{63}{392}} = 67.5$$

We can use the central limit theorem to approximate the probability that the sample mean differs from the population mean by more than 100 in absolute value; i.e.,

$P(|\overline{X} - \mu| > 100)$. First, from the symmetry of the normal distribution,

$$P(|\overline{X} - \mu| > 100) \approx 2P(\overline{X} - \mu > 100)$$

and

$$P(\overline{X} - \mu > 100) = 1 - P(\overline{X} - \mu < 100)$$

$$= 1 - P\left(\frac{\overline{X} - \mu}{\sigma_{\overline{X}}} < \frac{100}{\sigma_{\overline{X}}}\right)$$

$$\approx 1 - \Phi\left(\frac{100}{67.5}\right)$$

$$= .069$$

Thus the probability that the sample mean differs from the population mean by more than 100 is approximately .14. In fact, among the 500 samples of size 64 in Example A in Section 7.3.1, 82, or 16.4%, differed by more than 100 from the population mean. Similarly, the central limit theorem approximation gives .026 as the probability of deviations of more than 150 from the population mean. In the simulation in Example A in Section 7.3.1, 11 of 500, or 2.2%, differed by more than 150. If we are not too finicky, the central limit theorem gives us reasonable and useful approximations. ∎

E X A M P L E **B**   For a sample of size 50, the standard error of the sample mean number of discharges is

$$\sigma_{\overline{X}} = 78$$

For the particular sample of size 50 discussed in Example A in Section 7.3.2, we found $\overline{X} = 938.35$, so $\overline{X} - \mu = 123.9$. We now calculate an approximation of the probability of an error this large or larger:

$$P(|\overline{X} - \mu| \geq 123.9) = 1 - P(|\overline{X} - \mu| < 123.9)$$

$$\approx 1 - \left[2\Phi\left(\frac{123.9}{78}\right) - 1\right]$$

$$= 2 - 2\Phi(1.59)$$

$$= .11$$

Thus, we can expect an error this large or larger to occur about 11% of the time. ∎

E X A M P L E **C**   In Example C in Section 7.3.2, we found from the sample of size 50 an estimate $\hat{p} = .52$ of the proportion of hospitals that discharged fewer than 1000 patients; in fact, the actual proportion in the population is .65. Thus, $|\hat{p} - p| = .13$. What is the probability that an estimate will be off by an amount this large or larger?

We have

$$\sigma_{\hat{p}} = \sqrt{\frac{p(1-p)}{n}}\sqrt{1 - \frac{n-1}{N-1}}$$

$$= .068 \times .94 = .064$$

We can therefore calculate

$$P(|p - \hat{p}| > .13) = 1 - P(|p - \hat{p}| \leq .13)$$

$$= 1 - P\left(\frac{|p - \hat{p}|}{\sigma_{\hat{p}}} \leq \frac{.13}{\sigma_{\hat{p}}}\right)$$

$$\approx 2[1 - \Phi(2.03)] = .04$$

We see that the sample was rather "unlucky"—an error this large or larger would occur only about 4% of the time. ∎

We can now derive a **confidence interval** for the population mean, μ. A confidence interval for a population parameter, θ, is a random interval, calculated from the sample, that contains θ with some specified probability. For example, a 95% confidence interval for μ is a random interval that contains μ with probability .95; if we were to take many random samples and form a confidence interval from each one, about 95% of these intervals would contain μ. If the coverage probability is $1 - \alpha$, the interval is called a $100(1 - \alpha)\%$ confidence interval. Confidence intervals are frequently used in conjunction with point estimates to convey information about the uncertainty of the estimates.

For $0 \leq \alpha \leq 1$, let $z(\alpha)$ be that number such that the area under the standard normal density function to the right of $z(\alpha)$ is α (Figure 7.3). Note that the symmetry of the standard normal density function about zero implies that $z(1 - \alpha) = -z(\alpha)$. If Z follows a standard normal distribution, then, by definition of $z(\alpha)$,

$$P(-z(\alpha/2) \leq Z \leq z(\alpha/2)) = 1 - \alpha$$

From the central limit theorem, $(\overline{X} - \mu)/\sigma_{\overline{X}}$ has approximately a standard normal distribution, so

$$P\left(-z(\alpha/2) \leq \frac{\overline{X} - \mu}{\sigma_{\overline{X}}} \leq z(\alpha/2)\right) \approx 1 - \alpha$$

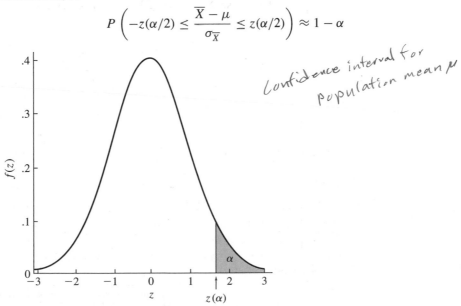

FIGURE 7.3   A standard normal density showing α and $z(\alpha)$.

Elementary manipulation of the inequalities gives

$$P(\overline{X} - z(\alpha/2)\sigma_{\overline{X}} \leq \mu \leq \overline{X} + z(\alpha/2)\sigma_{\overline{X}}) \approx 1 - \alpha$$

That is, the probability that μ lies in the interval $\overline{X} \pm z(\alpha/2)\sigma_{\overline{X}}$ is approximately $1 - \alpha$. The interval is thus called a $100(1 - \alpha)\%$ **confidence interval.** It is important to understand that this interval is random and that the preceding equation states that the probability that this random interval covers μ is $1 - \alpha$. In practice, α is assigned a small value, such as .1, .05, or .01, so that the probability that the interval covers μ will be large. Also, since the population variance is typically not known, $s_{\overline{X}}$ is substituted for $\sigma_{\overline{X}}$. For large samples, it can be shown that the effect of this substitution is practically negligible. It is impossible to give a precise answer to the question "How large is large?" As a rule of thumb, a value of n greater than 25 or 30 is usually adequate.

To illustrate the concept of a confidence interval, 20 samples each of size $n = 25$ were drawn from the population of hospital discharges. From each of these 20 samples, an approximate 95% confidence interval for μ, the mean number of discharges, was computed. These 20 confidence intervals are displayed as vertical lines in Figure 7.4; the dashed line in the figure is drawn at the true value, $\mu = 814.6$. Notice that it so

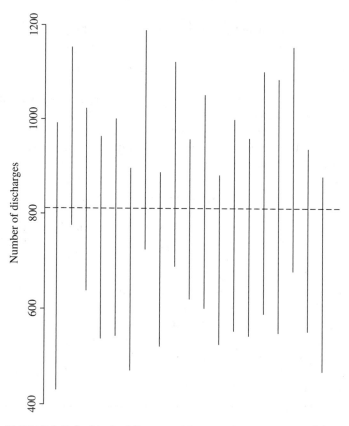

FIGURE 7.4 Vertical lines are 20 approximate 95% confidence intervals for μ. The horizontal line is the true value of μ.

happened that all the confidence intervals included μ; since these are 95% intervals, on the average 5%, or 1 out of 20, would not include μ.

The following example illustrates the procedure for calculating confidence intervals.

EXAMPLE **D** A particular area contains 8000 condominium units. In a survey of the occupants, a simple random sample of size 100 yields the information that the average number of motor vehicles per unit is 1.6, with a sample standard deviation of .8. The estimated standard error of $\overline{X}$ is thus

$$s_{\overline{X}} = \frac{s}{\sqrt{n}}\sqrt{1 - \frac{n}{N}}$$

$$= \frac{.8}{10}\sqrt{1 - \frac{100}{8000}}$$

$$= .08$$

Note that the finite population correction makes almost no difference. Since $z(.025) = 1.96$, a 95% confidence interval for the population average is $\overline{X} \pm 1.96s_{\overline{X}}$, or (1.44, 1.76).

An estimate of the total number of motor vehicles is $T = 8000 \times 1.6 = 12{,}800$. The estimated standard error of T is

$$s_T = Ns_{\overline{X}} = 640$$

A 95% confidence interval for the total number of motor vehicles is $T \pm 1.96s_T$, or (11,546, 14,054).

In the same survey, 12% of the respondents said they planned to sell their condos within the next year; $\hat{p} = .12$ is an estimate of the population proportion p. The estimated standard error is

$$s_{\hat{p}} = \sqrt{\frac{\hat{p}(1-\hat{p})}{n-1}}\sqrt{1 - \frac{100}{8000}} = .03$$

A 95% confidence interval for p is $\hat{p} \pm 1.96s_{\hat{p}}$, or (.06, .18).

The total number of owners planning to sell is estimated as $T = N\hat{p} = 960$. The estimated standard error of T is $s_T = Ns_{\hat{p}} = 240$. A 95% confidence interval for the number in the population planning to sell is $T \pm 1.96s_T$, or (490, 1430). The proper interpretation of this interval, (490, 1430), is a little subtle. We cannot state that the probability is 0.95 and that the number of owners planning to sell is between 490 and 1430, because that number is either in this interval or not. What is true is that 95% of intervals formed in this way will contain the true number in the long run. This interval is like one of those shown in Figure 7.4; in the long run, 95% of those intervals will contain the true number of discharges, but in the figure any particular interval either does or doesn't contain the true number. ∎

The width of a confidence interval is determined by the sample size n and the population standard deviation σ. If σ is known approximately, perhaps from earlier samples of the population, n can be chosen so as to obtain a confidence interval close to some desired length. Such analysis is usually an important aspect of planning the design of a sample survey.

E X A M P L E **E**   The interval for the total number of owners planning to sell in Example D might be considered too wide for practical purposes; reducing its width would require a larger sample size. Suppose that an interval with a half-width of 200 is desired. Neglecting the finite population correction, the half-width is

$$1.96s_T = 1.96N\sqrt{\frac{\hat{p}(1-\hat{p})}{n-1}} = \frac{5095}{\sqrt{n-1}}$$

Setting the last expression equal to 200 and solving for n yields $n = 650$ as the necessary sample size.   ■

Let us summarize: The fundamental result of this section is that the sampling distribution of the sample mean is approximately Gaussian. This approximation can be used to quantify the error committed in estimating the population mean by the sample mean, thus giving us a good understanding of the accuracy of estimates produced by a simple random sample. We next introduced the idea of a confidence interval, a random interval that contains a population parameter with a specified probability and thus provides an assessment of the accuracy of the corresponding estimate of that parameter. We have seen in our examples that the width of the confidence interval is a multiple of the estimated standard deviation of the estimate; for example, a confidence interval for μ is $\overline{X} \pm ks_{\overline{X}}$, where the constant k depends on the coverage probability of the interval.

7.4 Estimation of a Ratio

The foundations of the theory of survey sampling have been laid in the preceding sections on simple random sampling. This and the next section build on that foundation, developing some advanced topics in survey sampling.

In this section, we consider the estimation of a ratio. Suppose that for each member of a population, two values, x and y, may be measured. The ratio of interest is

$$r = \frac{\sum\limits_{i=1}^{N} y_i}{\sum\limits_{i=1}^{N} x_i} = \frac{\mu_y}{\mu_x}$$

Ratios arise frequently in sample surveys; for example, if households are sampled, the following ratios might be calculated:

- If y is the number of unemployed males aged 20–30 in a household and x is the number of males aged 20–30 in a household, then r is the proportion of unemployed males aged 20–30.

- If y is weekly food expenditure and x is number of inhabitants, then r is weekly food cost per inhabitant.
- If y is the number of motor vehicles and x is the number of inhabitants of driving age, then r is the number of motor vehicles per inhabitant of driving age.

In a survey of farms, y might be the acres of wheat planted and x the total acreage. In an inventory audit, y might be the audited value of an item and x the book value.

In this section, we first consider directly the problem of estimating a ratio. Later, we will use the estimation of a ratio as a technique for estimating μ_y. We will produce a new estimate, the ratio estimate, which we will compare to the ordinary estimate, $\overline{Y}$.

Before continuing, we note the elementary but sometimes overlooked fact that

$$r \neq \frac{1}{N} \sum_{i=1}^{N} \frac{y_i}{x_i}$$

Suppose that a sample is drawn consisting of the pairs (X_i, Y_i); the natural estimate of r is $R = \overline{Y}/\overline{X}$. We wish to derive expressions for $E(R)$ and $\mathrm{Var}(R)$, but since R is a nonlinear function of the random variables $\overline{X}$ and $\overline{Y}$, we cannot do this in closed form. We will therefore employ the approximate methods of Section 4.6.

In order to calculate the approximate variance of R, we need to know $\mathrm{Var}(\overline{X})$, $\mathrm{Var}(\overline{Y})$, and $\mathrm{Cov}(\overline{X}, \overline{Y})$. The first two quantities we know from Theorem B of Section 7.3.1. For the last quantity, we define the **population covariance** of x and y to be

$$\sigma_{xy} = \frac{1}{N} \sum_{i=1}^{N} (x_i - \mu_x)(y_i - \mu_y)$$

It can then be shown, in a manner entirely analogous to the proof of Theorem B in Section 7.3.1, that

$$\mathrm{Cov}(\overline{X}, \overline{Y}) = \frac{\sigma_{xy}}{n} \left(1 - \frac{n-1}{N-1}\right)$$

From Example C in Section 4.6, we have the following theorem.

THEOREM A

With simple random sampling, the approximate variance of $R = \overline{Y}/\overline{X}$ is

$$\mathrm{Var}(R) \approx \frac{1}{\mu_x^2} \left(r^2 \sigma_{\overline{X}}^2 + \sigma_{\overline{Y}}^2 - 2r\sigma_{\overline{XY}}\right)$$

$$= \frac{1}{n} \left(1 - \frac{n-1}{N-1}\right) \frac{1}{\mu_x^2} \left(r^2 \sigma_x^2 + \sigma_y^2 - 2r\sigma_{xy}\right) \qquad \blacksquare$$

The **population correlation coefficient** is defined as

$$\rho = \frac{\sigma_{xy}}{\sigma_x \sigma_y}$$

and is used as a measure of the strength of the linear relationship between the x and y values in the population. It can be shown that $-1 \leq \rho \leq 1$; large values of ρ

indicate a strong positive relationship between x and y, and small values indicate a strong negative relationship. (See Figure 4.7 for some illustrations of correlation.) The equation in Theorem A can be expressed in terms of the population correlation coefficient as follows:

$$\text{Var}(R) \approx \frac{1}{n} \left(1 - \frac{n-1}{N-1} \right) \frac{1}{\mu_x^2} \left(r^2 \sigma_x^2 + \sigma_y^2 - 2r\rho\sigma_x\sigma_y \right)$$

From this expression, we see that strong correlation of the same sign as r decreases the variance. We also note that the variance is affected by the size of μ_x—if μ_x is small, the variance is large, essentially because small values of $\overline{X}$ in the ratio $R = \overline{Y}/\overline{X}$ cause R to fluctuate wildly.

We now consider the approximate expectation of R. From Example C in Section 4.6 and the preceding calculations, we have the following theorem.

THEOREM B

With simple random sampling, the expectation of R is given approximately by

$$E(R) \approx r + \frac{1}{n} \left(1 - \frac{n-1}{N-1} \right) \frac{1}{\mu_x^2} \left(r\sigma_x^2 - \rho\sigma_x\sigma_y \right) \qquad \blacksquare$$

From the equation in Theorem B, we see that strong correlation of the same sign as r decreases the bias and that the bias is large if μ_x is small. Furthermore, note that the bias is of the order $1/n$, so its contribution to the mean squared error is of the order $1/n^2$. In comparison, the contribution of the variance is of the order $1/n$. Therefore, for large samples, the bias is negligible compared to the standard error of the estimate.

For large samples, truncating the Taylor series after the linear term provides a good approximation, since the deviations $\overline{X} - \mu_X$ and $\overline{Y} - \mu_Y$ are likely to be small. To this order of approximation, R is expressed as a linear combination of $\overline{X}$ and $\overline{Y}$, and an argument based on the central limit theorem can be used to show that R is approximately normally distributed. Approximate confidence intervals can thus be formed for r by using the normal distribution.

In order to estimate the standard error of R, we substitute R for r in the formula of Theorem A. The x and y population variances are estimated by s_x^2 and s_y^2. The population covariance is estimated by

$$s_{xy} = \frac{1}{n-1} \sum_{i=1}^{n} (X_i - \overline{X})(Y_i - \overline{Y})$$

$$= \frac{1}{n-1} \left(\sum_{i=1}^{n} X_i Y_i - n \overline{X}\overline{Y} \right)$$

(as can be seen by expanding the product), and the population correlation is estimated by

$$\hat{\rho} = \frac{s_{xy}}{s_x s_y}$$

The estimated variance of R is thus

$$s_R^2 = \frac{1}{n}\left(1 - \frac{n-1}{N-1}\right)\frac{1}{\overline{X}^2}(R^2 s_x^2 + s_y^2 - 2Rs_{xy})$$

An approximate $100(1-\alpha)\%$ confidence interval for r is $R \pm z(\alpha/2)s_R$.

EXAMPLE A   Suppose that 100 people who recently bought houses are surveyed, and the monthly mortgage payment and gross income of each buyer are determined. Let y denote the mortgage payment and x the gross income. Suppose that

$$\overline{X} = \$3100 \qquad \overline{Y} = \$868$$
$$s_y = \$250 \qquad s_x = \$1200$$
$$\hat{\rho} = .85 \qquad R = .28$$

Neglecting the finite population correction, the estimated standard error of R is

$$s_R = \frac{1}{10}\left(\frac{1}{3100}\right)\sqrt{.28^2 \times 1200^2 + 250^2 - 2 \times .28 \times .85 \times 250 \times 1200}$$
$$= .006$$

An approximate 95% confidence interval for r is $.28 \pm (1.96) \times (.006)$, or $.28 \pm .012$. Note that the high correlation between x and y causes the standard error of R to be small. We can use the observed values for the variances, covariances, and means to gauge the order of magnitude of the bias by substituting them in place of the population parameters in the formula of Theorem B. Doing so, and again neglecting the finite population correction, gives the value .00015 for the bias, which is negligible relative to s_R. Note that the large value of $\overline{X}$ and the large positive correlation coefficient cause the bias to be small. ∎

Ratios may also be used as tools for estimating population means and totals. To illustrate the concept, we return to the example of hospital discharges. For this population, the number of beds in each hospital is also known; let us denote the number of beds in the ith hospital by x_i and the number of discharges by y_i. Suppose that all the x_i are known, perhaps from an earlier enumeration, before a sample has been taken to estimate the number of discharges, and that we would like to take advantage of this information. One way to do this is to form a **ratio estimate** of μ_y:

$$\overline{Y}_R = \frac{\mu_x}{\overline{X}}\overline{Y} = \mu_x R$$

where $\overline{X}$ is the average number of beds and $\overline{Y}$ is the average number of discharges in the sample. The idea is fairly simple: We expect x_i and y_i to be closely related in the population, since a hospital with a large number of beds should tend to have a large number of discharges. This is borne out by Figure 7.5, a scatterplot of the number of discharges versus the number of beds. If $\overline{X} < \mu_x$, the sample underestimates the number of beds and probably the number of discharges as well; multiplying $\overline{Y}$ by $\mu_x/\overline{X}$ increases $\overline{Y}$ to $\overline{Y}_R$.

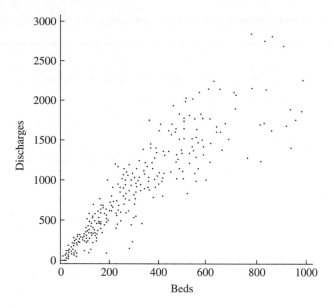

F I G U R E **7.5** Scatterplot of the number of discharges versus the number of beds for the 393 hospitals.

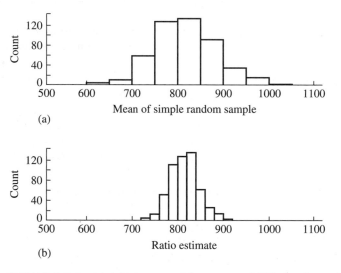

F I G U R E **7.6** (a) A histogram of the means of 500 simple random samples of size 64 from the population of discharges; (b) a histogram of the values of 500 ratio estimates of the mean number of discharges from samples of size 64.

To see how this ratio estimate works in practice, it was simulated from 500 samples of size 64 from the population of hospitals. The histogram of the results is shown in Figure 7.6 along with the histogram of the means of 500 simple random samples of size 64. The comparison shows dramatically how effective the ratio estimate is at reducing variability.

Two more examples will illustrate the scope of the ratio estimation method.

E X A M P L E **B**   Suppose that we want to estimate the total number of unemployed males aged 20–30 from a sample of households and that we know τ_x, the total number of males aged 20–30, from census data. The ratio estimate is

$$T_R = \tau_x \frac{\overline{Y}}{\overline{X}}$$

where $\overline{Y}$ is the average number of unemployed males aged 20–30 per household in the sample, and $\overline{X}$ is the sample average number of males aged 20–30 per household. ■

E X A M P L E **C**   A sample of items in an inventory is taken to estimate the total value of the inventory. Let Y_i be the audited value of the ith sample item, and let X_i be its book value. We assume that τ_x, the total book value of the inventory, is known, and we estimate the total audited value by

$$T_R = \tau_x \frac{\overline{Y}}{\overline{X}}$$ ■

We will now analyze the observed success of the ratio estimate. Since $\overline{Y}_R = \mu_X R$, $\text{Var}(\overline{Y}_R) = \mu_X^2 \text{Var}(R)$. From Theorem A, we thus have the following.

COROLLARY **A**

The approximate variance of the ratio estimate of μ_y is

$$\text{Var}(\overline{Y}_R) \approx \frac{1}{n}\left(1 - \frac{n-1}{N-1}\right)\left(r^2\sigma_x^2 + \sigma_y^2 - 2r\rho\sigma_x\sigma_y\right)$$ ■

Similarly, from Theorem B, we have another corollary.

COROLLARY **B**

The approximate bias of the ratio estimate of μ_y is

$$E(\overline{Y}_R) - \mu_Y \approx \frac{1}{n}\left(1 - \frac{n-1}{N-1}\right)\frac{1}{\mu_x}\left(r\sigma_x^2 - \rho\sigma_x\sigma_y\right)$$ ■

When will the ratio estimate Y_R be better than the ordinary estimate $\overline{Y}$? In the following, the finite population correction is neglected for simplicity. Since the variance of the ordinary estimate $\overline{Y}$ is

$$\text{Var}(\overline{Y}) = \frac{\sigma_y^2}{n}$$

the ratio estimate has a smaller variance if

$$r^2\sigma_x^2 - 2r\rho\sigma_x\sigma_y < 0$$

or (provided $r > 0$, for example)

$$2\rho\sigma_y > r\sigma_x$$

Letting $C_x = \sigma_x/\mu_x$ and $C_y = \sigma_y/\mu_y$, this last inequality is equivalent to

$$\rho > \frac{1}{2}\left(\frac{C_x}{C_y}\right)$$

C_x and C_y are called **coefficients of variation** and give the standard deviation as a proportion of the mean. (Coefficients of variation are often more meaningful than standard deviations. For example, a standard deviation of 10 means one thing if the true value of the quantity being measured is 100 and something entirely different if the true value is 10,000.)

In order to assess the accuracy of $\overline{Y}_R$, $\text{Var}(\overline{Y}_R)$ can be estimated from the sample.

COROLLARY C

The variance of $\overline{Y}_R$ can be estimated by

$$s_{\overline{Y}_R}^2 = \frac{1}{n}\left(1 - \frac{n-1}{N-1}\right)\left(R^2 s_x^2 + s_y^2 - 2R s_{xy}\right)$$

and an approximate $100(1-\alpha)\%$ confidence interval for μ_y is $(\overline{Y}_R \pm z(\frac{\alpha}{2})s_{\overline{Y}_R})$. ■

E X A M P L E **D**   For the population of 393 hospitals, we have

$$\begin{array}{ll} \mu_x = 274.8 & \sigma_x = 213.2 \\ \mu_y = 814.6 & \sigma_y = 589.7 \\ r\ \ = 2.96 & \rho\ \ = .91 \end{array}$$

Thus,

$$\begin{aligned} \text{Var}(\overline{Y}_R) &\approx \frac{1}{n}(2.96^2 \times 213.2^2 + 589.7^2 - 2 \times 2.96 \times .91 \times 213.2 \times 589.7) \\ &= \frac{68,697.4}{n} \end{aligned}$$

and

$$\sigma_{\overline{Y}_R} \approx \frac{262.1}{\sqrt{n}}$$

Including the finite population correction, the linearized approximation predicts that, with $n = 64$,

$$\sigma_{\overline{Y}_R} = \frac{1}{8}(262.1)\sqrt{1 - \frac{63}{392}} = 30.0$$

The actual standard deviation of the 500 sample values displayed in Figure 7.6 is 29.9, which is remarkably close. The mean of the 500 values is 816.2, compared to the population mean of 814.6; the slight apparent bias is consistent with Corollary B.

In contrast, the standard deviation of $\overline{Y}$ from a simple random sample of size $n = 64$ is

$$\sigma_{\overline{Y}} = \frac{\sigma}{\sqrt{n}}\sqrt{1 - \frac{n-1}{N-1}}$$

$$= \frac{589.7}{8}\sqrt{1 - \frac{63}{392}}$$

$$= 66.3$$

The comparison of $\sigma_{\overline{Y}}$ to $\sigma_{\overline{Y}_R}$ is consistent with the substantial reduction in variability accomplished by using a ratio estimate of μ_y shown in Figure 7.6.

The following is another way of interpreting this comparison. If a simple random sample of size n_1 is taken, the variance of the estimate is $\mathrm{Var}(\overline{Y}) = 589.7^2/n_1$. A ratio estimate from a sample of size n_2 will have the same variance if

$$\frac{262.1^2}{n_2} = \frac{589.7^2}{n_1}$$

or

$$n_2 = n_1\left(\frac{262.1}{589.7}\right)^2 = .1975n_1$$

Thus, in this case, we can obtain the same precision from a ratio estimate *using a sample about 80% smaller* than the simple random sample. Note that this comparison neglects the bias of the ratio estimate, which is justifiable in this case because the bias is quite small. Here is a case in which a biased estimate performs substantially better than an unbiased estimate, the bias being quite small and the reduction in variance being quite large. ∎

7.5 Stratified Random Sampling

7.5.1 Introduction and Notation

In stratified random sampling, the population is partitioned into subpopulations, or **strata,** which are then independently sampled. The results from the strata are then combined to estimate population parameters, such as the mean.

Following are some examples that suggest the range of situations in which stratification is natural:

- In auditing financial transactions, the transactions may be grouped into strata on the basis of their nominal values. For example, high-value, medium-value, and low-value strata might be formed.
- In samples of human populations, geographical areas often form natural strata.
- In a study of records of shipments of household goods by motor carriers, the carriers were grouped into three strata: large carriers, medium carriers, and small carriers.

Stratified samples are used for a variety of reasons. We are often interested in obtaining information about each of a number of natural subpopulations in addition to information about the population as a whole. The subpopulations might be defined by geographical areas or age groups. In an industrial application in which the population consists of items produced by a manufacturing process, relevant subpopulations might consist of items produced during different shifts or from different lots of raw material. The use of a stratified random sample guarantees a prescribed number of observations from each subpopulation, whereas the use of a simple random sample can result in underrepresentation of some subpopulations. A second reason for using stratification is that, as will be shown below, the stratified sample mean can be considerably more precise than the mean of a simple random sample, especially if the population members within each stratum are relatively homogeneous and if there is considerable variation between strata.

In the next section, properties of the stratified sample mean are derived. Since a simple random sample is taken within each stratum, the results will follow easily from the derivations of earlier sections. The section after that takes up the problem of how to allocate the total number of observations, n, among the various strata. Comparisons will be made of the efficiencies of different allocation schemes and also of the precisions of these allocation schemes relative to that of a simple random sample of the same total size.

7.5.2   Properties of Stratified Estimates

Suppose there are L strata in all. Let the number of population elements in stratum 1 be denoted by N_1, the number in stratum 2 be N_2, etc. The total population size is $N = N_1 + N_2 + \ldots + N_L$. The population mean and variance of the lth stratum are denoted by μ_l and σ_l^2. The overall population mean can be expressed in terms of the μ_l as follows. Let x_{il} denote the ith population value in the lth stratum and let $W_l = N_l/N$ denote the fraction of the population in the lth stratum. Then

$$\mu = \frac{1}{N} \sum_{l=1}^{L} \sum_{i=1}^{N_l} x_{il}$$

$$= \frac{1}{N} \sum_{l=1}^{L} N_l \mu_l$$

$$= \sum_{l=1}^{L} W_l \mu_l$$

Within each stratum, a simple random sample of size n_l is taken. The sample mean in stratum l is denoted by

$$\overline{X}_l = \frac{1}{n_l} \sum_{i=1}^{n_l} X_{il}$$

Here X_{il} denotes the ith sample value in the lth stratum. Note that $\overline{X}_l$ is the mean of a simple random sample from the population consisting of the lth stratum, so from Theorem A of Section 7.3.1, $E(\overline{X}_l) = \mu_l$. By analogy with the preceding relationship between the overall population mean and the population means of the various strata, the obvious estimate of μ is

$$\overline{X}_s = \sum_{l=1}^{L} \frac{N_l \overline{X}_l}{N}$$

$$= \sum_{l=1}^{L} W_l \overline{X}_l$$

THEOREM A

The stratified estimate, $\overline{X}_s$, of the population mean is unbiased.

Proof

$$E(\overline{X}_s) = \sum_{l=1}^{L} W_l E(\overline{X}_l)$$

$$= \frac{1}{N} \sum_{l=1}^{L} N_l \mu_l$$

$$= \mu$$

Since we assume that the samples from different strata are independent of one another and that within each stratum a simple random sample is taken, the variance of $\overline{X}_s$ can be easily calculated.

THEOREM B

The variance of the stratified sample mean is given by

$$\text{Var}(\overline{X}_s) = \sum_{l=1}^{L} W_l^2 \left(\frac{1}{n_l}\right) \left(1 - \frac{n_l - 1}{N_l - 1}\right) \sigma_l^2$$

Proof

Since the $\overline{X}_l$ are independent,

$$\text{Var}(\overline{X}_s) = \sum_{l=1}^{L} W_l^2 \text{Var}(\overline{X}_l)$$

From Theorem B of Section 7.3.1, we have

$$\text{Var}(\overline{X}_l) = \frac{1}{n_l} \left(1 - \frac{n_l - 1}{N_l - 1}\right) \sigma_l^2$$

Therefore, the desired result follows.

If the sampling fractions within all strata are small,

$$\text{Var}(\overline{X}_s) \approx \sum_{l=1}^{L} \frac{W_l^2 \sigma_l^2}{n_l}$$

E X A M P L E **A**   We again consider the population of hospitals. As we did in the discussion of ratio estimates, we assume that the number of beds in each hospital is known but that the number of discharges is not. We will try to make use of this knowledge by stratifying the hospitals according to the number of beds. Let stratum A consist of the 98 smallest hospitals, stratum B of the 98 next larger, stratum C of the 98 next larger, and stratum D of the 99 largest. The following table shows the results of this stratification of hospitals by size:

Stratum	N_l	W_l	μ_l	σ_l
A	98	.249	182.9	103.4
B	98	.249	526.5	204.8
C	98	.249	956.3	243.5
D	99	.251	1591.2	419.2

Suppose that we use a sample of total size n and let

$$n_1 = n_2 = n_3 = n_4 = \frac{n}{4}$$

so that we have equal sample sizes in each stratum. Then, from Theorem B, neglecting the finite population corrections and using the numerical values in the preceding table, we have

$$\text{Var}(\overline{X}_s) = \sum_{l=1}^{4} \frac{W_l^2 \sigma_l^2}{n_1}$$

$$= \frac{4}{n} \sum_{l=1}^{4} W_l^2 \sigma_l^2$$

$$= \frac{72,042.6}{n}$$

and

$$\sigma_{\overline{X}_s} = \frac{268.4}{\sqrt{n}}$$

The standard deviation of the mean of a simple random sample is

$$\sigma_{\overline{X}} = \frac{589.7}{\sqrt{n}}$$

Comparing the two standard deviations, we see that a tremendous gain in precision has resulted from the stratification. The ratio of the variances is .20; thus a stratified estimate based on a total sample size of $n/5$ is as precise as a simple random sample of size n. The reduction in variance due to stratification is comparable to that achieved by using a ratio estimate (Example D in Section 7.4). In later parts of this section, we will look more analytically at why the stratification done here produced such dramatic improvement. ∎

Let us next consider the stratified estimate of the population total, $T_s = N\overline{X}_s$. From Theorem B, we have the following corollary.

COROLLARY **A**

The expectation and variance of the stratified estimate of the population total are

$$E(T_s) = \tau$$

and

$$\text{Var}(T_s) = N^2 \text{Var}(\overline{X}_s)$$
$$= \sum_{l=1}^{L} N_l^2 \left(\frac{1}{n_l}\right) \left(1 - \frac{n_l - 1}{N_l - 1}\right) \sigma_l^2 \quad ∎$$

In order to estimate the standard errors of $\overline{X}_s$ and T_s, the variances of the individual strata must be separately estimated and substituted into the preceding formulae. The estimate of σ_l^2 is given by

$$s_l^2 = \frac{1}{n_l - 1} \sum_{i=1}^{n_l} (X_{il} - \overline{X}_l)^2$$

$\text{Var}(\overline{X}_s)$ is estimated by

$$s_{\overline{X}_s}^2 = \sum_{l=1}^{L} W_l^2 \left(\frac{1}{n_l}\right) \left(1 - \frac{n_l}{N_l}\right) s_l^2$$

The next example illustrates how this variance estimate can be used to find approximate confidence intervals for μ based on $\overline{X}_s$.

EXAMPLE **B**   A sample of size 10 was drawn from each of the four strata of hospitals described in Example A, yielding the following:

$$\overline{X}_1 = 240.6 \qquad s_1^2 = 6827.6$$
$$\overline{X}_2 = 507.4 \qquad s_2^2 = 23{,}790.7$$
$$\overline{X}_3 = 865.1 \qquad s_3^2 = 42{,}573.0$$
$$\overline{X}_4 = 1716.5 \qquad s_4^2 = 152{,}099.6$$

Therefore, $\overline{X}_s = 832.5$. The variance of the stratified sample mean is estimated by

$$s_{\overline{X}_s}^2 = \frac{1}{10} \sum_{l=1}^{4} W_l^2 \left(1 - \frac{n_l - 1}{N_l - 1}\right) s_l^2$$

$$= 1282.0$$

Thus,

$$s_{\overline{X}_s} = 35.8$$

An approximate 95% confidence interval for the population mean number of discharges is $\overline{X}_s \pm 1.96 s_{\overline{x}_s}$, or (762.4, 902.7).

The total number of discharges is estimated by $T_s = 393\overline{X}_s = 327{,}172$. The standard error of T_s is estimated by $s_{T_s} = 393 s_{\overline{X}_s} = 14{,}069$. An approximate 95% confidence interval for the population total is $T_s \pm 1.96 s_{T_s}$, or (299,596, 354, 748). ∎

7.5.3 Methods of Allocation

In Section 7.5.2, it was shown that, neglecting the finite population correction,

$$\text{Var}(\overline{X}_s) = \sum_{l=1}^{L} \frac{W_l^2 \sigma_l^2}{n_l}$$

If the resources of a survey allow only a total of n units to be sampled, the question arises of how to choose $n_1, \ldots, n_L$ to minimize $\text{Var}(\overline{X}_s)$ subject to the constraint $n_1 + \cdots + n_L = n$.

For the sake of simplicity, the calculations in this section ignore the finite population correction within each stratum. The analysis may be extended to include these corrections, but at the cost of some additional algebra. More complete results are contained in Cochran (1977).

THEOREM A

The sample sizes $n_1, \ldots, n_L$ that minimize $\text{Var}(\overline{X}_s)$ subject to the constraint $n_1 + \cdots + n_L = n$ are given by

$$n_l = n \frac{W_l \sigma_l}{\displaystyle\sum_{k=1}^{L} W_k \sigma_k}$$

where $l = 1, \ldots, L$.

Proof

We introduce a Lagrange multiplier, and we must then minimize

$$L(n_1, \ldots, n_L, \lambda) = \sum_{l=1}^{L} \frac{W_l^2 \sigma_l^2}{n_l} + \lambda \left(\sum_{l=1}^{L} n_l - n \right)$$

For $l = 1, \ldots, L$, we have

$$\frac{\partial L}{\partial n_l} = -\frac{W_l^2 \sigma_l^2}{n_l^2} + \lambda$$

Setting these partial derivatives equal to zero, we have the system of equations

$$n_l = \frac{W_l \sigma_l}{\sqrt{\lambda}}$$

for $l = 1, \ldots, L$. To determine λ, we first sum these equations over l:

$$n = \frac{1}{\sqrt{\lambda}} \sum_{l=1}^{L} W_l \sigma_l$$

Thus,

$$\frac{1}{\sqrt{\lambda}} = \frac{n}{\sum_{l=1}^{L} W_l \sigma_l}$$

and

$$n_l = n \frac{W_l \sigma_l}{\sum_{l=1}^{L} W_l \sigma_l}$$

which proves the theorem. ∎

This theorem shows that those strata for which $W_l \sigma_l$ is large should be sampled heavily. This makes sense intuitively. If W_l is large, the stratum contains a large fraction of the population; if σ_l is large, the population values in the stratum are quite variable, and in order to obtain a good determination of the stratum's mean, a relatively large sample size must be used. This optimal allocation scheme is called **Neyman allocation.**

Substituting the optimal values of n_l as given in Theorem A into the equation for $\text{Var}(\overline{X}_s)$ given in Theorem B in Section 7.5.2 gives us the following corollary.

COROLLARY **A**

Denoting by $\overline{X}_{so}$, the stratified estimate using the optimal allocations as given in Theorem A and neglecting the finite population correction,

$$\text{Var}(\overline{X}_{so}) = \frac{\left(\sum_{l=1}^{L} W_l \sigma_l \right)^2}{n}$$

∎

EXAMPLE A   For the population of hospitals, the weights for optimal allocation, $W_l \sigma_l / \sum W_l \sigma_l$, are, from the table of Example A of Section 7.5.2,

	Stratum			
	A	B	C	D
Weight	.106	.210	.250	.434

Note that, because of its larger standard deviation, stratum D is sampled more than four times as heavily as stratum A.   ■

The optimal allocations depend on the individual variances of the strata, which generally will not be known. Furthermore, if a survey measures several attributes for each population member, it is usually impossible to find an allocation that is simultaneously optimal for all of those variables. A simple and popular alternative method of allocation is to use the same sampling fraction in each stratum,

$$\frac{n_1}{N_1} = \frac{n_2}{N_2} = \cdots = \frac{n_L}{N_L}$$

which holds if

$$n_l = n \frac{N_l}{N} = n W_l$$

for $l = 1, \ldots, L$. This method is called **proportional allocation.** The estimate of the population mean based on proportional allocation is

$$\overline{X}_{sp} = \sum_{l=1}^{L} W_l \overline{X}_l$$

$$= \sum_{l=1}^{L} W_l \frac{1}{n_l} \sum_{i=1}^{n_l} X_{il}$$

$$= \frac{1}{n} \sum_{l=1}^{L} \sum_{i=1}^{n_l} X_{il}$$

since $W_l / n_l = 1/n$. This estimate is simply the unweighted mean of the sample values.

THEOREM **B**

With stratified sampling based on proportional allocation, ignoring the finite population correction,

$$\text{Var}(\overline{X}_{sp}) = \frac{1}{n} \sum_{l=1}^{L} W_l \sigma_l^2$$

Proof

From Theorem B of Section 7.5.2, we have

$$\text{Var}(\overline{X}_{sp}) = \sum_{l=1}^{L} W_l^2 \text{Var}(\overline{X}_l)$$

$$= \sum_{l=1}^{L} W_l^2 \frac{\sigma_l^2}{n_l}$$

Using $n_l = nW_l$, the result follows.   ∎

We now compare $\text{Var}(\overline{X}_{sp})$ and $\text{Var}(\overline{X}_{so})$ in order to discover the circumstances under which optimal allocation is substantially better than proportional allocation.

THEOREM C

With stratified random sampling, the difference between the variance of the estimate of the population mean based on proportional allocation and the variance of that estimate based on optimal allocation is, ignoring the finite population correction,

$$\text{Var}(\overline{X}_{sp}) - \text{Var}(\overline{X}_{so}) = \frac{1}{n} \sum_{l=1}^{L} W_l (\sigma_l - \bar{\sigma})^2$$

where

$$\bar{\sigma} = \sum_{l=1}^{L} W_l \sigma_l$$

Proof

$$\text{Var}(\overline{X}_{sp}) - \text{Var}(\overline{X}_{so}) = \frac{1}{n} \left[\sum_{l=1}^{L} W_l \sigma_l^2 - \left(\sum_{l=1}^{L} W_l \sigma_l \right)^2 \right]$$

The term within the large brackets equals $\sum_{l=1}^{L} W_l (\sigma_l - \bar{\sigma})^2$, which may be verified by expanding the square and collecting terms.   ∎

According to Theorem C, if the variances of the strata are all the same, proportional allocation yields the same results as optimal allocation. The more variable these variances are, the better it is to use optimal allocation.

E X A M P L E **B**   Let us calculate how much better optimal allocation is than proportional allocation for the population of hospitals. From Theorem C and Corollary A, we have

$$\text{Var}(\overline{X}_{sp}) = \text{Var}(\overline{X}_{so}) + \frac{1}{n}\sum W_l(\sigma_l - \bar{\sigma})^2$$

Therefore,

$$\frac{\text{Var}(\overline{X}_{sp})}{\text{Var}(\overline{X}_{so})} = 1 + \frac{\frac{1}{n}\sum W_l(\sigma_l - \bar{\sigma})^2}{\text{Var}(\overline{X}_{so})}$$

$$= 1 + \frac{\sum W_l(\sigma_l - \bar{\sigma})^2}{(\sum W_l\sigma_l)^2}$$

$$= 1 + .218$$

Thus, under proportional allocation, the variance of the mean is about 20% larger than it is under optimal allocation. ∎

We can also compare the variance under simple random sampling with the variance under proportional allocation. The variance under simple random sampling is, neglecting the finite population correction,

$$\text{Var}(\overline{X}) = \frac{\sigma^2}{n}$$

In order to compare this equation with that for the variance under proportional allocation, we need a relationship between the overall population variance, σ^2, and the strata variances, σ_l^2. The overall population variance may be expressed as

$$\sigma^2 = \frac{1}{N}\sum_{l=1}^{L}\sum_{i=1}^{N_l}(x_{il} - \mu)^2$$

Also,

$$(x_{il} - \mu)^2 = [(x_{il} - \mu_l) + (\mu_l - \mu)]^2$$
$$= (x_{il} - \mu_l)^2 + 2(x_{il} - \mu_l)(\mu_l - \mu) + (\mu_l - \mu)^2$$

When both sides of this last equation are summed over l, the middle term on the right-hand side becomes zero since $N_l\mu_l = \sum_{i=1}^{N_l} x_{il}$, so we have

$$\sum_{i=1}^{N_l}(x_{il} - \mu)^2 = \sum_{i=1}^{N_l}(x_{il} - \mu_l)^2 + N_l(\mu_l - \mu)^2$$
$$= N_l\sigma_l^2 + N_l(\mu_l - \mu)^2$$

Dividing both sides by N and summing over l, we have

$$\sigma^2 = \sum_{l=1}^{L} W_l\sigma_l^2 + \sum_{l=1}^{L} W_l(\mu_l - \mu)^2$$

Substituting this expression for σ^2 into $\mathrm{Var}(\overline{X}) = \sigma^2/n$ and using the formula for $\mathrm{Var}(\overline{X}_{sp})$ given in Theorem B completes a proof of the following theorem.

THEOREM D

The difference between the variance of the mean of a simple random sample and the variance of the mean of a stratified random sample based on proportional allocation is, neglecting the finite population correction,

$$\mathrm{Var}(\overline{X}) - \mathrm{Var}(\overline{X}_{sp}) = \frac{1}{n}\sum_{l=1}^{L} W_l(\mu_l - \mu)^2 \qquad \blacksquare$$

Thus, stratified random sampling with proportional allocation always gives a smaller variance than does simple random sampling, providing that the finite population correction is ignored. Comparing the equations for the variances under simple random sampling, proportional allocation, and optimal allocation, we see that stratification with proportional allocation is better than simple random sampling if the strata means are quite variable and that stratification with optimal allocation is even better than stratification with proportional allocation if the strata standard deviations are variable.

E X A M P L E C   We calculate the improvement that would result from using stratification with proportional allocation rather than simple random sampling for the population of hospitals. From Theorems B and D, we have

$$\frac{\mathrm{Var}(\overline{X}_{srs})}{\mathrm{Var}(\overline{X}_{sp})} = 1 + \frac{\sum W_l(\mu_l - \bar{\mu})^2}{\sum W_l\sigma_l^2}$$

$$= 1 + 3.83$$

As is frequently the case, the gain from using stratification with proportional allocation rather than simple random sampling is much greater than the gain from using optimal allocation rather than proportional allocation. Furthermore, proportional allocation requires knowledge only of the sizes of the strata, whereas optimal allocation requires knowledge of the standard deviations of the strata, and such knowledge is usually unavailable.   ■

Typically, stratified random sampling can result in substantial increases in precision for populations containing values that vary greatly in size. For example, a population of transactions, a sample of which is to be audited for errors, might contain transactions in the hundreds of thousands of dollars and transactions in the hundreds of dollars. If such a population were divided into several strata according to the dollar amounts of the transactions, there might well be considerable variation in the mean transaction errors between the strata, since there may be rather large errors on large transactions and small errors on small transactions. The variability of the errors might also be larger in the former strata as well.

We have not addressed the question of how many strata to form and how to define the strata. In order to construct the optimal number of strata, the population values themselves, which are of course unknown, would have to be used. Stratification must therefore be done on the basis of some related variable that is known (such as transaction amount in the preceding paragraph) or on the results of earlier samples. In practice, it usually turns out that such relationships are not strong enough to make it worthwhile constructing more than a few strata.

7.6 Concluding Remarks

This chapter introduced survey sampling. It first covered the most elementary method of probability sampling—simple random sampling. The theory of this method underlies the theory of more complex sampling techniques. Stratified sampling was also introduced and shown to increase the precision of estimates substantially in many cases.

Several concepts and techniques introduced here recur throughout statistics: the concept of a random estimate of a population parameter, such as the population mean; bias; the standard error of an estimate; confidence intervals based on the central limit theorem; and linearization, or propagation of error.

The theory and technique of survey sampling go far beyond the material in this introduction. One method that deserves mention because of its widespread use is **systematic sampling.** The population members are given in a list. If, say, a 10% sample is desired, every tenth member of the list is sampled starting from some random point among the first ten. If the list is in totally random order, this method is similar to simple random sampling. If, however, there is some correlation or relationship between successive members, the method is more similar to stratified sampling. The clear danger of this method is that there may be some periodic structure in the list, in which case bias can ensue.

Another commonly used method is **cluster sampling.** In sampling residential households, a survey might choose blocks randomly and then either sample every dwelling on each chosen block or further subsample the dwellings. Because one would expect dwellings within a single block to be relatively homogeneous, this method is typically less precise than a simple random sample of the same size.

We have developed a mathematical model for survey sampling and have deduced consequences of that model, including probabilistic error bounds for the estimates. As is always the case, reality never quite matches the mathematical model. The basic assumptions of the model are (1) that every population member appears in the sample with a specified probability and (2) that an exact measurement or response is obtained from every sample member. In practice, neither assumption will hold precisely. Converse and Traugott (1986) provide an interesting discussion of the practical difficulties of polls and surveys and consequences for the variability of the estimates.

The first assumption may fail because of the difficulty of obtaining an exact enumeration of the population or because of imprecision in its definition. For example, political surveys can be putatively based on all adults, all registered voters, or all "likely" voters. However, the most serious problem with respect to the first assumption is that of nonresponse. Response levels of only 60% to 70% are common in surveys of human populations. The possibility of substantial bias clearly arises if there is a relationship of potential answers to survey questions to the propensity to respond to those questions. For example, adults living in families are easier to contact by a telephone survey than those living alone, and the opinions of these two groups may well differ on certain issues. It is important to realize that the standard errors of estimates that we have developed earlier in this chapter account only for random variability in sample composition, not for systematic biases.

The *Literary Digest* poll of 1936, which predicted a 57% to 43% victory for Republican Alfred Landon over incumbent president Franklin Roosevelt, is one of the most famous of flawed surveys. Questionnaires were mailed to about 10 million voters, who were selected from lists such as telephone books and club memberships, and approximately 2.4 million of the questionnaires were returned. There were two intrinsic problems: (1) nonresponse—those who did not respond may have voted differently from those who did—and (2) selection bias—even if all 10 million voters had responded, they would not have constituted a random sample; those in lower socioeconomic classes (who were more likely to vote for Roosevelt) were less likely to have telephone service or belong to clubs and thus less likely to be included in the sample than were wealthier voters. The assumption that an exact measurement is obtained from every member of the sample may also be in error. In surveys conducted by interviewers, the interviewer's approach and personality may affect the response. In surveys that use questionnaires, the wording of the questions and the context within which they are lodged can have an effect. An interesting example is a poll conducted by Stanley Presser, (*New Yorker,* Oct 18, 2004). Half of the sample was asked, "Do you think the United States should allow public speeches against democracy?" The other half was asked, "Do you think the United States should forbid public speeches against democracy?" 56% said no to the first question, and 39% said yes to the second. The interesting paper by Hansen in Tanur et al. (1972) reports on efforts of the U.S. Bureau of the Census to investigate these sorts of problems.

7.7 Problems

1. Consider a population consisting of five values—1, 2, 2, 4, and 8. Find the population mean and variance. Calculate the sampling distribution of the mean of a sample of size 2 by generating all possible such samples. From them, find the mean and variance of the sampling distribution, and compare the results to Theorems A and B in Section 7.3.1.

2. Suppose that a sample of size $n = 2$ is drawn from the population of the preceding problem and that the proportion of the sample values that are greater than 3 is recorded. Find the sampling distribution of this statistic by listing all possible such samples. Find the mean and variance of the sampling distribution.

3. Which of the following is a random variable?
   a. The population mean
   b. The population size, N

    c. The sample size, n

    d. The sample mean

    e. The variance of the sample mean

    f. The largest value in the sample

    g. The population variance

    h. The estimated variance of the sample mean

4. Two populations are surveyed with simple random samples. A sample of size n_1 is used for population I, which has a population standard deviation σ_1; a sample of size $n_2 = 2n_1$ is used for population II, which has a population standard deviation $\sigma_2 = 2\sigma_1$. Ignoring finite population corrections, in which of the two samples would you expect the estimate of the population mean to be more accurate?

5. How would you respond to a friend who asks you, "How can we say that the sample mean is a random variable when it is just a number, like the population mean? For example, in Example A of Section 7.3.2, a simple random sample of size 50 produced $\bar{x} = 938.5$; how can the number 938.5 be a random variable?"

6. Suppose that two populations have equal population variances but are of different sizes: $N_1 = 100,000$ and $N_2 = 10,000,000$. Compare the variances of the sample means for a sample of size $n = 25$. Is it substantially easier to estimate the mean of the smaller population?

7. Suppose that a simple random sample is used to estimate the proportion of families in a certain area that are living below the poverty level. If this proportion is roughly .15, what sample size is necessary so that the standard error of the estimate is .02?

8. A sample of size $n = 100$ is taken from a population that has a proportion $p = 1/5$.

    a. Find δ such that $P(|\hat{p} - p| \geq \delta) = 0.025$.

    b. If, in the sample, $\hat{p} = 0.25$, will the 95% confidence interval for p contain the true value of p?

9. In a simple random sample of 1,500 voters, 55% said they planned to vote for a particular proposition, and 45% said they planned to vote against it. The estimated margin of victory for the proposition is thus 10%. What is the standard error of this estimated margin? What is an approximate 95% confidence interval for the margin?

10. True or false (and state why):

If a sample from a population is large, a histogram of the values in the sample will be approximately normal, even if the population is not normal.

11. Consider a population of size four, the members of which have values x_1, x_2, x_3, x_4.

    a. If simple random sampling were used, how many samples of size two are there?

    b. Suppose that rather than simple random sampling, the following sampling scheme is used. The possible samples of size two are

$$\{x_1, x_2\}, \{x_2, x_3\}, \{x_3, x_4\}, \{x_1, x_4\}$$

and the sampling is done in such a way that each of these four possible samples is equally likely. Is the sample mean unbiased?

12. Consider simple random sampling *with* replacement.

    a. Show that

    $$s^2 = \frac{1}{n-1} \sum_{i=1}^{n} (X_i - \overline{X})^2$$

    is an unbiased estimate of σ^2.

    b. Is s an unbiased estimate of σ?

    c. Show that $n^{-1}s^2$ is an unbiased estimate of $\sigma_{\overline{X}}^2$.

    d. Show that $n^{-1}N^2s^2$ is an unbiased estimate of σ_T^2.

    e. Show that $\hat{p}(1-\hat{p})/(n-1)$ is an unbiased estimate of $\sigma_{\hat{p}}^2$.

13. Suppose that the total number of discharges, τ, in Example A of Section 7.2 is estimated from a simple random sample of size 50. Denoting the estimate by T, use the central limit theorem to sketch the approximate probability density of the error $T - \tau$.

14. The proportion of hospitals in Example A of Section 7.2 that had fewer than 1000 discharges is $p = .654$. Suppose that the total number of hospitals having fewer than 1000 discharges is estimated from a simple random sample of size 25. Use the central limit theorem to sketch the approximate sampling distribution of the estimate.

15. Consider estimating the mean of the population of hospital discharges (Example A of Section 7.2) from a simple random sample of size n. Use the normal approximation to the distribution of $\overline{X}$ in answering the following:

    a. Sketch $P(|\overline{X} - \mu| > 200)$ as a function of n for $20 \le n \le 100$.

    b. For $n = 20, 40$, and 80, find Δ such that $P(|\overline{X} - \mu| > \Delta) \approx .10$. Similarly, find Δ such that $P(|\overline{X} - \mu| > \Delta) \approx .50$.

16. True or false?

    a. The center of a 95% confidence interval for the population mean is a random variable.

    b. A 95% confidence interval for μ contains the sample mean with probability .95.

    c. A 95% confidence interval contains 95% of the population.

    d. Out of one hundred 95% confidence intervals for μ, 95 will contain μ.

17. A 90% confidence interval for the average number of children per household based on a simple random sample is found to be $(.7, 2.1)$. Can we conclude that 90% of households have between .7 and 2.1 children?

18. From independent surveys of two populations, 90% confidence intervals for the population means are constructed. What is the probability that neither interval contains the respective population mean? That both do?

19. This problem introduces the concept of a *one-sided confidence interval*. Using the central limit theorem, how should the constant k be chosen so that the interval

$(-\infty, \overline{X} + k s_{\overline{X}})$ is a 90% confidence interval for μ—i.e., so that $P(\mu \leq \overline{X} + k s_{\overline{X}}) = .9$? This is called a one-sided confidence interval. How should k be chosen so that $(\overline{X} - k s_{\overline{X}}, \infty)$ is 95% one-sided confidence interval?

20. In Example D of Section 7.3.3, a 95% confidence interval for μ was found to be (1.44, 1.76). Because μ is some fixed number, it either lies in this interval or it doesn't, so it doesn't make any sense to claim that $P(1.44 \leq \mu \leq 1.76) = .95$. What do we mean, then, by saying this is a "95% confidence interval?"

21. In order to halve the width of a 95% confidence interval for a mean, by what factor should the sample size be increased? Ignore the finite population correction.

22. An investigator quantifies her uncertainty about the estimate of a population mean by reporting $\overline{X} \pm s_{\overline{X}}$. What size confidence interval is this?

23. a. Show that the standard error of an estimated proportion is largest when $p = 1/2$.

    b. Use this result and Corollary B of Section 7.3.2 to conclude that the quantity

    $$\frac{1}{2} \sqrt{\frac{N - n}{N(n - 1)}}$$

    is a conservative estimate of the standard error of $\hat{p}$ no matter what the value of p may be.

    c. Use the central limit theorem to conclude that the interval

    $$\hat{p} \pm \sqrt{\frac{N - n}{N(n - 1)}}$$

    contains p with probability at least .95.

24. For a random sample of size n from a population of size N, consider the following as an estimate of μ:

    $$\overline{X}_c = \sum_{i=1}^{n} c_i X_i$$

    where the c_i are fixed numbers and $X_1, \ldots, X_n$ is the sample.

    a. Find a condition on the c_i such that the estimate is unbiased.

    b. Show that the choice of c_i that minimizes the variances of the estimate subject to this condition is $c_i = 1/n$, where $i = 1, \ldots, n$.

25. Here is an alternative proof of Lemma B in Section 7.3.1. Consider a random permutation $Y_1, Y_2, \ldots, Y_N$ of $x_1, x_2, \ldots, x_N$. Argue that the joint distribution of any subcollection, $Y_{i_1}, \ldots, Y_{i_n}$, of the Y_i is the same as that of a simple random sample, $X_1, \ldots, X_n$. In particular,

$$\text{Var}(Y_i) = \text{Var}(X_k) = \sigma^2$$

and

$$\text{Cov}(Y_i, Y_j) = \text{Cov}(X_k, X_l) = \gamma$$

if $i \neq j$ and $k \neq l$. Since $Y_1 + Y_2 + \cdots + Y_N = \tau$,

$$\text{Var}\left(\sum_{i=1}^{N} Y_i\right) = 0$$

(Why?) Express $\text{Var}(\sum_{i=1}^{N} Y_i)$ in terms of σ^2 and the unknown covariance, γ. Solve for γ, and conclude that

$$\gamma = -\frac{\sigma^2}{N-1}$$

for $i \neq j$.

26. This is another proof of Lemma B in Section 7.3.1. Let U_i be a random variable with $U_i = 1$ if the ith population member is in the sample and equal to 0 otherwise.

    a. Show that the sample mean $\overline{X} = n^{-1} \sum_{i=1}^{N} U_i x_i$.
    b. Show that $P(U_i = 1) = n/N$. Find $E(U_i)$, using the fact that U_i is a Bernoulli random variable.
    c. What is the variance of the Bernoulli random variable U_i?
    d. Noting that $U_i U_j$ is a Bernoulli random variable, find $E(U_i U_j)$, $i \neq j$. (Be careful to take into account that the sample is drawn without replacement.)
    e. Find $\text{Cov}(U_i, U_j)$, $i \neq j$.
    f. Using the representation of $\overline{X}$ above, find $\text{Var}(\overline{X})$.

27. Suppose that the population size N is not known, but it is known that $n \leq N$. Show that the following procedure will generate a simple random sample of size n. Imagine that the population is arranged in a long list that you can read sequentially.

    a. Let the sample initially consist of the the first n elements in the list.
    b. For $k = 1, 2, \ldots$, as long as the end of the list has not been encountered:

       i. Read the $(n+k)$-th element in the list.
       ii. Place it in the sample with probability $n/(n+k)$ and, if it is placed in the sample, randomly drop one of the exisiting sample members.

28. In surveys, it is difficult to obtain accurate answers to sensitive questions such as "Have you ever used heroin?" or "Have you ever cheated on an exam?" Warner (1965) introduced the method of **randomized response** to deal with such situations. A respondent spins an arrow on a wheel or draws a ball from an urn containing balls of two colors to determine which of two statements to respond to: (1) "I have characteristic A," or (2) "I do not have characteristic A." The interviewer does not know which statement is being responded to but merely records a yes or a no. The hope is that an interviewee is more likely to answer truthfully if he or she realizes that the interviewer does not know which statement is being responded to. Let R be the proportion of a sample answering Yes. Let p be the probability that statement 1 is responded to (p is known from the structure of the randomizing device), and let q be the proportion of the population that has characteristic A. Let r be the probability that a respondent answers Yes.

    a. Show that $r = (2p-1)q + (1-p)$. [*Hint:* $P(\text{yes}) = P(\text{yes given question 1}) \times P(\text{question 1}) + P(\text{yes given question 2}) \times P(\text{question 2}).$]

b. If r were known, how could q be determined?

c. Show that $E(R) = r$, and propose an estimate, Q, for q. Show that the estimate is unbiased.

d. Ignoring the finite population correction, show that

$$\text{Var}(R) = \frac{r(1-r)}{n}$$

where n is the sample size.

e. Find an expression for $\text{Var}(Q)$.

29. A variation of the method described in Problem 28 has been proposed. Instead of responding to statement 2, the respondent answers an unrelated question for which the probability of a "yes" response is known, for example, "Were you born in June?"

    a. Propose an estimate of q for this method.

    b. Show that the estimate is unbiased.

    c. Obtain an expression for the variance of the estimate.

30. Compare the accuracies of the methods of Problems 28 and 29 by comparing their standard deviations. You may do this by substituting some plausible numerical values for p and q.

31. Referring to Example D in Section 7.3.3, how large should the sample be in order that the 95% confidence interval for the total number of owners planning to sell will have a width of 500?

32. Referring again to Example D in Section 7.3.3, suppose that a survey is done of another condominium project of 12,000 units. The sample size is 200, and the proportion planning to sell in this sample is .18.

    a. What is the standard error of this estimate? Give a 90% confidence interval.

    b. Suppose we use the notation $\hat{p}_1 = .12$ and $\hat{p}_2 = .18$ to refer to the proportions in the two samples. Let $\hat{d} = \hat{p}_1 - \hat{p}_2$ be an estimate of the difference, d, of the two population proportions p_1 and p_2. Using the fact that $\hat{p}_1$ and $\hat{p}_2$ are independent random variables, find expressions for the variance and standard error of $\hat{d}$.

    c. Because $\hat{p}_1$ and $\hat{p}_2$ are approximately normally distributed, so is $\hat{d}$. Use this fact to construct 99%, 95%, and 90% confidence intervals for d. Is there clear evidence that p_1 is really different from p_2?

33. Two populations are independently surveyed using simple random samples of size n, and two proportions, p_1 and p_2, are estimated. It is expected that both population proportions are close to .5. What should the sample size be so that the standard error of the difference, $\hat{p}_1 - \hat{p}_2$, will be less than .02?

34. In a survey of a very large population, the incidences of two health problems are to be estimated from the same sample. It is expected that the first problem will affect about 3% of the population and the second about 40%. Ignore the finite population correction in answering the following questions.

a. How large should the sample be in order for the standard errors of both estimates to be less than .01? What are the actual standard errors for this sample size?

b. Suppose that instead of imposing the same limit on both standard errors, the investigator wants the standard error to be less than 10% of the true value in each case. What should the sample size be?

35. A simple random sample of a population of size 2000 yields the following 25 values:

104	109	111	109	87
86	80	119	88	122
91	103	99	108	96
104	98	98	83	107
79	87	94	92	97

a. Calculate an unbiased estimate of the population mean.

b. Calculate unbiased estimates of the population variance and $\text{Var}(\overline{X})$.

c. Give approximate 95% confidence intervals for the population mean and total.

36. With simple random sampling, is $\overline{X}^2$ an unbiased estimate of μ^2? If not, what is the bias?

37. Two surveys were independently conducted to estimate a population mean, μ. Denote the estimates and their standard errors by $\overline{X}_1$ and $\overline{X}_2$ and $\sigma_{\overline{X}_1}$ and $\sigma_{\overline{X}_2}$. Assume that $\overline{X}_1$ and $\overline{X}_2$ are unbiased. For some α and β, the two estimates can be combined to give a better estimator:

$$X = \alpha \overline{X}_1 + \beta \overline{X}_2$$

a. Find the conditions on α and β that make the combined estimate unbiased.

b. What choice of α and β minimizes the variances, subject to the condition of unbiasedness?

38. Let $X_1, \ldots, X_n$ be a simple random sample. Show that $\dfrac{1}{n}\sum_{i=1}^{n} X_i^3$ is an unbiased estimate of $\dfrac{1}{N}\sum_{i=1}^{N} x_i^3$.

39. Suppose that of a population of N items, k are defective in some way. For example, the items might be documents, a small proportion of which are fraudulent. How large should a sample be so that with a specified probability it will contain at least one of the defective items? For example, if $N = 10,000, k = 50$, and $p = .95$, what should the sample size be? Such calculations are useful in planning sample sizes for acceptance sampling.

40. This problem presents an algorithm for drawing a simple random sample from a population in a sequential manner. The members of the population are considered for inclusion in the sample one at a time in some prespecified order (for example, the order in which they are listed). The ith member of the population is included

in the sample with probability

$$\frac{n - n_i}{N - i + 1}$$

where n_i is the number of population members already in the sample before the ith member is examined. Show that the sample selected in this way is in fact a simple random sample; that is, show that every possible sample occurs with probability

$$\frac{1}{\binom{N}{n}}$$

41. In accounting and auditing, the following sampling method is sometimes used to estimate a population total. In estimating the value of an inventory, suppose that a book value exists for each item and is readily accessible. For each item in the sample, the difference D, audited value minus book value, is determined. The inventory value is estimated by the sum of the book values of the population and $N\overline{D}$, where N is the population size.

    a. Show that the estimate is unbiased.
    b. Find an expression for the variance of the estimate.
    c. Compare the expression obtained in part (b) to the variance of the usual estimate, which is the product of N and the average audited value. Under what circumstances would the proposed method be more accurate?
    d. How could a ratio estimate be employed in this situation? Would there be any advantage or disadvantage to using a ratio estimate rather than the proposed method?

42. Show that the population correlation coefficient is less than or equal to 1 in absolute value.

43. Suppose that for Example D in Section 7.3.3, the average number of occupants per condominium unit in the sample is 2.2 with a sample standard deviation of .7 and the sample correlation coefficient between the number of occupants and the number of motor vehicles is .85. Estimate the population ratio of the number of motor vehicles per occupant and its standard error. Find an approximate 95% confidence interval for the estimate.

44. Show that

$$\frac{\text{Var}(\overline{Y}_R)}{\text{Var}(\overline{Y})} \approx 1 + \frac{C_x}{C_y}\left(\frac{C_x}{C_y} - 2\rho\right)$$

    Sketch the graph of this ratio as a function of C_x/C_y.

45. In the population of hospitals, the correlation of the number of beds and the number of discharges is $\rho = .91$ (Example D of Section 7.4). To see how $\text{Var}(\overline{Y}_R)$ would be different if the correlation were different, plot $\text{Var}(\overline{Y}_R)$ for $n = 64$ as a function of ρ for $-1 < \rho < 1$.

46. Use the central limit theorem to sketch the approximate sampling distribution of $\overline{Y}_R$ for $n = 64$ for the population of hospitals. Compare to the approximate sampling distribution of $\overline{Y}$.

47. For the population of hospitals and a sample size of $n = 64$, find the approximate bias of $\overline{Y}_R$ by applying Corollary B of Section 7.4 and compare it to the approximate standard deviation of the estimate. Repeat for $n = 128$.

48. A simple random sample of 100 households located in a city recorded the number of people living in the household, X, and the weekly expenditure for food, Y. It is known that there are 100,000 households in the city. In the sample

$$\sum X_i = 320$$

$$\sum Y_i = 10,000$$

$$\sum X_i^2 = 1250$$

$$\sum Y_i^2 = 1,100,000$$

$$\sum X_i Y_i = 36,000$$

Neglect the finite population correction in answering the following.

   a. Estimate the ratio $r = \mu_y/\mu_x$.
   b. Form an approximate 95% confidence interval for μ_y/μ_x.
   c. Using only the data on Y estimate the total weekly food expenditure, τ, for households in the city and form a 90% confidence interval.

49. In a wildlife survey, an area of desert land was divided into 1000 squares, or "quadrats," a simple random sample of 50 of which were surveyed. In each surveyed quadrat, the number of birds, Y, and the area covered by vegetation, X, were determined. It was found that

$$\sum X_i = 3000$$

$$\sum Y_i = 150$$

$$\sum X_i^2 = 225,000$$

$$\sum Y_i^2 = 650$$

$$\sum X_i Y_i = 11,000$$

   a. Estimate the ratio of the average number of birds per quadrat to the average vegetation cover per quadrat.
   b. Estimate the standard error of your estimate and find an approximate 90% confidence interval for the population average.
   c. Estimate the total number of birds and find an approximate 95% confidence interval for the population total.
   d. Suppose that from an aerial survey, the total area covered by vegetation could easily be determined. How could this information be used to provide another

estimate of the number of birds? Would you expect this estimate to be better than or worse than that found in part (c)?

50. Hartley and Ross (1954) derived the following exact bound on the relative size of the bias and standard error of a ratio estimate:

$$\frac{|E(R) - r|}{\sigma_R} \leq \frac{\sigma_{\overline{X}}}{\mu_x} = \frac{\sigma_x}{\mu_x} \sqrt{\frac{1}{n} \left(1 - \frac{n-1}{N-1}\right)}$$

   a. Derive this bound from the relation

$$\text{Cov}(R, \overline{X}) = E\left(\frac{\overline{Y}}{\overline{X}}\overline{X}\right) - E\left(\frac{\overline{Y}}{\overline{X}}\right) E(\overline{X})$$

   b. Apply the bound to Problem 43 using sample estimates in place of the given population parameters.

51. This problem introduces a technique called the "jackknife," originally proposed by Quenouille (1956) for reducing bias. Many nonlinear estimates, including the ratio estimator, have the property that

$$E(\hat{\theta}) = \theta + \frac{b_1}{n} + \frac{b_2}{n^2} + \cdots$$

   where $\hat{\theta}$ is an estimate of θ. The jackknife forms an estimate $\hat{\theta}_J$, which has a leading bias term of the order n^{-2} rather than n^{-1}. Thus, for sufficiently large n, the bias of $\hat{\theta}_J$ is substantially smaller than that of $\hat{\theta}$. The technique involves splitting the sample into several subsamples, computing the estimate for each subsample, and then combining the several estimates. The sample is split into p groups of size m, where $n = mp$. For $j = 1, \ldots, p$, the estimate $\hat{\theta}_j$ is calculated from the $m(p - 1)$ observations left after the jth group has been deleted. From the preceding expression,

$$E(\hat{\theta}_j) = \theta + \frac{b_1}{m(p-1)} + \frac{b_2}{[m(p-1)]^2} + \cdots$$

   Now, p "pseudovalues" are defined:

$$V_j = p\hat{\theta} - (p - 1)\hat{\theta}_j$$

   The jackknife estimate, $\hat{\theta}_J$, is defined as the average of the pseudovalues:

$$\hat{\theta}_J = \frac{1}{p} \sum_{j=1}^{p} V_j$$

   Show that the bias of $\hat{\theta}_J$ is of the order n^{-2}.

52. A population consists of three strata with $N_1 = N_2 = 1000$ and $N_3 = 500$. A stratified random sample with 10 observations in each stratum yields the

following data:

Stratum 1	94	99	106	106	101	102	122	104	97	97
Stratum 2	183	183	179	211	178	179	192	192	201	177
Stratum 3	343	302	286	317	289	284	357	288	314	276

Estimate the population mean and total and give a 90% confidence interval.

53. The following table (Cochran 1977) shows the stratification of all farms in a county by farm size and the mean and standard deviation of the number of acres of corn in each stratum.

Farm Size	N_l	μ_l	σ_l
0–40	394	5.4	8.3
41–80	461	16.3	13.3
81–120	391	24.3	15.1
121–160	334	34.5	19.8
161–200	169	42.1	24.5
201–240	113	50.1	26.0
241 +	148	63.8	35.2

    a. For a sample size of 100 farms, compute the sample sizes from each stratum for proportional and optimal allocation, and compare them.

    b. Calculate the variances of the sample mean for each allocation and compare them to each other and to the variance of an estimate formed from simple random sampling.

    c. What are the population mean and variance?

    d. Suppose that ten farms are sampled per stratum. What is $\text{Var}(\overline{X}_s)$? How large a simple random sample would have to be taken to attain the same variance? Ignore the finite population correction.

    e. Repeat part (d) using proportional allocation of the 70 samples.

54. a. Suppose that the cost of a survey is $C = C_0 + C_1 n$, where C_0 is a startup cost and C_1 is the cost per observation. For a given cost C, find the allocation $n_1, \ldots, n_L$ to L strata that is optimal in the sense that it minimizes the variance of the estimate of the population mean subject to the cost constraint.

    b. Suppose that the cost of an observation varies from stratum to stratum—in some strata the observations might be relatively cheap and in others relatively expensive. The cost of a survey with an allocation $n_1, \ldots, n_L$ is

$$C = C_0 + \sum_{l=1}^{L} C_l n_l$$

For a fixed total cost C, what choice of $n_1, \cdots, n_L$ minimizes the variance?

    c. Assuming that the cost function is as given in part (b), for a fixed variance, find n_l to minimize cost.

55. The designer of a sample survey stratifies a population into two strata, H and L. H contains 100,000 people, and L contains 500,000. He decides to allocate 100 samples to stratum H and 200 to stratum L, taking a simple random sample in each stratum.

  a. How should the designer estimate the population mean?

  b. Suppose that the population standard deviation in stratum H is 20 and the standard deviation in stratum L is 10. What will be the standard error of his estimate?

  c. Would it be better to allocate 200 samples to stratum H and 100 to stratum L?

  d. Would it be better to use proportional allocation?

56. How might stratification be used in each of the following sampling problems?

  a. A survey of household expenditures in a city.

  b. A survey to examine the lead concentration in the soil in a large plot of land.

  c. A survey to estimate the number of people who use elevators in a large building with a single bank of elevators.

  d. A survey of programs on a television station, taken to estimate the proportion of time taken up by advertising on Monday through Friday from 6 P.M. until 10 P.M. Assume that 52 weeks of recorded broadcasts are available for analysis.

57. Consider stratifying the population of Problem 1 into two strata: (1, 2, 2) and (4, 8). Assuming that one observation is taken from each stratum, find the sampling distribution of the estimate of the population mean and the mean and standard deviation of the sampling distribution. Compare to Theorems A and B in Section 7.5.2 and the results of Problem 1.

58. (Computer Exercise) Construct a population consisting of the integers from 1 to 100. Simulate the sampling distribution of the sample mean of a sample of size 12 by drawing 100 samples of size 12 and making a histogram of the results.

59. (Computer Exercise) Continuing with Problem 58, divide the population into two strata of equal size, allocate six observations per stratum, and simulate the distribution of the stratified estimate of the population mean. Do the same thing with four strata. Compare the results to each other and to the results of Problem 58.

60. A population consists of two strata, H and L, of sizes 100,000 and 500,000 and standard deviations 20 and 12, respectively. A stratified sample of size 100 is to be taken.

  a. Find the optimal allocation for estimating the population mean.

  b. Find the optimal allocation for estimating the difference of the means of the strata, $\mu_H - \mu_L$.

61. The value of a population mean increases linearly through time: $\mu(t) = \alpha + \beta t$ while the variance remains constant. Independent simple random samples of size n are taken at times $t = 1, 2,$ and 3.

  a. Find conditions on w_1, w_2, and w_3 such that

$$\hat{\beta} = w_1 \overline{X}_1 + w_2 \overline{X}_2 + w_3 \overline{X}_3$$

is an unbiased estimate of the rate of change, β. Here $\overline{X}_i$ denotes the sample mean at time t_i.

b. What values of the w_i minimize the variance subject to the constraint that the estimate is unbiased?

62. In Example B of Section 7.5.2, the standard error of $\overline{X}_s$ was estimated to be $s_{\overline{X}_s} = 35.8$. How good is this estimate—what is the actual standard error of $\overline{X}_s$?

63. (Open-ended) Monte Carlo evaluation of an integral was introduced in Example A of Section 5.2. Refer to that example for the following notation. Try to interpret that method from the point of view of survey sampling by considering an "infinite population" of numbers in the interval $[0, 1]$, each population member x having a value $f(x)$. Interpret $\hat{I}(f)$ as the mean of a simple random sample. What is the standard error of $\hat{I}(f)$? How could it be estimated? How could a confidence interval for $I(f)$ be formed? Do you think that anything could be gained by stratifying the "population?" For example, the strata could be the intervals $[0, .5)$ and $[.5, 1]$. You might find it helpful to consider some examples.

64. The value of an inventory is to be estimated by sampling. The items are stratified by book value in the following way:

Stratum	N_l	μ_l	σ_l
$1000+	70	3000	1250
$200–1000	500	500	100
$1–200	10,000	90	30

a. What should the relative sampling fraction in each stratum be for proportional and for optimal allocation? Ignore the finite population correction.

b. How do the variances under each type of allocation compare to each other and to the variance under simple random sampling?

65. The disk file `cancer` contains values for breast cancer mortality from 1950 to 1960 (y) and the adult white female population in 1960 (x) for 301 counties in North Carolina, South Carolina, and Georgia.

a. Make a histogram of the population values for cancer mortality.

b. What are the population mean and total cancer mortality? What are the population variance and standard deviation?

c. Simulate the sampling distribution of the mean of a sample of 25 observations of cancer mortality.

d. Draw a simple random sample of size 25 and use it to estimate the mean and total cancer mortality.

e. Estimate the population variance and standard deviation from the sample of part (d).

f. Form 95% confidence intervals for the population mean and total from the sample of part (d). Do the intervals cover the population values?

g. Repeat parts (d) through (f) for a sample of size 100.

h. Suppose that the size of the total population of each county is known and that this information is used to improve the cancer mortality estimates by forming a ratio estimator. Do you think this will be effective? Why or why not?

    i. Simulate the sampling distribution of ratio estimators of mean cancer mortality based on a simple random sample of size 25. Compare this result to that of part (c).

    j. Draw a simple random sample of size 25 and estimate the population mean and total cancer mortality by calculating ratio estimates. How do these estimates compare to those formed in the usual way in part (d) from the same data?

    k. Form confidence intervals about the estimates obtained in part (j).

    l. Stratify the counties into four strata by population size. Randomly sample six observations from each stratum and form estimates of the population mean and total mortality.

    m. Stratify the counties into four strata by population size. What are the sampling fractions for proportional allocation and optimal allocation? Compare the variances of the estimates of the population mean obtained using simple random sampling, proportional allocation, and optimal allocation.

    n. How much better than those in part (m) will the estimates of the population mean be if 8, 16, 32, or 64 strata are used instead?

66. A photograph of a large crowd on a beach is taken from a helicopter. The photo is of such high resolution that when sections are magnified, individual people can be identified, but to count the entire crowd in this way would be very time-consuming. Devise a plan to estimate the number of people on the beach by using a sampling procedure.

67. The data set `families` contains information about 43,886 families living in the city of Cyberville. The city has four regions: the Northern region has 10,149 families, the Eastern region has 10,390 families, the Southern region has 13,457 families, and the Western region has 9,890. For each family, the following information is recorded:

    1. Family type
        1: Husband-wife family
        2: Male-head family
        3: Female-head family
    2. Number of persons in family
    3. Number of children in family
    4. Family income
    5. Region
        1: North
        2: East
        3: South
        4: West
    6. Education level of head of household
        31: Less than 1st grade
        32: 1st, 2nd, 3rd, or 4th grade
        33: 5th or 6th grade
        34: 7th or 8th grade
        35: 9th grade
        36: 10th grade
        37: 11th grade

38: 12th grade, no diploma
39: High school graduate, high school diploma, or equivalent
40: Some college but no degree
41: Associate degree in college (occupation/vocation program)
42: Associate degree in college (academic program)
43: Bachelor's degree (e.g., B.S., B.A., A.B.)
44: Master's degree (e.g., M.S., M.A., M.B.A.)
45: Professional school degree (e.g., M.D., D.D.S., D.V.M., LL.B., J.D.)
46: Doctoral degree (e.g., Ph.D., Ed.D.)

In these exercises, you will try to learn about the families of Cyberville by using sampling.

a. Take a simple random sample of 500 families. Estimate the following population parameters, calculate the estimated standard errors of these estimates, and form 95% confidence intervals:

i. The proportion of female-headed families
ii. The average number of children per family
iii. The proportion of heads of households who did not receive a high school diploma
iv. The average family income

Repeat the preceding parameters for five different simple random samples of size 500 and compare the results.

b. Take 100 samples of size 400.

i. For each sample, find the average family income.
ii. Find the average and standard deviation of these 100 estimates and make a histogram of the estimates.
iii. Superimpose a plot of a normal density with that mean and standard deviation of the histogram and comment on how well it appears to fit.
iv. Plot the empirical cumulative distribution function (see Section 10.2). On this plot, superimpose the normal cumulative distribution function with mean and standard deviation as earlier. Comment on the fit.
v. Another method for examining a normal approximation is via a normal probability plot (Section 9.9). Make such a plot and comment on what it shows about the approximation.
vi. For each of the 100 samples, find a 95% confidence interval for the population average income. How many of those intervals actually contain the population target?
vii. Take 100 samples of size 100. Compare the averages, standard deviations, and histograms to those obtained for a sample of size 400 and explain how the theory of simple random sampling relates to the comparisons.

c. For a simple random sample of 500, compare the incomes of the three family types by comparing histograms and boxplots (see Chapter 10.6).

d. Take simple random samples of size 400 from each of the four regions.

i. Compare the incomes by region by making parallel boxplots.
ii. Does it appear that some regions have larger families than others?
iii. Are there differences in education level among the four regions?

e. Formulate a question of your choice and attempt to answer it with a simple random sample of size 400.

f. Does stratification help in estimating the average family income? From a simple random sample of size 400, estimate the average income and also the standard error of your estimate. Form a 95% confidence interval. Next, allocate the 400 observations proportionally to the four regions and estimate the average income from the stratified sample. Estimate the standard error and form a 95% confidence interval. Compare your results to the results of the simple random sample.

CHAPTER **8**

Estimation of Parameters and Fitting of Probability Distributions

The Poisson for [handwritten]

8.1 Introduction

In this chapter, we discuss fitting probability laws to data. Many families of probability laws depend on a small number of parameters; for example, the Poisson family depends on the parameter λ (the mean number of counts), and the Gaussian family depends on two parameters, μ and σ. Unless the values of parameters are known in advance, they must be estimated from data in order to fit the probability law.

After parameter values have been chosen, the model should be compared to the actual data to see if the fit is reasonable; Chapter 9 is concerned with measures and tests of goodness of fit.

In order to introduce and illustrate some of the ideas and to provide a concrete basis for later theoretical discussions, we will first consider a classical example—the fitting of a Poisson distribution to radioactive decay. The concepts introduced in this example will be elaborated in this and the next chapter.

The Gaussian Family depend on μ, σ^2 [handwritten]

8.2 Fitting the Poisson Distribution to Emissions of Alpha Particles

Records of emissions of alpha particles from radioactive sources show that the number of emissions per unit of time is not constant but fluctuates in a seemingly random fashion. If the underlying rate of emission is constant over the period of observation (which will be the case if the half-life is much longer than the time period of observation) and if the particles come from a very large number of independent sources (atoms), the Poisson model seems appropriate. For this reason, the Poisson distribution is frequently used as a model for radioactive decay. You should recall that the

Poisson distribution as a model for random counts in space or time rests on three assumptions: (1) the underlying rate at which the events occur is constant in space or time, (2) events in disjoint intervals of space or time occur independently, and (3) there are no multiple events.

Berkson (1966) conducted a careful analysis of data obtained from the National Bureau of Standards. The source of the alpha particles was americium 241. The experimenters recorded 10,220 times between successive emissions. The observed mean emission rate (total number of emissions divided by total time) was .8392 emissions per sec. The clock used to record the times was accurate to .0002 sec.

The first two columns of the following table display the counts, n, that were observed in 1207 intervals, each of length 10 sec. In 18 of the 1207 intervals, there were 0, 1, or 2 counts; in 28 of the intervals there were 3 counts, etc.

n	Observed	Expected
0–2	18	12.2
3	28	27.0
4	56	56.5
5	105	94.9
6	126	132.7
7	146	159.1
8	164	166.9
9	161	155.6
10	123	130.6
11	101	99.7
12	74	69.7
13	53	45.0
14	23	27.0
15	15	15.1
16	9	7.9
17+	5	7.1
	1207	1207

In fitting a Poisson distribution to the counts shown in the table, we view the 1207 counts as 1207 independent realizations of Poisson random variables, each of which has the probability mass function

$$\pi_k = P(X = k) = \frac{\lambda^k e^{-\lambda}}{k!}$$

In order to fit the Poisson distribution, we must estimate a value for λ from the observed data. Since the average count in a 10-second interval was 8.392, we take this as an estimate of λ (recall that the $E(X) = \lambda$) and denote it by $\hat{\lambda}$.

Before continuing, we want to mention some issues that will be explored in depth in subsequent sections of this chapter. First, observe that if the experiment were to be repeated, the counts would be different and the estimate of λ would be different; it is thus appropriate to regard the estimate of λ as a random variable which has a probability distribution referred to as its **sampling distribution.** The situation is entirely analogous to tossing a coin 10 times and regarding the number of heads as a binomially distributed random variable. Doing so and observing 6 heads generates one realization of this random variable; in the same sense 8.392 is a realization of a random variable. The question thus arises: what is the sampling distribution? This is of some practical interest, since the spread of the sampling distribution reflects the variability of the estimate. We could ask crudely, to what decimal place is the estimate 8.392 accurate? Second, later in this chapter we will discuss the rationale for choosing to estimate λ as we have done. Although estimating λ as the observed mean count is quite reasonable on its face, in principle there might be better procedures.

We now turn to assessing goodness of fit, a subject that will be taken up in depth in the next chapter. Consider the 16 cells into which the counts are grouped. Under the hypothesized model, the probability that a random count falls in any one of the cells may be calculated from the Poisson probability law. The probability that an observation falls in the first cell (0, 1, or 2 counts) is

$$p_1 = \pi_0 + \pi_1 + \pi_2$$

The probability that an observation falls in the second cell is $p_2 = \pi_3$. The probability that an observation falls in the 16th cell is

$$p_{16} = \sum_{k=17}^{\infty} \pi_k$$

Under the assumption that $X_1, \ldots, X_{1207}$ are independent Poisson random variables, the number of observations out of 1207 falling in a given cell follows a binomial distribution with a mean, or expected value, of $1207\,p_k$, and the joint distribution of the counts in all the cells is multinomial with $n = 1207$ and probabilities $p_1, p_2, \ldots, p_{16}$. The third column of the preceding table gives the expected number of counts in each cell; for example, because $p_4 = .0786$, the expected count in the corresponding cell is $1207 \times .0786 = 94.9$. Qualitatively, there is good agreement between the expected and observed counts. Quantitative measures will be presented in Chapter 9.

8.3 Parameter Estimation

As was illustrated in the example of alpha particle emissions, in order to fit a probability law to data, one typically has to estimate parameters associated with the probability law from the data. The following examples further illustrate this point.

E X A M P L E **A**   *Normal Distribution*
The normal, or Gaussian, distribution involves two parameters, μ and σ, where μ is the mean of the distribution and σ^2 is the variance:

$$f(x|\mu, \sigma) = \frac{1}{\sigma\sqrt{2\pi}} e^{-\frac{1}{2}\frac{(x-\mu)^2}{\sigma^2}}, \qquad -\infty < x < \infty$$

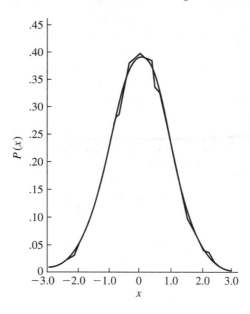

FIGURE **8.1**   Gaussian fit of current flow across a cell membrane to a frequency polygon.

The use of the normal distribution as a model is usually justified using some version of the central limit theorem, which says that the sum of a large number of independent random variables is approximately normally distributed. For example, Bevan, Kullberg, and Rice (1979) studied random fluctuations of current across a muscle cell membrane. The cell membrane contained a large number of channels, which opened and closed at random and were assumed to operate independently. The net current resulted from ions flowing through open channels and was therefore the sum of a large number of roughly independent currents. As the channels opened and closed, the net current fluctuated randomly. Figure 8.1 shows a smoothed histogram of values obtained from 49,152 observations of the net current and an approximating Gaussian curve. The fit of the Gaussian distribution is quite good, although the smoothed histogram seems to show a slight skewness. In this application, information about the characteristics of the individual channels, such as conductance, was extracted from the estimated parameters μ and σ^2.   ■

E X A M P L E **B**   *Gamma Distribution*
The gamma distribution depends on two parameters, α and λ:

$$f(x|\alpha, \lambda) = \frac{1}{\Gamma(\alpha)}\lambda^\alpha x^{\alpha-1}e^{-\lambda x}, \qquad 0 \le x \le \infty$$

The family of gamma distributions provides a flexible set of densities for nonnegative random variables.

Figure 8.2 shows how the gamma distribution fits to the amounts of rainfall from different storms (Le Cam and Neyman 1967). Gamma distributions were fit

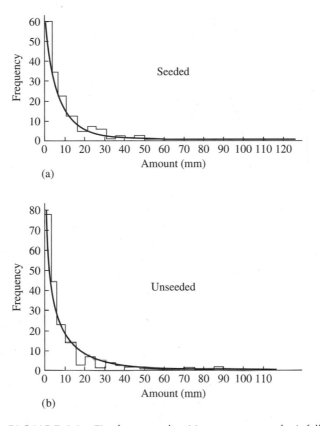

F I G U R E **8.2**   Fit of gamma densities to amounts of rainfall for (a) seeded and
(b) unseeded storms.

$$f(x \mid \alpha, \lambda) = \frac{1}{\Gamma(\alpha)} \lambda^{\alpha} x^{\alpha-1} e^{-\lambda x} \quad 0 \le x \le \infty$$

to rainfall amounts from storms that were seeded and unseeded in an experiment to
determine the effects, if any, of seeding. Differences in the distributions between the
seeded and unseeded conditions should be reflected in differences in the parameters
α and λ. ∎

As these examples illustrate, there are a variety of reasons for fitting probability
laws to data. A scientific theory may suggest the form of a probability distribution
and the parameters of that distribution may be of direct interest to the scientific investigation; the examples of alpha particle emission and Example A are of this character.
Example B is typical of situations in which a model is fit for essentially descriptive
purposes as a method of data summary or compression. A probability model may
play a role in a complex modeling situation; for example, utility companies interested
in projecting patterns of consumer demand find it useful to model daily temperatures
as random variables from a distribution of a particular form. This distribution may
then be used in simulations of the effects of various pricing and generation schemes.
In a similar way, hydrologists planning uses of water resources use stochastic models
of rainfall in simulations.

We will take the following basic approach to the study of parameter estimation. The observed data will be regarded as realizations of random variables $X_1, X_2, \ldots, X_n$, whose joint distribution depends on an unknown parameter θ. Note that θ may be a vector, such as (α, λ) in Example B. Usually the X_i will be modeled as independent random variables all having the same distribution $f(x|\theta)$, in which case their joint distribution is $f(x_1|\theta)f(x_2|\theta)\cdots f(x_n|\theta)$. We will refer to such X_i as independent and identically distributed, or i.i.d. An estimate of θ will be a function of $X_1, X_2, \ldots, X_n$ and will hence be a random variable with a probability distribution called its **sampling distribution.** We will use approximations to the sampling distribution to assess the variability of the estimate, most frequently through its standard deviation, which is commonly called its **standard error.**

It is desirable to have general procedures for forming estimates so that each new problem does not have to be approached *ab initio*. We will develop two such procedures, the method of moments and the method of maximum likelihood, concentrating primarily on the latter, because it is the more generally useful.

The advanced theory of statistics is heavily concerned with "optimal estimation," and we will touch lightly on this topic. The essential idea is that given a choice of many different estimation procedures, we would like to use that estimate whose sampling distribution is most concentrated around the true parameter value.

Before going on to the method of moments, let us note that there are strong similarities of the subject matter of this and the previous chapter. In Chapter 7 we were concerned with estimating population parameters, such as the mean and total, and the process of random sampling created random variables whose probability distributions depended on those parameters. We were concerned with the sampling distributions of the estimates and with assessing variability via standard errors and confidence intervals. In this chapter we consider models in which the data are generated from a probability distribution. This distribution usually has a more hypothetical status than that of Chapter 7, where the distribution was induced by deliberate randomization. In this chapter we will also be concerned with sampling distributions and with assessing variability through standard errors and confidence intervals.

8.4 The Method of Moments

The kth moment of a probability law is defined as

$$\mu_k = E(X^k)$$

where X is a random variable following that probability law (of course, this is defined only if the expectation exists). If $X_1, X_2, \ldots, X_n$ are i.i.d. random variables from that distribution, the kth **sample moment** is defined as

$$\hat{\mu}_k = \frac{1}{n}\sum_{i=1}^{n} X_i^k$$

We can view $\hat{\mu}_k$ as an estimate of μ_k. The method of moments estimates parameters by finding expressions for them in terms of the lowest possible order moments and then substituting sample moments into the expressions.

Suppose, for example, that we wish to estimate two parameters, θ_1 and θ_2. If θ_1 and θ_2 can be expressed in terms of the first two moments as

$$\theta_1 = f_1(\mu_1, \mu_2)$$
$$\theta_2 = f_2(\mu_1, \mu_2)$$

then the method of moments estimates are

$$\hat{\theta}_1 = f_1(\hat{\mu}_1, \hat{\mu}_2)$$
$$\hat{\theta}_2 = f_2(\hat{\mu}_1, \hat{\mu}_2)$$

The construction of a method of moments estimate involves three basic steps:

1. Calculate low order moments, finding expressions for the moments in terms of the parameters. Typically, the number of low order moments needed will be the same as the number of parameters.
2. Invert the expressions found in the preceding step, finding new expressions for the parameters in terms of the moments.
3. Insert the sample moments into the expressions obtained in the second step, thus obtaining estimates of the parameters in terms of the sample moments.

To illustrate this procedure, we consider some examples.

EXAMPLE A   *Poisson Distribution*

The first moment for the Poisson distribution is the parameter $\lambda = E(X)$. The first sample moment is

$$\hat{\mu}_1 = \overline{X} = \frac{1}{n} \sum_{i=1}^{n} X_i$$

which is, therefore, the method of moments estimate of λ: $\hat{\lambda} = \overline{X}$.

As a concrete example, let us consider a study done at the National Institute of Science and Technology (Steel et al. 1980). Asbestos fibers on filters were counted as part of a project to develop measurement standards for asbestos concentration. Asbestos dissolved in water was spread on a filter, and 3-mm diameter punches were taken from the filter and mounted on a transmission electron microscope. An operator counted the number of fibers in each of 23 grid squares, yielding the following counts:

31	29	19	18	31	28
34	27	34	30	16	18
26	27	27	18	24	22
28	24	21	17	24	

The Poisson distribution would be a plausible model for describing the variability from grid square to grid square in this situation and could be used to characterize the inherent variability in future measurements. The method of moments estimate of λ is simply the arithmetic mean of the counts listed above, these or $\hat{\lambda} = 24.9$.

If the experiment were to be repeated, the counts—and therefore the estimate— would not be exactly the same. It is thus natural to ask how stable this estimate is.

A standard statistical technique for addressing this question is to derive the sampling distribution of the estimate or an approximation to that distribution. The statistical model stipulates that the individual counts X_i are independent Poisson random variables with parameter λ_0. Letting $S = \sum X_i$, the parameter estimate $\hat{\lambda} = S/n$ is a random variable, the distribution of which is called its sampling distribution. Now from Example E in Section 4.5, the distribution of the sum of independent Poisson random variables is Poisson distributed, so the distribution of S is Poisson $(n\lambda_0)$. Thus the probability mass function of $\hat{\lambda}$ is

$$P(\hat{\lambda} = v) = P(S = nv)$$
$$= \frac{(n\lambda_0)^{nv} e^{-n\lambda_0}}{(nv)!}$$

for v such that nv is a nonnegative integer.

Since S is Poisson, its mean and variance are both $n\lambda_0$, so

$$E(\hat{\lambda}) = \frac{1}{n} E(S) = \lambda_0$$
$$\text{Var}(\hat{\lambda}) = \frac{1}{n^2} \text{Var}(S) = \frac{\lambda_0}{n}$$

From Example A in Section 5.3, if $n\lambda_0$ is large, the distribution of S is approximately normal; hence, that of $\hat{\lambda}$ is approximately normal as well, with mean and variance given above. Because $E(\hat{\lambda}) = \lambda_0$, we say that the estimate is **unbiased:** the sampling distribution is centered at λ_0. The second equation shows that the sampling distribution becomes more concentrated about λ_0 as n increases. The standard deviation of this distribution is called the **standard error** of $\hat{\lambda}$ and is

$$\sigma_{\hat{\lambda}} = \sqrt{\frac{\lambda_0}{n}}$$

Of course, we can't know the sampling distribution or the standard error of $\hat{\lambda}$ because they depend on λ_0, which is unknown. However, we can derive an approximation by substituting $\hat{\lambda}$ for λ_0 and use it to assess the variability of our estimate. In particular, we can calculate the **estimated standard error** of $\hat{\lambda}$ as

$$s_{\hat{\lambda}} = \sqrt{\frac{\hat{\lambda}}{n}}$$

For this example, we find

$$s_{\hat{\lambda}} = \sqrt{\frac{24.9}{23}} = 1.04$$

At the end of this section, we will present a justification for using $\hat{\lambda}$ in place of λ_0.

In summary, we have found that the sampling distribution of $\hat{\lambda}$ is approximately normal, centered at the true value λ_0 with standard deviation 1.04. This gives us a reasonable assessment of the variability of our parameter estimate. For example, because a normally distributed random variable is unlikely to be more than two standard deviations away from its mean, the error in our estimate of λ is unlikely to be more than 2.08. We thus have not only an estimate of λ_0, but also an understanding of the inherent variability of that estimate.

In Chapter 9, we will address the question of whether the Poisson distribution really fits these data. Clearly, we could calculate the average of any batch of numbers, whether or not they were well fit by the Poisson distribution. ∎

EXAMPLE B   *Normal Distribution*
The first and second moments for the normal distribution are

$$\mu_1 = E(X) = \mu$$
$$\mu_2 = E(X^2) = \mu^2 + \sigma^2$$

Therefore,

$$\mu = \mu_1$$
$$\sigma^2 = \mu_2 - \mu_1^2$$

The corresponding estimates of μ and σ^2 from the sample moments are

$$\hat{\mu} = \overline{X}$$

$$\hat{\sigma}^2 = \frac{1}{n}\sum_{i=1}^{n} X_i^2 - \overline{X}^2 = \frac{1}{n}\sum_{i=1}^{n}(X_i - \overline{X})^2$$

From Section 6.3, the sampling distribution of $\overline{X}$ is $N(\mu, \sigma^2/n)$ and $n\hat{\sigma}^2/\sigma^2 \sim \chi_{n-1}^2$. Furthermore, $\overline{X}$ and $\hat{\sigma}^2$ are independently distributed. We will return to these sampling distributions later in the chapter. ∎

EXAMPLE C   *Gamma Distribution*
The first two moments of the gamma distribution are

$$\mu_1 = \frac{\alpha}{\lambda}$$

$$\mu_2 = \frac{\alpha(\alpha + 1)}{\lambda^2}$$

(see Example B in Section 4.5). To apply the method of moments, we must express α and λ in terms of μ_1 and μ_2. From the second equation,

$$\mu_2 = \mu_1^2 + \frac{\mu_1}{\lambda}$$

or

$$\lambda = \frac{\mu_1}{\mu_2 - \mu_1^2}$$

Also, from the equation for the first moment given here,

$$\alpha = \lambda\mu_1 = \frac{\mu_1^2}{\mu_2 - \mu_1^2}$$

The method of moments estimates are, since $\hat{\sigma}^2 = \hat{\mu}_2 - \hat{\mu}_1^2$,

$$\hat{\lambda} = \frac{\overline{X}}{\hat{\sigma}^2}$$

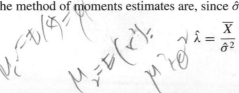

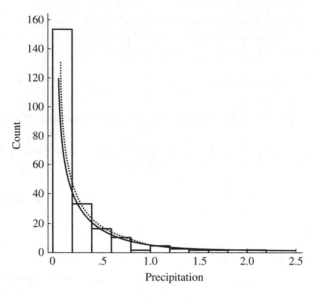

F I G U R E **8.3** Gamma densities fit by the methods of moments and by the method of maximum likelihood to amounts of precipitation; the solid line shows the method of moments estimate and the dotted line the maximum likelihood estimate.

and

$$\hat{\alpha} = \frac{\overline{X}^2}{\hat{\sigma}^2}$$

As a concrete example, let us consider the fit of the amounts of precipitation during 227 storms in Illinois from 1960 to 1964 to a gamma distribution (Le Cam and Neyman 1967). The data, listed in Problem 42 at the end of Chapter 10, were gathered and analyzed in an attempt to characterize the natural variability in precipitation from storm to storm. A histogram shows that the distribution is quite skewed, so a gamma distribution is a natural candidate for a model. For these data, $\overline{X} = .224$ and $\hat{\sigma}^2 = .1338$, and therefore $\hat{\alpha} = .375$ and $\hat{\lambda} = 1.674$.

The histogram with the fitted density is shown in Figure 8.3. Note that, in order to make visual comparison easy, the density was normalized to have a total area equal to the total area under the histogram, which is the number of observations times the bin width of the histogram, or $227 \times .2 = 45.4$. Alternatively, the histogram could have been normalized to have a total area of 1. Qualitatively, the fit in Figure 8.3 looks reasonable; we will examine it in more detail in Example C in Section 9.9. ■

We now turn to a discussion of the sampling distributions of $\hat{\alpha}$ and $\hat{\lambda}$. In the previous two examples, we were able to use known theoretical results in deriving sampling distributions, but it appears that it would be difficult to derive the exact forms of the sampling distributions of $\hat{\lambda}$ and $\hat{\alpha}$, because they are each rather complicated functions of the sample values $X_1, X_2, \ldots, X_n$. However, the problem can be approached by simulation. Imagine for the moment that we knew the true values λ_0 and α_0. We could generate many, many samples of size $n = 227$ from the gamma distribution with these parameter values, and from each of these samples we could calculate estimates of λ and α. A histogram of the values of the estimates of λ, for example, should then give us a good idea of the sampling distribution of $\hat{\lambda}$.

The only problem with this idea is that it requires knowing the true parameter values. (Notice that we faced a problem very much like this in Example A.) So we substitute our estimates of λ and α for the true values; that is we draw many, many samples of size $n = 227$ from a gamma distribution with parameters $\alpha = .375$ and $\lambda = 1.674$. The results of drawing 1000 such samples of size $n = 227$ are displayed in Figure 8.4. Figure 8.4(a) is a histogram of the 1000 estimates of α so obtained and Figure 8.4(b) shows the corresponding histogram for λ. These histograms indicate the variability that is inherent in estimating the parameters from a sample of this size. For example, we see that if the true value of α is .375, then it would not be very unusual for the estimate to be in error by .1 or more. Notice that the shapes of the histograms suggest that they might be approximated by normal densities.

The variability shown by the histograms can be summarized by calculating the standard deviations of the 1000 estimates, thus providing estimated standard errors of $\hat{\alpha}$ and $\hat{\lambda}$. To be precise, if the 1000 estimates of α are denoted by $\alpha_i^*, i = 1, 2, \ldots, 1000$, the standard error of $\hat{\alpha}$ is estimated as

$$s_{\hat{\alpha}} = \sqrt{\frac{1}{1000} \sum_{i=1}^{1000} (\alpha_i^* - \bar{\alpha})^2}$$

where $\bar{\alpha}$ is the mean of the 1000 values. The results of this calculation and the corresponding one for $\hat{\lambda}$ are $s_{\hat{\alpha}} = .06$ and $s_{\hat{\lambda}} = .34$. These standard errors are concise quantifications of the amount of variability of the estimates $\hat{\alpha} = .375$ and $\hat{\lambda} = 1.674$ displayed in Figure 8.4.

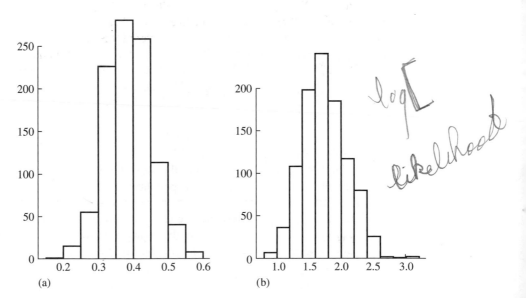

FIGURE **8.4**   Histogram of 1000 simulated method of moment estimates of (a) α and (b) λ.

Our use of simulation (or Monte Carlo) here is an example of what in statistics is called the **bootstrap.** We will see more examples of this versatile method later.

EXAMPLE D   *An Angular Distribution*

The angle θ at which electrons are emitted in muon decay has a distribution with the density

$$f(x|\alpha) = \frac{1 + \alpha x}{2}, \qquad -1 \le x \le 1 \qquad \text{and} \qquad -1 \le \alpha \le 1$$

where $x = \cos\theta$. The parameter α is related to polarization. Physical considerations dictate that $|\alpha| \le \frac{1}{3}$, but we note that $f(x|\alpha)$ is a probability density for $|\alpha| \le 1$. The method of moments may be applied to estimate α from a sample of experimental measurements, $X_1, \ldots, X_n$. The mean of the density is

$$\mu = \int_{-1}^{1} x \frac{1 + \alpha x}{2}\, dx = \frac{\alpha}{3}$$

Thus, the method of moments estimate of α is $\hat{\alpha} = 3\overline{X}$. Consideration of the sampling distribution of $\hat{\alpha}$ is left as an exercise (Problem 13).   ■

Under reasonable conditions, method of moments estimates have the desirable property of consistency. An estimate, $\hat{\theta}$, is said to be a **consistent** estimate of a parameter, θ, if $\hat{\theta}$ approaches θ as the sample size approaches infinity. The following states this more precisely.

DEFINITION

Let $\hat{\theta}_n$ be an estimate of a parameter θ based on a sample of size n. Then $\hat{\theta}_n$ is said to be consistent in probability if $\hat{\theta}_n$ converges in probability to θ as n approaches infinity; that is, for any $\epsilon > 0$,

$$P(|\hat{\theta}_n - \theta| > \epsilon) \to 0 \qquad \text{as } n \to \infty$$ ■

The weak law of large numbers implies that the sample moments converge in probability to the population moments. If the functions relating the estimates to the sample moments are continuous, the estimates will converge to the parameters as the sample moments converge to the population moments.

The consistency of method of moments estimates can be used to provide a justification for a procedure that we used in estimating standard errors in the previous examples. We were interested in the variance (or its square root—the standard error) of a parameter estimate $\hat{\theta}$. Denoting the true parameter by θ_0, we had a relationship of the form

$$\sigma_{\hat{\theta}} = \frac{1}{\sqrt{n}} \sigma(\theta_0)$$

(In Example A, $\sigma_{\hat{\lambda}} = \sqrt{\lambda_0/n}$, so that $\sigma(\lambda) = \sqrt{\lambda}$.) We approximated this by the

estimated standard error

$$s_{\hat{\theta}} = .\frac{1}{\sqrt{n}}\sigma(\hat{\theta})$$

We now claim that the consistency of $\hat{\theta}$ implies that $s_{\hat{\theta}} \approx \sigma_{\hat{\theta}}$. More precisely,

$$\lim_{n\to\infty} \frac{s_{\hat{\theta}}}{\sigma_{\hat{\theta}}} = 1$$

provided that the function $\sigma(\theta)$ is continuous in θ. The result follows since if $\hat{\theta} \to \theta_0$, then $\sigma(\hat{\theta}) \to \sigma(\theta_0)$. Of course, this is just a limiting result and we always have a finite value of n in practice, but it does provide some hope that the ratio will be close to 1 and that the estimated standard error will be a reasonable indication of variability.

Let us summarize the results of this section. We have shown how the method of moments can provide estimates of the parameters of a probability distribution based on a "sample" (an i.i.d. collection) of random variables from that distribution. We addressed the question of variability or reliability of the estimates by observing that if the sample is random, the parameter estimates are random variables having distributions that are referred to as their sampling distributions. The standard deviation of the sampling distribution is called the *standard error of the estimate.* We then faced the problem of how to ascertain the variability of an estimate from the sample itself. In some cases the sampling distribution was of an explicit form depending upon the unknown parameters (Examples A and B); in these cases we could substitute our estimates for the unknown parameters in order to approximate the sampling distribution. In other cases the form of the sampling distribution was not so obvious, but we realized that even if we didn't know it explicitly, we could simulate it. By using the bootstrap we avoided doing perhaps difficult analytic calculations by sitting back and instructing a computer to generate random numbers.

8.5 The Method of Maximum Likelihood

As well as being a useful tool for parameter estimation in our current context, the method of maximum likelihood can be applied to a great variety of other statistical problems, such as curve fitting, for example. This general utility is one of the major reasons for the importance of likelihood methods in statistics. We will later see that maximum likelihood estimates have nice theoretical properties as well.

Suppose that random variables $X_1, \ldots, X_n$ have a joint density or frequency function $f(x_1, x_2, \ldots, x_n | \theta)$. Given observed values $X_i = x_i$, where $i = 1, \ldots, n$, the likelihood of θ as a function of $x_1, x_2, \ldots, x_n$ is defined as

$$\text{lik}(\theta) = f(x_1, x_2, \ldots, x_n | \theta)$$

Note that we consider the joint density as a function of θ rather than as a function of the x_i. If the distribution is discrete, so that f is a frequency function, the likelihood function gives the probability of observing the given data as a function of the parameter θ. The **maximum likelihood estimate (mle)** of θ is that value of θ that maximizes the likelihood—that is, makes the observed data "most probable" or "most likely."

If the X_i are assumed to be i.i.d., their joint density is the product of the marginal densities, and the likelihood is

$$\text{lik}(\theta) = \prod_{i=1}^{n} f(X_i|\theta)$$

Rather than maximizing the likelihood itself, it is usually easier to maximize its natural logarithm (which is equivalent since the logarithm is a monotonic function). For an i.i.d. sample, the **log likelihood** is

$$l(\theta) = \sum_{i=1}^{n} \log[f(X_i|\theta)]$$

(In this text, "log" will always mean the natural logarithm.)

Let us find the maximum likelihood estimates for the examples first considered in Section 8.4.

EXAMPLE A   *Poisson Distribution*

If X follows a Poisson distribution with parameter λ, then

$$P(X = x) = \frac{\lambda^x e^{-\lambda}}{x!}$$

If $X_1, \ldots, X_n$ are i.i.d. and Poisson, their joint frequency function is the product of the marginal frequency functions. The log likelihood is thus

$$l(\lambda) = \sum_{i=1}^{n} (X_i \log \lambda - \lambda - \log X_i!)$$

$$= \log \lambda \sum_{i=1}^{n} X_i - n\lambda - \sum_{i=1}^{n} \log X_i!$$

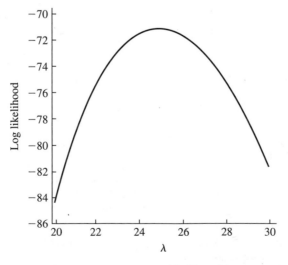

FIGURE 8.5   Plot of the log likelihood function of λ for asbestos data.

Figure 8.5 is a graph of $l(\lambda)$ for the asbestos counts of Example A in Section 8.4. Setting the first derivative of the log likelihood equal to zero, we find

$$l'(\lambda) = \frac{1}{\lambda} \sum_{i=1}^{n} X_i - n = 0$$

The mle is then

$$\hat{\lambda} = \overline{X}$$

We can check that this is indeed a maximum (in fact, $l(\lambda)$ is a concave function of λ; see Figure 8.5). The maximum likelihood estimate agrees with the method of moments for this case and thus has the same sampling distribution. ∎

EXAMPLE B   *Normal Distribution*

If $X_1, X_2, \ldots, X_n$ are i.i.d. $N(\mu, \sigma^2)$, their joint density is the product of their marginal densities:

$$f(x_1, x_2, \ldots, x_n | \mu, \sigma) = \prod_{i=1}^{n} \frac{1}{\sigma \sqrt{2\pi}} \exp\left(-\frac{1}{2}\left[\frac{x_i - \mu}{\sigma}\right]^2 \right)$$

Regarded as a function of μ and σ, this is the likelihood function. The log likelihood is thus

$$l(\mu, \sigma) = -n \log \sigma - \frac{n}{2} \log 2\pi - \frac{1}{2\sigma^2} \sum_{i=1}^{n} (X_i - \mu)^2$$

The partials with respect to μ and σ are

$$\frac{\partial l}{\partial \mu} = \frac{1}{\sigma^2} \sum_{i=1}^{n} (X_i - \mu)$$

$$\frac{\partial l}{\partial \sigma} = -\frac{n}{\sigma} + \sigma^{-3} \sum_{i=1}^{n} (X_i - \mu)^2$$

Setting the first partial equal to zero and solving for the mle, we obtain

$$\hat{\mu} = \overline{X}$$

Setting the second partial equal to zero and substituting the mle for μ, we find that the mle for σ is

$$\hat{\sigma} = \sqrt{\frac{1}{n} \sum_{i=1}^{n} (X_i - \overline{X})^2}$$

Again, these estimates and their sampling distributions are the same as those obtained by the method of moments. ∎

EXAMPLE C   *Gamma Distribution*
Since the density function of a gamma distribution is

$$f(x|\alpha, \lambda) = \frac{1}{\Gamma(\alpha)} \lambda^\alpha x^{\alpha-1} e^{-\lambda x}, \qquad 0 \le x < \infty$$

the log likelihood of an i.i.d. sample, $X_i, \ldots, X_n$, is

$$l(\alpha, \lambda) = \sum_{i=1}^{n} [\alpha \log \lambda + (\alpha - 1) \log X_i - \lambda X_i - \log \Gamma(\alpha)]$$

$$= n\alpha \log \lambda + (\alpha - 1) \sum_{i=1}^{n} \log X_i - \lambda \sum_{i=1}^{n} X_i - n \log \Gamma(\alpha)$$

The partial derivatives are

$$\frac{\partial l}{\partial \alpha} = n \log \lambda + \sum_{i=1}^{n} \log X_i - n \frac{\Gamma'(\alpha)}{\Gamma(\alpha)}$$

$$\frac{\partial l}{\partial \lambda} = \frac{n\alpha}{\lambda} - \sum_{i=1}^{n} X_i$$

Setting the second partial equal to zero, we find

$$\hat{\lambda} = \frac{n\hat{\alpha}}{\sum_{i=1}^{n} X_i} = \frac{\hat{\alpha}}{\overline{X}}$$

But when this solution is substituted into the equation for the first partial, we obtain a nonlinear equation for the mle of α:

$$n \log \hat{\alpha} - n \log \overline{X} + \sum_{i=1}^{n} \log X_i - n \frac{\Gamma'(\hat{\alpha})}{\Gamma(\hat{\alpha})} = 0$$

This equation cannot be solved in closed form; an iterative method for finding the roots has to be employed. To start the iterative procedure, we could use the initial value obtained by the method of moments.

For this example, the two methods do not give the same estimates. The mle's are computed from the precipitation data of Example C in Section 8.4 by an iterative procedure (a combination of the secant method and the method of bisection) using the method of moments estimates as starting values. The resulting estimates are $\hat{\alpha} = .441$ and $\hat{\lambda} = 1.96$. In Example C in Section 8.4, the method of moments estimates were found to be $\hat{\alpha} = .375$ and $\hat{\lambda} = 1.674$. Figure 8.3 shows fitted densities from both types of estimates of α and λ. There is clearly little practical difference, especially if we keep in mind that the gamma distribution is only a possible model and should not be taken as being literally true.

Because the maximum likelihood estimates are not given in closed form, obtaining their exact sampling distribution would appear to be intractable. We thus use the bootstrap to approximate these distributions, just as we did to approximate the sampling distributions of the method of moments estimates. The underlying rationale is the same: If we knew the "true" values, α_0 and λ_0, say, we could approximate

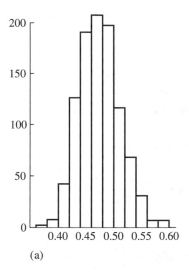
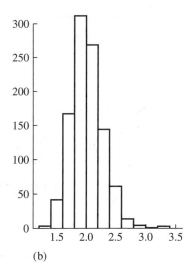

(a)   (b)

FIGURE **8.6**   Histograms of 1000 simulated maximum likelihood estimates of (a) α and (b) λ.

the sampling distribution of their maximum likelihood estimates by generating many, many samples of size $n = 227$ from a gamma distribution with parameters α_0 and λ_0, forming the maximum likelihood estimates from each sample, and displaying the results in histograms. Since, of course, we don't know the true values, we let our maximum likelihood estimates play their role: We generated 1000 samples each of size $n = 227$ of gamma distributed random variables with $\alpha = .441$ and $\lambda = 1.96$. For each of these samples, the maximum likelihood estimates of α and λ were calculated. Histograms of these 1000 estimates are shown in Figure 8.6; we regard these histograms as approximations to the sampling distribution of the maximum likelihood estimates $\hat{\alpha}$ and $\hat{\lambda}$.

Comparison of Figures 8.6 and 8.4 is interesting. We see that the sampling distributions of the maximum likelihood estimates are substantially less dispersed than those of the method of moments estimates, which indicates that in this situation, the method of maximum likelihood is more precise than the method of moments. The standard deviations of the values displayed in the histograms are the estimated standard errors of the maximum likelihood estimates; we find $s_{\hat{\alpha}} = .03$ and $s_{\hat{\lambda}} = .26$. Recall that in Example C of Section 8.4 the corresponding estimated standard errors for the method of moments estimates were found to be .06 and .34.   ■

EXAMPLE **D**   *Muon Decay*
From the form of the density given in Example D in Section 8.4, the log likelihood is

$$l(\alpha) = \sum_{i=1}^{n} \log(1 + \alpha X_i) - n \log 2$$

Setting the derivative equal to zero, we see that the mle of α satisfies the following nonlinear equation:

$$\sum_{i=1}^{n} \frac{X_i}{1 + \hat{\alpha} X_i} = 0$$

Again, we would have to use an iterative technique to solve for $\hat{\alpha}$. The method of moments estimate could be used as a starting value.   ∎

In Examples C and D, in order to find the maximum likelihood estimate, we would have to solve a nonlinear equation. In general, in some problems involving several parameters, systems of nonlinear equations must be solved to find the mle's. We will not discuss numerical methods here; a good discussion is found in Chapter 6 of Dahlquist and Bjorck (1974).

8.5.1   Maximum Likelihood Estimates of Multinomial Cell Probabilities

The method of maximum likelihood is often applied to problems involving multinomial cell probabilities. Suppose that $X_1, \ldots, X_m$, the counts in cells $1, \ldots, m$, follow a multinomial distribution with a total count of n and cell probabilities $p_1, \ldots, p_m$. We wish to estimate the p's from the x's. The joint frequency function of $X_1, \ldots, X_m$ is

$$f(x_1, \ldots, x_m | p_1, \ldots, p_m) = \frac{n!}{\prod\limits_{i=1}^{m} x_i!} \prod_{i=1}^{m} p_i^{x_i}$$

Note that the marginal distribution of each X_i is binomial (n, p_i), and that since the X_i are not independent (they are constrained to sum to n), their joint frequency function is not the product of the marginal frequency functions, as it was in the examples considered in the preceding section. We can, however, still use the method of maximum likelihood since we can write an expression for the joint distribution. We assume n is given, and we wish to estimate $p_1, \ldots, p_m$ with the constraint that the p_i sum to 1. From the joint frequency function just given, the log likelihood is

$$l(p_1, \ldots, p_m) = \log n! - \sum_{i=1}^{m} \log x_i! + \sum_{i=1}^{m} x_i \log p_i$$

To maximize this likelihood subject to the constraint, we introduce a Lagrange multiplier and maximize

$$L(p_1, \ldots, p_m, \lambda) = \log n! - \sum_{i=1}^{m} \log x_i! + \sum_{i=1}^{m} x_i \log p_i + \lambda \left(\sum_{i=1}^{m} p_i - 1 \right)$$

Setting the partial derivatives equal to zero, we have the following system of equations:

$$\hat{p}_j = -\frac{x_j}{\lambda}, \qquad j = 1, \ldots, m$$

Summing both sides of this equation, we have

$$1 = \frac{-n}{\lambda}$$

or

$$\lambda = -n$$

Therefore,

$$\hat{p}_j = \frac{x_j}{n}$$

which is an obvious set of estimates. The sampling distribution of $\hat{p}_j$ is determined by the distribution of x_j, which is binomial.

In some situations, such as frequently occur in the study of genetics, the multinomial cell probabilities are functions of other unknown parameters θ; that is, $p_l = p_i(\theta)$. In such cases, the log likelihood of θ is

$$l(\theta) = \log n! - \sum_{i=1}^{m} \log x_i! + \sum_{i=1}^{m} x_i \log p_i(\theta)$$

EXAMPLE A   *Hardy-Weinberg Equilibrium*
If gene frequencies are in equilibrium, the genotypes AA, Aa, and aa occur in a population with frequencies $(1 - \theta)^2$, $2\theta(1 - \theta)$, and θ^2, according to the Hardy-Weinberg law. In a sample from the Chinese population of Hong Kong in 1937, blood types occurred with the following frequencies, where M and N are erythrocyte antigens:

	Blood Type			
	M	MN	N	Total
Frequency	342	500	187	1029

There are several possible ways to estimate θ from the observed frequencies. For example, if we equate θ^2 with $187/1029$, we obtain .4263 as an estimate of θ. Intuitively, however, it seems that this procedure ignores some of the information in the other cells. If we let X_1, X_2, and X_3 denote the counts in the three cells and let $n = 1029$, the log likelihood of θ is (you should check this):

$$l(\theta) = \log n! - \sum_{i=1}^{3} \log X_i! + X_1 \log(1 - \theta)^2 + X_2 \log 2\theta(1 - \theta) + X_3 \log \theta^2$$

$$= \log n! - \sum_{i=1}^{3} \log X_i! + (2X_1 + X_2) \log(1 - \theta)$$

$$+ (2X_3 + X_2) \log \theta + X_2 \log 2$$

In maximizing $l(\theta)$, we do not need to explicitly incorporate the constraint that the cell probabilities sum to 1 since the functional form of $p_i(\theta)$ is such that $\sum_{i=1}^{3} p_i(\theta) = 1$.

Setting the derivative equal to zero, we have

$$-\frac{2X_1 + X_2}{1 - \theta} + \frac{2X_3 + X_2}{\theta} = 0$$

Solving this, we obtain the mle:

$$\hat{\theta} = \frac{2X_3 + X_2}{2X_1 + 2X_2 + 2X_3}$$

$$= \frac{2X_3 + X_2}{2n}$$

$$= \frac{2 \times 187 + 500}{2 \times 1029} = .4247$$

How precise is this estimate? Do we have faith in the accuracy of the first, second, third, or fourth decimal place? We will address these questions by using the bootstrap to estimate the sampling distribution and the standard error of $\hat{\theta}$. The bootstrap logic is as follows: If θ were known, then the three multinomial cell probabilities, $(1 - \theta)^2$, $2\theta(1 - \theta)$, and θ^2, would be known. To find the sampling distribution of $\hat{\theta}$, we could simulate many multinomial random variables with these probabilities and $n = 1029$, and for each we could form an estimate of θ. A histogram of these estimates would be an approximation to the sampling distribution. Since, of course, we don't know the actual value of θ to use in such a simulation, the bootstrap principle tells us to use $\hat{\theta} = .4247$ in its place. With this estimated value of θ the three cell probabilities (M, MN, N) are .331, .489, and .180. One thousand multinomial random counts, each with total count 1029, were simulated with these probabilities (see problem 35 at the end of the chapter for the method of generating these random counts). From each of these 1000 computer "experiments," a value θ^* was determined. A histogram of the estimates (Figure 8.7) can be regarded as an estimate of the sampling distribution of $\hat{\theta}$. The estimated standard error of $\hat{\theta}$ is the standard deviation of these 1000 values: $s_{\hat{\theta}} = .011$.   ∎

8.5.2 Large Sample Theory for Maximum Likelihood Estimates

In this section we develop approximations to the sampling distribution of maximum likelihood estimates by using limiting arguments as the sample size increases. The theory we shall sketch shows that under reasonable conditions, maximum likelihood estimates are consistent. We also develop a useful and important approximation for the variance of a maximum likelihood estimate and argue that for large sample sizes, the sampling distribution is approximately normal.

The rigorous development of this large sample theory is quite technical; we will simply state some results and give very rough, heuristic arguments for the case of an i.i.d. sample and a one-dimensional parameter. (The arguments for Theorems A and B may be skipped without loss of continuity. Rigorous proofs may be found in Cramér (1946).)

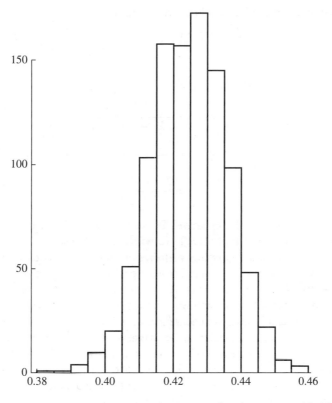

FIGURE 8.7 Histogram of 1000 simulated maximum likelihood estimates of θ described in Example A.

For an i.i.d. sample of size n, the log likelihood is

$$l(\theta) = \sum_{i=1}^{n} \log f(x_i|\theta)$$

We denote the true value of θ by θ_0. It can be shown that under reasonable conditions $\hat{\theta}$ is a consistent estimate of θ_0; that is, $\hat{\theta}$ converges to θ_0 in probability as n approaches infinity.

THEOREM A

Under appropriate smoothness conditions on f, the mle from an i.i.d. sample is consistent.

Proof

The following is merely a sketch of the proof. Consider maximizing

$$\frac{1}{n}l(\theta) = \frac{1}{n}\sum_{i=1}^{n} \log f(X_i|\theta)$$

As n tends to infinity, the law of large numbers implies that

$$\frac{1}{n} l(\theta) \to E \log f(X|\theta)$$

$$= \int \log f(x|\theta) f(x|\theta_0) \, dx$$

It is thus plausible that for large n, the θ that maximizes $l(\theta)$ should be close to the θ that maximizes $E \log f(X|\theta)$. (An involved argument is necessary to establish this.) To maximize $E \log f(X|\theta)$, we consider its derivative:

$$\frac{\partial}{\partial \theta} \int \log f(x|\theta) f(x|\theta_0) \, dx = \int \frac{\frac{\partial}{\partial \theta} f(x|\theta)}{f(x|\theta)} f(x|\theta_0) \, dx$$

If $\theta = \theta_0$, this equation becomes

$$\int \frac{\partial}{\partial \theta} f(x|\theta_0) \, dx = \frac{\partial}{\partial \theta} \int f(x|\theta_0) \, dx = \frac{\partial}{\partial \theta}(1) = 0$$

which shows that θ_0 is a stationary point and hopefully a maximum. Note that we have interchanged differentiation and integration and that the assumption of smoothness on f must be strong enough to justify this. ∎

We will now derive a useful intermediate result.

LEMMA A

Define $I(\theta)$ by

$$I(\theta) = E \left[\frac{\partial}{\partial \theta} \log f(X|\theta) \right]^2$$

Under appropriate smoothness conditions on f, $I(\theta)$ may also be expressed as

$$I(\theta) = -E \left[\frac{\partial^2}{\partial \theta^2} \log f(X|\theta) \right]$$

Proof

First, we observe that since $\int f(x|\theta) \, dx = 1$,

$$\frac{\partial}{\partial \theta} \int f(x|\theta) \, dx = 0$$

Combining this with the identity

$$\frac{\partial}{\partial \theta} f(x|\theta) = \left[\frac{\partial}{\partial \theta} \log f(x|\theta) \right] f(x|\theta)$$

we have

$$0 = \frac{\partial}{\partial \theta} \int f(x|\theta)\, dx = \int \left[\frac{\partial}{\partial \theta} \log f(x|\theta) \right] f(x|\theta)\, dx$$

where we have interchanged differentiation and integration (some assumptions must be made in order to do this). Taking second derivatives of the preceding expressions, we have

$$0 = \frac{\partial}{\partial \theta} \int \left[\frac{\partial}{\partial \theta} \log f(x|\theta) \right] f(x|\theta)\, dx$$

$$= \int \left[\frac{\partial^2}{\partial \theta^2} \log f(x|\theta) \right] f(x|\theta)\, dx + \int \left[\frac{\partial}{\partial \theta} \log f(x|\theta) \right]^2 f(x|\theta)\, dx$$

From this, the desired result follows. ■

The large sample distribution of a maximum likelihood estimate is approximately normal with mean θ_0 and variance $1/[nI(\theta_0)]$. Since this is merely a limiting result, which holds as the sample size tends to infinity, we say that the mle is **asymptotically unbiased** and refer to the variance of the limiting normal distribution as the **asymptotic variance of the mle.**

$$I(\theta) = E\left[\frac{\partial}{\partial \theta} \log f(N\theta) \right]^2$$
$$I(\theta) = -E\left[\frac{\partial^2}{\partial \theta^2} \log f(N\theta) \right]$$

THEOREM B

Under smoothness conditions on f, the probability distribution of $\sqrt{nI(\theta_0)}(\hat{\theta} - \theta_0)$ tends to a standard normal distribution.

Proof

The following is merely a sketch of the proof; the details of the argument are beyond the scope of this book. From a Taylor series expansion,

$$0 = l'(\hat{\theta}) \approx l'(\theta_0) + (\hat{\theta} - \theta_0)l''(\theta_0)$$

$$(\hat{\theta} - \theta_0) \approx \frac{-l'(\theta_0)}{l''(\theta_0)}$$

mle is asympt

$$n^{1/2}(\hat{\theta} - \theta_0) \approx \frac{-n^{-1/2}l'(\theta_0)}{n^{-1}l''(\theta_0)}$$

First, we consider the numerator of this last expression. Its expectation is

$$E[n^{-1/2}l'(\theta_0)] = n^{-1/2} \sum_{i=1}^{n} E\left[\frac{\partial}{\partial \theta} \log f(X_i|\theta_0) \right]$$

$$= 0$$

as in Theorem A. Its variance is

$$\text{Var}[n^{-1/2}l'(\theta_0)] = \frac{1}{n}\sum_{i=1}^{n} E\left[\frac{\partial}{\partial\theta}\log f(X_i|\theta_0)\right]^2$$

$$= I(\theta_0)$$

Next, we consider the denominator:

$$\frac{1}{n}l''(\theta_0) = \frac{1}{n}\sum_{i=1}^{n}\frac{\partial^2}{\partial\theta^2}\log f(x_i|\theta_0)$$

By the law of large numbers, the latter expression converges to

$$E\left[\frac{\partial^2}{\partial\theta^2}\log f(X|\theta_0)\right] = -I(\theta_0)$$

from Lemma A.

We thus have

$$n^{1/2}(\hat{\theta} - \theta_0) \approx \frac{n^{-1/2}l'(\theta_0)}{I(\theta_0)}$$

Therefore,

$$E[n^{1/2}(\hat{\theta} - \theta_0)] \approx 0$$

Furthermore,

$$\text{Var}[n^{1/2}(\hat{\theta} - \theta_0)] \approx \frac{I(\theta_0)}{I^2(\theta_0)}$$

$$= \frac{1}{I(\theta_0)}$$

and thus

$$\text{Var}(\hat{\theta} - \theta_0) \approx \frac{1}{nI(\theta_0)}$$

The central limit theorem may be applied to $l'(\theta_0)$, which is a sum of i.i.d. random variables:

$$l'(\theta_0) = \sum_{i=1}^{n}\frac{\partial}{\partial\theta_0}\log f(X_i|\theta) \qquad \blacksquare$$

Another interpretation of the result of Theorem B is as follows. For an i.i.d. sample, the maximum likelihood estimate is the maximizer of the log likelihood function,

$$l(\theta) = \sum_{i=1}^{n}\log f(X_i|\theta)$$

The asymptotic variance is

$$\frac{1}{nI(\theta_0)} = -\frac{1}{El''(\theta_0)}$$

when $El''(\theta_0)$ is large, $l(\theta)$ is, on average, changing very rapidly in a vicinity of θ_0 and the variance of the maximizer is small.

A corresponding result can be proved from the multidimensional case. The vector of maximum likelihood estimates is asymptotically normally distributed. The mean of the asymptotic distribution is the vector of true parameters, θ_0. The covariance of the estimates $\hat{\theta}_i$ and $\hat{\theta}_j$ is given by the ij entry of the matrix $n^{-1}I^{-1}(\theta_0)$, where $I(\theta)$ is the matrix with ij component

$$E\left[\frac{\partial}{\partial\theta_i}\log f(X|\theta)\frac{\partial}{\partial\theta_j}\log f(X|\theta)\right] = -E\left[\frac{\partial^2}{\partial\theta_i\partial\theta_j}\log f(X|\theta)\right]$$

Since we do not wish to delve deeply into technical details, we do not specify the conditions under which the results obtained in this section hold. It is worth mentioning, however, that the true parameter value, θ_0, is required to be an interior point of the set of all parameter values. Thus the results would not be expected to apply in Example D of Section 8.5 if $\alpha_0 = 1$, for example. It is also required that the support of the density or frequency function $f(x|\theta)$ [the set of values for which $f(x|\theta) > 0$] does not depend on θ. Thus, for example, the results would not be expected to apply to estimating θ from a sample of random variables that were uniformly distributed on the interval $[0, \theta]$.

The following sections will apply these results in several examples.

8.5.3 Confidence Intervals from Maximum Likelihood Estimates

In Chapter 7, confidence intervals for the population mean μ were introduced. Recall that the confidence interval for μ was a random interval that contained μ with some specified probability. In the current context, we are interested in estimating the parameter θ of a probability distribution. We will develop confidence intervals for θ based on $\hat{\theta}$; these intervals serve essentially the same function as they did in Chapter 7 in that they express in a fairly direct way the degree of uncertainty in the estimate $\hat{\theta}$. A confidence interval for θ is an interval based on the sample values used to estimate θ. Since these sample values are random, the interval is random and the probability that it contains θ is called the coverage probability of the interval. Thus, for example, a 90% confidence interval for θ is a random interval that contains θ with probability .9. A confidence interval quantifies the uncertainty of a parameter estimate.

We will discuss three methods for forming confidence intervals for maximum likelihood estimates: exact methods, approximations based on the large sample properties of maximum likelihood estimates, and bootstrap confidence intervals. The construction of confidence intervals for parameters of a normal distribution illustrates the use of exact methods.

EXAMPLE A   We found in Example B of Section 8.5 that the maximum likelihood estimates of μ and σ^2 from an i.i.d. normal sample are

$$\hat{\mu} = \overline{X}$$

$$\hat{\sigma}^2 = \frac{1}{n}\sum_{i=1}^{n}(X_i - \overline{X})^2$$

A confidence interval for μ is based on the fact that

$$\frac{\sqrt{n}(\overline{X} - \mu)}{S} \sim t_{n-1}$$

where t_{n-1} denotes the t distribution with $n - 1$ degrees of freedom and

$$S^2 = \frac{1}{n-1} \sum_{i=1}^{n} (X_i - \overline{X})^2$$

(see Section 6.3). Let $t_{n-1}(\alpha/2)$ denote that point beyond which the t distribution with $n - 1$ degrees of freedom has probability $\alpha/2$. Since the t distribution is symmetric about 0, the probability to the left of $-t_{n-1}(\alpha/2)$ is also $\alpha/2$. Then, by definition,

$$P\left(-t_{n-1}(\alpha/2) \leq \frac{\sqrt{n}(\overline{X} - \mu)}{S} \leq t_{n-1}(\alpha/2)\right) = 1 - \alpha$$

The inequality can be manipulated to yield

$$P\left(\overline{X} - \frac{S}{\sqrt{n}} t_{n-1}(\alpha/2) \leq \mu \leq \overline{X} + \frac{S}{\sqrt{n}} t_{n-1}(\alpha/2)\right) = 1 - \alpha$$

According to this equation, the probability that μ lies in the interval $\overline{X} \pm St_{n-1}(\alpha/2)/\sqrt{n}$ is $1 - \alpha$. Note that this interval is *random:* The center is at the random point $\overline{X}$ and the width is proportional to S, which is also random.

Now let us turn to a confidence interval for σ^2. From Section 6.3,

$$\frac{n\hat{\sigma}^2}{\sigma^2} \sim \chi_{n-1}^2$$

where χ_{n-1}^2 denotes the chi-squared distribution with $n - 1$ degrees of freedom. Let $\chi_m^2(\alpha)$ denote the point beyond which the chi-square distribution with m degrees of freedom has probability α. It then follows by definition that

$$P\left(\chi_{n-1}^2(1 - \alpha/2) \leq \frac{n\hat{\sigma}^2}{\sigma^2} \leq \chi_{n-1}^2(\alpha/2)\right) = 1 - \alpha$$

Manipulation of the inequalities yields

$$P\left(\frac{n\hat{\sigma}^2}{\chi_{n-1}^2(\alpha/2)} \leq \sigma^2 \leq \frac{n\hat{\sigma}^2}{\chi_{n-1}^2(1 - \alpha/2)}\right) = 1 - \alpha$$

Therefore, a $100(1 - \alpha)\%$ confidence interval for σ^2 is

$$\left(\frac{n\hat{\sigma}^2}{\chi_{n-1}^2(\alpha/2)}, \frac{n\hat{\sigma}^2}{\chi_{n-1}^2(1 - \alpha/2)}\right)$$

Note that this interval is not symmetric about $\hat{\sigma}^2$—it is not of the form $\hat{\sigma}^2 \pm c$, unlike the previous example.

A simulation illustrates these ideas: The following experiment was done on a computer 20 times. A random sample of size $n = 11$ from normal distribution with mean $\mu = 10$ and variance $\sigma^2 = 9$ was generated. From the sample, $\overline{X}$ and $\hat{\sigma}^2$ were

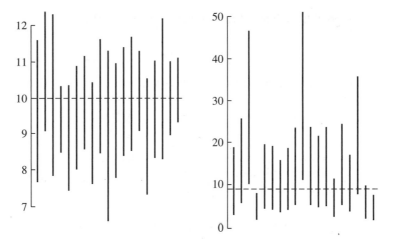

FIGURE **8.8**   20 confidence intervals for μ (left panel) and for σ^2 (right panel) as described in Example A. Horizontal lines indicate the true values.

calculated and 90% confidence intervals for μ and σ^2 were constructed, as described before. Thus at the end there were 20 intervals for μ and 20 intervals for σ^2. The 20 intervals for μ are shown as vertical lines in the left panel of Figure 8.8 and the 20 intervals for σ^2 are shown in the right panel. Horizontal lines are drawn at the true values $\mu = 10$ and $\sigma^2 = 9$. Since these are 90% confidence intervals, we expect the true parameter values to fall outside the intervals 10% of the time; thus on the average we would expect 2 of 20 intervals to fail to cover the true parameter value. From the figure, we see that all the intervals for μ actually cover μ, whereas four of the intervals of σ^2 failed to contain σ^2.   ■

Exact methods such as that illustrated in the previous example are the exception rather than the rule in practice. To construct an exact interval requires detailed knowledge of the sampling distribution as well as some cleverness. A second method of constructing confidence intervals is based on the large sample theory of the previous section. According to the results of that section, the distribution of $\sqrt{nI(\theta_0)}(\hat{\theta} - \theta_0)$ is approximately the standard normal distribution. Since θ_0 is unknown, we will use $I(\hat{\theta})$ in place of $I(\theta_0)$; we have employed similar substitutions a number of times before—for example, in finding an approximate standard error in Example A of Section 8.4. It can be further argued that the distribution of $\sqrt{nI(\hat{\theta})}(\hat{\theta} - \theta_0)$ is also approximately standard normal. Since the standard normal distribution is symmetric about 0,

$$P\left(-z(\alpha/2) \leq \sqrt{nI(\hat{\theta})}(\hat{\theta} - \theta_0) \leq z(\alpha/2)\right) \approx 1 - \alpha$$

Manipulation of the inequalities yields

$$\hat{\theta} \pm z(\alpha/2)\frac{1}{\sqrt{nI(\hat{\theta})}}$$

as an approximate $100(1 - \alpha)\%$ confidence interval. We now illustrate this procedure with an example.

EXAMPLE B   *Poisson Distribution*
The mle of λ from a sample of size n from a Poisson distribution is

$$\hat{\lambda} = \overline{X}$$

Since the sum of independent Poisson random variables follows a Poisson distribution, the parameter of which is the sum of the parameters of the individual summands, $n\hat{\lambda} = \sum_{i=1}^{n} X_i$ follows a Poisson distribution with mean $n\lambda$. Also, the sampling distribution of $\hat{\lambda}$ is known, although it depends on the true value of λ, which is unknown. Exact confidence intervals for λ may be obtained by using this fact, and special tables are available (Pearson and Hartley 1966).

For large samples, confidence intervals may be derived as follows. First, we need to calculate $I(\lambda)$. Let $f(x|\lambda)$ denote the probability mass function of a Poisson random variable with parameter λ. There are two ways to do this. We may use the definition

$$I(\lambda) = E\left[\frac{\partial}{\partial\lambda} \log f(X|\lambda)\right]^2$$

We know that

$$\log f(x|\lambda) = x \log \lambda - \lambda - \log x!$$

and thus

$$I(\lambda) = E\left(\frac{X}{\lambda} - 1\right)^2$$

Rather than evaluate this quantity, we may use the alternative expression for $I(\lambda)$ given by Lemma A of Section 8.5.2:

$$I(\lambda) = -E\left[\frac{\partial^2}{\partial\lambda^2} \log f(X|\lambda)\right]$$

Since

$$\frac{\partial^2}{\partial\lambda^2} \log f(X|\lambda) = -\frac{X}{\lambda^2}$$

$I(\lambda)$ is simply

$$\frac{E(X)}{\lambda^2} = \frac{1}{\lambda}$$

Thus, an approximate $100(1 - \alpha)\%$ confidence interval for λ is

$$\overline{X} \pm z(\alpha/2)\sqrt{\frac{\overline{X}}{n}}$$

Note that the asymptotic variance is in fact the exact variance in this case. The confidence interval, however, is only approximate, since the sampling distribution of $\overline{X}$ is only approximately normal.

As a concrete example, let us return to the study that involved counting asbestos fibers on filters, discussed earlier. In Example A in Section 8.4, we found $\hat{\lambda} = 24.9$. The estimated standard error of $\hat{\lambda}$ is thus ($n = 23$)

$$s_{\hat{\lambda}} = \sqrt{\frac{\hat{\lambda}}{n}} = 1.04$$

An approximate 90% confidence interval for λ is

$$\hat{\lambda} \pm 1.65 s_{\hat{\lambda}}$$

or $(23.2, 26.6)$. This interval gives a good indication of the uncertainty inherent in the determination of the average asbestos level using the model that the counts in the grid squares are independent Poisson random variables.   ∎

In a similar way, approximate confidence intervals can be obtained for parameters estimated from random multinomial counts. The counts are not i.i.d., so the variance of the parameter estimate is not of the form $1/[nI(\theta)]$. However, it can be shown that

$$\text{Var}(\hat{\theta}) \approx \frac{1}{E[l'(\theta_0)^2]} = -\frac{1}{E[l''(\theta_0)]}$$

and the maximum likelihood estimate is approximately normally distributed. Example C illustrates this concept.

EXAMPLE C   *Hardy-Weinberg Equilibrium*
Let us return to the example of Hardy-Weinberg equilibrium discussed in Example A in Section 8.5.1. There we found $\hat{\theta} = .4247$. Now,

$$l'(\theta) = -\frac{2X_1 + X_2}{1 - \theta} + \frac{2X_3 + X_2}{\theta}$$

In order to calculate $E[l'(\theta)^2]$, we would have to deal with the variances and covariances of the X_i. This does not look too inviting; it turns out to be easier to calculate $E[l''(\theta)]$.

$$l''(\theta) = -\frac{2X_1 + X_2}{(1 - \theta)^2} - \frac{2X_3 + X_2}{\theta^2}$$

Since the X_i are binomially distributed, we have

$$E(X_1) = n(1 - \theta)^2$$
$$E(X_2) = 2n\theta(1 - \theta)$$
$$E(X_3) = n\theta^2$$

We find, after some algebra, that

$$E[l''(\theta)] = -\frac{2n}{\theta(1 - \theta)}$$

Since θ is unknown, we substitute $\hat{\theta}$ in its place and obtain the estimated standard error of $\hat{\theta}$:

$$s_{\hat{\theta}} = \frac{1}{\sqrt{-I''(\hat{\theta})}}$$

$$= \sqrt{\frac{\hat{\theta}(1-\hat{\theta})}{2n}} = .011$$

An approximate 95% confidence interval for θ is $\hat{\theta} \pm 1.96 s_{\hat{\theta}}$, or (.403, .447). (Note that this estimated standard error of $\hat{\theta}$ agrees with that obtained by the bootstrap in Example 8.5.1A.)   ∎

Finally, we describe the use of the bootstrap for finding approximate confidence intervals. Suppose that $\hat{\theta}$ is an estimate of a parameter θ—the true, unknown value of which is θ_0—and suppose for the moment that the distribution of $\Delta = \hat{\theta} - \theta_0$ is known. Denote the $\alpha/2$ and $1 - \alpha/2$ quantiles of this distribution by $\underline{\delta}$ and $\overline{\delta}$; i.e.,

$$P(\hat{\theta} - \theta_0 \le \underline{\delta}) = \frac{\alpha}{2}$$

$$P(\hat{\theta} - \theta_0 \le \overline{\delta}) = 1 - \frac{\alpha}{2}$$

Then

$$P(\underline{\delta} \le \hat{\theta} - \theta_0 \le \overline{\delta}) = 1 - \alpha$$

and from manipulation of the inequalities,

$$P(\hat{\theta} - \overline{\delta} \le \theta_0 \le \hat{\theta} - \underline{\delta}) = 1 - \alpha$$

The preceding assumed that the distribution of $\hat{\theta} - \theta_0$ was known, which is typically not the case. If θ_0 were known, this distribution could be approximated arbitrarily well by simulation: Many, many samples of observations could be randomly generated on a computer with the true value θ_0; for each sample, the difference $\hat{\theta} - \theta_0$ could be recorded; and the two quantiles $\underline{\delta}$ and $\overline{\delta}$ could, consequently, be determined as accurately as desired. Since θ_0 is not known, the bootstrap principle suggests using $\hat{\theta}$ in its place: Generate many, many samples (say, B in all) from a distribution with value $\hat{\theta}$; and for each sample construct an estimate of θ, say θ_j^*, $j = 1, 2, \ldots, B$. The distribution of $\hat{\theta} - \theta_0$ is then approximated by that of $\theta^* - \hat{\theta}$, the quantiles of which are used to form an approximate confidence interval. Examples may make this clearer.

E X A M P L E **D**   We first apply this technique to the Hardy-Weinberg equilibrium problem; we will find an approximate 95% confidence interval based on the bootstrap and compare the result to the interval obtained in Example C, where large-sample theory for maximum likelihood estimates was used. The 1000 bootstrap estimates of θ of Example A of Section 8.5.1 provide an estimate of the distribution of θ^*; in particular the 25th largest is .403 and the 975th largest is .446, which are our estimates of the .025 and

.975 quantiles of the distribution. The distribution of $\theta^* - \hat{\theta}$ is approximated by subtracting $\hat{\theta} = .425$ from each θ_i^*, so the .025 and .975 quantiles of this distribution are estimated as

$$\underline{\delta} = .403 - .425 = -.022$$
$$\overline{\delta} = .446 - .425 = .021$$

Thus our approximate 95% confidence interval is

$$(\hat{\theta} - \overline{\delta}, \hat{\theta} - \underline{\delta}) = (.404, .447)$$

Since the uncertainty in $\hat{\theta}$ is in the second decimal place, this interval and that found in Example C are identical for all practical purposes. ∎

EXAMPLE **E**   Finally, we apply the bootstrap to find approximate confidence intervals for the parameters of the gamma distribution fit in Example C of Section 8.5. Recall that the estimates were $\hat{\alpha} = .471$ and $\hat{\lambda} = 1.97$. Of the 1000 bootstrap values of $\alpha^*, \alpha_1^*, \alpha_2^*, \ldots, \alpha_{1000}^*$, the 50th largest was .419 and the 950th largest was .538; the .05 and .95 quantiles of the distribution of $\alpha^* - \hat{\alpha}$ are approximated by subtracting $\hat{\alpha}$ from these values, giving

$$\underline{\delta} = .419 - .471 = -.052$$
$$\overline{\delta} = .538 - .471 = .067$$

Our approximate 90% confidence interval for α_0 is thus

$$(\hat{\alpha} - \overline{\delta}, \hat{\alpha} - \underline{\delta}) = (.404, .523)$$

The 50th and 950th largest values of λ^* were 1.619 and 2.478, and the corresponding approximate 90% confidence interval for λ_0 is (1.462, 2.321). ∎

We caution the reader that there are a number of different methods of using the bootstrap to find approximate confidence intervals. We have chosen to present the preceding method largely because the reasoning leading to its development is fairly direct. Another popular method, the *bootstrap percentile method,* uses the quantiles of the bootstrap distribution of $\hat{\theta}$ directly. Using this method in the previous example, the confidence interval for α would be (.419, .538). Although this direct equation of quantiles of the bootstrap sampling distribution with confidence limits may seem initially appealing, its rationale is somewhat obscure. If the bootstrap distribution is symmetric, the two methods are equivalent (see Problem 38).

8.6 The Bayesian Approach to Parameter Estimation

A preview of the Bayesian approach was given in Example E of Section 3.5.2, which should be reviewed before continuing.

In the Bayesian approach, the unknown parameter θ is treated as a random variable, with "prior distribution" $f_\Theta(\theta)$ representing what we know about the parameter before observing data, X. In the following, we assume Θ is a continuous random variable; the discrete case is entirely analogous. This model is in contrast to the approaches described in the previous sections, in which θ was treated as an unknown constant. For a given value, $\Theta = \theta$, the data have the probability distribution (density or probability mass function) $f_{X|\Theta}(x|\theta)$. The joint distribution of X and Θ is thus

$$f_{X,\Theta}(x, \theta) = f_{X|\Theta}(x|\theta) f_\Theta(\theta)$$

and the marginal distribution of X is

$$f_X(x) = \int f_{X,\Theta}(x, \theta)d\theta$$

$$= \int f_{X|\Theta}(x|\theta) f_\Theta(\theta)d\theta$$

The distribution of Θ given the data X is thus

$$f_{\Theta|X}(\theta|x) = \frac{f_{X,\Theta}(x, \theta)}{f_X(x)}$$

$$= \frac{f_{X|\Theta}(x|\theta) f_\Theta(\theta)}{\int f_{X|\Theta}(x|\theta) f_\Theta(\theta)d\theta}$$

This is called the **posterior distribution;** it represents what is known about Θ having observed data X. Note that the likelihood is $f_{X|\Theta}(x|\theta)$, viewed as a function of θ, and we may usefully summarize the preceding result as

$$f_{\Theta|X}(\theta|x) \propto f_{X|\Theta}(x|\theta) \times f_\Theta(\theta)$$

Posterior density $\propto$ Likelihood $\times$ Prior density

The Bayes paradigm has an appealing formal simplicity as it involves elementary probability operations. We will now see what it amounts to for examples we considered earlier.

EXAMPLE A   *Fitting a Poisson Distribution*
Here the unknown parameter is λ, which has a prior distribution $f_\Lambda(\lambda)$, and the data are n i.i.d. observations $X_1, X_2, \ldots, X_n$, which for a given value λ are Poisson random variables with

$$f_{X_i|\Lambda}(x_i|\lambda) = \frac{\lambda^{x_i} e^{-\lambda}}{x_i!}, \qquad x_i = 0, 1, 2, \ldots$$

Their joint distribution given λ is (from independence) the product of their marginal distributions given λ

$$f_{X|\Lambda}(x|\lambda) = \frac{\lambda^{\sum_{i=1}^n x_i} e^{-n\lambda}}{\prod_{i=1}^n x_i!}$$

where X denotes $(X_1, X_2, \ldots, X_n)$. The posterior distribution of Λ given X is then

$$f_{\Lambda|X}(\lambda|x) = \frac{\lambda^{\sum_{i=1}^{n} x_i} e^{-n\lambda} f_\Lambda(\lambda)}{\int \lambda^{\sum_{i=1}^{n} x_i} e^{-n\lambda} f_\Lambda(\lambda) \, d\lambda}$$

(the term $\prod_{i=1}^{n} x_i!$ has cancelled out).

Thus, to evaluate the posterior distribution, we have to do two things: specify the prior distribution $f_\Lambda(\lambda)$ and carry out the integration in the denominator of the preceding expression. For illustration, we consider the data of Examples 8.4A and 8.5A.

We will consider two approaches to specifying the prior distribution. The first is that of an orthodox Bayesian who takes very seriously the model that the prior distribution specifies his prior opinion. Note that this specification should be done *before* seeing the data, X, and he is required to provide the probability density $f_\Lambda(\lambda)$ through introspection. This is not an easy task to carry out, and even the orthodox often compromise for convenience. He thus decides to quantify his opinion by specifying a prior mean $\mu_1 = 15$ and standard deviation $\sigma = 5$ and to use, because the math works out nicely as we will see, a gamma density with that mean and standard deviation. This choice could be aided by plotting gamma densities for various parameter values. The prior density is shown in Figure 8.9. Using the relationships developed in Example C in Section 8.4, the second moment is $\mu_2 = \mu_1^2 + \sigma^2 = 250$ and the parameters of the gamma density are

$$\nu = \frac{\mu_1}{\mu_2 - \mu_1^2} = 0.6$$

$$\alpha = \nu\mu_1 = 9$$

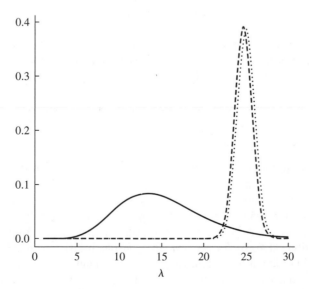

FIGURE **8.9** First statistician's prior (solid) and posterior (dashed). Second statistician's posterior (dotted).

(We denote the parameter by ν rather than by the usual λ since λ has already been used for the parameter of the Poisson distribution.) The prior distribution for Λ is then

$$f_\Lambda(\lambda) = \frac{\nu^\alpha}{\Gamma(\alpha)} \lambda^{\alpha-1} e^{-\nu\lambda}$$

After some cancellation, the posterior density is

$$f_{\Lambda|X}(\lambda|x) = \frac{\lambda^{\Sigma x_i + \alpha - 1} e^{-(n+\nu)\lambda}}{\int_0^\infty \lambda^{\Sigma x_i + \alpha - 1} e^{-(n+\nu)\lambda} d\lambda}$$

Now, consider this an important trick that is used time and again in Bayesian calculations: the denominator is a constant that makes the expression integrate to 1. We can deduce from the form of the numerator that the ratio *must* be a gamma density with parameters

$$\alpha' = \sum x_i + \alpha = 582$$
$$\nu' = n + \nu = 23.6$$

This standard trick allows the statistician to avoid having to do any explicit integration. (Make sure you understand it, because it will occur again several times.) The posterior density is shown in Figure 8.9. Compare it to the prior distribution to observe how observation of the data, X, has drastically changed his state of knowledge about Λ. Notice that the posterior density is much more symmetric and looks like a normal density (that this is no accident will be shown later).   ■

According to the Bayesian paradigm, all the information about Λ is contained in the posterior distribution. The mean of this distribution (the **posterior mean**) is

$$\mu_{\text{post}} = \frac{\alpha'}{\nu'} = 24.7$$

The most probable value of Λ, the **posterior mode,** is 24.6. (Verify that the gamma density is maximized at $(\alpha - 1)/\nu$.) Either of these two values could be used as a point estimate of the unknown mean of the Poisson distribution, if a single number is required.

The variance of the posterior distribution is

$$\sigma_{\text{post}}^2 = \frac{\alpha'}{\nu'^2} = 1.04$$

and the posterior standard deviation is $\sigma_{\text{post}} = 1.02$, which is a simple measure of variability—the posterior distribution of Λ has mean 24.7 and standard deviation 1.02. A Bayesian analogue of a 90% confidence interval is the interval from the 5th percentile to the 95th percentile of the posterior, which can be found numerically to be [23.02, 26.34]. A common alternative to this interval is a **high posterior density (HPD) interval,** formed as follows: Imagine placing a horizontal line at the high point of the posterior density and moving it downward until the interval of λ formed below where the line cuts the density contained 90% probability. If the posterior density is symmetric and unimodal, as is nearly the case in Figure 8.9, the HPD interval will coincide with the interval between the percentiles.

The second statistician takes a more utilitarian, noncommittal approach. She believes that it is implausible that the mean count λ could be larger than 100, and uses a simple prior that is uniform on [0, 100], without trying to quantify her opinion more precisely. The posterior density is thus

$$f_{\Lambda|X}(\lambda|x) = \frac{\lambda^{\sum_{i=1}^{n} x_i} e^{-n\lambda} \frac{1}{100}}{\frac{1}{100} \int_0^{100} \lambda^{\sum_{i=1}^{n} x_i} e^{-n\lambda} d\lambda}, \qquad 0 \leq \lambda \leq 100$$

The denominator has to be integrated numerically, but this is easy to do for such a smooth function. The resulting posterior density is shown in Figure 8.9. Using numerical evaluations, she finds that the posterior mode is 24.9, the posterior mean is 25.0, and the posterior standard deviation is 1.04. The interval from the 5th to the 95th percentile is [23.3, 26.7].

We now compare these two results to each other and to the results of maximum likelihood analysis.

Estimate	Bayes 1	Bayes 2	Maximum Likelihood
mode	24.6	24.9	24.9
mean	24.7	25.0	—
standard deviation	1.02	1.04	1.04
upper limit	26.3	26.7	26.6
lower limit	23.0	23.3	23.2

Comparing the results of the second Bayesian to those of maximum likelihood, it is important to realize that her posterior density is directly proportional to the likelihood for $0 \leq \lambda \leq 100$, because the prior is flat over this range and the posterior is proportional to the prior times the likelihood. Thus, her posterior mode and the maximum likelihood estimate are identical. There is no such guarantee that her posterior standard deviation and the approximate standard error of the maximum likelihood estimate are identical, but they turn out to be, to the number of significant figures displayed in the table. The two 90% intervals are very close.

Now compare the results of the first and second Bayesians. Observe that although his prior opinion was not in accord with the data, the data strongly modified the prior, to produce a posterior that is close to hers. Even though they start with quite different assumptions, the data forces them to very similar conclusions. His prior opinion has indeed influenced the results: his posterior mean and mode are less than hers, but the influence has been mild. (If there had been less data or if his prior opinions had been much more biased to low values, the results would have been in greater conflict.) The fundamental result that the posterior is proportional to the prior times the likelihood helps us to understand the difference: the likelihood is substantial only in the region approximately between $\lambda = 22$ and $\lambda = 28$. (This can be seen in the figure, because the second statistician's posterior is proportional to the likelihood. See Figure 8.5, also). In this region, his prior decreases slowly, so the posterior is proportional to a weighted version of the likelihood, with slowly decreasing weight.

The first Bayesian's posterior thus differs from the second by being pushed up slightly on the left and pulled down on the right.

Although they are very similar numerically, there is an important difference between the Bayesian and frequentist interpretation of the confidence intervals. In the Bayesian framework, Λ is a random variable and it makes perfect sense to say, "Given the observations, the probability that Λ is in the interval [23.3, 26.7] is 0.90." Under the frequentist framework, such a statement makes no sense, because λ is a constant, albeit unknown, and it either lies in the interval [23.3, 26.7] or doesn't—no probability is involved. Before the data are observed, the interval is random, and it makes sense to state that the probability that the interval contains the true parameter value is 0.90, but after the data are observed, nothing is random anymore. One way to understand the difference of interpretation is to realize that in the Bayesian analysis the interval refers to the state of knowledge about λ and not to λ itself.

Finally, we note that an alternative for the second statistician would have been to use a gamma prior because of its analytical convenience, but to make the prior very flat. This can be accomplished by setting α and λ to be very small.

E X A M P L E B   *Normal Distribution*
It is convenient to reparametrize the normal distribution, replacing σ^2 by $\xi = 1/\sigma^2$; ξ is called the **precision**. We will also use θ in place of μ. The density is then

$$f(x|\theta, \xi) = \left(\frac{\xi}{2\pi}\right)^{1/2} \exp\left(-\frac{1}{2}\xi(x-\theta)^2\right)$$

The normal distribution has two parameters, and we will consider cases of Bayesian analysis depending on which of them are known and unknown.   ■

Case of Unknown Mean and Known Variance
We first consider the case in which the precision is known, $\xi = \xi_0$ and the mean, θ, is unknown. In the Bayesian treatment, the mean is a random variable, Θ. It is mathematically convenient to use a prior distribution for Θ, which is $N(\theta_0, \xi_{prior}^{-1})$. This prior is very flat, or uninformative, when ξ_{prior} is very small, i.e., when the prior variance is very large. Thus, if $X = (X_1, X_2, \ldots, X_n)$ are independent given θ

$$f_{\Theta|X}(\theta|x) \propto f_{X|\Theta}(x|\theta) \times f_\Theta(\theta)$$

$$= \left(\frac{\xi_0}{2\pi}\right)^{n/2} \prod_{i=1}^{n} \exp\left(\frac{-\xi_0}{2}(x_i - \theta)^2\right) \times \left(\frac{\xi_{prior}}{2\pi}\right)^{1/2}$$

$$\times \exp\left(\frac{-\xi_{prior}}{2}(\theta - \theta_0)^2\right)$$

$$\propto \exp\left(-\frac{1}{2}\left[\xi_0 \sum_{i=1}^{n}(x_i - \theta)^2 + \xi_{prior}(\theta - \theta_0)^2\right]\right)$$

Here we have exhibited only the terms in the posterior density that depend upon θ; the last expression above shows the shape of the posterior density as a function of θ. The posterior density itself is proportional to this expression, with a proportionality constant that is determined by the requirement that the posterior density integrates to 1.

We will now manipulate the expression for the numerator to cast it in a form so that we can recognize that the posterior density is normal. Expressing $\sum(x_i - \theta)^2 = \sum(x_i - \bar{x})^2 + n(\theta - \bar{x})^2$, and absorbing more terms that do not depend on θ into the constant of proportionality (a typical move in Bayesian calculations), we find

$$f_{\Theta|X}(\theta|x) \propto \exp\left(-\frac{1}{2}[n\xi_0(\theta - \bar{x})^2 + \xi_{\text{prior}}(\theta - \theta_0)^2]\right)$$

Now, observe that this is of the form $\exp(-(1/2)Q(\theta))$, where $Q(\theta)$ is a quadratic polynomial. We can find expressions ξ_{post} and θ_{post}, and write

$$Q(\theta) = \xi_{\text{post}}(\theta - \theta_{\text{post}})^2 + \text{terms that do not depend on } \theta$$

and conclude that the posterior density is normal with posterior mean θ_{post} and posterior precision ξ_{post}. Again, terms that do not depend on θ do not affect the shape of the posterior density and are absorbed in the normalization constant that makes the posterior density integrate to 1. Thus we expand $Q(\theta)$ and identify the coefficient of θ^2 as the posterior precision and the coefficient of $-\theta$ as twice the posterior mean times the posterior precision. Doing so, we find

$$\xi_{\text{post}} = n\xi_0 + \xi_{\text{prior}}$$

$$\theta_{\text{post}} = \frac{n\xi_0\bar{x} + \theta_0\xi_{\text{prior}}}{n\xi_0 + \xi_{\text{prior}}}$$

$$= \bar{x}\frac{n\xi_0}{n\xi_0 + \xi_{\text{prior}}} + \theta_0\frac{\xi_{\text{prior}}}{n\xi_0 + \xi_{\text{prior}}}$$

The posterior density of θ is thus normal with this mean and precision. Note that the precision has increased and that the posterior mean is a weighted combination of the sample mean and the prior mean.

To interpret these results, consider what happens when $\xi_{\text{prior}} \ll n\xi_0$, which would be the case if n were sufficiently large of if ξ_{prior} were small (as for a very flat prior). Then the posterior mean would be

$$\theta_{\text{post}} \approx \bar{x}$$

which is the maximum likelihood estimate, and

$$\xi_{\text{post}} \approx n\xi_0$$

This last equation can be written as $\sigma_{\text{post}}^2 = \sigma_0^2/n$, which is just the variance of $\overline{X}$ in the non-Bayesian setting. In summary, if the flat prior with very small ξ_{prior} is used, the posterior density of θ is very close to normal with mean $\bar{x}$ and variance σ_0^2/n. ∎

Case of Known Mean and Unknown Variance
In this case, the precision is unknown and is treated as a random variable Ξ, with prior distribution $f_\Xi(\xi)$. Given ξ, the X_i are independent $N(\theta_0, \xi^{-1})$. Let $X = (X_1, X_2, \ldots, X_n)$. Then

$$f_{\Xi|X}(\xi|x) \propto f_{X|\Xi}(x|\xi)f_\Xi(\xi)$$

$$\propto \xi^{n/2}\exp\left(-\frac{1}{2}\xi\sum(x_i - \theta_0)^2\right)f_\Xi(\xi)$$

Observing how the density depends on ξ, we realize that it is analytically convenient to specify the prior to be a gamma density: $\Xi \sim \Gamma(\alpha, \lambda)$. Then

$$f_{\Xi|X}(\xi|x) \propto \xi^{n/2} \exp\left(-\frac{1}{2}\xi \sum(x_i - \theta_0)^2\right) \xi^{\alpha-1} e^{-\lambda\xi}$$

which is a gamma density with parameters,

$$\alpha_{\text{post}} = \alpha + \frac{n}{2}$$

$$\lambda_{\text{post}} = \lambda + \frac{1}{2}\sum(x_i - \theta_0)^2$$

In the case of a flat prior (small α and λ), the posterior mean and mode are

$$\text{Posterior mean} \approx \frac{1}{n}\sum(x_i - \theta_0)^2$$

$$\text{Posterior mode} \approx \frac{1}{n-2}\sum(x_i - \theta_0)^2$$

The former is the maximum likelihood estimate of σ^2. In the limit, $\lambda \to 0$, $\alpha \to 0$,

$$f_{\Xi|X}(\xi|x) \propto \xi^{n/2-1} \exp\left(-\frac{1}{2}\xi \sum(x_i - \theta_0)^2\right) \qquad \blacksquare$$

Case of Unknown Mean and Unknown Variance

In this case, there are two unknown parameters, and a Bayesian approach requires the specification of a joint two-dimensional prior distribution. We follow a path of mathematical convenience and take the priors to be independent:

$$\Theta \sim N\left(\theta_0, \xi_{\text{prior}}^{-1}\right)$$

$$\Xi \sim \Gamma(\alpha, \lambda)$$

We then have

$$f_{\Theta, \Xi|X}(\theta, \xi|x) \propto f_{X|\Theta, \Xi}(x|\theta, \xi) f_\Theta(\theta) f_\Xi(\xi)$$

$$\propto \xi^{n/2} \exp\left(-\frac{\xi}{2}\sum(x_i - \theta)^2\right)$$

$$\times \exp\left(-\frac{\xi_{\text{prior}}}{2}(\theta - \theta_0)^2\right) \xi^{\alpha-1} \exp(-\lambda\xi)$$

From the manner in which θ and ξ occur in the first exponential, it appears that the two variables are not independent in the posterior even though they were in the prior. To evaluate this joint posterior density, we would have to find the constant of proportionality that makes it integrate to 1—the normalization constant. Two dimensional numerical integration could be used.

Often the primary interest is in the mean, θ, and one useful aspect of Bayesian analysis is that information about θ can be "marginalized" by integrating out ξ:

$$f_{\Theta|X}(\theta|x) = \int_0^\infty f_{\Theta, \Xi|X}(\theta, \xi|x) d\xi$$

Examining the preceding expression for $f_{\Theta, \Xi|X}(\theta, \xi|x)$ as a function of ξ, we see that it is of the form of a gamma density, with parameters $\tilde{\alpha} = \alpha + n/2$ and $\tilde{\lambda} = \lambda + (1/2)\sum(x_i - \theta)^2$, so we can evaluate the integral. We thus find

$$f_{\Theta|X}(\theta|x) \propto \exp\left(-\frac{\xi_{\text{prior}}}{2}(\theta - \theta_0)^2\right) \frac{\Gamma(\alpha + n/2)}{[\lambda + \frac{1}{2}\sum(x_i - \theta)^2]^{\alpha + n/2}}$$

This is not a density that we recognize, but it could be evaluated numerically. Doing so would again entail finding the normalizing constant, which could be done by numerical integration. Some simplifications occur when n is large or when the prior is quite flat ($\alpha, \lambda, \xi_{\text{prior}}$ are small). Then

$$f_{\Theta|X}(\theta|x) \propto \left(\sum(x_i - \theta)^2\right)^{-n/2}$$

This posterior is maximized when $\sum(x_i - \theta)^2$ is minimized, which occurs at $\theta = \bar{x}$. We can relate this to the result we found for maximum likelihood analysis by expressing

$$\sum(x_i - \theta)^2 = \sum(x_i - \bar{x})^2 + n(\theta - \bar{x})^2$$
$$= (n-1)s^2 + n(\theta - \bar{x})^2$$
$$= (n-1)s^2\left(1 + \frac{n(\theta - \bar{x})^2}{(n-1)s^2}\right)$$

Substituting this above and absorbing terms that do not depend on θ into the proportionality constant, we find

$$f_{\Theta|X}(\theta|x) \propto \left(1 + \frac{1}{n-1}\frac{n(\theta - \bar{x})^2}{s^2}\right)^{-n/2}$$

Now comparing this to the definition of the t distribution (Section 6.2), we see that

$$\frac{\sqrt{n}(\Theta - \bar{x})}{s} \sim t_{n-1}$$

corresponding to the result from maximum likelihood analysis.

The interval $\bar{x} \pm t_{n-1}(\alpha/2)s/\sqrt{n}$ was earlier derived as a $100(1-\alpha)\%$ confidence interval centered about the maximum likelihood estimate, and here it has reappeared in the Bayesian analysis as an interval with posterior probability $1 - \alpha$. There are differences of interpretation, however, just as there were for the earlier Poisson case. The Bayesian interval is a probability statement referring to the state of knowledge about θ given the observed data, regarding θ as a random variable. The frequentist confidence interval is based on a probability statement about the possible values of the observations, regarding θ as a constant, albeit unknown. ∎

E X A M P L E C   *Hardy-Weinberg Equilibrium*

We now turn to a Bayesian treatment of Example A in Section 8.5.1. We use the multinomial likelihood function and a prior for θ, which is uniform on $[0, 1]$. The posterior density is thus proportional to the likelihood, and is shown in Figure 8.10. Note that it looks very much like a normal density, a phenomenon that will be explored in a later section. Since $f_{X|\Theta}(x|\theta)$ is a polynomial in θ (of high degree),

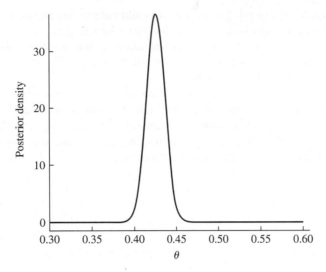

FIGURE **8.10**   Posterior distribution of Θ.

the normalization constant can in principle be computed analytically. (Alternatively, all the computations can be done numerically.)

Because the prior is flat, the posterior is directly proportional to the likelihood and the maximum of the posterior density is the maximum likelihood estimate, $\hat{\theta} = 0.4247$. The 0.025 percentile of the density is 0.404, and the 0.975 percentile is 0.446. These results agree with the approximate confidence interval found for the maximum likelihood estimate in Example C in Section 8.5.3.   ■

8.6.1   Further Remarks on Priors

In the previous section, we saw that if the prior for a Poisson parameter is chosen to be a gamma density, then the posterior is also a gamma density. Similarly, when the prior for a normal mean with known variance is chosen to be normal, then the posterior is normal as well. Earlier, in Example E in Section 3.5.2, a beta prior was used for a binomial parameter, and the posterior turned out to be beta as well. These are examples of **conjugate priors:** if the prior distribution belongs to a family G and, conditional on the parameters of G, the data have a distribution H, then G is said to be conjugate to H if the posterior is in the family G. Other conjugate priors will be the subject of problems at the end of the chapter. Conjugate priors are used for mathematical convenience (required integrations can be done in closed form) and because they can assume a variety of shapes as the parameters of the prior are varied.

In scientific applications, it is usually desirable to use a flat, or "uninformative," prior so that the data can speak for themselves. Even if a scientific investigator actually had a strong prior opinion, he or she might want to present an "objective" analysis. This is accomplished by using a flat prior so that the conclusions, as summarized in the posterior density, are those of one who is initially unopinionated or unprejudiced.

If an informative prior were used, it would have to be justified to the larger scientific community. The objective prior thus has a hypothetical, or "what if," status: if one was initially indifferent to parameter values in the range in which the likelihood is large, then one's opinion after observing the data would be expressed as a posterior proportional to the likelihood.

Attempts have been made to formalize more precisely what the notion of an uninformative prior means. One problem that is addressed is caused by reparametrization. For example, suppose that the prior density of the precision ξ is taken to be uniform on an interval $[a, b]$, which might seem to be a reasonable way to quantify the notion of being uniformative. However, if the variance $\sigma^2 = 1/\xi$, rather than the precison, was used, the prior density of σ^2 would not be uniform on $[b^{-1}, a^{-1}]$. We will not delve further into these issues here, except to note that the parametrization θ or $g(\theta)$ would make a difference only if the difference in the shapes of the priors was substantial in the region in which the likelihood was large.

We saw in the Poisson example that if α and ν are very small, the gamma prior is quite flat and the posterior is proportional to the likelihood function. Formally, if α and ν are set equal to zero, then the prior is

$$f_{\Lambda|\alpha,\nu}(\lambda) = \lambda^{-1}, \;\; 0 \le \lambda < \infty$$

But this function does not integrate to 1—it is not a probability density. A similar phenomena occurs in the normal case with unknown mean and known precision, if the prior precision is set equal to 0. The prior is then

$$f_\Theta(\theta) \propto 1, \;\; -\infty < \theta < \infty$$

and not a probability density either. Such priors are called **improper priors** (priors that lack propriety).

In general, if an improper prior is formally used, the posterior may not be a density either, because the denominator of the expression for the posterior density, $\int f_{X|\Theta}(x|\theta) f_\Theta(\theta)\, d\theta$ may not converge. (Note that it is integrated with respect to θ, not x.) This has not been the case in our examples. For the Poisson example, if $f_\Lambda(\lambda) \propto \lambda^{-1}$, then the denominator is

$$\int_0^\infty \lambda^{\sum x_i - 1} e^{-n\lambda} d\lambda < \infty$$

In the normal case, too, the integral is defined, and thus there is a well-defined posterior density.

Let us revisit some examples using the device of an improper prior. In the Poisson example, using the improper prior $f_\Lambda(\lambda) = \lambda^{-1}$ results in a (proper) posterior

$$f_{\Lambda|X}(\lambda|x) \propto \lambda^{\sum x_i - 1} e^{-n\lambda}$$

which can be recognized as a gamma density.

In the normal example with unknown mean and variance, we can take θ and ξ to be independent with improper priors $f_\Theta(\theta) = 1$ and $f_\Xi(\xi) = \xi^{-1}$. The joint posterior of θ and ξ is then

$$f_{\Theta,\Xi|X}(\theta, \xi|x) \propto \xi^{n/2-1} \exp\left(-\frac{\xi}{2}\sum(x_i - \theta)^2\right)$$

Expressing $\sum_{i=1}^{n}(x_i - \theta)^2 = (n-1)s^2 + n(\theta - \bar{x})^2$, we have

$$f_{\Theta, \Xi | X}(\theta, \xi | x) \propto \xi^{n/2-1} \exp\left(-\frac{\xi}{2}(n-1)s^2\right) \exp\left(-\frac{n\xi}{2}(\theta - \bar{x})^2\right)$$

For fixed ξ, this expression is proportional to the conditional density of θ given ξ. (Why?) From the form of the dependence on θ, we see that conditional on ξ, θ is normal with mean $\bar{x}$ and precision $n\xi$. By integrating out ξ, we can find the marginal distribution of θ and relate it to the t distribution as was done earlier.

Since improper priors are not actually probability densities, they are difficult to interpret literally. However, the resulting posteriors can be viewed as approximations to those that would have arisen with extreme values of the parameters of proper priors. The priors corresponding to such extreme values are very flat, so the posterior is dominated by the likelihood. Then it is only in the range in which the likelihood is large that the prior makes any practical difference—truncating the improper prior well outside this range to produce a proper prior will not appreciably change the posterior.

8.6.2 Large Sample Normal Approximation to the Posterior

We have seen in several examples that the posterior distribution is nearly normal with the mean equal to the maximum likelihood estimate, and that the posterior standard deviation is close to the asymptotic standard deviation of the maximum likelihood estimate. The two methods thus often give quite comparable results. We will not give a formal proof here, but rather will sketch an argument that the posterior distribution is approximately normal with the mean equal the the maximum likelihood estimate, $\hat{\theta}$, and variance approximately equal to $-[l''(\hat{\theta})]^{-1}$.

Denoting the observations generically by x, the posterior distribution is

$$\begin{aligned}
f_{\Theta | X}(\theta | x) &\propto f_{\Theta}(\theta) f_{X | \Theta}(x | \theta) \\
&= \exp[\log f_{\Theta}(\theta)] \exp[\log f_{X | \Theta}(x | \theta)] \\
&= \exp[\log f_{\Theta}(\theta)] \exp[l(\theta)]
\end{aligned}$$

Now, if the sample is large, the posterior is dominated by the likelihood, and in the region where the likelihood is large, the prior is nearly constant. Thus, to an approximation,

$$\begin{aligned}
f_{\Theta | X}(\theta | x) &\propto \exp\left[l(\hat{\theta}) + (\theta - \hat{\theta})l'(\hat{\theta}) + \frac{1}{2}(\theta - \hat{\theta})^2 l''(\hat{\theta})\right] \\
&\propto \exp\left[\frac{1}{2}(\theta - \hat{\theta})^2 l''(\hat{\theta})\right]
\end{aligned}$$

In the last step, we used the fact that since $\hat{\theta}$ is the maximum likelihood estimate $l'(\hat{\theta}) = 0$. The term $l(\hat{\theta})$ was absorbed into a proportionality constant, since we are evaluating the posterior as a function of θ. Finally, observe that the last expression is proportional to a normal density with mean $\hat{\theta}$ and variance $-[l''(\hat{\theta})]^{-1}$.

8.6.3 Computational Aspects

Contemporary computational resources have had an enormous impact on Bayesian inference. As we have seen in several examples, the computationally difficult part of Bayesian inference is the calculation of the normalizing constant that makes the posterior density integrate to 1. Traditionally, such calculations were performed analytically, often using conjugate priors so that the integrations could be done explicitly. The numerical integration of a well-behaved function of a small number of variables is now trivial.

Difficulties do arise in high dimensional problems, however, and the integrations are often done by sophisticated Monte Carlo methods. We will not go into these sorts of methods in this book, but will hint at their nature in the following example of a method called **Gibbs Sampling.** Consider, as a simple example, inference for a normal distribution with unknown mean and variance. From Example B in Section 8.6

$$f_{\Theta, \Xi|X}(\theta, \xi|x) \propto \xi^{n/2} \exp\left(-\frac{\xi}{2} \sum (x_i - \theta)^2\right)$$
$$\times \exp\left(-\frac{\xi_{\text{prior}}}{2}(\theta - \theta_0)^2\right) \xi^{\alpha-1} \exp(-\lambda \xi)$$

For simplicity, suppose that an improper prior is used: $\xi_{\text{prior}} \to 0, \alpha \to 0, \lambda \to 0$. Then

$$f_{\Theta, \Xi|X}(\theta, \xi|x) \propto \xi^{n/2-1} \exp\left(-\frac{\xi}{2} \sum (x_i - \theta)^2\right)$$
$$\propto \xi^{n/2-1} \exp\left(\frac{n\xi}{2}(\theta - \bar{x})^2\right)$$

Here we expressed

$$\sum (x_l - \theta)^2 = \sum (x_i - \bar{x})^2 + n(\theta - \bar{x})^2$$

and absorbed terms that do not involve θ into the constant of proportionality. To study the posterior distribution of ξ and θ by Monte Carlo, we would draw many pairs (ξ_k, θ_k) from this joint density; the problem is how to actually do this.

Gibbs Sampling would accomplish this in the following way. Observe that the expression $f_{\Theta, \Xi|X}(\theta, \xi|x)$ shows that for given ξ, θ is normally distributed with mean $\bar{x}$ and precision $n\xi$. (Fix ξ in the expression and recognize a normal density in θ.) Also, if θ is fixed, the density of ξ is a gamma density. Gibbs Sampling alternates back and forth between the two conditional distributions:

1. Choose an initial value θ_0; $\bar{x}$ would be a natural choice.
2. Generate ξ_0 from a gamma density with parameter θ_0.
3. Generate θ_1 from a normal distribution with parameter ξ_0.
4. Generate ξ_1 from a gamma density with parameter θ_1.
5. Continue on in this fashion.

The analysis of the algorithm and why it works is beyond the scope of this book. A "burn-in" period is required so that we might run this scheme for a few hundred steps before beginning to record pairs (ξ_k, θ_k), $k = 1, \ldots, N$, which would be regarded as simulated pairs from the posterior. A further complication is that these pairs are not independent of one another. But, nonetheless, a histogram of the collection of θ_k could be used as an estimate of the marginal posterior distribution of Θ. The posterior mean of Θ can be estimated as

$$E(\Theta|X) \approx \frac{1}{N} \sum_{k=1}^{N} \theta_k$$

8.7 Efficiency and the Cramér-Rao Lower Bound

In most statistical estimation problems, there are a variety of possible parameter estimates. For example, in Chapter 7 we considered both the sample mean and a ratio estimate, and in this chapter we considered the method of moments and the method of maximum likelihood. Given a variety of possible estimates, how would we choose which to use? Qualitatively, it would be sensible to choose that estimate whose sampling distribution was most highly concentrated about the true parameter value. To define this aim operationally, we would need to specify a quantitative measure of such concentration. Mean squared error is the most commonly used measure of concentration, largely because of its analytic simplicity. The mean squared error of $\hat{\theta}$ as an estimate of θ_0 is

$$MSE(\hat{\theta}) = E(\hat{\theta} - \theta_0)^2$$
$$= \text{Var}(\hat{\theta}) + (E(\hat{\theta}) - \theta_0)^2$$

(See Theorem A of Section 4.2.1.) If the estimate $\hat{\theta}$ is unbiased [$E(\hat{\theta}) = \theta_0$], $MSE(\hat{\theta}) = \text{Var}(\hat{\theta})$. When the estimates under consideration are unbiased, comparison of their mean squared errors reduces to comparison of their variances, or equivalently, standard errors.

Given two estimates, $\hat{\theta}$ and $\tilde{\theta}$, of a parameter θ, the **efficiency** of $\hat{\theta}$ relative to $\tilde{\theta}$ is defined to be

$$\text{eff}(\hat{\theta}, \tilde{\theta}) = \frac{\text{Var}(\tilde{\theta})}{\text{Var}(\hat{\theta})}$$

Thus, if the efficiency is smaller than 1, $\hat{\theta}$ has a larger variance than $\tilde{\theta}$ has. This comparison is most meaningful when both $\hat{\theta}$ and $\tilde{\theta}$ are unbiased or when both have the same bias. Frequently, the variances of $\hat{\theta}$ and $\tilde{\theta}$ are of the form

$$\text{Var}(\hat{\theta}) = \frac{c_1}{n}$$
$$\text{Var}(\tilde{\theta}) = \frac{c_2}{n}$$

where n is the sample size. If this is the case, the efficiency can be interpreted as the ratio of sample sizes necessary to obtain the same variance for both $\hat{\theta}$ and $\tilde{\theta}$. (In Chapter 7, we compared the efficiencies of estimates of a population mean from a simple random sample, a stratified random sample with proportional allocation, and a stratified random sample with optimal allocation.)

EXAMPLE A *Muon Decay*

Two estimates have been derived for α in the problem of muon decay. The method of moments estimate is

$$\tilde{\alpha} = 3\overline{X}$$

The maximum likelihood estimate is the solution of the nonlinear equation

$$\sum_{i=1}^{n} \frac{X_i}{1 + \hat{\alpha} X_i} = 0$$

We need to find the variances of these two estimates.

Since the variance of a sample mean is σ^2/n, we compute σ^2:

$$\sigma^2 = E(X^2) - [E(X)]^2$$
$$= \int_{-1}^{1} x^2 \frac{1 + \alpha x}{2} \, dx - \frac{\alpha^2}{9}$$
$$= \frac{1}{3} - \frac{\alpha^2}{9}$$

Thus, the variance of the method of moments estimate is

$$\text{Var}(\tilde{\alpha}) = 9 \, \text{Var}(\overline{X}) = \frac{3 - \alpha^2}{n}$$

The exact variance of the mle, $\hat{\theta}$, cannot be computed in closed form, so we approximate it by the asymptotic variance,

$$\text{Var}(\hat{\alpha}) \approx \frac{1}{nI(\alpha)}$$

and then compare this asymptotic variance to the variance of $\tilde{\alpha}$. The ratio of the former to the latter is called the **asymptotic relative efficiency.** By definition,

$$I(\alpha) = E \left[\frac{\partial}{\partial \alpha} \log f(x|\alpha) \right]^2$$
$$= \int_{-1}^{1} \frac{x^2}{(1 + \alpha x)^2} \left(\frac{1 + \alpha x}{2} \right) dx$$
$$= \frac{\log \left(\dfrac{1+\alpha}{1-\alpha} \right) - 2\alpha}{2\alpha^3}, \qquad -1 < \alpha < 1, \alpha \neq 0$$
$$= \frac{1}{3}, \qquad \alpha = 0$$

The asymptotic relative efficiency is thus (for $\alpha \neq 0$)

$$\frac{\text{Var}(\hat{\alpha})}{\text{Var}(\tilde{\alpha})} = \frac{2\alpha^3}{3 - \alpha^2} \left[\frac{1}{\log \left(\dfrac{1+\alpha}{1-\alpha} \right) - 2\alpha} \right]$$

The following table gives this efficiency for various values of α between 0 and 1; symmetry would yield the values between -1 and 0.

α	Efficiency
0.0	1.0
.1	.997
.2	.989
.3	.975
.4	.953
.5	.931
.6	.878
.7	.817
.8	.727
.9	.582
.95	.464

As α tends to 1, the efficiency tends to 0. Thus, the mle is not much better than the method of moments estimate for α close to 0 but does increasingly better as α tends to 1.

It must be kept in mind that we used the asymptotic variance of the mle, so we calculated an asymptotic relative efficiency, viewing this as an approximation to the actual relative efficiency. To gain more precise information for a given sample size, a simulation of the sampling distribution of the mle could be conducted. This might be especially interesting for $\alpha = 1$, a case for which the formula for the asymptotic variance given above does not appear to make much sense. With a simulation study, the behavior of the bias as n and α vary could be analyzed (we showed that the mle is asymptotically unbiased, but there may be bias for a finite sample size), and the actual distribution could be compared to the approximating normal. ∎

In searching for an optimal estimate, we might ask whether there is a lower bound for the MSE of *any* estimate. If such a lower bound existed, it would function as a benchmark against which estimates could be compared. If an estimate achieved this lower bound, we would know that it could not be improved upon. In the case in which the estimate is unbiased, the Cramér-Rao inequality provides such a lower bound. We now state and prove the Cramér-Rao inequality.

THEOREM A   *Cramér-Rao Inequality*

Let $X_1, \ldots, X_n$ be i.i.d. with density function $f(x|\theta)$. Let $T = t(X_1, \ldots, X_n)$ be an unbiased estimate of θ. Then, under smoothness assumptions on $f(x|\theta)$,

$$\mathrm{Var}(T) \geq \frac{1}{nI(\theta)}$$

Proof

Let

$$Z = \sum_{i=1}^{n} \frac{\partial}{\partial \theta} \log f(X_i|\theta)$$

$$= \sum_{i=1}^{n} \frac{\frac{\partial}{\partial \theta} f(X_i|\theta)}{f(X_i|\theta)}$$

In Section 8.5.2, we showed that $E(Z) = 0$. Because the correlation coefficient of Z and T is less than or equal to 1 in absolute value

$$\text{Cov}^2(Z, T) \leq \text{Var}(Z)\text{Var}(T)$$

It was also shown in Section 8.5.2 that

$$\text{Var}\left[\frac{\partial}{\partial \theta} \log f(X|\theta)\right] = I(\theta)$$

Therefore,

$$\text{Var}(Z) = nI(\theta)$$

The proof will be completed by showing that $\text{Cov}(Z, T) = 1$. Since Z has mean 0,

$$\text{Cov}(Z, T) = E(ZT)$$

$$= \int \cdots \int t(x_1, \ldots, x_n) \left[\sum_{i=1}^{n} \frac{\frac{\partial}{\partial \theta} f(x_i|\theta)}{f(x_i|\theta)}\right] \prod_{j=1}^{n} f(x_j|\theta) \, dx_j$$

Noting that

$$\sum_{i=1}^{n} \frac{\frac{\partial}{\partial \theta} f(x_i|\theta)}{f(x_i|\theta)} \prod_{j=1}^{n} f(x_j|\theta) = \frac{\partial}{\partial \theta} \prod_{i=1}^{n} f(x_i|\theta)$$

we rewrite the expression for the covariance of Z and T as

$$\text{Cov}(Z, T) = \int \cdots \int t(x_1, \ldots, x_n) \frac{\partial}{\partial \theta} \prod_{i=1}^{n} f(x_i|\theta) \, dx_i$$

$$= \frac{\partial}{\partial \theta} \int \cdots \int t(x_1, \ldots, x_n) \prod_{i=1}^{n} f(x_i|\theta) \, dx_i$$

$$= \frac{\partial}{\partial \theta} E(T) = \frac{\partial}{\partial \theta}(\theta) = 1$$

which proves the inequality. [Note the interchange of differentiation and integration that must be justified by the smoothness assumptions on $f(x|\theta)$.] ∎

Theorem A gives a lower bound on the variance of *any* unbiased estimate. An unbiased estimate whose variance achieves this lower bound is said to be **efficient.** Since the asymptotic variance of a maximum likelihood estimate is equal to the lower bound, maximum likelihood estimates are said to be **asymptotically efficient.** For a finite sample size, however, a maximum likelihood estimate may not be efficient, and maximum likelihood estimates are not the only asymptotically efficient estimates.

E X A M P L E **B**   *Poisson Distribution*

In Example B in Section 8.5.3, we found that for the Poisson distribution

$$I(\lambda) = \frac{1}{\lambda}$$

Therefore, by Theorem A, for any unbiased estimate T of λ, based on a sample of independent Poisson random variables, $X_1, \ldots, X_n$,

$$\text{Var}(T) \geq \frac{\lambda}{n}$$

The mle of λ was found to be $\overline{X} = S/n$, where $S = X_1 + \cdots + X_n$. Since S follows a Poisson distribution with parameter $n\lambda$, $\text{Var}(S) = n\lambda$ and $\text{Var}(\overline{X}) = \lambda/n$. Therefore, $\overline{X}$ attains the Cramér-Rao lower bound, and we know that no unbiased estimator of λ can have a smaller variance. In this sense, $\overline{X}$ is optimal for the Poisson distribution. But note that the theorem does not preclude the possibility that there is a biased estimator of λ that has a smaller mean squared error than $\overline{X}$ does.   ■

8.7.1   An Example: The Negative Binomial Distribution

The Poisson distribution is often the first model considered for random counts; it has the property that the mean of the distribution is equal to the variance. When it is found that the variance of the counts is substantially larger than the mean, the negative binomial distribution is sometimes instead considered as a model. We consider a reparametrization and generalization of the negative binomial distribution introduced in Section 2.1.3, which is a discrete distribution on the nonnegative integers with a frequency function depending on the parameters m and k:

$$f(x|m, k) = \left(1 + \frac{m}{k}\right)^{-k} \frac{\Gamma(k + x)}{x!\,\Gamma(k)} \left(\frac{m}{m + k}\right)^x$$

The mean and variance of the negative binomial distribution can be shown to be

$$\mu = m$$
$$\sigma^2 = m + \frac{m^2}{k}$$

It is apparent that this distribution is overdispersed ($\sigma^2 > \mu$) relative to the Poisson. We will not derive the mean and variance. (They are most easily obtained by using moment-generating functions.)

The negative binomial distribution can be used as a model in several cases:

- If k is an integer, the distribution of the number of successes up to the kth failure in a sequence of independent Bernoulli trials with probability of success $p = m/(m+k)$ is negative binomial.
- Suppose that Λ is a random variable following a gamma distribution and that for λ, a given value of Λ, X follows a Poisson distribution with mean λ. It can be shown that the unconditional distribution of X is negative binomial. Thus, for situations in which the rate varies randomly over time or space, the negative binomial distribution might tentatively be considered as a model.
- The negative binomial distribution also arises with a particular type of clustering. Suppose that counts of colonies, or clusters, follow a Poisson distribution and that each colony has a random number of individuals. If the probability distribution of the number of individuals per colony is of a particular form (the logarithmic series distribution), it can be shown that the distribution of counts of individuals is negative binomial. The negative binomial distribution might be a plausible model for the distribution of insect counts if the insects hatch from depositions, or clumps, of larvae.
- The negative binomial distribution can be applied to model population size in a certain birth/death process, the assumption being that the birth rate and death rate per individual are constant and that there is a constant rate of immigration.

Anscombe (1950) discusses estimation of the parameters m and k and compares the efficiencies of several methods of estimation. The simplest method is the method of moments; from the relations of m and k to μ and σ^2 given previously, the method of moments estimates of m and k are

$$\hat{m} = \overline{X}$$
$$\hat{k} = \frac{\overline{X}^2}{\hat{\sigma}^2 - \overline{X}}$$

Another relatively simple method of estimation of m and k is based on the number of zeros. The probability of the count being zero is

$$p_0 = \left(1 + \frac{m}{k}\right)^{-k}$$

If m is estimated by the sample mean and there are n_0 zeros out of a sample size of n, then k is estimated by $\hat{k}$, where $\hat{k}$ satisfies

$$\frac{n_0}{n} = \left(1 + \frac{\overline{X}}{\hat{k}}\right)^{-\hat{k}}$$

Although the solution cannot be obtained in closed form, it is not difficult to find by iteration.

Figure 8.11, from Anscombe (1950), shows the asymptotic efficiencies of the two methods of estimation of the negative binomial parameters relative to the maximum likelihood estimate. In the figure, the method of moments is method 1 and the method based on the number of zeros is method 2. Method 2 is quite efficient when the mean is small—that is, when there are a large number of zeros. Method 1 becomes more efficient as k increases.

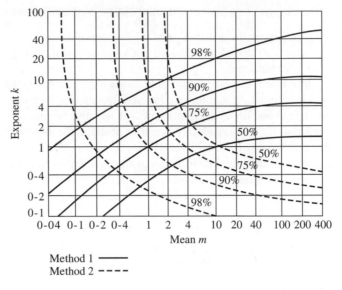

FIGURE **8.11**   Asymptotic efficiencies of estimates of negative binomial parameters.

The maximum likelihood estimate is asymptotically efficient but is somewhat more difficult to compute. The equations will not be written out here. Bliss and Fisher (1953) discuss computational methods and give several examples. The maximum likelihood estimate of m is the sample mean, but that of k is the solution of a nonlinear equation.

EXAMPLE **A**   *Insect Counts*
Let us consider an example from Bliss and Fisher (1953). From each of 6 apple trees in an orchard that was sprayed, 25 leaves were selected. On each of the leaves, the number of adult female red mites was counted. Intuitively, we might conclude that this situation was too heterogeneous for a Poisson model to fit; the rates of infestation might be different on different trees and at different locations on the same tree. The following table shows the observed counts and the expected counts from fitting Poisson and negative binomial distributions. The mle's for k and m were $\hat{k} = 1.025$ and $\hat{m} = 1.146$.

Number per Leaf	Observed Count	Poisson Distribution	Negative Binomial Distribution
0	70	47.7	69.5
1	38	54.6	37.6
2	17	31.3	20.1
3	10	12.0	10.7
4	9	3.4	5.7
5	3	.75	3.0
6	2	.15	1.6
7	1	.03	.85
8+	0	.00	.95

Casual inspection of this table makes it clear that the Poisson does not fit; there are many more small and large counts observed than are expected for a Poisson distribution.   ∎

A recursive relation is useful in fitting the negative binomial distribution:

$$p_0 = \left(1 + \frac{m}{k}\right)^{-k}$$

$$p_n = \frac{k+n-1}{n}\left(\frac{m}{k+m}\right)p_{n-1}$$

8.8 Sufficiency

This section introduces the concept of sufficiency and some of its theoretical implications. Suppose that $X_1, \ldots, X_n$ is a sample from a probability distribution with the density or frequency function $f(x|\theta)$. The concept of sufficiency arises as an attempt to answer the following question: Is there a statistic, a function $T(X_1, \ldots, X_n)$, that contains all the information in the sample about θ? If so, a reduction of the original data to this statistic without loss of information is possible. For example, consider a sequence of independent Bernoulli trials with unknown probability of success, θ. We may have the intuitive feeling that the total number of successes contains all the information about θ that there is in the sample, that the order in which the successes occurred, for example, does not give any additional information. The following definition formalizes this idea.

> DEFINITION
>
> A statistic $T(X_1, \ldots, X_n)$ is said to be **sufficient** for θ if the conditional distribution of $X_1, \ldots, X_n$, given $T = t$, does not depend on θ for any value of t.   ∎

In other words, given the value of T, which is called a **sufficient statistic,** we can gain no more knowledge about θ from knowing more about the probability distribution of $X_1, \ldots, X_n$. (Formally, we could envision keeping only T and throwing away all the X_i without any loss of information. Informally, and more realistically, this would make no sense at all. The values of the X_i might indicate that the model did not fit or that something was fishy about the data. What would you think, for example, if you saw 50 ones followed by 50 zeros in a sequence of supposedly independent Bernoulli trials?)

EXAMPLE A Let $X_1, \ldots, X_n$ be a sequence of independent Bernoulli random variables with $P(X_i = 1) = \theta$. We will verify that $T = \sum_{i=1}^{n} X_i$ is sufficient for θ.

$$P(X_1 = x_1, \ldots, X_n = x_n | T = t) = \frac{P(X_1 = x_1, \ldots, X_n = x_n, T = t)}{P(T = t)}$$

Bearing in mind that the X_i can take on only the values 0s and 1s, the probability in the numerator is the probability that some particular set of t X_i are equal to 1s and the other $n - t$ are 0s. Since the X_i are independent, the probability of this is the product of the marginal probabilities, or $\theta^t (1 - \theta)^{n-t}$. To find the denominator note that the distribution of T, the total number of ones, is binomial with n trials and probability of success θ. The ratio in question is thus

$$\frac{\theta^t (1 - \theta)^{n-t}}{\binom{n}{t} \theta^t (1 - \theta)^{n-t}} = \frac{1}{\binom{n}{t}}$$

The conditional distribution thus does not involve θ at all. Given the total number of ones, the probability that they occur on any particular set of t trials is the same for any value of θ so that set of trials contains no additional information about θ. ∎

8.8.1 A Factorization Theorem

The preceding definition of sufficiency is hard to work with, because it does not indicate how to go about finding a sufficient statistic, and given a candidate statistic, T, it would typically be very hard to conclude whether it was sufficient because of the difficulty in evaluating the conditional distribution. The following factorization theorem provides a convenient means of identifying sufficient statistics.

THEOREM A

A necessary and sufficient condition for $T(X_1, \ldots, X_n)$ to be sufficient for a parameter θ is that the joint probability function (density function or frequency function) factors in the form

$$f(x_1, \ldots, x_n | \theta) = g[T(x_1, \ldots, x_n), \theta] h(x_1, \ldots, x_n)$$

Proof

We give a proof for the discrete case. (The proof for the general case is more subtle and requires regularity conditions, but the basic ideas are the same.) First, suppose that the frequency function factors as given in the theorem. To simplify notation, we will let $\mathbf{X}$ denote $(X_1, \ldots, X_n)$ and $\mathbf{x}$ denote $(x_1, \ldots, x_n)$. We have

$$P(T = t) = \sum_{T(x)=t} P(\mathbf{X} = \mathbf{x})$$

$$= g(t, \theta) \sum_{T(x)=t} h(\mathbf{x})$$

Here the notation indicates that the sum is over all $\mathbf{x}$ such that $T(\mathbf{x}) = t$. We then have

$$P(\mathbf{X} = \mathbf{x} | T = t) = \frac{P(\mathbf{X} = \mathbf{x}, T = t)}{P(T = t)}$$

$$= \frac{h(\mathbf{x})}{\sum_{T(\mathbf{X})=t} h(\mathbf{x})}$$

This conditional distribution does not depend on θ, as was to be shown.

To show that the conclusion holds in the other direction, suppose that the conditional distribution of $\mathbf{X}$ given T is independent of θ. Let

$$g(t, \theta) = P(T = t | \theta)$$
$$h(\mathbf{x}) = P(\mathbf{X} = \mathbf{x} | T = t)$$

We then have

$$P(\mathbf{X} = \mathbf{x} | \theta) = P(T = t | \theta) P(\mathbf{X} = \mathbf{x} | T = t)$$
$$= g(t, \theta) h(\mathbf{x})$$

as was to be shown. ∎

We can demonstrate the utility of Theorem A by applying it to some examples. More examples are included in the problems at the end of this chapter.

E X A M P L E **A**   Consider a sequence of independent Bernoulli random variables, $X_1, \ldots, X_n$, where

$$P(X_i = x) = \theta^x (1 - \theta)^{1-x}, \qquad x = 0 \text{ or } x = 1$$

then

$$f(\mathbf{x} | \theta) = \prod_{i=1}^{n} \theta^{x_i} (1 - \theta)^{1-x_i}$$

$$= \theta^{\sum_{i=1}^{n} x_i} (1 - \theta)^{n - \sum_{i=1}^{n} x_i}$$

$$= \left(\frac{\theta}{1 - \theta} \right)^{\sum_{i=1}^{n} x_i} (1 - \theta)^n$$

We see that $f(\mathbf{x}|\theta)$ depends only on $x_1, x_2, \ldots, x_n$ through the sufficient statistic $t = \sum_{i=1}^{n} x_i$ and $f(x|\theta)$ is of the form $g(\sum_{i=1}^{n} x_i, \theta)h(\mathbf{x})$, where $h(\mathbf{x}) = 1$ and

$$g(t, \theta) = \left(\frac{\theta}{1-\theta}\right)^t (1-\theta)^n$$

∎

E X A M P L E B   Consider a random sample from a normal distribution that has an unknown mean and variance. We have

$$f(\mathbf{x}|\mu, \sigma) = \prod_{i=1}^{n} \frac{1}{\sigma\sqrt{2\pi}} \exp\left[\frac{-1}{2\sigma^2}(x_i - \mu)^2\right]$$

$$= \frac{1}{\sigma^n (2\pi)^{n/2}} \exp\left[\frac{-1}{2\sigma^2} \sum_{i=1}^{n} (x_i - \mu)^2\right]$$

$$= \frac{1}{\sigma^n (2\pi)^{n/2}} \exp\left[\frac{-1}{2\sigma^2} \left(\sum_{i=1}^{n} x_i^2 - 2\mu \sum_{i=1}^{n} x_i + n\mu^2\right)\right]$$

This expression is just a function of $\sum_{i=1}^{n} x_i$ and $\sum_{i=1}^{n} x_i^2$, which are therefore sufficient statistics. In this example we have a two-dimensional sufficient statistic. Although Theorem A was stated explicitly for a one-dimensional sufficient statistic, the multidimensional analogue holds also.   ∎

Because the likelihood,

$$f(x_1, \ldots, x_n; \theta) = g[T(x_1, \ldots, x_n), \theta]h(x_1, \ldots, x_n)$$

it depends only on the data through $T(x_1, \ldots, x_n)$. The maximum likelihood estimate is found by maximizing $g[T(x_1, \ldots, x_n), \theta]$. In Example A, the likelihood is a function of $t = \sum_{i=1}^{n} x_i$, and the maximum likelihood estimate is $\hat{\theta} = t/n$.

Similarly, in a Bayesian framework, the posterior distribution of θ is proportional to the product of the prior distribution of θ and the likelihood. As a function of θ, the posterior distribution thus depends only on the data through $g[T(x_1, \ldots, x_n), \theta]$—the posterior probability of θ is the same for all $\{x_1, \ldots, x_n\}$ which have a common value of $T(x_1, \ldots, x_n)$. The sufficient statistic carries all the information about θ that is contained in the data $x_1, x_2, \ldots, x_n$.

A study of the properties of probability distributions that have sufficient statistics of the same dimension as the parameter space regardless of sample size led to the development of what is called the **exponential family** of probability distributions. Many common distributions, including the normal, the binomial, the Poisson, and the gamma, are members of this family. One-parameter members of the exponential family have density or frequency functions of the form

$$f(x|\theta) = \exp[c(\theta)T(x) + d(\theta) + S(x)], \qquad x \in A$$

$$= 0, \qquad x \notin A$$

where the set A does not depend on θ. Suppose that $X_1, \ldots, X_n$ is a sample from a member of the exponential family; the joint probability function is

$$f(\mathbf{x}|\theta) = \prod_{i=1}^{n} \exp[c(\theta)T(x_i) + d(\theta) + S(x_i)]$$

$$= \exp\left[c(\theta)\sum_{i=1}^{n} T(x_i) + nd(\theta)\right] \exp\left[\sum_{i=1}^{n} S(x_i)\right]$$

From this result, it is apparent by the factorization theorem that $\sum_{i=1}^{n} T(X_i)$ is a sufficient statistic.

EXAMPLE C   The frequency function of the Bernoulli distribution is

$$P(X = x) = \theta^x(1 - \theta)^{1-x}, \qquad x = 0 \text{ or } x = 1$$

$$= \exp\left[x \log\left(\frac{\theta}{1 - \theta}\right) + \log(1 - \theta)\right]$$

This is a member of the exponential family with $T(x) = x$, and we have already seen that $\sum_{i=1}^{n} X_i$, is a sufficient statistic for a sample from the Bernoulli distribution.   ■

A k-parameter member of the exponential family has a density or frequency function of the form

$$f(x|\theta) = \exp\left[\sum_{i=1}^{k} c_i(\theta)T_i(x) + d(\theta) + S(x)\right], \qquad x \in A$$

$$= 0, \qquad x \notin A$$

where the set A does not depend on θ.

The normal distribution is of this form. A great deal of theoretical work has centered around the exponential family; further discussion of this family can be found in Bickel and Doksum (2001).

We conclude this section with the following corollary of Theorem A.

COROLLARY A

If T is sufficient for θ, the maximum likelihood estimate is a function of T.

Proof

From Theorem A, the likelihood is $g(T, \theta)h(\mathbf{x})$, which depends on θ only through T. To maximize this quantity, we need only maximize $g(T, \theta)$.   ■

Corollary A and the Rao-Blackwell theorem of the next section may be interpreted as giving some theoretical support to the use of maximum likelihood estimates.

8.8.2 The Rao-Blackwell Theorem

In the preceding section, we argued for the importance of sufficient statistics on essentially qualitative grounds. The Rao-Blackwell theorem gives a quantitative rationale for basing an estimator of a parameter θ on a sufficient statistic if one exists.

THEOREM **A**   *Rao-Blackwell Theorem*

Let $\hat{\theta}$ be an estimator of θ with $E(\hat{\theta}^2) < \infty$ for all θ. Suppose that T is sufficient for θ, and let $\tilde{\theta} = E(\hat{\theta}|T)$. Then, for all θ,

$$E(\tilde{\theta} - \theta)^2 \le E(\hat{\theta} - \theta)^2$$

The inequality is strict unless $\hat{\theta} = \tilde{\theta}$.

Proof

We first note that, from the property of iterated conditional expectation (Theorem A of Section 4.4.1),

$$E(\tilde{\theta}) = E[E(\hat{\theta}|T)] = E(\hat{\theta})$$

Therefore, to compare the mean squared error of the two estimators, we need only compare their variances. From Theorem B of Section 4.4.1, we have

$$\text{Var}(\hat{\theta}) = \text{Var}[E(\hat{\theta}|T)] + E[\text{Var}(\hat{\theta}|T)]$$

or

$$\text{Var}(\hat{\theta}) = \text{Var}(\tilde{\theta}) + E[\text{Var}(\hat{\theta}|T)]$$

Thus, $\text{Var}(\hat{\theta}) > \text{Var}(\tilde{\theta})$ unless $\text{Var}(\hat{\theta}|T) = 0$, which is the case only if $\hat{\theta}$ is a function of T, which would imply $\hat{\theta} = \tilde{\theta}$. ∎

Since $E(\hat{\theta}|T)$ is a function of the sufficient statistic T, the Rao-Blackwell theorem gives a strong rationale for basing estimators on sufficient statistics if they exist. If an estimator is not a function of a sufficient statistic, it can be improved.

Suppose that there are two estimates, $\hat{\theta}_1$ and $\hat{\theta}_2$, having the same expectation. Assuming that a sufficient statistic T exists, we may construct two other estimates, $\tilde{\theta}_1$ and $\tilde{\theta}_2$, by conditioning on T. The theory we have developed so far gives no clues as to which one of these two is better. If the probability distribution of T has the property called *completeness*, $\tilde{\theta}_1$ and $\tilde{\theta}_2$ are identical, by a theorem of Lehmann and Scheffé. We will not define completeness or pursue this topic further; Lehmann and Casella (1998) and Bickel and Doksum (2001) discuss this concept.

8.9 Concluding Remarks

Certain key ideas first introduced in the context of survey sampling in Chapter 7 have recurred in this chapter. We have viewed an estimate as a random variable having a probability distribution called its sampling distribution. In Chapter 7, the estimate was of a parameter, such as the mean, of a finite population; in this chapter, the estimate was of a parameter of a probability distribution. In both cases, characteristics of the sampling distribution, such as the bias and the variance and the large sample approximate form, have been of interest. In both chapters, we studied confidence intervals for the true value of the unknown parameter. The method of propagation of error, or linearization, has been a useful tool in both chapters. These key ideas will be important in other contexts in later chapters as well.

Important concepts and techniques in estimation theory were introduced in this chapter. We discussed two general methods of estimation—the method of moments and the method of maximum likelihood. The latter especially has great general utility in statistics. We developed and applied some approximate distribution theory for maximum likelihood estimates. Other theoretical developments included the concept of efficiency, the Cramér-Rao lower bound, and the concept of sufficiency and some of its consequences.

Bayesian inference was introduced in this chapter. The point of view contrasts rather sharply with that of frequentist inference in that the Bayesian formalism allows uncertainty statements about parameter values to be probabilistic, for example, "After seeing the data, the probability is 95% that $1.8 \leq \theta \leq 6.3$." In frequentist inference, θ is not a random variable, and a statement like this would literally make no sense; it would be replaced by, "A 95% confidence interval for θ is $[1.8, 6.3]$," perhaps followed by a long convoluted explication of the meaning of a confidence interval. Despite this apparently sharp philosophical difference, Bayesian and frequentist procedures have a great deal in common and typically lead to similar conclusions. Despite the distinction between the two statements above, the statements may well mean essentially the same thing operationally to a practitioner who has analyzed the data. The likelihood function is fundamental for both frequentist and Bayesian inference. In an application, the choice of a model, that is, the choice of a likelihood function, will typically be much more important than whether on subsequently multiplies it by a prior or just maximizes it. This is especially true if flat priors are used; in fact, one might regard a flat prior as a device that allows the likelihood to be treated as a probability density.

In this chapter, we introduced the bootstrap method for assessing the variability of an estimate. Such uses of simulation have become increasingly widespread as computers have become faster and cheaper; the bootstrap as a general method has been developed only quite recently and has rapidly become one of the most important statistical tools. We will see other situations in which the bootstrap is useful in later chapters. Efron and Tibshirani (1993) give an excellent introduction to the theory and applications of the bootstrap.

The context in which we have introduced the bootstrap is often referred to as the **parametric bootstrap.** The nonparametric bootstrap will be introduced in Chapter 10. The parametric bootstrap can be thought about somewhat abstractly in the following way. We have data $\mathbf{x}$ that we regard as being generated by a probability distribution $F(\mathbf{x}|\theta)$, which depends on a parameter θ. We wish to know $Eh(\mathbf{X}, \theta)$ for some function $h(\ )$. For example, if θ itself is estimated from the data as $\hat{\theta}(\mathbf{x})$ and $h(\mathbf{X}, \theta) = [\hat{\theta}(\mathbf{X}) - \theta]^2$, then $Eh(\mathbf{X}, \theta)$ is the mean square error of the estimate. As another example, if

$$h(\mathbf{X}, \theta) = \begin{cases} 1 \text{ if } |\hat{\theta}(\mathbf{X}) - \theta| > \Delta \\ 0 \text{ otherwise} \end{cases}$$

then $Eh(\mathbf{X}, \theta)$ is the probability that $|\hat{\theta}(\mathbf{X}) - \theta| > \Delta$. We realize that if θ were known, we could use the computer to generate independent random variables $\mathbf{X}_1, \mathbf{X}_2, \ldots, \mathbf{X}_B$ from $F(\mathbf{x}|\theta)$ and then appeal to the law of large numbers:

$$Eh(\mathbf{X}, \theta) \approx \frac{1}{B} \sum_{i=1}^{B} h(\mathbf{X}_i, \theta)$$

This approximation could be made arbitrarily precise by choosing B sufficiently large. The parametric bootstrap principle is to perform this Monte Carlo simulation using $\hat{\theta}$ in place of the unknown θ—that is, using $F(\mathbf{x}|\hat{\theta})$ to generate the $\mathbf{X}_i$. It is difficult to give a concise answer to the natural question: How much error is introduced by using $\hat{\theta}$ in place of θ? The answer depends on the continuity of $Eh(\mathbf{X}, \theta)$ as a function of θ—if small changes in θ can give rise to large changes in $Eh(\mathbf{X}, \theta)$, the parametric bootstrap will not work well.

8.10 Problems

1. The following table gives the observed counts in 1-second intervals for Berkson's data (Section 8.2). What are the expected counts from a Poisson distribution? Do they match the observed counts?

n	Observed
0	5267
1	4436
2	1800
3	534
4	111
5+	21

2. The Poisson distribution has been used by traffic engineers as a model for light traffic, based on the rationale that if the rate is approximately constant and the traffic is light (so the individual cars move independently of each other), the distribution of counts of cars in a given time interval or space area should be nearly Poisson (Gerlough and Schuhl 1955). The following table shows the number of right turns during 300 3-min intervals at a specific intersection. Fit a Poisson distribution. Comment on the fit by comparing observed and expected counts. It is useful to know that the 300 intervals were distributed over various hours of the day and various days of the week.

n	Frequency
0	14
1	30
2	36
3	68
4	43
5	43
6	30
7	14
8	10
9	6
10	4
11	1
12	1
13+	0

3. One of the earliest applications of the Poisson distribution was made by Student (1907) in studying errors made in counting yeast cells or blood corpuscles with a haemacytometer. In this study, yeast cells were killed and mixed with water and gelatin; the mixture was then spread on a glass and allowed to cool. Four different concentrations were used. Counts were made on 400 squares, and the data are summarized in the following table:

Number of Cells	Concentration 1	Concentration 2	Concentration 3	Concentration 4
0	213	103	75	0
1	128	143	103	20
2	37	98	121	43
3	18	42	54	53
4	3	8	30	86
5	1	4	13	70
6	0	2	2	54
7	0	0	1	37
8	0	0	0	18
9	0	0	1	10
10	0	0	0	5
11	0	0	0	2
12	0	0	0	2

   a. Estimate the parameter λ for each of the four sets of data.
   b. Find an approximate 95% confidence interval for each estimate.
   c. Compare observed and expected counts.

4. Suppose that X is a discrete random variable with

$$P(X = 0) = \frac{2}{3}\theta$$

$$P(X = 1) = \frac{1}{3}\theta$$

$$P(X = 2) = \frac{2}{3}(1 - \theta)$$

$$P(X = 3) = \frac{1}{3}(1 - \theta)$$

where $0 \le \theta \le 1$ is a parameter. The following 10 independent observations were taken from such a distribution: (3, 0, 2, 1, 3, 2, 1, 0, 2, 1).

a. Find the method of moments estimate of θ.

b. Find an approximate standard error for your estimate.

c. What is the maximum likelihood estimate of θ?

d. What is an approximate standard error of the maximum likelihood estimate?

e. If the prior distribution of Θ is uniform on [0, 1], what is the posterior density? Plot it. What is the mode of the posterior?

5. Suppose that X is a discrete random variable with $P(X = 1) = \theta$ and $P(X = 2) = 1 - \theta$. Three independent observations of X are made: $x_1 = 1$, $x_2 = 2$, $x_3 = 2$.

a. Find the method of moments estimate of θ.

b. What is the likelihood function?

c. What is the maximum likelihood estimate of θ?

d. If Θ has a prior distribution that is uniform on [0, 1], what is its posterior density?

6. Suppose that $X \sim \text{bin}(n, p)$.

a. Show that the mle of p is $\hat{p} = X/n$.

b. Show that mle of part (a) attains the Cramér-Rao lower bound.

c. If $n = 10$ and $X = 5$, plot the log likelihood function.

7. Suppose that X follows a geometric distribution,

$$P(X = k) = p(1 - p)^{k-1}$$

and assume an i.i.d. sample of size n.

a. Find the method of moments estimate of p.

b. Find the mle of p.

c. Find the asymptotic variance of the mle.

d. Let p have a uniform prior distribution on [0, 1]. What is the posterior distribution of p? What is the posterior mean?

8. In an ecological study of the feeding behavior of birds, the number of hops between flights was counted for several birds. For the following data, (a) fit a geometric distribution, (b) find an approximate 95% confidence interval for p, (c) examine goodness of fit. (d) If a uniform prior is used for p, what is the posterior distribution and what are the posterior mean and standard deviation?

Number of Hops	Frequency
1	48
2	31
3	20
4	9
5	6
6	5
7	4
8	2
9	1
10	1
11	2
12	1

9. How would you respond to the following argument? This talk of sampling distributions is ridiculous! Consider Example A of Section 8.4. The experimenter found the mean number of fibers to be 24.9. How can this be a "random variable" with an associated "probability distribution" when it's just a number? The author of this book is guilty of deliberate mystification!

10. Use the normal approximation of the Poisson distribution to sketch the approximate sampling distribution of $\hat{\lambda}$ of Example A of Section 8.4. According to this approximation, what is $P(|\lambda_0 - \hat{\lambda}| > \delta)$ for $\delta = .5, 1, 1.5, 2$, and 2.5, where λ_0 denotes the true value of λ?

11. In Example A of Section 8.4, we used knowledge of the exact form of the sampling distribution of $\hat{\lambda}$ to estimate its standard error by

$$s_{\hat{\lambda}} = \sqrt{\frac{\hat{\lambda}}{n}}$$

This was arrived at by realizing that $\sum X_i$ follows a Poisson distribution with parameter $n\lambda_0$. Now suppose we hadn't realized this but had used the bootstrap, letting the computer do our work for us by generating B samples of size $n = 23$ of Poisson random variables with parameter $\lambda = 24.9$, forming the mle of λ from each sample, and then finally computing the standard deviation of the resulting collection of estimates and taking this as an estimate of the standard error of $\hat{\lambda}$. Argue that as $B \to \infty$, the standard error estimated in this way will tend to $s_{\hat{\lambda}}$.

12. Suppose that you had to choose either the method of moments estimates or the maximum likelihood estimates in Example C of Section 8.4 and C of Section 8.5. Which would you choose and why?

13. In Example D of Section 8.4, the method of moments estimate was found to be $\hat{\alpha} = 3\overline{X}$. In this problem, you will consider the sampling distribution of $\hat{\alpha}$.

   a. Show that $E(\hat{\alpha}) = \alpha$—that is, that the estimate is unbiased.

b. Show that $\text{Var}(\hat{\alpha}) = (3 - \alpha^2)/n$. [*Hint:* What is $\text{Var}(\overline{X})$?]

c. Use the central limit theorem to deduce a normal approximation to the sampling distribution of $\hat{\alpha}$. According to this approximation, if $n = 25$ and $\alpha = 0$, what is the $P(|\hat{\alpha}| > .5)$?

14. In Example C of Section 8.5, how could you use the bootstrap to estimate the following measures of the accuracy of $\hat{\alpha}$: (a) $P(|\hat{\alpha} - \alpha_0| > .05)$, (b) $E(|\hat{\alpha} - \alpha_0|)$, (c) that number Δ such that $P(|\hat{\alpha} - \alpha_0| > \Delta) = .5$.

15. The upper quartile of a distribution with cumulative distribution F is that point $q_{.25}$ such that $F(q_{.25}) = .75$. For a gamma distribution, the upper quartile depends on α and λ, so denote it as $q(\alpha, \lambda)$. If a gamma distribution is fit to data as in Example C of Section 8.5 and the parameters α and λ are estimated by $\hat{\alpha}$ and $\hat{\lambda}$, the upper quartile could then be estimated by $\hat{q} = q(\hat{\alpha}, \hat{\lambda})$. Explain how to use the bootstrap to estimate the standard error of $\hat{q}$.

16. Consider an i.i.d. sample of random variables with density function

$$f(x|\sigma) = \frac{1}{2\sigma} \exp\left(-\frac{|x|}{\sigma}\right).$$

a. Find the method of moments estimate of σ.

b. Find the maximum likelihood estimate of σ.

c. Find the asymptotic variance of the mle.

d. Find a sufficient statistic for σ.

17. Suppose that $X_1, X_2, \ldots, X_n$ are i.i.d. random variables on the interval $[0, 1]$ with the density function

$$f(x|\alpha) = \frac{\Gamma(2\alpha)}{\Gamma(\alpha)^2}[x(1 - x)]^{\alpha-1}$$

where $\alpha > 0$ is a parameter to be estimated from the sample. It can be shown that

$$E(X) = \frac{1}{2}$$

$$\text{Var}(X) = \frac{1}{4(2\alpha + 1)}$$

a. How does the shape of the density depend on α?

b. How can the method of moments be used to estimate α?

c. What equation does the mle of α satisfy?

d. What is the asymptotic variance of the mle?

e. Find a sufficient statistic for α.

18. Suppose that $X_1, X_2, \ldots, X_n$ are i.i.d. random variables on the interval $[0, 1]$ with the density function

$$f(x|\alpha) = \frac{\Gamma(3\alpha)}{\Gamma(\alpha)\Gamma(2\alpha)}x^{\alpha-1}(1 - x)^{2\alpha-1}$$

where $\alpha > 0$ is a parameter to be estimated from the sample. It can be shown that

$$E(X) = \frac{1}{3}$$

$$\text{Var}(X) = \frac{2}{9(3\alpha + 1)}$$

a. How could the method of moments be used to estimate α?
b. What equation does the mle of α satisfy?
c. What is the asymptotic variance of the mle?
d. Find a sufficient statistic for α.

19. Suppose that $X_1, X_2, \ldots, X_n$ are i.i.d. $N(\mu, \sigma^2)$.

  a. If μ is known, what is the mle of σ?
  b. If σ is known, what is the mle of μ?
  c. In the case above (σ known), does any other unbiased estimate of μ have smaller variance?

20. Suppose that $X_1, X_2, \ldots, X_{25}$ are i.i.d. $N(\mu, \sigma^2)$, where $\mu = 0$ and $\sigma = 10$. Plot the sampling distributions of $\overline{X}$ and $\hat{\sigma}^2$.

21. Suppose that $X_1, X_2, \ldots, X_n$ are i.i.d. with density function

$$f(x|\theta) = e^{-(x-\theta)}, \qquad x \geq \theta$$

and $f(x|\theta) = 0$ otherwise.

  a. Find the method of moments estimate of θ.
  b. Find the mle of θ. (*Hint:* Be careful, and don't differentiate before thinking. For what values of θ is the likelihood positive?)
  c. Find a sufficient statistic for θ.

22. The Weibull distribution was defined in Problem 67 of Chapter 2. This distribution is sometimes fit to lifetimes. Describe how to fit this distribution to data and how to find approximate standard errors of the parameter estimates.

23. A company has manufactured certain objects and has printed a serial number on each manufactured object. The serial numbers start at 1 and end at N, where N is the number of objects that have been manufactured. One of these objects is selected at random, and the serial number of that object is 888. What is the method of moments estimate of N? What is the mle of N?

24. Find a very new shiny penny. Hold it on its edge and spin it. Do this 20 times and count the number of times it comes to rest heads up. Letting π denote the probability of a head, graph the log likelihood of π. Next, repeat the experiment in a slightly different way: This time spin the coin until 10 heads come up. Again, graph the log likelihood of π.

25. If a thumbtack is tossed in the air, it can come to rest on the ground with either the point up or the point touching the ground. Find a thumbtack. Before doing any experiment, what do you think π, the probability of it landing point up, is? Next, toss the thumbtack 20 times and graph the log likelihood of π. Then do

another experiment: Toss the thumbtack until it lands point up 5 times, and graph the log likelihood of π based on this experiment.

Find and graph the posterior distribution arising from a uniform prior on π. Find the posterior mean and standard deviation and compare the posterior with a normal distribution with that mean and standard deviation. Finally, toss the thumbtack 20 more times and compare the posterior distribution based on all 40 tosses to that based on the first 20.

26. In an effort to determine the size of an animal population, 100 animals are captured and tagged. Some time later, another 50 animals are captured, and it is found that 20 of them are tagged. How would you estimate the population size? What assumptions about the capture/recapture process do you need to make? (See Example I of Section 1.4.2.)

27. Suppose that certain electronic components have lifetimes that are exponentially distributed: $f(t|\tau) = (1/\tau)\exp(-t/\tau), t \geq 0$. Five new components are put on test, the first one fails at 100 days, and no further observations are recorded.

   a. What is the likelihood function of τ?
   b. What is the mle of τ?
   c. What is the sampling distribution of the mle?
   d. What is the standard error of the mle?

   (*Hint:* See Example A of Section 3.7.)

28. Why do the intervals in the left panel of Figure 8.8 have different centers? Why do they have different lengths?

29. Are the estimates of σ^2 at the centers of the confidence intervals shown in the right panel of Figure 8.8? Why are some intervals so short and others so long? For which of the samples that produced these confidence intervals was $\hat{\sigma}^2$ smallest?

30. The exponential distribution is $f(x; \lambda) = \lambda e^{-\lambda x}$ and $E(X) = \lambda^{-1}$. The cumulative distribution function is $F(x) = P(X \leq x) = 1 - e^{-\lambda x}$. Three observations are made by an instrument that reports $x_1 = 5$ and $x_2 = 3$, but x_3 is too large for the instrument to measure and it reports only that $x_3 > 10$. (The largest value the instrument can measure is 10.0.)

   a. What is the likelihood function?
   b. What is the mle of λ?

31. George spins a coin three times and observes no heads. He then gives the coin to Hilary. She spins it until the first head occurs, and ends up spinning it four times total. Let θ denote the probability the coin comes up heads.

   a. What is the likelihood of θ?
   b. What is the MLE of θ?

32. The following 16 numbers came from normal random number generator on a computer:

5.3299	4.2537	3.1502	3.7032	1.6070	6.3923	3.1181
6.5941	3.5281	4.7433	0.1077	1.5977	5.4920	1.7220
4.1547	2.2799					

a. What would you guess the mean and variance (μ and σ^2) of the generating normal distribution were?

b. Give 90%, 95%, and 99% confidence intervals for μ and σ^2.

c. Give 90%, 95%, and 99% confidence intervals for σ.

d. How much larger a sample do you think you would need to halve the length of the confidence interval for μ?

33. Suppose that $X_1, X_2, \ldots, X_n$ are i.i.d. $N(\mu, \sigma^2)$, where μ and σ are unknown. How should the constant c be chosen so that the interval $(-\infty, \overline{X} + c)$ is a 95% confidence interval for μ; that is, c should be chosen so that $P(-\infty < \mu \leq \overline{X} + c) = .95$.

34. Suppose that $X_1, X_2, \ldots, X_n$ are i.i.d. $N(\mu_0, \sigma_0^2)$ and μ and σ^2 are estimated by the method of maximum likelihood, with resulting estimates $\hat{\mu}$ and $\hat{\sigma}^2$. Suppose the bootstrap is used to estimate the sampling distribution of $\hat{\mu}$.

   a. Explain why the bootstrap estimate of the distribution of $\hat{\mu}$ is $N(\hat{\mu}, \frac{\hat{\sigma}^2}{n})$.

   b. Explain why the bootstrap estimate of the distribution of $\hat{\mu} - \mu_0$ is $N(0, \frac{\hat{\sigma}^2}{n})$.

   c. According to the result of the previous part, what is the form of the bootstrap confidence interval for μ, and how does it compare to the exact confidence interval based on the t distribution?

35. (Bootstrap in Example A of Section 8.5.1) Let $U_1, U_2, \ldots, U_{1029}$ be independent uniformly distributed random variables. Let X_1 equal the number of U_i less than .331, X_2 equal the number between .331 and .820, and X_3 equal the number greater than .820. Why is the joint distribution of X_1, X_2, and X_3 multinomial with probabilities .331, .489, and .180 and $n = 1029$?

36. How do the approximate 90% confidence intervals in Example E of Section 8.5.3 compare to those that would be obtained approximating the sampling distributions of $\hat{\alpha}$ and $\hat{\lambda}$ by normal distributions with standard deviations given by $s_{\hat{\alpha}}$ and $s_{\hat{\lambda}}$ as in Example C of Section 8.5?

37. Using the notation of Section 8.5.3, suppose that $\underline{\theta}$ and $\overline{\theta}$ are lower and upper quantiles of the distribution of θ^*. Show that the bootstrap confidence interval for θ can be written as $(2\hat{\theta} - \overline{\theta}, 2\hat{\theta} - \underline{\theta})$.

38. Continuing Problem 37, show that if the sampling distribution of θ^* is symmetric about $\hat{\theta}$, then the bootstrap confidence interval is $(\underline{\theta}, \overline{\theta})$.

39. In Section 8.5.3, the bootstrap confidence interval was derived from consideration of the sampling distribution of $\hat{\theta} - \theta_0$. Suppose that we had started with considering the distribution of $\hat{\theta}/\theta$. How would the argument have proceeded, and would the bootstrap interval that was finally arrived at have been different?

40. In Example A of Section 8.5.1, how could you use the bootstrap to estimate the following measures of the accuracy of $\hat{\theta}$: (a) $P(|\hat{\theta} - \theta_0| > .01)$, (b) $E(|\hat{\theta} - \theta_0|)$, (c) that number Δ such that $P(|\hat{\theta} - \theta_0| > \Delta) = .5$?

41. What are the relative efficiencies of the method of moments and maximum likelihood estimates of α and λ in Example C of Section 8.4 and Example C of Section 8.5?

42. The file `gamma-ray` contains a small quantity of data collected from the Compton Gamma Ray Observatory, a satellite launched by NASA in 1991 (`http://cossc.gsfc.nasa.gov/`). For each of 100 sequential time intervals of variable lengths (given in seconds), the number of gamma rays originating in a particular area of the sky was recorded. Assuming a model that the arrival times are a Poisson process with constant emission rate ($\lambda =$ events per second), estimate λ. What is the estimated standard error? How might you informally check the assumption that the emission rate is constant? What is the posterior distribution of Λ if an improper gamma prior is used?

43. The file `gamma-arrivals` contains another set of gamma-ray data, this one consisting of the times between arrivals (interarrival times) of 3,935 photons (units are seconds).

   a. Make a histogram of the interarrival times. Does it appear that a gamma distribution would be a plausible model?

   b. Fit the parameters by the method of moments and by maximum likelihood. How do the estimates compare?

   c. Plot the two fitted gamma densities on top of the histogram. Do the fits look reasonable?

   d. For both maximum likelihood and the method of moments, use the bootstrap to estimate the standard errors of the parameter estimates. How do the estimated standard errors of the two methods compare?

   e. For both maximum likelihood and the method of moments, use the bootstrap to form approximate confidence intervals for the parameters. How do the confidence intervals for the two methods compare?

   f. Is the interarrival time distribution consistent with a Poisson process model for the arrival times?

44. The file `bodytemp` contains normal body temperature readings (degrees Fahrenheit) and heart rates (beats per minute) of 65 males (coded by 1) and 65 females (coded by 2) from Shoemaker (1996). Assuming that the population distributions are normal (an assumption that will be investigated in a later chapter), estimate the means and standard deviations of the males and females. Form 95% confidence intervals for the means. Standard folklore is that the average body temperature is 98.6 degrees Fahrenheit. Does this appear to be the case?

45. A Random Walk Model for Chromatin. A human chromosome is a very large molecule, about 2 or 3 centimeters long, containing 100 million base pairs (Mbp). The cell nucleus, where the chromosome is contained, is in contrast only about a thousandth of a centimeter in diameter. The chromosome is packed in a series of coils, called *chromatin*, in association with special proteins (histones), forming a string of microscopic beads. It is a mixture of DNA and protein. In the G0/G1 phase of the cell cycle, between mitosis and the onset of DNA replication, the mitotic chromosomes diffuse into the interphase nucleus. At this stage, a number of important processes related to chromosome function take place. For example, DNA is made accessible for transcription and is duplicated, and repairs are made of DNA strand breaks. By the time of the next mitosis, the chromosomes have been duplicated. The complexity of these and other processes raises many questions about the large-scale spatial organization of chromosomes and how this organization relates to cell function. Fundamentally, it is puzzling how these processes can unfold in such a spatially restricted environment.

At a scale of about 10^{-3} Mbp, the DNA forms a chromatin fiber about 30 nm in diameter; at a scale of about 10^{-1} Mbp the chromatin may form loops. Very little is known about the spatial organization beyond this scale. Various models have been proposed, ranging from highly random to highly organized, including irregularly folded fibers, giant loops, radial loop structures, systematic organization to make the chromatin readily accessible to transcription and replication machinery, and stochastic configurations based on random walk models for polymers.

A series of experiments (Sachs et al., 1995; Yokota et al., 1995) were conducted to learn more about spatial organization on larger scales. Pairs of small DNA sequences (size about 40 kbp) at specified locations on human chromosome 4 were flourescently labeled in a large number of cells. The distances between the members of these pairs were then determined by flourescence microscopy. (The distances measured were actually two-dimensional distances between the projections of the paired locations onto a plane.) The empirical distribution of these distances provides information about the nature of large-scale organization.

There has long been a tradition in chemistry of modeling the configurations of polymers by the theory of random walks. As a consequence of such a model, the two-dimensional distance should follow a Rayleigh distribution

$$f(r|\theta) = \frac{r}{\theta^2} \exp\left(\frac{-r^2}{2\theta^2}\right)$$

Basically, the reason for this is as follows: The random walk model implies that the joint distribution of the locations of the pair in R^3 is multivariate Gaussian; by properties of the multivariate Gaussian, it can be shown the joint distribution of the locations of the projections onto a plane is bivariate Gaussian. As in Example A of Section 3.6.2 of the text, it can be shown that the distance between the points follows a Rayleigh distribution.

In this exercise, you will fit the Rayleigh distribution to some of the experimental results and examine the goodness of fit. The entire data set comprises 36 experiments in which the separation between the pairs of flourescently tagged locations ranged from 10 Mbp to 192 Mbp. In each such experimental condition, about 100–200 measurements of two-dimensional distances were determined. This exercise will be concerned just with the data from three experiments (short, medium, and long separation). The measurements from these experiments is contained in the files `Chromatin/short`, `Chromatin/medium`, `Chromatin/long`.

a. What is the maximum likelihood estimate of θ for a sample from a Rayleigh distribution?

b. What is the method of moments estimate?

c. What are the approximate variances of the mle and the method of moments estimate?

  d. For each of the three experiments, plot the likelihood functions and find the mle's and their approximate variances.

  e. Find the method of moments estimates and the approximate variances.

  f. For each experiment, make a histogram (with unit area) of the measurements and plot the fitted densities on top. Do the fits look reasonable? Is there any appreciable difference between the maximum likelihood fits and the method of moments fits?

  g. Does there appear to be any relationship between your estimates and the genomic separation of the points?

  h. For one of the experiments, compare the asymptotic variances to the results obtained from a parametric bootstrap. In order to do this, you will have to generate random variables from a Rayleigh distribution with parameter θ.

   Show that if X follows a Rayleigh distribution with $\theta = 1$, then $Y = \theta X$ follows a Rayleigh distribution with parameter θ. Thus it is sufficient to figure out how to generate random variables that are Rayleigh, $\theta = 1$. Show how Proposition D of Section 2.3 of the text can be applied to accomplish this.

   $B = 100$ bootstrap samples should suffice for this problem. Make a histogram of the values of the θ^*. Does the distribution appear roughly normal? Do you think that the large sample theory can be reasonably applied here? Compare the standard deviation calculated from the bootstrap to the standard errors you found previously.

  i. For one of the experiments, use the bootstrap to construct an approximate 95% confidence interval for θ using $B = 1000$ bootstrap samples. Compare this interval to that obtained using large sample theory.

46. The data of this exercise were gathered as part of a study to estimate the population size of the bowhead whale (Raftery and Zeh 1993). The statistical procedures for estimating the population size along with an assessment of the variability of the estimate were quite involved, and this problem deals with only one aspect of the problem—a study of the distribution of whale swimming speeds. Pairs of sightings and corresponding locations that could be reliably attributed to the same whale were collected, thus providing an estimate of velocity for each whale. The velocities, $v_1, v_2, \ldots, v_{210}$ (km/h), were converted into times $t_1, t_2, \ldots, t_{210}$ to swim 1 km—$t_i = 1/v_i$. The distribution of the t_i was then fit by a gamma distribution. The times are contained in the file `whales`.

  a. Make a histogram of the 210 values of t_i. Does it appear that a gamma distribution would be a plausible model to fit?

  b. Fit the parameters of the gamma distribution by the method of moments.

  c. Fit the parameters of the gamma distribution by maximum likelihood. How do these values compare to those found before?

  d. Plot the two gamma densities on top of the histogram. Do the fits look reasonable?

  e. Estimate the sampling distributions and the standard errors of the parameters fit by the method of moments by using the bootstrap.

  f. Estimate the sampling distributions and the standard errors of the parameters fit by maximum likelihood by using the bootstrap. How do they compare to the results found previously?

g. Find approximate confidence intervals for the parameters estimated by maximum likelihood.

47. The Pareto distribution has been used in economics as a model for a density function with a slowly decaying tail:

$$f(x|x_0, \theta) = \theta x_0^\theta x^{-\theta-1}, \qquad x \geq x_0, \quad \theta > 1$$

Assume that $x_0 > 0$ is given and that $X_1, X_2, \ldots, X_n$ is an i.i.d. sample.

a. Find the method of moments estimate of θ.
b. Find the mle of θ.
c. Find the asymptotic variance of the mle.
d. Find a sufficient statistic for θ.

48. Consider the following method of estimating λ for a Poisson distribution. Observe that

$$p_0 = P(X = 0) = e^{-\lambda}$$

Letting Y denote the number of zeros from an i.i.d. sample of size n, λ might be estimated by

$$\tilde{\lambda} = -\log\left(\frac{Y}{n}\right)$$

Use the method of propagation of error to obtain approximate expressions for the variance and the bias of this estimate. Compare the variance of this estimate to the variance of the mle, computing relative efficiencies for various values of λ. Note that $Y \sim \text{bin}(n, p_0)$.

49. For the example on muon decay in Section 8.4, suppose that instead of recording $x = \cos\theta$, only whether the electron goes backward $(x < 0)$ or forward $(x > 0)$ is recorded.

a. How could α be estimated from n independent observations of this type? (*Hint:* Use the binomial distribution.)
b. What is the variance of this estimate and its efficiency relative to the method of moments estimate and the mle for $\alpha = 0, .1, .2, .3, .4, .5, .6, .7, .8, .9$?

50. Let $X_1, \ldots, X_n$ be an i.i.d. sample from a Rayleigh distribution with parameter $\theta > 0$:

$$f(x|\theta) = \frac{x}{\theta^2} e^{-x^2/(2\theta^2)}, \qquad x \geq 0$$

(This is an alternative parametrization of that of Example A in Section 3.6.2.)

a. Find the method of moments estimate of θ.
b. Find the mle of θ.
c. Find the asymptotic variance of the mle.

51. The double exponential distribution is

$$f(x|\theta) = \frac{1}{2} e^{-|x-\theta|}, \qquad -\infty < x < \infty$$

For an i.i.d. sample of size $n = 2m + 1$, show that the mle of θ is the median of the sample. (The observation such that half of the rest of the observations are smaller and half are larger.) [*Hint:* The function $g(x) = |x|$ is not differentiable. Draw a picture for a small value of n to try to understand what is going on.]

52. Let $X_1, \ldots, X_n$ be i.i.d. random variables with the density function

$$f(x|\theta) = (\theta + 1)x^\theta, \qquad 0 \le x \le 1$$

   a. Find the method of moments estimate of θ.
   b. Find the mle of θ.
   c. Find the asymptotic variance of the mle.
   d. Find a sufficient statistic for θ.

53. Let $X_1, \ldots, X_n$ be i.i.d. uniform on $[0, \theta]$.

   a. Find the method of moments estimate of θ and its mean and variance.
   b. Find the mle of θ.
   c. Find the probability density of the mle, and calculate its mean and variance. Compare the variance, the bias, and the mean squared error to those of the method of moments estimate.
   d. Find a modification of the mle that renders it unbiased.

54. Suppose that an i.i.d. sample of size 15 from a normal distribution gives $\overline{X} = 10$ and $s^2 = 25$. Find 90% confidence intervals for μ and σ^2.

55. For two factors—starchy or sugary, and green base leaf or white base leaf—the following counts for the progeny of self-fertilized heterozygotes were observed (Fisher 1958):

Type	Count
Starchy green	1997
Starchy white	906
Sugary green	904
Sugary white	32

According to genetic theory, the cell probabilities are $.25(2 + \theta)$, $.25(1 - \theta)$, $.25(1 - \theta)$, and $.25\theta$, where $\theta (0 < \theta < 1)$ is a parameter related to the linkage of the factors.

   a. Find the mle of θ and its asymptotic variance.
   b. Form an approximate 95% confidence interval for θ based on part (a).
   c. Use the bootstrap to find the approximate standard deviation of the mle and compare to the result of part (a).
   d. Use the bootstrap to find an approximate 95% confidence interval and compare to part (b).

56. Referring to Problem 55, consider two other estimates of θ. (1) The expected number of counts in the first cell is $n(2 + \theta)/4$; if this expected number is equated to the count X_1, the following estimate is obtained:

$$\tilde{\theta}_1 = \frac{4X_1}{n} - 2$$

(2) The same procedure done for the last cell yields

$$\tilde{\theta}_2 = \frac{4X_4}{n}$$

Compute these estimates. Using that X_1 and X_4 are binomial random variables, show that these estimates are unbiased, and obtain expressions for their variances. Evaluate the estimated standard errors and compare them to the estimated standard error of the mle.

57. This problem is concerned with the estimation of the variance of a normal distribution with unknown mean from a sample $X_1, \ldots, X_n$ of i.i.d. normal random variables. In answering the following questions, use the fact that (from Theorem B of Section 6.3)

$$\frac{(n-1)s^2}{\sigma^2} \sim \chi_{n-1}^2$$

and that the mean and variance of a chi-square random variable with r df are r and $2r$, respectively.

a. Which of the following estimates is unbiased?

$$s^2 = \frac{1}{n-1}\sum_{i=1}^{n}(X_i - \overline{X})^2 \qquad \hat{\sigma}^2 = \frac{1}{n}\sum_{i=1}^{n}(X_i - \overline{X})^2$$

b. Which of the estimates given in part (a) has the smaller MSE?
c. For what value of ρ does $\rho \sum_{i=1}^{n}(X_i - \overline{X})^2$ have the minimal MSE?

58. If gene frequencies are in equilibrium, the genotypes AA, Aa, and aa occur with probabilities $(1-\theta)^2$, $2\theta(1-\theta)$, and θ^2, respectively. Plato et al. (1964) published the following data on haptoglobin type in a sample of 190 people:

	Haptoglobin Type	
Hp1-1	Hp1-2	Hp2-2
10	68	112

a. Find the mle of θ.
b. Find the asymptotic variance of the mle.
c. Find an approximate 99% confidence interval for θ.
d. Use the bootstrap to find the approximate standard deviation of the mle and compare to the result of part (b).
e. Use the bootstrap to find an approximate 99% confidence interval and compare to part (c).

59. Suppose that in the population of twins, males (M) and females (F) are equally likely to occur and that the probability that twins are identical is α. If twins are not identical, their genes are independent.

a. Show that

$$P(MM) = P(FF) = \frac{1+\alpha}{4} \qquad P(MF) = \frac{1-\alpha}{2}$$

b. Suppose that n twins are sampled. It is found that n_1 are *MM*, n_2 are *FF*, and n_3 are *MF*, but it is not known which twins are identical. Find the mle of α and its variance.

60. Let $X_1, \ldots, X_n$ be an i.i.d. sample from an exponential distribution with the density function

$$f(x|\tau) = \frac{1}{\tau} e^{-x/\tau}, \qquad 0 \le x < \infty$$

   a. Find the mle of τ.
   b. What is the exact sampling distribution of the mle?
   c. Use the central limit theorem to find a normal approximation to the sampling distribution.
   d. Show that the mle is unbiased, and find its exact variance. (*Hint:* The sum of the X_i follows a gamma distribution.)
   e. Is there any other unbiased estimate with smaller variance?
   f. Find the form of an approximate confidence interval for τ.
   g. Find the form of an exact confidence interval for τ.

61. Laplace's rule of succession. Laplace claimed that when an event happens n times in a row and never fails to happen, the probability that the event will occur the next time is $(n+1)/(n+2)$. Can you suggest a rationale for this claim?

62. Show that the gamma distribution is a conjugate prior for the exponential distribution. Suppose that the waiting time in a queue is modeled as an exponential random variable with unknown parameter λ, and that the average time to serve a random sample of 20 customers is 5.1 minutes. A gamma distribution is used as a prior. Consider two cases: (1) the mean of the gamma is 0.5 and the standard deviation is 1, and (2) the mean is 10 and the standard deviation is 20. Plot the two posterior distributions and compare them. Find the two posterior means and compare them. Explain the differences.

63. Suppose that 100 items are sampled from a manufacturing process and 3 are found to be defective. A beta prior is used for the unknown proportion θ of defective items. Consider two cases: (1) $a = b = 1$, and (2) $a = 0.5$ and $b = 5$. Plot the two posterior distributions and compare them. Find the two posterior means and compare them. Explain the differences.

64. This is a continuation of the previous problem. Let $X = 0$ or 1 according to whether an item is defective. For each choice of the prior, what is the marginal distribution of X before the sample is taken? What are the marginal distributions after the sample is taken? (*Hint:* for the second question, use the posterior distribution of θ.)

65. Suppose that a random sample of size 20 is taken from a normal distribution with unknown mean and known variance equal to 1, and the mean is found to be $\bar{x} = 10$. A normal distribution was used as the prior for the mean, and it was found that the posterior mean was 15 and the posterior standard deviation was 0.1. What were the mean and standard deviation of the prior?

66. Let the unknown probability that a basketball player makes a shot successfully be θ. Suppose your prior on θ is uniform on $[0, 1]$ and that she then makes two shots in a row. Assume that the outcomes of the two shots are independent.

    a. What is the posterior density of θ?

    b. What would you estimate the probability that she makes a third shot to be?

67. Evans (1953) considered fitting the negative binomial distribution and other distributions to a number of data sets that arose in ecological studies. Two of these sets will be used in this problem. The first data set gives frequency counts of *Glaux maritima* made in 500 contiguous 20-cm^2 quadrants. For the second data set, a plot of potato plants 48 rows wide and 96 ft long was examined. The area was split into 2304 sampling units consisting of 2-ft lengths of row and in each unit the number of potato beetles was counted. Fit Poisson and negative binomial distributions, and comment on the goodness of fit. For these data, the method of moments should be fairly efficient.

Count	*Glaux maritima*	Potato Beetles
0	1	190
1	15	264
2	27	304
3	42	260
4	77	294
5	77	219
6	89	183
7	57	150
8	48	104
9	24	90
10	14	60
11	16	46
12	9	29
13	3	36
14	1	19
15		12
16		11
17		6
18		10
19		2
20		4
21		1
22		3
23		4
24		1
25		1
26		0
27		0
28		1

68. Let $X_1, \ldots, X_n$ be an i.i.d. sample from a Poisson distribution with mean λ, and let $T = \sum_{i=1}^{n} X_i$.

   a. Show that the distribution of $X_1, \ldots, X_n$ given T is independent of λ, and conclude that T is sufficient for λ.

   b. Show that X_1 is not sufficient.

   c. Use Theorem A of Section 8.8.1 to show that T is sufficient. Identify the functions g and h of that theorem.

69. Use the factorization theorem (Theorem A in Section 8.8.1) to conclude that $T = \sum_{i=1}^{n} X_i$ is a sufficient statistic when the X_i are an i.i.d. sample from a geometric distribution.

70. Use the factorization theorem to find a sufficient statistic for the exponential distribution.

71. Let $X_1, \ldots, X_n$ be an i.i.d. sample from a distribution with the density function

$$f(x|\theta) = \frac{\theta}{(1+x)^{\theta+1}}, \qquad 0 < \theta < \infty \text{ and } 0 \le x < \infty$$

Find a sufficient statistic for θ.

72. Show that $\prod_{i=1}^{n} X_i$ and $\sum_{i=1}^{n} X_i$ are sufficient statistics for the gamma distribution.

73. Find a sufficient statistic for the Rayleigh density,

$$f(x|\theta) = \frac{x}{\theta^2} e^{-x^2/(2\theta^2)}, \qquad x \ge 0$$

74. Show that the binomial distribution belongs to the exponential family.

75. Show that the gamma distribution belongs to the exponential family.

CHAPTER 9

Testing Hypotheses and Assessing Goodness of Fit

9.1 Introduction

We will introduce some of the basic concepts of this chapter by means of a simple artificial example; it is important that you read the example carefully. Suppose that I have two coins, coin 0 has probability of heads equal to 0.5 and coin 1 has probability of heads equal to 0.7. I choose one of the coins, toss it 10 times and tell you the number of heads, but do not tell you whether it was coin 0 or coin 1. On the basis of the number of heads, your task is to decide which coin it was. How should your decision rule be?

Let X denote the number of heads. Figure 9.1 gives $p(x)$ for each of the coins.

x	0	1	2	3	4	5	6	7	8	9	10
coin 0	.0010	.0098	.0439	.1172	.2051	.2461	.2051	.1172	.0439	.0098	.0010
coin 1	.0000	.0001	.0014	.0090	.0368	.1029	.2001	.2668	.2335	.1211	.0282

FIGURE **9.1**

Suppose that you observed two heads. Then $P_0(2)/P_1(2)$ is about 30, which we will call the **likelihood ratio**—coin 0 was about 30 times more likely to produce this result than was coin 1. This result would favor coin 0. On the other hand, if there were 8 heads, the likelihood ratio would be $.0439/.2335 = 0.19$, which would favor coin 1. The likelihood ratio will play a central role in the procedures we develop.

We specify two hypotheses, H_0 and H_1, according to whether coin 0 or coin 1 was tossed. We first develop a Bayesian methodology for assessing the evidence for each of the hypotheses. This approach requires the specification of prior probabilities $P(H_0)$ and $P(H_1)$ for each of the hypotheses before observing any data. If you believed that I have no reason to choose coin 0 over coin 1, you would take $P(H_0) = P(H_1) = 1/2$.

After observing the number of heads, your posterior probabilities would be $P(H_0|x)$ and $P(H_1|x)$. The former would be

$$P(H_0|x) = \frac{P(H_0, x)}{P(x)}$$

$$= \frac{P(x|H_0)P(H_0)}{P(x)}$$

The ratio

$$\frac{P(H_0|x)}{P(H_1|x)} = \frac{P(H_0)}{P(H_1)} \frac{P(x|H_0)}{P(x|H_1)}$$

is the product of the ratio of prior probabilities and the likelihood ratio. Thus, the evidence provided by the data is contained in the likelihood ratio, which is multiplied by the ratio of prior probabilities to produce the ratio of posterior probabilities.

The likelihood ratio is evaluated in Figure 9.2.

x	0	1	2	3	4	5	6	7	8	9	10
$\frac{P(x|H_0)}{P(x|H_1)}$	165.4	70.88	30.38	13.02	5.579	2.391	1.025	0.4392	0.1882	0.0807	0.0346

FIGURE 9.2

(The numbers in Figure 9.2 do not precisely agree with the ratios of the numbers in the Figure 9.1 because the former are truncated to four decimal places.) The evidence x is increasingly favorable to H_0 as x decreases, i.e., the likelihood ratio is monotonic in x. If one's prior probabilities were equal, then for zero to six heads, H_0 would be more probable and for seven to ten heads, H_1 would be more probable. If the prior probabilities change, the breakpoint changes. If you were asked to choose H_0 or H_1 on the basis of the data x, it seems reasonable that you would choose the hypothesis which had larger posterior probability. You would choose H_0 if

$$\frac{P(H_0|x)}{P(H_1|x)} = \frac{P(H_0)}{P(H_1)} \frac{P(x|H_0)}{P(x|H_1)} > 1$$

or equivalently if

$$\frac{P(x|H_0)}{P(x|H_1)} > c$$

where the critical value c depends upon your prior probability. Your decision would be based on the likelihood ratio: you accept H_0 if the likelihood ratio is greater than c, and you reject H_0 if the likelihood ratio is less than c.

Let us now further examine the consequences of a particular decision rule, i.e., a particular specification of the constant c. First suppose that $c = 1$; then H_0 is accepted as long as $X \leq 6$ and is rejected in favor of H_1 if $X > 6$. We can make two possible errors: reject H_0 when it is true, or accept H_0 when it is false. The probabilities of these two possible errors can be evaluated as follows:

$$P(\text{reject } H_0|H_0) = P(X > 6|H_0)$$
$$= 0.18$$

Here we used the binomial probabilities above corresponding to H_0. Similarly, the probability of the other kind of error is

$$P(\text{accept } H_0|H_1) = P(X \leq 6|H_1)$$
$$= 0.35$$

Now suppose that $c = 0.1$, which corresponds to $P(H_0)/P(H_1) = 10$. Then from Figure 9.2, H_0 is accepted when $X \leq 8$. Compared to equal odds, more extreme evidence is required to reject H_0 because the prior probabilities greatly favor H_0. Then the probabilities of the two types of errors are

$$P(\text{reject } H_0|H_0) = P(X > 8|H_0)$$
$$= 0.01$$
$$P(\text{accept } H_0|H_1) = P(X \leq 8|H_1)$$
$$= 0.85$$

In this way, we see that there is a correspondence between the prior probabilities and the probabilities of the two types of errors. The constant c controls the tradeoff between the probabilities of the two types of errors.

9.2 The Neyman-Pearson Paradigm

Rather than using a Bayesian approach, Neyman and Pearson formulated their theory of hypothesis testing by casting it as a decision problem and making the probabilities of the two types of errors central, thus bypassing the necessity of specifying prior probabilities. In doing so, this approach introduced an asymmetry: one hypothesis is singled out as the **null hypothesis** and the other as the **alternative hypothesis,** the former usually denoted by H_0 and the latter by H_1 or H_A. We will see later through examples how this specification is naturally made, but for now we will continue with the example of the previous section and arbitrarily declare H_0 to the null hypothesis. The following terminology is standard:

- Rejecting H_0 when it is true is called a **type I error.**
- The probability of a type I error is called the **significance level** of the test and is usually denoted by α.
- Accepting the null hypothesis when it is false is called a **type II error** and its probability is usually denoted by β.
- The probability that the null hypothesis is rejected when it is false is called the **power** of the test, and equals $1 - \beta$.
- We have seen in this example how rejecting H_0 when the likelihood ratio is less than a constant c is equivalent to rejecting when the number of heads is greater than some value x_0. The likelihood ratio, or equivalently, the number of heads, is called the **test statistic.**
- The set of values of the test statistic that leads to rejection of the null hypothesis is called the **rejection region,** and the set of values that leads to acceptance is called the **acceptance region.**
- The probability distribution of the test statistic when the null hypothesis is true is called the **null distribution.**

In the example in the introduction to this chapter, the null and alternative hypotheses each completely specify the probability distribution of the number of heads, as binomial(10,0.5) or binomial(10,0.7), respectively. These are called **simple hypotheses.** The Neyman-Pearson Lemma shows that basing the test on the likelihood ratio as we did is optimal:

NEYMAN-PEARSON LEMMA

Suppose that H_0 and H_1 are simple hypotheses and that the test that rejects H_0 whenever the likelihood ratio is less than c has significance level α. Then *any other test* for which the significance level is less than or equal to α has power less than or equal to that of the likelihood ratio test. ∎

The point is that there are many possible tests. Any partition of the set of possible outcomes of the observations into a set that has probability less than or equal to α when the null hypothesis is true and its complement, and that rejects when the observations are in the complement has significance level less than or equal to α by construction. Among all such possible partitions, that based on the likelihood ratio maximizes the power.

Proof

Let $f(x)$ denote the probability density function or frequency function of the observations. A test of $H_0 : f(x) = f_0(x)$ versus $H_1 : f(x) = f_A(x)$ amounts to using a decision function $d(x)$, where $d(x) = 0$ if H_0 is accepted and $d(x) = 1$ if H_0 is rejected. Since $d(X)$ is a Bernoulli random variable, $E(d(X)) = P(d(X) = 1)$. The significance level of the test is thus $\alpha = P_0(d(X) = 1) = E_0(d(X))$, and the power is $P_A(d(X) = 0) = E_A(d(X))$. Here E_0 denotes expectation under the probability law specified by H_0, etc.

Let $d(X)$ correspond to the likelihood ratio test: $d(x) = 1$ if $f_0(x) < cf_A(x)$ and $E_0(d(X)) = \alpha$. Let $d^*(x)$ be the decision function of another test satisfying $E_0(d^*(X)) \leq E_0(d(X)) = \alpha$. We will show that $E_A(d^*(X)) \leq E_A(d(X))$. This will follow from the key inequality

$$d^*(x)[cf_A(x) - f_0(x)] \leq d(x)[cf_A(x) - f_0(x)]$$

which holds since if $d(x) = 1$, $cf_A(x) - f_0(x) > 0$ and if $d(x) = 0$, $cf_A(x) - f_0(x) \leq 0$. Now integrating (or summing) both sides of the inequality above with respect to x gives

$$cE_A(d^*(X)) - E_0(d^*(X)) \leq cE_A(d(X)) - E_0(d(X))$$

and thus

$$E_0(d(X)) - E_0(d^*(X)) \leq c[E_A(d(X)) - E_A(d^*(X))]$$

The conclusion follows since the left-hand side of this inequality is nonnegative by assumption. ∎

E X A M P L E   A   Let $X_1, \ldots, X_n$ be a random sample from a normal distribution having known variance σ^2. Consider two simple hypotheses:

$$H_0: \mu = \mu_0$$
$$H_A: \mu = \mu_1$$

where μ_0 and μ_1 are given constants. Let the significance level α be prescribed. The Neyman-Pearson Lemma states that among all tests with significance level α, the test that rejects for small values of the likelihood ratio is most powerful. We thus calculate the likelihood ratio, which is

$$\frac{f_0(\mathbf{X})}{f_1(\mathbf{X})} = \frac{\exp\left[\frac{-1}{2\sigma^2} \sum_{i=1}^{n} (X_i - \mu_0)^2\right]}{\exp\left[\frac{-1}{2\sigma^2} \sum_{i=1}^{n} (X_i - \mu_1)^2\right]}$$

since the multipliers of the exponentials cancel. Small values of this statistic correspond to small values of $\sum_{i=1}^{n}(X_i - \mu_1)^2 - \sum_{i=1}^{n}(X_i - \mu_0)^2$. Expanding the squares, we see that the latter expression reduces to

$$2n\overline{X}(\mu_0 - \mu_1) + n\mu_1^2 - n\mu_0^2$$

Now, if $\mu_0 - \mu_1 > 0$, the likelihood ratio is small if $\overline{X}$ is small; if $\mu_0 - \mu_1 < 0$, the likelihood ratio is small if $\overline{X}$ is large. To be concrete, let us assume the latter case. We then know that the likelihood ratio is a function of $\overline{X}$ and is small when $\overline{X}$ is large. The Neyman-Pearson lemma thus tells us that the most powerful test rejects for $\overline{X} > x_0$ for some x_0, and we choose x_0 so as to give the test the desired level α. That is, x_0 is chosen so that $P(\overline{X} > x_0) = \alpha$ if H_0 is true. Under H_0 in this example, the null distribution of $\overline{X}$ is a normal distribution with mean μ_0 and variance σ^2/n, so x_0 can be chosen from tables of the standard normal distribution. Since

$$P(\overline{X} > x_0) = P\left(\frac{\overline{X} - \mu_0}{\sigma/\sqrt{n}} > \frac{x_0 - \mu_0}{\sigma/\sqrt{n}}\right)$$

we can solve

$$\frac{x_0 - \mu_0}{\sigma/\sqrt{n}} = z(\alpha)$$

for x_0 in order to find the rejection region for a level α test. Here, as usual, $z(\alpha)$ denotes the upper α point of the standard normal distribution; that is, if Z is a standard normal random variable, $P(Z > z(\alpha)) = \alpha$.   ∎

This example is typical of the way that the Neyman-Pearson Lemma is used. We write down the likelihood ratio and observe that small values of it correspond in a one-to-one manner with extreme values of a test statistic, in this case $\overline{X}$. Knowing the null distribution of the test statistic makes it possible to choose a critical level that produces a desired significance level α.

Unfortunately, the Neyman-Pearson Lemma is of little direct utility in most practical problems, because the case of testing a simple null hypothesis versus a simple alternative is rarely encountered. If a hypothesis does not completely specify the probability distribution, the hypothesis is called a **composite hypothesis.** Here are some examples:

EXAMPLE B   *Goodness-of-Fit Test*

Let $X_1, X_2, \ldots, X_n$ be a sample from a discrete probability distribution. The null hypothesis could be that the distribution is Poisson with some unspecified mean, and the alternative could be that the distribution is not Poisson. For example, we might want to test whether a Poisson model is reasonable for the data of Example A in Section 8.4. Since the null hypothesis does not completely specify the distribution of the X_i's, it is composite. If the null hypothesis were refined to state that the distribution was Poisson with some specified mean, then it would be simple. The alternative hypothesis does not completely specify the distribution, so it is composite. We will take up the subject of testing for goodness of fit later in this chapter. ■

EXAMPLE C   *Testing for ESP*

Consider a hypothetical experiment in which a subject is asked to identify, without looking, the suits of 20 cards drawn randomly with replacement from a 52 card deck. Let T be the number of correct identifications. The null hypothesis states that the person is purely guessing, and the alternative states that the person has extrasensory ability. The null hypothesis is simple because then T is binomial(20,0.25). The alternative does not completely specify the distribution of T, so it is composite; note that it does not even specify that the distribution is binomial. ■

This example is useful for further illustrating two other issues that arise in hypothesis testing: the specification of the significance level and the choice of the null hypothesis.

9.2.1   Specification of the Significance Level and the Concept of a *p*-value

One of the strengths of the Neyman-Pearson approach is that only the distribution under the null hypothesis is needed in order to construct a test. In Example C above, it would be conventional and convenient to take the null hypothesis to be that of pure guessing; we discuss this further in the next section. In this case, the null distribution of T is binomial(20,0.25). Because large values of T tend to lend credence to the alternative, the rejection region would be of the form $\{T \geq t_0\}$ where t_0 is chosen so that $P(T \geq t_0 | H_0) = \alpha$, the desired significance level of the test. For example, from calculating binomial probabilities, we find $P(T \geq 12) = .0009$, so for this choice of the critical region, the null hypothesis of no ESP ability would be falsely rejected only with probability about one in a thousand. Note that we did not need to specify the form of the probability distribution under the alternative, but used only the notion that if the alternative is true, the subject would be expected to correctly identify more suits than if purely guessing. In comparison, a fully Bayesian treatment would have to specify the distribution under the alternative as well as prior probabilities.

The theory requires the specification of the significance level, α, in advance of analyzing the data, but gives no guidance about how to make this choice. In practice it is almost always the case that the choice of α is essentially arbitrary, but is heavily influenced by custom. Small values, such as 0.01 and 0.05, are commonly used. Another criticism of the paradigm is that it is built on the assumption that one must either reject or not reject a hypothesis, when typically no such decision is actually required. The theory is thus often applied in a hypothetical manner. For example, suppose that the subject above guessed the suit correctly nine times. Since $P(T \geq 9|H_0) = .041$, the null hypothesis would have been "rejected" at the significance level $\alpha = .05$, if one were actually "rejecting" or "not rejecting." Thus, the evidence is often summarized as a *p-value*, which is defined to be the smallest significance level at which the null hypothesis would be rejected. If nine suits were identified correctly, the p-value would be 0.041. If ten suits were identified, it would be 0.014, since $P(T \geq 10|H_0) = .014$, etc.

The use of a p-value to summarize the evidence against the null hypothesis was advocated by the eminent statistician Sir Ronald Fisher. But rather than casting it within a hypothetical framework of "rejection," he thought of the p-value as being the probability under the null hypothesis of a result as or more extreme than that actually observed. So for example, in the case that ten suits are identified, the p-value is the chance of someone getting at least that many correct by purely guessing. The smaller the p-value, the stronger the evidence against the null hypothesis.

The Bayesian paradigm summarizes the evidence for and against the null hypothesis as a posterior probability. Its application depends on specifying probability models under both the null and the alternative and on assigning meaningful prior probabilities. It is important to understand that a p-value is *not* the posterior probability that the null hypothesis is true. To reiterate, the p-value is the probability of a result as or more extreme than that actually observed if the null hypothesis were true. This is a probability, but it is not the posterior probability that the null hypothesis is true; the latter depends on the specification of prior probabilities. Consider the example of Section 9.1. If $x = 8$ heads are observed, the p-value is $.0439 + .0098 + .0010 = .0546$, or about 5%. Suppose that the prior probabilities were equal. The likelihood ratio is $0.1882 = P(H_0|x)/(1 - P(H_0|x))$ from which it follows that $P(H_0|x) = 0.1584$, or about 16%.

9.2.2 The Null Hypothesis

As should be clear by now, there is an asymmetry in the Neyman-Pearson paradigm between the null and alternative hypotheses. The decision as to which is the null and which is the alternative hypothesis is not a mathematical one, and depends on scientific context, custom, and convenience. This will gradually become clearer as we see more real examples in this and later chapters, and for now we will make only the following remarks:

- In Example B of Section 9.2, we chose as the null hypothesis the hypothesis that the distribution was Poisson and as the alternative hypothesis the hypothesis that the distribution was not Poisson. In this case, the null hypothesis is simpler than the alternative, which in a sense contains more distributions than does the null. It is conventional to choose the simpler of two hypotheses as the null.
- The consequences of incorrectly rejecting one hypothesis may be graver than those of incorrectly rejecting the other. In such a case, the former should be chosen as the null hypothesis, because the probability of falsely rejecting it could be controlled by choosing α. Examples of this kind arise in screening new drugs; frequently, it must be documented rather conclusively that a new drug has a positive effect before it is accepted for general use.
- In scientific investigations, the null hypothesis is often a simple explanation that must be discredited in order to demonstrate the presence of some physical phenomenon or effect. The hypothetical ESP experiment referred to earlier falls in this category; the null hypothesis states that the subject is merely guessing, that there is no ESP. The validity of the null hypothesis would not be cast in doubt unless the results would be extremely unlikely under the null. We will see other examples of this type beginning in Chapter 11.

9.2.3   Uniformly Most Powerful Tests

The optimality result of the Neyman-Pearson Lemma requires that both hypotheses be simple. In some cases, the theory can be extended to include composite hypotheses. If the alternative H_1 is composite, a test that is most powerful for *every* simple alternative in H_1 is said to be **uniformly most powerful.**

E X A M P L E A   Continuing with Example A of Section 9.2, consider testing $H_0 : \mu = \mu_0$ versus $H_1 : \mu > \mu_0$. In Example A, we saw that for a particular alternative $\mu_1 > \mu_0$, the most powerful test rejects for $\overline{X} > x_0$, where x_0 depends on μ_0, σ, n, but *not* on μ_1. Because this test is most powerful and is the same for every alternative, it is uniformly most powerful.   ■

It can also be argued that the test is uniformly most powerful for testing $H_0 : \mu \leq \mu_0$ versus $H_1 : \mu > \mu_0$. But it is not uniformly most powerful for testing $H_0 : \mu = \mu_0$ versus $H_1 : \mu \neq \mu_0$. This follows from further examination of the example, which shows that the test that is most powerful against the alternative that $\mu > \mu_0$ rejects for large values of $\overline{X} - \mu_0$, whereas the test that is most powerful against the alternative $\mu < \mu_0$ rejects for small values of $\overline{X} - \mu_0$. The most powerful test is thus not the same for every alternative.

In typical composite situations, there is no uniformly most powerful test. The alternatives $H_1 : \mu < \mu_0$ and $H_1 : \mu > \mu_0$ are called **one-sided alternatives.** The alternative $H_1 : \mu \neq \mu_0$ is a **two-sided alternative.**

9.3 The Duality of Confidence Intervals and Hypothesis Tests

There is a duality between confidence intervals (more generally, confidence sets) and hypothesis tests. In this section, we will show that a confidence set can be obtained by "inverting" a hypothesis test, and vice versa. Before presenting the general structure, we consider an example.

E X A M P L E A   Let $X_1, \ldots, X_n$ be a random sample from a normal distribution having unknown mean μ and known variance σ^2. We consider testing the following hypotheses:

$$H_0: \mu = \mu_0$$
$$H_A: \mu \neq \mu_0$$

Consider a test at a specific level α that rejects for $|\overline{X} - \mu_0| > x_0$, where x_0 is determined so that $P(|\overline{X} - \mu_0| > x_0) = \alpha$ if H_0 is true: $x_0 = \sigma_{\overline{X}} z(\alpha/2)$. Here the standard deviation of $\overline{X}$ is denoted by $\sigma_{\overline{X}} = \sigma/\sqrt{n}$. The test thus accepts when

$$|\overline{X} - \mu_0| < \sigma_{\overline{X}} z(\alpha/2)$$

or

$$-\sigma_{\overline{X}} z(\alpha/2) < \overline{X} - \mu_0 < \sigma_{\overline{X}} z(\alpha/2)$$

or

$$\overline{X} - \sigma_{\overline{X}} z(\alpha/2) < \mu_0 < \overline{X} + \sigma_{\overline{X}} z(\alpha/2)$$

A $100(1-\alpha)\%$ confidence interval for μ_0 is

$$\left[\overline{X} - \sigma_{\overline{X}} z(\alpha/2), \overline{X} + \sigma_{\overline{X}} z(\alpha/2) \right]$$

Comparing the acceptance region of the test to the confidence interval, we see that μ_0 lies in the confidence interval for μ if and only if the hypothesis test accepts. In other words, *the confidence interval consists precisely of all those values of μ_0 for which the null hypothesis $H_0: \mu = \mu_0$ is accepted.* ∎

We now demonstrate that this duality holds more generally. Let θ be a parameter of a family of probability distributions, and denote the set of all possible values of θ by Θ. Denote the random variables constituting the data by $\mathbf{X}$.

THEOREM A

Suppose that for every value θ_0 in Θ there is a test at level α of the hypothesis $H_0: \theta = \theta_0$. Denote the acceptance region of the test by $A(\theta_0)$. Then the set

$$C(\mathbf{X}) = \{\theta: \mathbf{X} \in A(\theta)\}$$

is a $100(1 - \alpha)\%$ confidence region for θ.

Proof

Because A is the acceptance region of a test at level α,

$$P[\mathbf{X} \in A(\theta_0)|\theta = \theta_0] = 1 - \alpha$$

Now,

$$P[\theta_0 \in C(\mathbf{X})|\theta = \theta_0] = P[\mathbf{X} \in A(\theta_0)|\theta = \theta_0]$$
$$= 1 - \alpha$$

by the definition of $C(\mathbf{X})$.   ■

It is helpful to state Theorem A in words: A $100(1 - \alpha)\%$ confidence region for θ consists of all those values of θ_0 for which the hypothesis that θ equals θ_0 will not be rejected at level α.

THEOREM B

Suppose that $C(\mathbf{X})$ is a $100(1 - \alpha)\%$ confidence region for θ; that is, for every θ_0,

$$P[\theta_0 \in C(\mathbf{X})|\theta = \theta_0] = 1 - \alpha$$

Then an acceptance region for a test at level α of the hypothesis $H_0: \theta = \theta_0$ is

$$A(\theta_0) = \{\mathbf{X}|\theta_0 \in C(\mathbf{X})\}$$

Proof

The test has level α because

$$P(\mathbf{X} \in A(\theta_0)|\theta = \theta_0) = P(\theta_0 \in C(\mathbf{X})|\theta = \theta_0) = 1 - \alpha$$ ■

In words, Theorem B says that the hypothesis that θ equals θ_0 is accepted if θ_0 lies in the confidence region.

This duality can be quite useful. In some situations, it is possible to form confidence intervals for parameters of probability distributions and then use those intervals to test hypotheses about the values of those parameters. In other situations, it may be relatively easy to test hypotheses and then determine the acceptance regions for the test to form confidence intervals that might have been quite difficult to derive in a more direct manner. We will see examples of both types of situations in later chapters.

9.4 Generalized Likelihood Ratio Tests

The likelihood ratio test is optimal for testing a simple hypothesis versus a simple hypothesis. In this section, we will develop a generalization of this test for use in situations in which the hypotheses are not simple. Such tests are not generally optimal, but they are typically nonoptimal in situations for which no optimal test exists, and they usually perform reasonably well. Generalized likelihood ratio tests have wide utility; they play the same role in testing as mle's do in estimation.

It is frequently the case that the hypotheses under consideration specify, or partially specify, the values of parameters of the probability distribution that has generated the data. Specifically, suppose that the observations $\mathbf{X} = (X_1, \ldots, X_n)$ have a joint density or frequency function $f(\mathbf{x}|\theta)$. Then H_0 may specify that $\theta \in \omega_0$, where ω_0 is a subset of the set of all possible values of θ, and H_1 may specify that $\theta \in \omega_1$, where ω_1 is disjoint from ω_0. Let $\Omega = \omega_0 \cup \omega_1$. Based on the data, a plausible measure of the relative tenability of the hypotheses is the ratio of their likelihoods. If the hypotheses are composite, each likelihood is evaluated at that value of θ that maximizes it, yielding the generalized likelihood ratio

$$\Lambda^* = \frac{\max\limits_{\theta \in \omega_0}[\text{lik}(\theta)]}{\max\limits_{\theta \in \omega_1}[\text{lik}(\theta)]}$$

Small values of Λ^* tend to discredit H_0.

It is preferable for certain technical reasons to use the test statistic

$$\Lambda = \frac{\max\limits_{\theta \in \omega_0}[\text{lik}(\theta)]}{\max\limits_{\theta \in \Omega}[\text{lik}(\theta)]}$$

rather than Λ^*. Note that $\Lambda = \min(\Lambda^*, 1)$ so small values of Λ^* correspond to small values of Λ. The rejection region for a likelihood ratio test consists of small values of Λ, for example, all $\Lambda \leq \lambda_0$. The threshhold λ_0 is chosen so that $P(\Lambda \leq \lambda_0 | H_0) = \alpha$, the desired significance level of the test.

We now illustrate the construction of a likelihood ratio test with a simple example.

E X A M P L E A   *Testing a Normal Mean*

Let $X_1, \ldots, X_n$ be i.i.d. and normally distributed with mean μ and variance σ^2, where σ is known. We wish to test H_0: $\mu = \mu_0$ against H_1: $\mu \neq \mu_0$, where μ_0 is a prescribed number. The role of θ is played by μ, and $\omega_0 = \{\mu_0\}$, $\omega_1 = \{\mu | \mu \neq \mu_0\}$, and $\Omega = \{-\infty < \mu < \infty\}$.

Since ω_0 consists of only one point, the numerator of the likelihood ratio statistic is

$$\frac{1}{(\sigma\sqrt{2\pi})^n} e^{-\frac{1}{2\sigma^2}\sum_{i=1}^{n}(X_i-\mu_0)^2}$$

For the denominator, we have to maximize the likelihood for $\mu \in \Omega$, which is achieved when μ is the mle $\overline{X}$. The denominator is the likelihood of $X_1, X_2, \ldots, X_n$ evaluated with $\mu = \overline{X}$:

$$\frac{1}{(\sigma\sqrt{2\pi})^n} e^{-\frac{1}{2\sigma^2}\sum_{i=1}^{n}(X_i-\overline{X})^2}$$

The likelihood ratio statistic is, therefore,

$$\Lambda = \exp\left(-\frac{1}{2\sigma^2}\left[\sum_{i=1}^{n}(X_i-\mu_0)^2 - \sum_{i=1}^{n}(X_i-\overline{X})^2\right]\right)$$

Rejecting for small values of Λ is equivalent to rejecting for large values of

$$-2\log\Lambda = \frac{1}{\sigma^2}\left(\sum_{i=1}^{n}(X_i-\mu_0)^2 - \sum_{i=1}^{n}(X_i-\overline{X})^2\right)$$

Using the identity

$$\sum_{i=1}^{n}(X_i-\mu_0)^2 = \sum_{i=1}^{n}(X_i-\overline{X})^2 + n(\overline{X}-\mu_0)^2$$

we see that the likelihood ratio test rejects for large values of $-2\log\Lambda = n(\overline{X}-\mu_0)^2/\sigma^2$. The distribution of this statistic under H_0 is chi-square with 1 degree of freedom. This follows, since under H_0, $\overline{X} \sim N(\mu_0, \sigma^2/n)$, which implies that $\sqrt{n}(\overline{X}-\mu_0)/\sigma \sim N(0, 1)$ and hence its square, $-2\log\Lambda \sim \chi_1^2$. Knowing the null distribution of the test statistic makes possible the construction of a rejection region for any significance level α: The test rejects when

$$\frac{n}{\sigma^2}(\overline{X}-\mu_0)^2 > \chi_1^2(\alpha)$$

Again using the fact that a chi-square random variable with 1 degree of freedom is the square of a standard normal random variable, we can rewrite this relation to show that the rejection region for the test is

$$|\overline{X}-\mu_0| \geq \frac{\sigma}{\sqrt{n}}z(\alpha/2) \qquad \blacksquare$$

The preceding derivation has been rather formal, but upon examination the result looks perfectly reasonable or perhaps even so obvious as to make us doubt the value of the formal exercise: The test of H_0: $\mu = \mu_0$ versus H_1: $\mu \neq \mu_0$ rejects when $|\overline{X}-\mu_0|$ is large. The test does not reject when $-\sigma z(\alpha/2)/\sqrt{n} \leq \overline{X} - \mu_0 \leq \sigma z(\alpha/2)/\sqrt{n}$ or, equivalently, when $\overline{X} - \sigma z(\alpha/2)/\sqrt{n} \leq \mu_0 \leq \overline{X} + \sigma z(\alpha/2)/\sqrt{n}$. That is, the test does not reject when μ_0 lies in a $100(1-\alpha)\%$ confidence interval for μ. Compare to Example A of Section 9.3.

In order for the likelihood ratio test to have the significance level α, λ_0 must be chosen so that $P(\Lambda \leq \lambda_0) = \alpha$ if H_0 is true. If the sampling distribution of Λ under H_0 is known, we can determine λ_0. Generally, the sampling distribution is not of a simple form, but in many situations the following theorem provides the basis for an approximation to the null distribution.

THEOREM A

Under smoothness conditions on the probability density or frequency functions involved, the null distribution of $-2 \log \Lambda$ tends to a chi-square distribution with degrees of freedom equal to $\dim \Omega - \dim \omega_0$ as the sample size tends to infinity. ∎

The proof, which is beyond the scope of this book, is based on a second-order Taylor series expansion.

In the statement of Theorem A, $\dim \Omega$ and $\dim \omega_0$ are the numbers of free parameters under Ω and ω_0, respectively. In Example A, the null hypothesis completely specifies μ and σ^2; there are no free parameters under ω_0, so $\dim \omega_0 = 0$. Under Ω, σ is fixed but μ is free, so $\dim \Omega = 1$. For this example, the null distribution of $-2 \log \Lambda$ is exactly χ_1^2.

9.5 Likelihood Ratio Tests for the Multinomial Distribution

In this section we will develop a generalized likelihood ratio test of the goodness of fit of a model for multinomial cell probabilities. Under the model, the vector of cell probabilities p is described by a hypothesis H_0, which specifies that $p = p(\theta)$, $\theta \in \omega_0$, where θ is a parameter that may be unknown. For example, in Section 8.2 we considered fitting Poisson probabilities that depended on an unknown parameter (there called λ, which played the role of θ) to the cell counts in a table. We want to judge the plausibility of the model relative to a model H_1 in which the cell probabilities are free except for the constraints that they are nonnegative and sum to 1. If there are m cells, Ω is thus the set consisting of m nonnegative numbers that sum to 1. The numerator of the likelihood ratio is

$$\max_{p \in \omega_0} \left(\frac{n!}{x_1! \cdots x_m!} \right) p_1(\theta)^{x_1} \cdots p_m(\theta)^{x_m}$$

where the x_i are the observed counts in the m cells. By the definition of the maximum likelihood estimate, this likelihood is maximized when $\hat{\theta}$ is the maximum likelihood estimate of θ. The corresponding probabilities will be denoted by $p_i(\hat{\theta})$.

Since the probabilities are unrestricted under Ω, the denominator is maximized by the unrestricted mle's, or

$$\hat{p}_i = \frac{x_i}{n}$$

The likelihood ratio is, therefore,

$$\Lambda = \frac{\frac{n!}{x_1! \cdots x_m!} p_1(\hat{\theta})^{x_1} \cdots p_m(\hat{\theta})^{x_m}}{\frac{n!}{x_1! \cdots x_m!} \hat{p}_1^{x_1} \cdots \hat{p}_m^{x_m}}$$

$$= \prod_{i=1}^{m} \left(\frac{p_i(\hat{\theta})}{\hat{p}_i} \right)^{x_i}$$

Also, since $x_i = n\hat{p}_i$,

$$-2 \log \Lambda = -2n \sum_{i=1}^{m} \hat{p}_i \log \left(\frac{p_i(\hat{\theta})}{\hat{p}_i} \right)$$

$$= 2 \sum_{i=1}^{m} O_i \log \left(\frac{O_i}{E_i} \right)$$

where $O_i = n\hat{p}_i$ and $E_i = np_i(\hat{\theta})$ denote the observed and expected counts, respectively.

Under Ω, the cell probabilities are allowed to be free, with the constraint that they sum to 1, so dim $\Omega = m - 1$. If, under H_0, the probabilities $p_i(\hat{\theta})$ depend on a k-dimensional parameter θ that has been estimated from the data, dim $\omega_0 = k$. The large sample distribution of $-2 \log \Lambda$ is thus a chi-square distribution with $m - k - 1$ degrees of freedom (the number of cells minus the number of estimated parameters minus 1).

Pearson's chi-square statistic is commonly used to test for goodness of fit

$$X^2 = \sum_{i=1}^{m} \frac{[x_i - np_i(\hat{\theta})]^2}{np_i(\hat{\theta})}$$

Pearson's statistic and the likelihood ratio are asymptotically equivalent under H_0. To indicate heuristically why this is so, we will go through a Taylor series argument. To begin,

$$-2 \log \Lambda = 2n \sum_{i=1}^{m} \hat{p}_i \log \left(\frac{\hat{p}_i}{p_i(\hat{\theta})} \right)$$

If H_0 is true and n is large, $\hat{p}_i \approx p_i(\hat{\theta})$. The Taylor series expansion of the function

$$f(x) = x \log \left(\frac{x}{x_0} \right)$$

about x_0 is

$$f(x) = (x - x_0) + \frac{1}{2}(x - x_0)^2 \frac{1}{x_0} + \cdots$$

Thus,

$$-2\log \Lambda \approx 2n \sum_{i=1}^{m} [\hat{p}_i - p_i(\hat{\theta})] + n \sum_{i=1}^{m} \frac{[\hat{p}_i - p_i(\hat{\theta})]^2}{p_i(\hat{\theta})}$$

The first term on the right-hand side is equal to 0 since the probabilities sum to 1, and the second term on the right-hand side may be expressed as

$$\sum_{i=1}^{m} \frac{[x_i - np_i(\hat{\theta})]^2}{np_i(\hat{\theta})}$$

since x_i, the observed count, equals $n\hat{p}_i$ for $i = 1, \ldots, m$.

We have argued for the approximate equivalence of two test statistics. Pearson's test has been more commonly used than the likelihood ratio test, because it is somewhat easier to calculate without the use of a computer.

Let us consider some examples.

EXAMPLE A   *Hardy-Weinberg Equilibrium*
Hardy-Weinberg equilibrium was first introduced in Example A in Section 8.5.1. We will now test whether this model fits the observed data. Recall that the Hardy-Weinberg equilibrium model says that the cell probabilities are $(1 - \theta)^2$, $2\theta(1 - \theta)$, and θ^2. Using the maximum likelihood estimate for θ, $\hat{\theta} = .4247$, and multiplying the resulting probabilities by the sample size $n = 1029$, we calculate expected counts, which are compared with observed counts in the following table:

	Blood Type		
	M	MN	N
Observed	342	500	187
Expected	340.6	502.8	185.6

The null hypothesis will be that the multinomial distribution is as specified by the Hardy-Weinberg equilibrium frequencies, with unknown parameter θ. The alternative hypothesis will be that the multinomial distribution does not have probabilities of that specified form. We first choose a value for α, the significance level for the test (recall that the significance level is the probability of falsely rejecting the hypothesis that the multinomial cell probabilities are as specified by genetic theory). In this application, there is no compelling reason to choose any particular value of α, so we will follow convention and let $\alpha = .05$. This means that our decision rule will falsely reject H_0 in only 5% of the cases.

We will use Pearson's chi-square test, and therefore X^2 as our test statistic. The null distribution of X^2 is approximately chi-square with 1 degree of freedom. (There are two independent cells, and one parameter has been estimated from the data.) Since, from Table 3 in Appendix B, the point defining the upper 5% of the chi-square distribution with 1 degree of freedom is 3.84, the test rejects if $X^2 > 3.84$. We next calculate X^2:

$$X^2 = \sum \frac{(O - E)^2}{E} = .0319$$

Thus, the null hypothesis is not rejected.

There is a certain unnecessary rigidity in this procedure, because it is not clear that such a decision (to reject or not) has to be made at all. There is also a certain arbitrariness: There was no strong reason to let $\alpha = .05$, but that choice essentially determined our decision. If we had let $\alpha = .01$, the decision would have been the same since $\chi^2(.01) > \chi^2(.05)$, but what if we had let $\alpha = .10$, or $.20$? It is here that the concept of the p-value becomes useful. Recall that the p-value is the smallest significance level at which the null hypothesis would be rejected. From a computer calculation of the chi-square distribution (or from a table of the normal distribution, since a chi-square distribution with 1 degree of freedom is the square of a standard normal random variable), the probability that a chi-square random variable with 1 degree of freedom is greater than or equal to .0319 is .86, which is the p-value. Another interpretation of this p-value is that if the model were correct, deviations this large or larger would occur 86% of the time. Thus, the data give us no reason to doubt the model.

In comparison, the likelihood ratio test statistic is

$$-2 \log \Lambda = 2 \sum_{i=1}^{3} O_i \log \left(\frac{O_i}{E_i} \right) = .0319$$

The two tests lead to the same conclusion.

Finally, we note that the actual maximized likelihood ratio is $\Lambda = \exp(-.0319/2) = .98$. Thus the Hardy-Weinberg model is almost as likely as the most general possible model. ∎

EXAMPLE **B** *Bacterial Clumps*

In testing milk for bacterial contamination, 0.01 ml of milk is spread over an area of 1 cm^2 on a slide. The slide is mounted on a microscope, and counts of bacterial clumps within grid squares are made. The Poisson model appears quite reasonable for the distribution of the clumps at first glance: The clumps are presumably mixed uniformly throughout the milk, and there is no reason to suspect that the clumps bunch together. However, on closer examination, two possible problems are noted. First, bacteria held by surface tension on the lower surface of the drop may adhere to the glass slide on contact, producing increased concentrations in that area of the film. Second, the film is not of uniform thickness, being thicker in the center and thinner at the edges, giving rise to nonuniform concentrations of bacteria. The following table, taken from Bliss and Fisher (1953), summarizes the counts of clumps on 400 grid squares.

Number per Square	0	1	2	3	4	5	6	7	8	9	10	19
Frequency	56	104	80	62	42	27	9	9	5	3	2	1

To fit a Poisson distribution to these data, we compute the mle, $\hat{\lambda}$, which is the mean of the 400 counts:

$$\hat{\lambda} = \frac{0 \times 56 + 1 \times 104 + 2 \times 80 + \cdots + 19 \times 1}{400}$$

$$= 2.44$$

The following table shows the observed and expected counts and the components of chi-square test statistic. (The last several cells were grouped together so that the minimum expected count would be nearly 5.)

Observed	56	104	80	62	42	27	9	20
Expected	34.9	85.1	103.8	84.4	51.5	25.1	10.2	5.0
Component of X^2	12.8	4.2	5.5	5.9	1.8	.14	.14	45.0

The chi-square statistic is $X^2 = 75.4$. With 6 degrees of freedom (there are eight cells, and one parameter has been estimated from the data), the null hypothesis is conclusively rejected [$\chi_6^2(.005) = 18.55$, so the p-value is less than .005]. When a goodness-of-fit test rejects, it is instructive to find out why; where does the model fail to fit? This can be seen by looking at the cells that make large contributions to X^2 and the signs of the observed minus the expected counts for those cells. We see here that the greatest contributions to X^2 come from the first and last cells of the table—there are too many small counts and too many large counts relative to what is expected for a Poisson distribution. ∎

E X A M P L E   C   *Fisher's Reexamination of Mendel's Data*
In one of his famous experiments, Mendel crossed 556 smooth, yellow male peas with wrinkled, green female peas. According to now established genetic theory, the relative frequencies of the progeny should be as given in the following table.

Type	Frequency
Smooth yellow	$\frac{9}{16}$
Smooth green	$\frac{3}{16}$
Wrinkled yellow	$\frac{3}{16}$
Wrinkled green	$\frac{1}{16}$

The counts that Mendel recorded and the expected counts are given in the following table:

Type	Observed Count	Expected Count
Smooth yellow	315	312.75
Smooth green	108	104.25
Wrinkled yellow	102	104.25
Wrinkled green	31	34.75

Calculating the likelihood ratio test statistic, we obtain

$$-2 \log \Lambda = 2 \sum_{i=1}^{4} O_i \log \left(\frac{O_i}{E_i} \right) = .618$$

Comparing this value with the chi-square distribution with 3 degrees of freedom (three independent parameters are estimated under Ω and none under ω_0), we have a p-value of slightly less than .9. Pearson's chi-square statistic is .604, which is quite close to the value from the likelihood ratio test. We interpret the p-value as meaning that, even if the model were correct, discrepancies this large or larger would be expected to occur on the basis of chance about 90% of the time. There is thus no reason to reject the hypothesis that the counts come from a multinomial distribution with the prescribed probabilities. We would tend to doubt this hypothesis for only small p-values.

The p-value can also be interpreted to mean that on the basis of chance we would expect agreement this close or closer about only 10% of the time. There is some validity to the suggestion that the data agree with the model too well; if the p-value had been .999, for example, we would definitely be suspicious.

Fisher pooled the results of all of Mendel's experiments in the following way. Suppose that two independent experiments give chi-square statistics with p and r degrees of freedom, respectively. Then, under the null hypothesis that the models were correct, the sum of those two test statistics would follow a chi-square distribution with $p + r$ degrees of freedom. Fisher added the chi-square statistics for all the independent experiments and compared the result with the chi-square distribution with degrees of freedom equal to the sum of all the degrees of freedom. The resulting p-value was .99996. Such close agreement would only occur 4 times out of 100,000 on the basis of chance!

What happened? Did Mendel deliberately or unconsciously fudge the data? Did he have an overzealous lab technician who was hoping for a recommendation to medical school? Was there divine intervention? Perhaps the best explanation is that Mendel continued experimenting until the results looked good and then he stopped. The statistical analysis here assumes that the sample size is fixed before the data are collected.   ∎

Mendel is not the only scientist whose data are too good to be true. Cyril Burt was an English psychologist whose work had a great impact on the debate concerning the genetic basis for intelligence. His many papers and extensive data argue for such a basis. In 1946, Burt became the first psychologist to be knighted; however, during the 1970s, his work came under increasing attack, and he was accused of actually fabricating data. One of his most famous studies was of the intelligence and occupational status of 40,000 fathers and sons. Dorfman (1978) studied the goodness of fit of these intelligence scores to a normal distribution, using Pearson's chi-square test. The p-values for fathers and sons were greater than $1 - 10^{-7}$ and $1 - 10^{-6}$, respectively! Dorfman concluded that "it may well be that Burt's frequency distributions are the most normally distributed in the history of anthropometric measurement."

9.6 The Poisson Dispersion Test

The likelihood ratio test and Pearson's chi-square test are carried out with respect to the general alternative hypothesis that the cell probabilities are completely free. If one has a specific alternative hypothesis in mind, better power can usually be obtained by testing against that alternative rather than against a more general alternative. Such a test is developed in this section for the hypothesis that a distribution is Poisson. The test is quite useful, and its construction affords another illustration of a generalized likelihood ratio test.

The two key assumptions underlying the Poisson distribution are that the rate is constant and that the counts in one interval of time or space are independent of the counts in disjoint intervals. These assumptions are often not met. For example, suppose that insects are counted on leaves of plants. The leaves are of different sizes and occur at various locations on different plants; the rate of infestation may well not be constant over the different locations. Furthermore, if the insects hatched from eggs that were deposited in groups, there might be clustering of the insects and the independence assumption might fail. If counts occurring over time are being recorded, the underlying rate of the phenomenon being studied might not be constant. Motor vehicle counts for traffic studies, for example, typically vary cyclically over time.

Given counts $x_1, \ldots, x_n$, we consider testing the null hypothesis that the counts are Poisson with the common parameter λ versus the alternative hypothesis that they are Poisson but have different rates, $\lambda_1, \ldots, \lambda_n$. Under Ω, there are n different rates; $\omega_0 \subset \Omega$ is the special case that they are all equal. Under ω_0, the maximum likelihood estimate of λ is $\hat{\lambda} = \overline{X}$. Under Ω, the maximum likelihood estimates of the λ_i are $x_1, \ldots, x_n$; we denote these estimates by $\tilde{\lambda}_i$. The likelihood ratio is thus

$$\Lambda = \frac{\displaystyle\prod_{i=1}^{n} \hat{\lambda}^{x_i} e^{-\hat{\lambda}}/x_i!}{\displaystyle\prod_{i=1}^{n} \tilde{\lambda}_i^{x_i} e^{-\tilde{\lambda}_i}/x_i!}$$

$$= \prod_{i=1}^{n} \left(\frac{\bar{x}}{x_i}\right)^{x_i} e^{x_i - \bar{x}}$$

The likelihood ratio test statistic is

$$-2 \log \Lambda = -2 \sum_{i=1}^{n} \left[x_i \log \left(\frac{\bar{x}}{x_i} \right) + (x_i - \bar{x}) \right]$$

$$= 2 \sum_{i=1}^{n} x_i \log \left(\frac{x_i}{\bar{x}} \right)$$

A nearly equivalent form of this statistic is produced using the Taylor series argument given in Section 9.5:

$$-2 \log \Lambda \approx \frac{1}{\bar{x}} \sum_{i=1}^{n} (x_i - \bar{x})^2$$

Under Ω, there are n independent parameters, $\lambda_1, \ldots, \lambda_n$, so dim $\Omega = n$. Under ω_0, there is only one parameter, λ, so dim $\omega_0 = 1$, and the degrees of freedom are $n - 1$.

The last equation given for the test statistic may be interpreted as being the ratio of n times the estimated variance to the estimated mean. For the Poisson distribution, the variance equals the mean; for the types of alternatives discussed above, the variance is typically greater than the mean. For this reason the test is often called the **Poisson dispersion test.** It is sensitive to—that is, has high power against—alternatives that are overdispersed relative to the Poisson, such as the negative binomial distribution. The ratio $\hat{\sigma}^2 / \bar{x}$ is sometimes used as a measure of clustering.

E X A M P L E **A**   *Asbestos Fibers*

In Example A in Section 8.4, we considered whether counts of asbestos fibers on grid squares could be modeled as a Poisson distribution. Applying the Poisson dispersion test, we find that

$$\frac{1}{\bar{x}} \sum (x_i - \bar{x})^2 = 26.56$$

or, if the likelihood ratio statistic is used,

$$2 \sum x_i \log \left(\frac{x_i}{\bar{x}} \right) = 27.11$$

Since there are 23 observations, there are 22 degrees of freedom. From a computer calculation, the p-value for the likelihood ratio statistic is .21. The evidence against the null hypothesis is not persuasive; however, the sample size is small and the test may have low power. ∎

E X A M P L E **B**   *Bacterial Clumps*

In Example B in Section 9.5, we applied Pearson's chi-square test to test whether counts of bacteria clumps in milk were fit by a Poisson distribution. There we found

$\bar{x} = 2.44$. The sample variance is

$$\hat{\sigma}^2 = \frac{0^2 \times 56 + 1^2 \times 104 + \cdots + 19^2 \times 1}{400} - \bar{x}^2$$

$$= 4.59$$

The ratio of the variance to the mean is 1.88 rather than 1; the test statistic is

$$T = \frac{n\hat{\sigma}^2}{\bar{x}}$$

$$= \frac{400 \times 4.59}{2.40} = 752.7$$

Under the null hypothesis, the statistic approximately follows a chi-square distribution with 399 degrees of freedom. Since a chi-square random variable with m degrees of freedom is the sum of the squares of m independent $N(0, 1)$ random variables, the central limit theorem implies that for large values of m the chi-square distribution with m degrees of freedom is approximately normal. For a chi-square distribution, the mean equals the number of degrees of freedom and variance equals twice the number of degrees of freedom. The p-value can thus be found by standardizing the statistic and using tables of the standard normal distribution:

$$P(T \geq 752.7) = P\left(\frac{T - 399}{\sqrt{2 \times 399}} \geq \frac{752.7 - 399}{\sqrt{2 \times 399}}\right)$$

$$\approx 1 - \Phi(12.5) \approx 0$$

Thus, there is almost no doubt that the Poisson distribution fails to fit the data. ∎

9.7 Hanging Rootograms

In this and the next section, we develop additional informal techniques for assessing goodness of fit. The first of these is the hanging rootogram. **Hanging rootograms** are a graphical display of the differences between observed and fitted values in histograms. To illustrate the construction and interpretation of hanging rootograms, we will use a set of data from the field of clinical chemistry (Martin, Gudzinowicz, and Fanger 1975). The following table gives the empirical distribution of 152 serum potassium levels. In clinical chemistry, distributions such as this are often tabulated to establish a range of "normal" values against which the level of the chemical found in a patient can be compared to determine whether it is abnormal. The tabulated distributions are often fit to parametric forms such as the normal distribution.

Serum Potassium Levels	
Interval Midpoint	Frequency
3.2	2
3.3	1
3.4	3
3.5	2
3.6	7
3.7	8
3.8	8
3.9	14
4.0	14
4.1	18
4.2	16
4.3	15
4.4	10
4.5	8
4.6	8
4.7	6
4.8	4
4.9	1
5.0	1
5.1	1
5.2	4
5.3	1

Figure 9.3(a) is a histogram of the frequencies. The plot looks roughly bell-shaped, but the normal distribution is not the only bell-shaped distribution. In order to evaluate their distribution more exactly, we must compare the observed frequencies to frequencies fit by the normal distribution. This can be done in the following way. Suppose that the parameters μ and σ of the normal distribution are estimated from the data by $\bar{x}$ and $\hat{\sigma}$. If the jth interval has the left boundary x_{j-1} and the right boundary x_j, then according to the model, the probability that an observation falls in that interval is

$$\hat{p}_j = \Phi\left(\frac{x_j - \bar{x}}{\hat{\sigma}}\right) - \Phi\left(\frac{x_{j-1} - \bar{x}}{\hat{\sigma}}\right)$$

If the sample is of size n, the predicted, or fitted, count in the jth interval is

$$\hat{n}_j = n\hat{p}_j$$

which may be compared to the observed counts, n_j.

Figure 9.3(b) is a *hanging histogram* of the differences: observed count (n_j) minus expected count ($\hat{n}_j$). These differences are difficult to interpret since the variability is not constant from cell to cell. If we neglect the variability in the estimated expected counts, we have

$$\text{Var}(n_j - \hat{n}_j) = \text{Var}(n_j) = np_j(1 - p_j)$$
$$= np_j - np_j^2$$

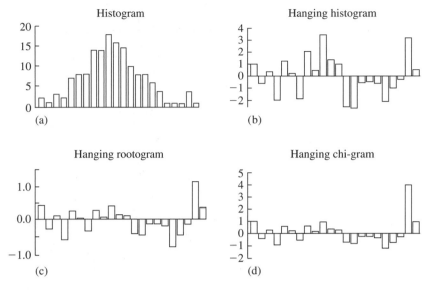

FIGURE **9.3**   (a) Histogram, (b) hanging histogram, (c) hanging rootogram, and (d) hanging chi-gram for normal fit to serum potassium data.

In this case, the p_j are small, so

$$\text{Var}(n_j - \hat{n}_j) \approx np_j$$

Thus, cells with large values of p_j (equivalent to large values of n_j if the model is at all close) have more variable differences, $n_j - \hat{n}_j$. In a hanging histogram, we expect larger fluctuations in the center than in the tails. This unequal variability makes it difficult to assess and compare the fluctuations, since a large deviation may indicate real misfit of the model or may be merely caused by large random variability.

To put the differences between observed and expected values on a scale on which they all have equal variability, a **variance-stabilizing transformation** may be used. (Such transformations will be used in later chapters as well.) Suppose that a random variable X has mean μ and variance $\sigma^2(\mu)$, which depends on μ. If $Y = f(X)$, the method of propagation of error (Section 4.6) shows that

$$\text{Var}(Y) \approx \sigma^2(\mu)[f'(\mu)]^2$$

Thus if f is chosen so that $\sigma^2(\mu)[\,f'(\mu)]^2$ is constant, the variance of Y will not depend on μ. The transformation f that accomplishes this is called a variance-stabilizing transformation.

Let us apply this idea to the case we have been considering:

$$E(n_j) = np_j = \mu$$
$$\text{Var}(n_j) \approx np_j = \sigma^2(\mu)$$

That is, $\sigma^2(\mu) = \mu$, so f will be a variance-stabilizing transformation if $\mu[f'(\mu)]^2$ is constant. The function $f(x) = \sqrt{x}$ does the job, and

$$E(\sqrt{n_j}) \approx \sqrt{np_j}$$
$$\text{Var}(\sqrt{n_j}) \approx \tfrac{1}{4}$$

if the model is correct.

Figure 9.3(c) shows a *hanging rootogram*, a display showing

$$\sqrt{n_j} - \sqrt{\hat{n}_j}$$

The advantage of the hanging rootogram is that the deviations from cell to cell have approximately the same statistical variability. To assess the deviations, we may use the rough rule of thumb that a deviation of more than 2 or 3 standard deviations (more than 1.0 or 1.5 in this case) is "large." The most striking feature of the hanging rootogram in Figure 9.3(c) is the large deviation in the right tail. Generally, deviations in the center have been down-weighted and those in the tails emphasized by the transformation. Also, it is noteworthy that although the deviations other than the one in the right tail are not especially large, they have a certain systematic character: Note the run of positive deviations followed by the run of negative deviations and then the large positive deviation in the extreme right tail. This may indicate some asymmetry in the distribution.

A possible alternative to the rootogram is what can be called a *hanging chi-gram*, a plot of the components of Pearson's chi-square statistic:

$$\frac{n_j - \hat{n}_j}{\sqrt{\hat{n}_j}}$$

Neglecting the variability in the expected counts, as before, $\text{Var}(n_j - \hat{n}_j) \approx np_j = \hat{n}_j$, so

$$\text{Var}\left(\frac{n_j - \hat{n}_j}{\sqrt{\hat{n}_j}}\right) \approx 1$$

so this technique also stabilizes variance. Figure 9.3(d) is a hanging chi-gram for the case we have been considering; it is quite similar in overall character to the hanging rootogram, but the deviation in the right tail is emphasized even more.

9.8 Probability Plots

Probability plots are an extremely useful graphical tool for qualitatively assessing the fit of data to a theoretical distribution. Consider a sample of size n from a uniform distribution on [0, 1]. Denote the *ordered* sample values by $X_{(1)} < X_{(2)} \cdots < X_{(n)}$. These values are called **order statistics.** It can be shown (see Problem 17 at the end of Chapter 4) that

$$E(X_{(j)}) = \frac{j}{n+1}$$

This suggests plotting the ordered observations $X_{(1)}, \ldots, X_{(n)}$ against their expected values $1/(n+1), \ldots, n/(n+1)$. If the underlying distribution is uniform, the plot

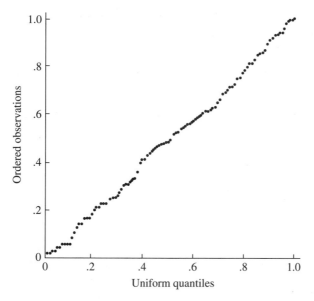

FIGURE **9.4** Uniform-uniform probability plot.

should look roughly linear. Figure 9.4 is such a plot for a sample of size 100 from a uniform distribution.

Now suppose that a sample $Y_1, \ldots, Y_{100}$ is generated in which each Y is half the sum of two independent uniform random variables. The distribution of Y is no longer uniform but triangular:

$$f(y) = \begin{cases} 4y, & 0 \le y \le \frac{1}{2} \\ 4 - 4y, & \frac{1}{2} \le y \le 1 \end{cases}$$

The ordered observations $Y_{(1)}, \ldots, Y_{(n)}$ are plotted against the points $1/(n+1), \ldots,$ $n/(n+1)$. The graph in Figure 9.5 shows a clear deviation from linearity and enables us to describe qualitatively the deviation of the distribution of the Y's from the uniform distribution. Note that in the left tail of the plotted distribution (near 0), the order statistics are larger than expected for a uniform distribution, and in the right tail (near 1), they are smaller, indicating that the tails of the distribution of the Y's decrease more quickly (are "lighter") than the tails of the uniform distribution.

The technique can be extended to other continuous probability laws by means of Proposition C of Section 2.3, which states that if X is a continuous random variable with a strictly increasing cumulative distribution function, F_X, and if $Y = F_X(X)$, then Y has a uniform distribution on [0, 1]. The transformation $Y = F_X(X)$ is known as the **probability integral transform.**

The following procedure is suggested by the proposition just referred to. Suppose that it is hypothesized that X follows a certain distribution, F. Given a sample $X_1, \ldots, X_n$, we plot

$$F(X_{(k)}) \qquad \text{vs.} \qquad \frac{k}{n+1}$$

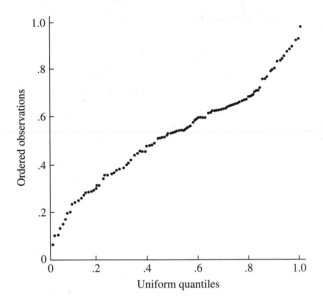

FIGURE **9.5**   Uniform-triangular probability plot.

or equivalently

$$X_{(k)} \quad \text{vs.} \quad F^{-1}\left(\frac{k}{n+1}\right)$$

In some cases, F is of the form

$$F(x) = G\left(\frac{x - \mu}{\sigma}\right)$$

where μ and σ are called location and scale parameters, respectively. The normal distribution is of this form. We could plot

$$\frac{X_{(k)} - \mu}{\sigma} \quad \text{vs.} \quad G^{-1}\left(\frac{k}{n+1}\right)$$

or if we plotted

$$X_{(k)} \quad \text{vs.} \quad G^{-1}\left(\frac{k}{n+1}\right)$$

the result would be approximately a straight line if the model were correct:

$$X_{(k)} \approx \sigma G^{-1}\left(\frac{k}{n+1}\right) + \mu$$

Slight modifications of this procedure are sometimes used. For example, rather than $G^{-1}[k/(n+1)]$, $E(X_{(k)})$, the expected value of the kth smallest observation can be used. But it can be argued that

$$E(X_{(k)}) \approx F^{-1}\left(\frac{k}{n+1}\right)$$
$$= \sigma G^{-1}\left(\frac{k}{n+1}\right) + \mu$$

so this modification yields very similar results to the original procedure.

The procedure can be viewed from another perspective. Recall from Section 2.2 that $F^{-1}[k/(n+1)]$ is the $k/(n+1)$ quantile of the distribution F; that is, it is the point such that the probability that a random variable with distribution function F is less than it is $k/(n+1)$. We are thus plotting the ordered observations (which may be viewed as the observed or empirical quantiles) versus the quantiles of the theoretical distribution.

EXAMPLE A   We can illustrate the procedure just described using a set of 100 observations, which are Michelson's determinations of the velocity of light made from June 5, 1879 to July 2, 1879; 299,000 has been subtracted from the determinations to give the values listed [data from Stigler (1977)]:

850	960	880	890	890	740
940	880	810	840	900	960
880	810	780	1070	940	860
820	810	930	880	720	800
760	850	800	720	770	810
950	850	620	760	790	980
880	860	740	810	980	900
970	750	820	880	840	950
760	850	1000	830	880	910
870	980	790	910	920	870
930	810	850	890	810	650
880	870	860	740	760	880
840	880	810	810	830	840
720	940	1000	800	850	840
950	1000	790	840	850	800
960	760	840	850	810	960
800	840	780	870		

Figure 9.6 shows the normal probability plot. The plot looks straight, showing that the normal distribution gives a reasonable fit.

A word of caution is in order here: Probability plots are by nature monotone-increasing and they all tend to look fairly straight. Some experience is necessary in gauging "straightness." Simulations, which are easily done, are very helpful in sharpening one's judgment. Some find it useful to hold the plot so that they are looking down the plotted line as if it were a roadway; this often makes curvature much more apparent.   ■

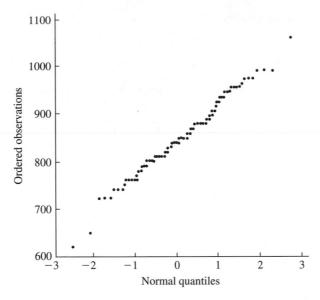

FIGURE **9.6**   Normal probability plot of Michelson's data.

EXAMPLE **B**   In order to be able to interpret probability plots, it is useful to see how they are shaped for samples from nonnormal distributions. Figure 9.7 is a normal probability plot of 500 pseudorandom variables from a double exponential distribution:

$$f(x) = \frac{1}{2}e^{-|x|}, \qquad -\infty < x < \infty$$

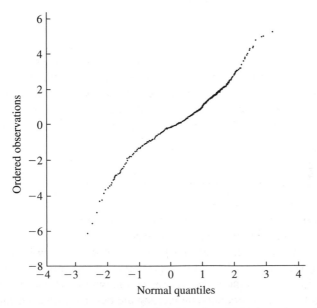

FIGURE **9.7**   Normal probability plot of 500 pseudorandom variables from a double exponential distribution.

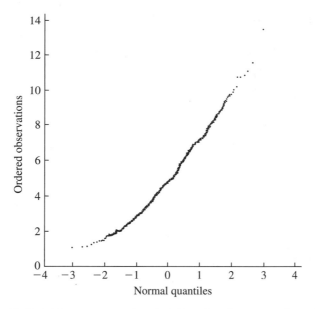

FIGURE **9.8**   Normal probability plot of 500 pseudorandom variables from a gamma distribution with the shape parameter $\alpha = 5$.

This density is symmetric about zero, but its tails die off at the rate $\exp(-|x|)$. This rate is slower than that for the tails of the normal distribution, which decay at the rate $\exp(-x^2)$. Note how the plot in Figure 9.7 bends down at the left and up at the right, indicating that the observations in the left tail were more negative than expected for a normal distribution and the observations in the right tail were more positive. In other words, the extreme observations were larger in magnitude than extreme observations from a normal distribution would be. This effect results because the tails of the double exponential are "heavier" than those of a normal distribution.

Figure 9.8 is a normal probability plot of 500 pseudorandom numbers from a gamma distribution with the shape parameter $\alpha = 5$ and the scale parameter $\lambda = 1$. As can be seen in Figure 2.11, the gamma density with $\alpha = 5$ is nonsymmetric, or skewed, and this is reflected by the bowlike appearance of the probability plot. ∎

EXAMPLE C   As an example for a nonnormal distribution, Figure 9.9 is a gamma probability plot of the precipitation amounts of Example C in Section 8.4.

The parameter λ of a gamma distribution is a scale parameter and so, as we have seen before, affects only the slope of a probability plot, not its straightness. Thus in constructing a probability plot we can take $\lambda = 1$ without loss. A computer was used to find the quantiles of a gamma distribution with parameter $\alpha = .471$ and $\lambda = 1$, and Figure 9.9 was produced by plotting the observed sorted values of precipitation versus the quantiles. Qualitatively, the fit looks reasonable, because there is no gross systematic deviation from a straight line. ∎

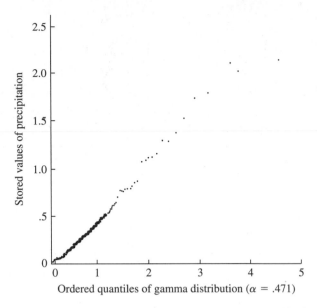

F I G U R E **9.9** Gamma probability plot of rainfall distribution.

Probability plots can also be constructed for grouped data, such as the data on serum potassium levels in Section 9.7. Because the ordered observations are not all available in such a case, the procedure must be modified. Suppose that the grouping gives the points $x_1, x_2, \ldots, x_{m+1}$ for the histogram's bin boundaries and that in the interval $[x_i, x_{i+1})$ there are n_i counts, where $i = 1, \ldots, m$. We denote the cumulative frequencies by $N_j = \sum_{i=1}^{j} n_i$. Then $N_1 < N_2 < \cdots < N_m$ and $N_m = n$, which is the total sample size. We thus plot

$$x_{j+1} \qquad \text{vs.} \qquad G^{-1}\left(\frac{N_j}{n+1}\right), \qquad j = 1, \ldots, m$$

E X A M P L E **D** Figure 9.10 shows a normal probability plot for the serum potassium data of Section 9.7. The cumulative frequencies are found by summing the frequencies in each bin. The deviations in the right tail are immediately apparent. ■

9.9 Tests for Normality

A wide variety of tests are available for testing goodness of fit to the normal distribution. We discuss some of them in this section; more discussion may be found in the works referred to.

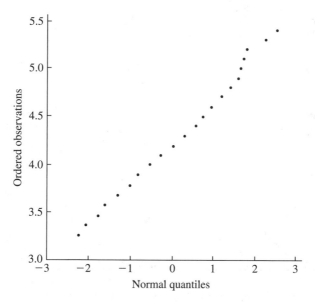

FIGURE **9.10**   Normal probability plot of serum potassium data.

If the data are grouped into bins, with several counts in each bin, Pearson's chi-square test for goodness of fit may be applied. But if the parameters are estimated from ungrouped data and the expected counts in each bin are calculated using the estimated parameters, the limiting distribution of the test statistic is no longer chi-square. In order for the limiting distribution to be chi-square, the parameters must be estimated from the grouped data. This was pointed out by Chernoff and Lehmann (1954) and is further discussed by Dahiya and Gurland (1972). Generally speaking, it seems rather artificial and wasteful of information to group continuous data.

Departures from normality often take the form of asymmetry, or skewness. For a normal distribution, the third central moment is $\int_{-\infty}^{\infty}(x-\mu)^3\varphi(x)dx$, which equals 0 since the density is symmetric about μ. Suppose that we wish to test the null hypothesis that $X_1, \ldots, X_n$ are independent normally distributed random variables with the same mean and variance. A goodness-of-fit test can be based on the **coefficient of skewness** for the sample,

$$b_1 = \frac{\frac{1}{n}\sum_{i=1}^{n}(X_i - \overline{X})^3}{s^3}$$

The test rejects for large values of $|b_1|$.

Symmetric distributions can depart from normality by being heavy-tailed or light-tailed or too peaked or too flat in the center. These forms of departures may be detected by the **coefficient of kurtosis** for the sample,

$$b_2 = \frac{\frac{1}{n}\sum_{i=1}^{n}(X_i - \overline{X})^4}{s^4}$$

If either of these measures is to be used as a test statistic, its sampling distribution when the distribution generating the data is normal must be determined. The hypothesis test rejects normality when the observed value of the statistic is in the tails of the sampling distribution. These sampling distributions are difficult to evaluate in closed form but can be approximated by simulations.

A goodness-of-fit test may also be based on the linearity of the probability plot, as measured by the correlation coefficient, r, of the x and y coordinates of the points of the probability plot. The test rejects for small values of r. The sampling distribution of r under normality has been approximated by simulations and is tabled in Filliben (1975). Ryan and Joiner (unpublished) give a short table for the null sampling distribution of r from normal probability plots with critical values for the correlation coefficient corresponding to significance levels .1, .05, and .01:

n	.1	.05	.01
4	.8951	.8734	.8318
5	.9033	.8804	.8320
10	.9347	.9180	.8804
15	.9506	.9383	.9110
20	.9600	.9503	.9290
25	.9662	.9582	.9408
30	.9707	.9639	.9490
40	.9767	.9715	.9597
50	.9807	.9764	.9664
60	.9836	.9799	.9710
75	.9865	.9835	.9752

They also report the results of some simulations of the power of r as the test statistic for certain alternative distributions. For example, the power against a uniform distribution with a significance level of .1 is .13 for $n = 10$ and .20 for $n = 20$. This is somewhat discouraging—the test rejects only 13% of the time and 20% of the time for the given sample sizes if the real underlying distribution is uniform. The moral is that it may be quite difficult to detect departure from normality in small samples. On a more positive note, the power of r against an exponential distribution is 53% for $n = 10$ and 89% for $n = 20$.

Pearson, D'Agostino, and Bowman (1977) report the results of quite extensive simulations of the power for several alternative distributions and give further references.

For Michelson's data (see Example A in Section 9.8), the correlation coefficient is .995. From the tables in Filliben (1975), this falls between the 50th and 75th percentiles of the null sampling distribution, giving no reason to reject the hypothesis of normality. It may not be very realistic, however, to model the 100 observations of the velocity of light as a sample of 100 independent random variables from some probability distribution and to use this model to test goodness of fit. We have little information about how these data were collected and processed. For example, since the observations were made sequentially, it is quite possible that the measurement process drifted in time or that successive errors were correlated. It is also possible that Michelson discarded some obviously bad data.

9.10 Concluding Remarks

Two very important concepts, estimation and hypothesis testing, have been introduced in this chapter and the last. They have been introduced here in the context of fitting probability distributions but will recur throughout the rest of this book in various other contexts. Generally, observations are taken from a probability law that depends on a parameter, θ. Estimation theory is concerned with estimating θ from the data; the theory of hypothesis testing is concerned with testing hypotheses about the value of θ. Methods based on likelihood, maximum likelihood estimation and likelihood ratio tests, have also been introduced. These methods are much more generally useful than has been demonstrated by the specific purposes to which they have been put in these chapters. The likelihood and the likelihood ratio are key concepts of statistics, from both Bayesian and frequentist perspectives.

The fundamental concepts and techniques of hypothesis testing have been introduced in this chapter. We have seen how to test a null hypothesis by choosing a test statistic and a rejection region such that, under the null hypothesis, the probability that the test statistic falls in the rejection region is α, the significance level of the test. The choice of this region is determined by knowing, at least approximately, the null distribution of the test statistic. The test statistic is frequently, but not always, a likelihood ratio; when the exact distribution of the likelihood ratio cannot be found, we can use the chi-square distribution as a large-sample approximation. We have also explored the relation of the p-value of the test statistic to the significance level. In some situations, the p-value is a less rigid summary of the evidence than is a decision whether to reject the null hypothesis.

With the increasing availability of flexible computer programs and inexpensive computers, graphical methods are being used more and more in statistics. The last part of this chapter introduced two graphical techniques: hanging rootograms and probability plots. Other graphical techniques will be introduced in Chapter 10. Such informal techniques are often of more practical use than are more formal techniques, such as hypothesis testing. Literally testing for goodness of fit is often rather artificial—a parametric distribution is usually entertained only as a model for the distribution of data values, and it is clear a priori that the data do not really come from that distribution. If enough data were available, the goodness-of-fit test would certainly reject. Rather than test a hypothesis that no one literally believes could hold, it is usually more useful to ascertain qualitatively where the model fits and where and how it fails to fit.

Some concepts introduced in Chapters 7 and 8 have been elaborated on in this chapter. In Chapter 7, we introduced confidence intervals for the parameters of finite populations; in Chapter 8, we considered confidence intervals for parameters of probability distributions. In this chapter, we have introduced hypothesis testing and developed a relation between hypothesis tests and confidence intervals. The method of propagation of error, used in Chapter 7 as a tool for analyzing the statistical behavior of ratio estimates, has been used in this chapter in connection with variance-stabilizing transformations.

9.11 Problems

1. A coin is thrown independently 10 times to test the hypothesis that the probability of heads is $\frac{1}{2}$ versus the alternative that the probability is not $\frac{1}{2}$. The test rejects if either 0 or 10 heads are observed.

   a. What is the significance level of the test?
   b. If in fact the probability of heads is .1, what is the power of the test?

2. Which of the following hypotheses are simple, and which are composite?

   a. X follows a uniform distribution on [0, 1].
   b. A die is unbiased.
   c. X follows a normal distribution with mean 0 and variance $\sigma^2 > 10$.
   d. X follows a normal distribution with mean $\mu = 0$.

3. Suppose that $X \sim \text{bin}(100, p)$. Consider the test that rejects H_0: $p = .5$ in favor of H_A: $p \neq .5$ for $|X - 50| > 10$. Use the normal approximation to the binomial distribution to answer the following:

   a. What is α?
   b. Graph the power as a function of p.

4. Let X have one of the following distributions:

X	H_0	H_A
x_1	.2	.1
x_2	.3	.4
x_3	.3	.1
x_4	.2	.4

   a. Compare the likelihood ratio, Λ, for each possible value X and order the x_i according to Λ.
   b. What is the likelihood ratio test of H_0 versus H_A at level $\alpha = .2$? What is the test at level $\alpha = .5$?
   c. If the prior probabilities are $P(H_0) = P(H_A)$, which outcomes favor H_0?
   d. What prior probabilities correspond to the decision rules with $\alpha = .2$ and $\alpha = .5$?

5. True or false, and state why:

   a. The significance level of a statistical test is equal to the probability that the null hypothesis is true.
   b. If the significance level of a test is decreased, the power would be expected to increase.
   c. If a test is rejected at the significance level α, the probability that the null hypothesis is true equals α.
   d. The probability that the null hypothesis is falsely rejected is equal to the power of the test.
   e. A type I error occurs when the test statistic falls in the rejection region of the test.

f. A type II error is more serious than a type I error.

g. The power of a test is determined by the null distribution of the test statistic.

h. The likelihood ratio is a random variable.

6. Consider the coin tossing example of Section 9.1. Suppose that instead of tossing the coin 10 times, the coin was tossed until a head came up and the total number of tosses, X, was recorded.

   a. If the prior probabilities are equal, which outcomes favor H_0 and which favor H_1?

   b. Suppose $P(H_0)/P(H_1) = 10$. What outcomes favor H_0?

   c. What is the significance level of a test that rejects H_0 if $X \geq 8$?

   d. What is the power of this test?

7. Let $X_1, \ldots, X_n$ be a sample from a Poisson distribution. Find the likelihood ratio for testing H_0: $\lambda = \lambda_0$ versus H_A: $\lambda = \lambda_1$, where $\lambda_1 > \lambda_0$. Use the fact that the sum of independent Poisson random variables follows a Poisson distribution to explain how to determine a rejection region for a test at level α.

8. Show that the test of Problem 7 is uniformly most powerful for testing H_0: $\lambda = \lambda_0$ versus H_A: $\lambda > \lambda_0$.

9. Let $X_1, \ldots, X_{25}$ be a sample from a normal distribution having a variance of 100. Find the rejection region for a test at level $\alpha = .10$ of H_0: $\mu = 0$ versus H_A: $\mu = 1.5$. What is the power of the test? Repeat for $\alpha = .01$.

10. Suppose that $X_1, \ldots, X_n$ form a random sample from a density function, $f(x|\theta)$, for which T is a sufficient statistic for θ. Show that the likelihood ratio test of H_0: $\theta = \theta_0$ versus H_A: $\theta = \theta_1$ is a function of T. Explain how, if the distribution of T is known under H_0, the rejection region of the test may be chosen so that the test has the level α.

11. Suppose that $X_1, \ldots, X_{25}$ form a random sample from a normal distribution having a variance of 100. Graph the power of the likelihood ratio test of H_0: $\mu = 0$ versus H_A: $\mu \neq 0$ as a function of μ, at significance levels .10 and .05. Do the same for a sample size of 100. Compare the graphs and explain what you see.

12. Let $X_1, \ldots, X_n$ be a random sample from an exponential distribution with the density function $f(x|\theta) = \theta \exp[-\theta x]$. Derive a likelihood ratio test of H_0: $\theta = \theta_0$ versus H_A: $\theta \neq \theta_0$, and show that the rejection region is of the form $\{\overline{X} \exp[-\theta_0 \overline{X}] \leq c\}$.

13. Suppose, to be specific, that in Problem 12, $\theta_0 = 1$, $n = 10$, and that $\alpha = .05$. In order to use the test, we must find the appropriate value of c.

   a. Show that the rejection region is of the form $\{\overline{X} \leq x_0\} \cup \{\overline{X} \geq x_1\}$, where x_0 and x_1 are determined by c.

   b. Explain why c should be chosen so that $P(\overline{X} \exp(-\overline{X}) \leq c) = .05$ when $\theta_0 = 1$.

   c. Explain why $\sum_{i=1}^{10} X_i$ and hence $\overline{X}$ follow gamma distributions when $\theta_0 = 1$. How could this knowledge be used to choose c?

d. Suppose that you hadn't thought of the preceding fact. Explain how you could determine a good approximation to c by generating random numbers on a computer (simulation).

14. Suppose that under H_0, a measurement X is $N(0, \sigma^2)$, and that under H_1, X is $N(1, \sigma^2)$ and that the prior probability $P(H_0) = 2 \times P(H_1)$. As in Section 9.1, the hypothesis H_0 will be chosen if $P(H_0|x) > P(H_1|x)$. For $\sigma^2 = 0.1, 0.5, 1.0, 5.0$:

   a. For what values of X will H_0 be chosen?
   b. In the long run, what proportion of the time will H_0 be chosen if H_0 is true $\frac{2}{3}$ of the time?

15. Suppose that under H_0, a measurement X is $N(0, \sigma^2)$, and that under H_1, X is $N(1, \sigma^2)$ and that the prior probability $P(H_0) = P(H_1)$. For $\sigma = 1$ and $x \in [0, 3]$, plot and compare (1) the p-value for the test of H_0 and (2) $P(H_0|x)$. Can the p-value be interpreted as the probability that H_0 is true? Choose another value of σ and repeat.

16. In the previous problem, with $\sigma = 1$, what is the probability that the p-value is less than 0.05 if H_0 is true? What is the probability if H_1 is true?

17. Let $X \sim N(0, \sigma^2)$, and consider testing $H_0 : \sigma = \sigma_0$ versus $H_A : \sigma = \sigma_1$, where $\sigma_1 > \sigma_0$. The values σ_0 and σ_1 are fixed.

   a. What is the likelihood ratio as a function of x? What values favor H_0? What is the rejection region of a level α test?
   b. For a sample, $X_1, X_2, \ldots, X_n$ distributed as above, repeat the previous question.
   c. Is the test in the previous question uniformly most powerful for testing $H_0 : \sigma = \sigma_0$ versus $H_1 : \sigma > \sigma_0$?

18. Let $X_1, X_2, \ldots, X_n$ be i.i.d. random variables from a double exponential distribution with density $f(x) = \frac{1}{2}\lambda \exp(-\lambda|x|)$. Derive a likelihood ratio test of the hypothesis $H_0 : \lambda = \lambda_0$ versus $H_1 : \lambda = \lambda_1$, where λ_0 and $\lambda_1 > \lambda_0$ are specified numbers. Is the test uniformly most powerful against the alternative $H_1 : \lambda > \lambda_0$?

19. Under H_0, a random variable has the cumulative distribution function $F_0(x) = x^2$, $0 \le x \le 1$; and under H_1, it has the cumulative distribution function $F_1(x) = x^3$, $0 \le x \le 1$.

   a. If the two hypotheses have equal prior probability, for what values of x is the posterior probability of H_0 greater than that of H_1?
   b. What is the form of the likelihood ratio test of H_0 versus H_1?
   c. What is the rejection region of a level α test?
   d. What is the power of the test?

20. Consider two probability density functions on $[0, 1]$: $f_0(x) = 1$, and $f_1(x) = 2x$. Among all tests of the null hypothesis $H_0 : X \sim f_0(x)$ versus the alternative $X \sim f_1(x)$, with significance level $\alpha = 0.10$, how large can the power possibly be?

21. Suppose that a single observation X is taken from a uniform density on $[0, \theta]$, and consider testing $H_0 : \theta = 1$ versus $H_1 : \theta = 2$.

a. Find a test that has significance level $\alpha = 0$. What is its power?

b. For $0 < \alpha < 1$, consider the test that rejects when $X \in [0, \alpha]$. What is its significance level and power?

c. What is the significance level and power of the test that rejects when $X \in [1 - \alpha, 1]$?

d. Find another test that has the same significance level and power as the previous one.

e. Does the likelihood ratio test determine a unique rejection region?

f. What happens if the null and alternative hypotheses are interchanged—$H_0 : \theta = 2$ versus $H_1 : \theta = 1$?

22. In Example A of Section 8.5.3 a confidence interval for the variance of a normal distribution was derived. Use Theorem B of Section 9.3 to derive an acceptance region for testing the hypothesis $H_0: \sigma^2 = \sigma_0^2$ at the significance level α based on a sample $X_1, X_2, \ldots, X_n$. Precisely describe the rejection region if $\sigma_0 = 1, n = 15, \alpha = .05$.

23. Suppose that a 99% confidence interval for the mean μ of a normal distribution is found to be $(-2.0, 3.0)$. Would a test of $H_0: \mu = -3$ versus $H_A: \mu \neq -3$ be rejected at the .01 significance level?

24. Let X be a binomial random variable with n trials and probability p of success.

a. What is the generalized likelihood ratio for testing $H_0: p = .5$ versus $H_A: p \neq .5$?

b. Show that the test rejects for large values of $|X - n/2|$.

c. Using the null distribution of X, show how the significance level corresponding to a rejection region $|X - n/2| > k$ can be determined.

d. If $n = 10$ and $k = 2$, what is the significance level of the test?

e. Use the normal approximation to the binomial distribution to find the significance level if $n = 100$ and $k = 10$.

This analysis is the basis of the **sign test**, a typical application of which would be something like this: An experimental drug is to be evaluated on laboratory rats. In n pairs of litter mates, one animal is given the drug and the other is given a placebo. A physiological measure of benefit is made after some time has passed. Let X be the number of pairs for which the animal receiving the drug benefited more than its litter mate. A simple model for the distribution of X if there is no drug effect is binomial with $p = .5$. This is then the null hypothesis that must be made untenable by the data before one could conclude that the drug had an effect.

25. Calculate the likelihood ratio for Example B of Section 9.5 and compare the results of a test based on the likelihood ratio to those of one based on Pearson's chi-square statistic.

26. True or false:

a. The generalized likelihood ratio statistic Λ is always less than or equal to 1.

b. If the p-value is .03, the corresponding test will reject at the significance level .02.

    c. If a test rejects at significance level .06, then the p-value is less than or equal to .06.

    d. The p-value of a test is the probability that the null hypothesis is correct.

    e. In testing a simple versus simple hypothesis via the likelihood ratio, the p-value equals the likelihood ratio.

    f. If a chi-square test statistic with 4 degrees of freedom has a value of 8.5, the p-value is less than .05.

27. What values of a chi-square test statistic with 7 degrees of freedom yield a p-value less than or equal to .10?

28. Suppose that a test statistic T has a standard normal null distribution.

    a. If the test rejects for large values of $|T|$, what is the p-value corresponding to $T = 1.50$?

    b. Answer the same question if the test rejects for large T.

29. Suppose that a level α test based on a test statistic T rejects if $T > t_0$. Suppose that g is a monotone-increasing function and let $S = g(T)$. Is the test that rejects if $S > g(t_0)$ a level α test?

30. Suppose that the null hypothesis is true, that the distribution of the test statistic, T say, is continuous with cdf F and that the test rejects for large values of T. Let V denote the p-value of the test.

    a. Show that $V = 1 - F(T)$.

    b. Conclude that the null distribution of V is uniform. (*Hint:* See Proposition C of Section 2.3.)

    c. If the null hypothesis is true, what is the probability that the p-value is greater than .1?

    d. Show that the test that rejects if $V < \alpha$ has significance level α.

31. What values of the generalized likelihood ratio Λ are necessary to reject the null hypothesis at the significance level $\alpha = .1$ if the degrees of freedom are 1, 5, 10, and 20?

32. The intensity of light reflected by an object is measured. Suppose there are two types of possible objects, A and B. If the object is of type A, the measurement is normally distributed with mean 100 and standard deviation 25; if it is of type B, the measurement is normally distributed with mean 125 and standard deviation 25. A single measurement is taken with the value $X = 120$.

    a. What is the likelihood ratio?

    b. If the prior probabilities of A and B are equal ($\frac{1}{2}$ each), what is the posterior probability that the item is of type B?

    c. Suppose that a decision rule has been formulated that declares the object to be of type B if $X > 125$. What is the significance level associated with this rule?

    d. What is the power of this test?

    e. What is the p-value when $X = 120$?

33. It has been suggested that dying people may be able to postpone their death until after an important occasion, such as a wedding or birthday. Phillips and King

(1988) studied the patterns of death surrounding Passover, an important Jewish holiday, in California during the years 1966–1984. They compared the number of deaths during the week before Passover to the number of deaths during the week after Passover for 1919 people who had Jewish surnames. Of these, 922 occurred in the week before Passover and 997, in the week after Passover. The significance of this discrepancy can be assessed by statistical calculations. We can think of the counts before and after as constituting a table with two cells. If there is no holiday effect, then a death has probability $\frac{1}{2}$ of falling in each cell. Thus, in order to show that there is a holiday effect, it is necessary to show that this simple model does not fit the data. Test the goodness of fit of the model by Pearson's X^2 test or by a likelihood ratio test. Repeat this analysis for a group of males of Chinese and Japanese ancestry, of whom 418 died in the week before Passover and 434 died in the week after. What is the relevance of this latter analysis to the former?

34. Test the goodness of fit of the data to the genetic model given in Problem 55 of Chapter 8.

35. Test the goodness of fit of the data to the genetic model given in Problem 58 of Chapter 8.

36. The National Center for Health Statistics (1970) gives the following data on distribution of suicides in the United States by month in 1970. Is there any evidence that the suicide rate varies seasonally, or are the data consistent with the hypothesis that the rate is constant? (*Hint:* Under the latter hypothesis, model the number of suicides in each month as a multinomial random variable with the appropriate probabilities and conduct a goodness-of-fit test. Look at the signs of the deviations, $O_i - E_i$, and see if there is a pattern.)

Month	Number of Suicides	Days/Month
Jan.	1867	31
Feb.	1789	28
Mar.	1944	31
Apr.	2094	30
May	2097	31
June	1981	30
July	1887	31
Aug.	2024	31
Sept.	1928	30
Oct.	2032	31
Nov.	1978	30
Dec.	1859	31

37. The following table gives the number of deaths due to accidental falls for each month during 1970. Is there any evidence for a departure from uniformity in the rate over time? That is, is there a seasonal pattern to this death rate? If so, describe its pattern and speculate as to causes.

Month	Number of Deaths
Jan.	1668
Feb.	1407
Mar.	1370
Apr.	1309
May	1341
June	1338
July	1406
Aug.	1446
Sept.	1332
Oct.	1363
Nov.	1410
Dec.	1526

38. Yip et al. (2000) studied seasonal variations in suicide rates in England and Wales during 1982–1996, collecting counts shown in the following table:

Month	Jan	Feb	Mar	Apr	May	June	July	Aug	Sept	Oct	Nov	Dec
Male	3755	3251	3777	3706	3717	3660	3669	3626	3481	3590	3605	3392
Female	1362	1244	1496	1452	1448	1376	1370	1301	1337	1351	1416	1226

Do either the male or female data show seasonality?

39. There is a great deal of folklore about the effects of the full moon on humans and other animals. Do animals bite humans more during a full moon? In an attempt to study this question, Bhattacharjee et al. (2000) collected data on admissions to a medical facility for treatment of bites by animals: cats, rats, horses, and dogs. 95% of the bites were by man's best friend, the dog. The lunar cycle was divided into 10 periods, and the number of bites in each period is shown in the following table. Day 29 is the full moon. Is there a temporal trend in the incidence of bites?

Lunar Day	16,17,18	19,20,21	22,23,24	25,26,27	28,29,1	2,3,4	5,6,7	8,9,10	11,12,13	14,15
Number of Bites	137	150	163	201	269	155	142	146	148	110

40. Consider testing goodness of fit for a multinomial distribution with two cells. Denote the number of observations in each cell by X_1 and X_2 and let the hypothesized probabilities be p_1 and p_2. Pearson's chi-square statistic is equal to

$$\sum_{i=1}^{2} \frac{(X_i - np_i)^2}{np_i}$$

Show that this may be expressed as

$$\frac{(X_1 - np_1)^2}{np_1(1 - p_1)}$$

Because X_1 is binomially distributed, the following holds approximately under the null hypothesis:

$$\frac{X_1 - np_1}{\sqrt{np_1(1 - p_1)}} \sim N(0, 1)$$

Thus, the square of the quantity on the left-hand side is approximately distributed as a chi-square random variable with 1 degree of freedom.

41. Let $X_i \sim \text{bin}(n_i, p_i)$, for $i = 1, \ldots, m$, be independent. Derive a likelihood ratio test for the hypothesis

$$H_0: p_1 = p_2 = \cdots = p_m$$

against the alternative hypothesis that the p_i are not all equal. What is the large-sample distribution of the test statistic?

42. Nylon bars were tested for brittleness (Bennett and Franklin 1954). Each of 280 bars was molded under similar conditions and was tested in five places. Assuming that each bar has uniform composition, the number of breaks on a given bar should be binomially distributed with five trials and an unknown probability p of failure. If the bars are all of the same uniform strength, p should be the same for all of them; if they are of different strengths, p should vary from bar to bar. Thus, the null hypothesis is that the p's are all equal. The following table summarizes the outcome of the experiment:

Breaks/Bar	Frequency
0	157
1	69
2	35
3	17
4	1
5	1

   a. Under the given assumption, the data in the table consist of 280 observations of independent binomial random variables. Find the mle of p.

   b. Pooling the last three cells, test the agreement of the observed frequency distribution with the binomial distribution using Pearson's chi-square test.

   c. Apply the test procedure derived in the previous problem.

43. a. In 1965, a newspaper carried a story about a high school student who reported getting 9207 heads and 8743 tails in 17,950 coin tosses. Is this a significant discrepancy from the null hypothesis $H_0: p = \frac{1}{2}$?

   b. Jack Youden, a statistician at the National Bureau of Standards, contacted the student and asked him exactly how he had performed the experiment (Youden

1974). To save time, the student had tossed groups of five coins at a time, and a younger brother had recorded the results, shown in the following table:

Number of Heads	Frequency
0	100
1	524
2	1080
3	1126
4	655
5	105

Are the data consistent with the hypothesis that all the coins were fair ($p = \frac{1}{2}$)?

c. Are the data consistent with the hypothesis that all five coins had the same probability of heads but that this probability was not necessarily $\frac{1}{2}$? (*Hint:* Use the binomial distribution.)

44. Derive and carry out a likelihood ratio test of the hypothesis $H_0: \theta = \frac{1}{2}$ versus $H_1: \theta \neq \frac{1}{2}$ for Problem 58 of Chapter 8.

45. In a classic genetics study, Geissler (1889) studied hospital records in Saxony and compiled data on the gender ratio. The following table shows the number of male children in 6115 families with 12 children. If the genders of successive children are independent and the probabilities remain constant over time, the number of males born to a particular family of 12 children should be a binomial random variable with 12 trials and an unknown probability p of success. If the probability of a male child is the same for each family, the table represents the occurrence of 6115 binomial random variables. Test whether the data agree with this model. Why might the model fail?

Number	Frequency
0	7
1	45
2	181
3	478
4	829
5	1112
6	1343
7	1033
8	670
9	286
10	104
11	24
12	3

46. Show that the transformation $Y = \sin^{-1} \sqrt{\hat{p}}$ is variance-stabilizing if $\hat{p} = X/n$, where $X \sim \text{bin}(n, p)$.

47. Let X follow a Poisson distribution with mean λ. Show that the transformation $Y = \sqrt{X}$ is variance-stabilizing.

48. Suppose that $E(X) = \mu$ and $\text{Var}(X) = c\mu^2$, where c is a constant. Find a variance-stabilizing transformation.

49. An English naturalist collected data on the lengths of cuckoo eggs, measuring to the nearest .5 mm. Examine the normality of this distribution by (a) constructing a histogram and superposing a normal density, (b) plotting on normal probability paper, and (c) constructing a hanging rootogram.

Length	Frequency
18.5	0
19.0	1
19.5	3
20.0	33
20.5	39
21.0	156
21.5	152
22.0	392
22.5	288
23.0	286
23.5	100
24.0	86
24.5	21
25.0	12
25.5	2
26.0	0
26.5	1

50. Burr (1974) gives the following data on the percentage of manganese in iron made in a blast furnace. For 24 days, a single analysis was made on each of five casts. Examine the normality of this distribution by making a normal probability plot and a hanging rootogram. (As a prelude to topics that will be taken up in later chapters, you might also informally examine whether the percentage of manganese is roughly constant from one day to the next or whether there are significant trends over time.)

Day 1	Day 2	Day 3	Day 4	Day 5	Day 6	Day 7	Day 8	Day 9	Day 10	Day 11	Day 12
1.40	1.40	1.80	1.54	1.52	1.62	1.58	1.62	1.60	1.38	1.34	1.50
1.28	1.34	1.44	1.50	1.46	1.58	1.64	1.46	1.44	1.34	1.28	1.46
1.36	1.54	1.46	1.48	1.42	1.62	1.62	1.38	1.46	1.36	1.08	1.28
1.38	1.44	1.50	1.52	1.58	1.76	1.72	1.42	1.38	1.58	1.08	1.18
1.44	1.46	1.38	1.58	1.70	1.68	1.60	1.38	1.34	1.38	1.36	1.28

(Continued)

Day 13	Day 14	Day 15	Day 16	Day 17	Day 18	Day 19	Day 20	Day 21	Day 22	Day 23	Day 24
1.26	1.52	1.50	1.42	1.32	1.16	1.24	1.30	1.30	1.48	1.32	1.44
1.50	1.50	1.42	1.32	1.40	1.34	1.22	1.48	1.52	1.46	1.22	1.28
1.52	1.46	1.38	1.48	1.40	1.40	1.20	1.28	1.76	1.48	1.72	1.10
1.38	1.34	1.36	1.36	1.26	1.16	1.30	1.18	1.16	1.42	1.18	1.06
1.50	1.40	1.38	1.38	1.26	1.54	1.36	1.28	1.28	1.36	1.36	1.10

51. Examine the probability plot in Figure 9.6 and explain why there are several sets of horizontal bands of points.

52. The following table gives values of two abundance ratios for different isotopes of potassium from several samples of minerals (H. Ku, private communication). Examine whether each of the ratios appears normally distributed by first making histograms and superposing normal densities and then making probability plots.

$^{39}K/^{41}K$	$^{41}K/^{40}K$	$^{39}K/^{41}K$	$^{41}K/^{40}K$	$^{39}K/^{41}K$	$^{41}K/^{40}K$
13.8645	576.369	13.8689	578.277	13.8724	576.017
13.8695	578.012	13.8593	574.708	13.8665	574.881
13.8659	575.597	13.8742	573.630	13.8566	578.508
13.8622	575.244	13.8703	576.069	13.8555	576.796
13.8696	575.567	13.8472	575.637	13.8534	580.394
13.8604	576.836	13.8555	575.971	13.8685	576.772
13.8672	576.236	13.8439	576.403	13.8694	576.501
13.8598	575.291	13.8646	576.179	13.8599	574.950
13.8641	576.478	13.8702	575.129	13.8605	577.614
13.8673	576.992	13.8606	577.084	13.8619	574.506
13.8597	578.335	13.8622	576.749	13.9641	576.317
13.8604	576.767	13.8588	576.669	13.8597	575.665
13.8591	576.571	13.8547	575.869	13.8617	575.815
13.8472	576.617	13.8597	577.793	13.861	576.109
13.863	575.885	13.8663	577.770	13.8615	576.144
13.8566	576.651	13.8597	577.697	13.8469	576.820
13.8503	575.974	13.8604	576.299	13.8582	576.672
13.8553	577.255	13.8634	575.903	13.8645	576.169
13.8642	574.664	13.8658	574.773	13.8713	575.390
13.8613	576.405	13.8547	577.391	13.8593	575.108
13.8706	574.306	13.8519	577.057	13.8522	576.663
13.8601	577.095	13.863	577.286	13.8489	578.358
13.866	576.957	13.8581	575.510	13.8609	575.371
13.8655	576.434	13.8644	576.509	13.857	575.851
13.8612	575.211	13.8665	574.300	13.8566	575.644
13.8598	576.630	13.8648	575.846	13.864	574.462

53. Hoaglin (1980) suggested a "Poissonness plot"—a simple visual method for assessing goodness of fit. The expected frequencies for a sample of size n from a Poisson distribution are

$$E_k = nP(X = k) = ne^{-\lambda}\frac{\lambda^k}{k!}$$

or

$$\log E_k = \log n - \lambda + k \log \lambda - \log k!$$

Thus, a plot of $\log(O_k) + \log k!$ versus k should yield nearly a straight line with a slope of $\log \lambda$ and an intercept of $\log n - \lambda$. Construct such plots for the data of Problems 1, 2, and 3 of Chapter 8. Comment on how straight they are.

54. A random variable X is said to follow a lognormal distribution if $Y = \log(X)$ follows a normal distribution. The lognormal is sometimes used as a model for heavy-tailed skewed distributions.

   a. Calculate the density function of the lognormal distribution.
   b. Examine whether the lognormal roughly fits the following data (Robson 1929), which are the dorsal lengths in millimeters of taxonomically distinct octopods.

110	15	60	54	19	115	73
190	57	43	44	18	37	43
55	19	23	82	175	50	80
65	63	36	16	10	17	52
43	70	22	95	20	41	17
15	12	11	29	29	61	22
40	17	26	30	16	116	28
32	33	29	27	16	55	8
11	49	82	85	20	67	27
44	16	6	35	17	26	32
76	150	21	5	6	51	75
23	29	64	22	47	9	10
28	18	84	52	130	50	45
12	21	73				

55. a. Generate samples of size 25, 50, and 100 from a normal distribution. Construct probability plots. Do this several times to get an idea of how probability plots behave when the underlying distribution is really normal.
   b. Repeat part (a) for a chi-square distribution with 10 df.
   c. Repeat part (a) for $Y = Z/U$, where $Z \sim N(0, 1)$ and $U \sim U[0, 1]$ and Z and U are independent.
   d. Repeat part (a) for a uniform distribution.
   e. Repeat part (a) for an exponential distribution.
   f. Can you distinguish between the normal distribution of part (a) and the subsequent nonnormal distributions?

56. Suppose that a sample is taken from a symmetric distribution whose tails decrease more slowly than those of the normal distribution. What would be the qualitative shape of a normal probability plot of this sample?

57. The Cauchy distribution has the probability density function

$$f(x) = \frac{1}{\pi}\left(\frac{1}{1+x^2}\right), \qquad -\infty < x < \infty$$

What would be the qualitative shape of a normal probability plot of a sample from this distribution?

58. Show how probability plots for the exponential distribution, $F(x) = 1 - e^{-\lambda x}$, may be constructed. Berkson (1966) recorded times between events and fit them to an exponential distribution. (The times between events in a Poisson process are exponentially distributed.) The following table comes from Berkson's paper. Make an exponential probability plot, and evaluate its "straightness."

Time Interval (sec)	Observed Frequency
0–60	115
60–120	104
120–181	99
181–243	106
243–306	113
306–369	104
369–432	101
432–497	106
497–562	104
562–628	96
628–698	512
689–1130	524
1130–1714	468
1714–2125	531
2125–2567	461
2567–3044	526
3044–3562	506
3562–4130	509
4130–4758	520
4758–5460	540
5460–6255	542
6255–7174	499
7174–8260	494
8260–9590	500
9590–11,304	550
11,304–13,719	465
13,719–14,347	104
14,347–15,049	97
15,049–15,845	101
15,845–16,763	104
16,763–17,849	92
17,849–19,179	102
19,179–20,893	103
20,893–23,309	110
23,309–27,439	112
27,439+	100

59. Construct a hanging rootogram from the data of the previous problem in order to compare the observed distribution to an exponential distribution.

60. The exponential distribution is widely used in studies of reliability as a model for lifetimes, largely because of its mathematical simplicity. Barlow, Toland, and Freeman (1984) analyzed data on the strength of Kevlar 49/epoxy, a material used in the space shuttle. The times to failure (in hours) of 76 strands tested at a stress level of 90% are given in the following table.

Times to Failure at 90% Stress Level				
.01	.01	.02	.02	.02
.03	.03	.04	.05	.06
.07	.07	.08	.09	.09
.10	.10	.11	.11	.12
.13	.18	.19	.20	.23
.24	.24	.29	.34	.35
.36	.38	.40	.42	.43
.52	.54	.56	.60	.60
.63	.65	.67	.68	.72
.72	.72	.73	.79	.79
.80	.80	.83	.85	.90
.92	.95	.99	1.00	1.01
1.02	1.03	1.05	1.10	1.10
1.11	1.15	1.18	1.20	1.29
1.31	1.33	1.34	1.40	1.43
1.45	1.50	1.51	1.52	1.53
1.54	1.54	1.55	1.58	1.60
1.63	1.64	1.80	1.80	1.81
2.02	2.05	2.14	2.17	2.33
3.03	3.03	3.24	4.20	4.69
7.89				

a. Construct a probability plot of the data against the quantiles of an exponential distribution to assess qualitatively whether the exponential is a reasonable model. Can you explain the peculiar appearance of the plot?

b. Compare the data to the exponential distribution by means of a hanging rootogram.

61. The files `haliburton` and `macdonalds` give the monthly returns on the stocks of these two companies from 1975 through 1999.

a. Make histograms of the returns and superimpose fitted normal densities. Comment on the quality of the fit. Which stock is more volatile?

b. Make normal probability plots and again comment on the quality of the fit.

62. Apply the Poisson dispersion test to the data on gamma-ray counts—Problem 42 of Chapter 8. You will have to modify the development of the likelihood ratio test in Section 9.5 to take account of the time intervals being of different lengths.

63. Construct a gamma probability plot for the data of Problem 46 of Chapter 8.

64. The file `bodytemp` contains normal body temperature readings (degrees Fahrenheit) and heart rates (beats per minute) of 65 males (coded by 1) and 65 females (coded by 2) from Shoemaker (1996).

   a. Assess the normality of the male and female body temperatures by making normal probability plots. In order to judge the inherent variability of these plots, simulate several samples from normal distributions with matching means and standard deviations, and make normal probability plots. What do you conclude?

   b. Repeat the preceding problem for heart rates.

   c. For the males, test the null hypothesis that the mean body temperature is 98.6° versus the alternative that the mean is not equal to 98.6°. Do the same for the females. What do you conclude?

65. This problem continues the analysis of the chromatin data from Problem 45 of Chapter 8 and is concerned with further examining goodness of fit.

   a. Goodness of fit can also be examined via probability plots in which the quantiles of a theoretical distribution are plotted against those of the empirical distribution. Following the discussion in Section 9.8, show that it is sufficient to plot the observed order statistics, $X_{(k)}$, versus the quantiles of the Rayleigh distribution with $\theta = 1$. Construct three such probability plots and comment on any systematic lack of fit that you observe. To get an idea of what sort of variability could be expected due to chance, simulate several sets of data from a Rayleigh distribution and make corresponding probability plots.

   b. Formally test goodness of fit by performing a chi-squared goodness of fit test, comparing histogram counts to those predicted from the Rayleigh model. You may need to combine cells of the histograms so that the expected counts in each cell are at least 5.

CHAPTER 10

Summarizing Data

10.1 Introduction

This chapter deals with methods of describing and summarizing data that are in the form of one or more samples, or batches. These procedures, many of which generate graphical displays, are useful in revealing the structure of data that are initially in the form of numbers printed in columns on a page or recorded on a tape or disk as a computer file. In the absence of a stochastic model, the methods are useful for purely descriptive purposes. If it is appropriate to entertain a stochastic model, the implications of that model for the method are of interest. For example, the arithmetic mean $\bar{x}$ is often used as a summary of a collection of numbers $x_1, x_2, \ldots, x_n$; it indicates a "typical value." (We discuss some of its strengths and weaknesses in this regard in Section 10.4.) In some situations, it may be useful to model the collection of numbers as a realization of n independent random variables $X_1, X_2, \ldots, X_n$ with common mean μ and variance σ^2. The question of variability of $\bar{x}$ can be addressed with such a model—the mean $\bar{x}$ is regarded as an estimate of μ, and we know from previous work that the stochastic model implies $E(\overline{X}) = \mu$ and $\text{Var}(\overline{X}) = \sigma^2/n$.

We will first discuss methods that are data analogues of the cumulative distribution function of a random variable. These methods are useful in displaying the distribution of data values. Next, we will discuss the histogram and related graphical displays that play the role for data that the probability density or frequency function plays for a random variable, giving a different view of the distribution of data values than that provided by the cumulative distribution function. We then discuss simpler numerical summaries of data, numbers that indicate a typical or central value of the data and a quantification of the spread. Such statistics provide a more condensed summary than do the cumulative distribution function and the histogram. We will pay particular attention to the effect of extreme data points on these measures. Next, we will introduce boxplots, graphical summaries that combine in a simple form information about the central values, spread, and shape of a distribution. Finally, scatterplots are introduced as a method for displaying information about relationships of variables.

10.2 Methods Based on the Cumulative Distribution Function

10.2.1 The Empirical Cumulative Distribution Function

Suppose that $x_1, \ldots, x_n$ is a batch of numbers. (The word *sample* is often used in the case that the x_i are independently and identically distributed with some distribution function; the word *batch* will imply no such commitment to a stochastic model.) The **empirical cumulative distribution function (ecdf)** is defined as

$$F_n(x) = \frac{1}{n}(\#x_i \leq x)$$

(With this definition, F_n is right-continuous; in the former Soviet Union and Eastern Europe, the ecdf is usually defined to be left-continuous.)

Denote the ordered batch of numbers by $x_{(1)} \leq x_{(2)} \leq \cdots \leq x_{(n)}$. Then if $x < x_{(1)}$, $F_n(x) = 0$, if $x_{(1)} \leq x < x_{(2)}$, $F_n(x) = 1/n$, if $x_{(k)} \leq x < x_{(k+1)}$, $F_n(x) = k/n$, and so on. If there is a single observation with value x, F_n has a jump of height $1/n$ at x; if there are r observations with the same value x, F_n has a jump of height r/n at x.

The ecdf is the data analogue of the cumulative distribution function of a random variable: $F(x)$ gives the probability that $X \leq x$ and $F_n(x)$ gives the proportion of the collection of numbers less than or equal to x.

EXAMPLE A   As an example of the use of the ecdf, let us consider data taken from a study by White, Riethof, and Kushnir (1960) of the chemical properties of beeswax. The aim of the study was to investigate chemical methods for detecting the presence of synthetic waxes that had been added to beeswax. For example, the addition of microcrystalline wax raises the melting point of beeswax. If all pure beeswax had the same melting point, its determination would be a reasonable way to detect dilutions. The melting point and other chemical properties of beeswax, however, vary from one beehive to another. The authors obtained samples of pure beeswax from 59 sources, measured several chemical properties, and examined the variability of the measurements. The 59 melting points (in °C) are listed here. As a summary of these measurements, the ecdf is plotted in Figure 10.1.

63.78	63.45	63.58	63.08	63.40	64.42	63.27	63.10
63.34	63.50	63.83	63.63	63.27	63.30	63.83	63.50
63.36	63.86	63.34	63.92	63.88	63.36	63.36	63.51
63.51	63.84	64.27	63.50	63.56	63.39	63.78	63.92
63.92	63.56	63.43	64.21	64.24	64.12	63.92	63.53
63.50	63.30	63.86	63.93	63.43	64.40	63.61	63.03
63.68	63.13	63.41	63.60	63.13	63.69	63.05	62.85
63.31	63.66	63.60					

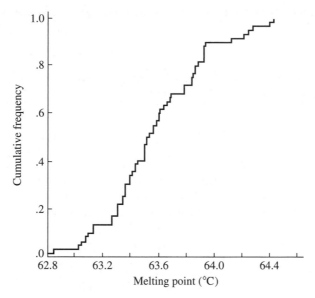

FIGURE **10.1**   The empirical cumulative distribution function of the melting points of beeswax.

Figure 10.1 conveniently summarizes the natural variability in melting points. For example, we can see from the graph that about 90% of the samples had melting points less than 64.2°C and that about 12% had melting points less than 63.2°C.

White, Riethof, and Kushnir showed that the addition of 5% microcrystalline wax raised the melting point of beeswax by .85°C and the addition of 10% raised the melting point by 2.22°C. From Figure 10.1, we can see that an addition of 5% microcrystalline wax might well be difficult to detect, especially if it was made to beeswax that had a low melting point, but that an addition of 10% would be detectable. In further calculations, the investigators modeled the distribution of melting points as Gaussian. How reasonable does this model appear to be?   ■

Let us briefly consider some of the elementary statistical properties of the ecdf in the case in which $X_1, \ldots, X_n$ is a random sample from a continuous distribution function, F. For purposes of analysis, it is convenient to express F_n in the following way:

$$F_n(x) = \frac{1}{n} \sum_{i=1}^{n} I_{(-\infty, x]}(X_i)$$

where

$$I_{(-\infty, x]}(X_i) = \begin{cases} 1, & \text{if } X_i \leq x \\ 0, & \text{if } X_i > x \end{cases}$$

The random variables $I_{(-\infty,x]}(X_i)$ are independent Bernoulli random variables:

$$I_{(-\infty,x]}(X_i) = \begin{cases} 1, & \text{with probability } F(x) \\ 0, & \text{with probability } 1 - F(x) \end{cases}$$

Thus, $nF_n(x)$ is a binomial random variable (n trials, probability $F(x)$ of success) and so

$$E[F_n(x)] = F(x)$$

$$\text{Var}[F_n(x)] = \frac{1}{n}F(x)[1 - F(x)]$$

As an estimate of $F(x)$, $F_n(x)$ is unbiased and has a maximum variance at that value of x such that $F(x) = .5$, that is, at the median. As x becomes very large or very small, the variance tends to zero.

In the preceding paragraph, we considered $F_n(x)$ for fixed x; the results can be applied to form a confidence interval for $F(x)$ for any given value of x. Much deeper analysis focuses on the stochastic behavior of F_n as a random function; that is, all values of x are considered simultaneously. It turns out, somewhat surprisingly, that the distribution of

$$\max_{-\infty<x<\infty} |F_n(x) - F(x)|$$

does not depend on F if F is continuous. This result makes possible the construction of a simultaneous confidence band about F_n, which can be used to test goodness of fit. [For further details, refer to Section 9.6 of Bickel and Doksum (1977).] It is important to realize the difference between the simultaneous confidence band and the individual confidence intervals that may be constructed using the binomial distribution. Each such individual confidence interval covers F at one point with a certain probability, say, $1 - \alpha$, but the probability that all such intervals cover F simultaneously is not necessarily $1 - \alpha$. We will encounter other phenomena of this type in later chapters.

10.2.2 The Survival Function

The **survival function** is equivalent to the cumulative distribution function and is defined as

$$S(t) = P(T > t) = 1 - F(t)$$

where T is a random variable with cdf F. In applications where the data consist of times until failure or death and are thus nonnegative, it is often customary to work with the survival function rather than the cumulative distribution function, although the two give equivalent information. Data of this type occur in medical and reliability studies. In these cases, $S(t)$ is simply the probability that the lifetime will be longer than t. We will be concerned with the sample analogue of S,

$$S_n(t) = 1 - F_n(t)$$

which gives the proportion of the data greater than t.

EXAMPLE **A**   As an example, let us consider the use of the survival function with a study of the lifetimes of guinea pigs infected with varying doses of tubercle bacilli (Bjerkdal 1960). In one study, five groups of 72 animals each were inoculated with the bacilli at increasing dosages, and a control group of 107 animals was used. We denote the inoculated groups by I, II, III, IV, and V, in order of increasing dose. The animals were observed over a 2-year period, and their times of death (in days) were recorded. The data are given here. Note that not all the animals in the lower-dosage regimens died.

Control Lifetimes							
18	36	50	52	86	87	89	91
102	105	114	114	115	118	119	120
149	160	165	166	167	167	173	178
189	209	212	216	273	278	279	292
341	355	367	380	382	421	421	432
446	455	463	474	506	515	546	559
576	590	603	607	608	621	634	634
637	638	641	650	663	665	688	725
735							

Dose I Lifetimes							
76	93	97	107	108	113	114	119
136	137	138	139	152	154	154	160
164	164	166	168	178	179	181	181
183	185	194	198	212	213	216	220
225	225	244	253	256	259	265	268
268	270	283	289	291	311	315	326
326	361	373	373	376	397	398	406
452	466	592	598				

Dose II Lifetimes							
72	72	78	83	85	99	99	110
113	113	114	114	118	119	123	124
131	133	135	137	140	142	144	145
154	156	157	162	162	164	165	167
171	176	177	181	182	187	192	196
211	214	216	216	218	228	238	242
248	256	257	262	264	267	267	270
286	303	309	324	326	334	335	358
409	473	550					

Dose III Lifetimes							
10	33	44	56	59	72	74	77
92	93	96	100	100	102	105	107
107	108	108	108	109	112	113	115
116	120	121	122	122	124	130	134
136	139	144	146	153	159	160	163
163	168	171	172	176	183	195	196
197	202	213	215	216	222	230	231
240	245	251	253	254	254	278	293
327	342	347	361	402	432	458	555

Dose IV Lifetimes							
43	45	53	56	56	57	58	66
67	73	74	79	80	80	81	81
81	82	83	83	84	88	89	91
91	92	92	97	99	99	100	100
101	102	102	102	103	104	107	108
109	113	114	118	121	123	126	128
137	138	139	144	145	147	156	162
174	178	179	184	191	198	211	214
243	249	329	380	403	511	522	598

Dose V Lifetimes							
12	15	22	24	24	32	32	33
34	38	38	43	44	48	52	53
54	54	55	56	57	58	58	59
60	60	60	60	61	62	63	65
65	67	68	70	70	72	73	75
76	76	81	83	84	85	87	91
95	96	98	99	109	110	121	127
129	131	143	146	146	175	175	211
233	258	258	263	297	341	341	376

A plot (Figure 10.2) of the empirical survival functions provides a convenient summary of the data. The proportions surviving beyond given times are plotted; it is not necessary to know the actual lifetimes of the animals that survived beyond the termination of the study. The graph is a much more effective presentation of the data than the tabular listings.

One of Bjerkdahl's primary interests was comparing the effect increased exposure had on guinea pigs that had different levels of resistivity. Comparing groups III and V, for example, we see that the difference in lifetimes of the weakest guinea pigs (say the 10% weakest) from the two groups was about 50 days, whereas the difference in lifetimes for stronger animals increases to about 100 days.   ■

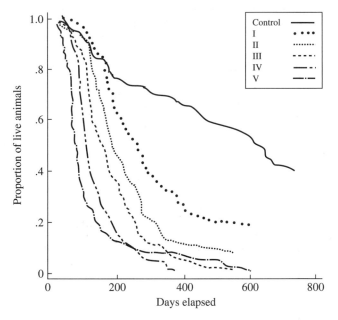

FIGURE **10.2** Survival functions for guinea pig lifetimes. For purposes of visual clarity, the points have been joined by lines: The solid line corresponds to the control group, the dotted line to group I, the short-dash line to group II, the long-dash line to group III, the dot-and-long-dash line to group IV, and the short-and-long-dash line to group V.

Survival plots may also be used for informal examinations of the **hazard function,** which may be interpreted as the instantaneous death rate for individuals who have survived up to a given time. If an individual is alive at time t, the probability that that individual will die in the time interval $(t, t + \delta)$ is, assuming that the density function f is continuous at t,

$$P(t \leq T \leq t + \delta | T \geq t) = \frac{P(t \leq T \leq t + \delta)}{P(T \geq t)}$$

$$= \frac{F(t + \delta) - F(t)}{1 - F(t)}$$

$$\approx \frac{\delta f(t)}{1 - F(t)}$$

The hazard function is defined as

$$h(t) = \frac{f(t)}{1 - F(t)}$$

and may be thought of as the instantaneous rate of mortality for an individual alive at time t. If T is the lifetime of a manufactured component, it may be natural to think of $h(t)$ as the instantaneous or age-specific failure rate. It may also be expressed as

$$h(t) = -\frac{d}{dt} \log[1 - F(t)] = -\frac{d}{dt} \log S(t)$$

which reveals that it is the negative of the slope of the log of the survival function.

Consider, for example, the exponential distribution:

$$F(t) = 1 - e^{-\lambda t}$$
$$S(t) = e^{-\lambda t}$$
$$f(t) = \lambda e^{-\lambda t}$$
$$h(t) = \lambda$$

The instantaneous mortality rate is constant. If the exponential distribution were used as a model for the time until failure of a component, it would imply that the probability of the component failing did not depend on its age. This is a consequence of the "memoryless" property of the exponential distribution (Section 2.2.1). An alternative model might have a hazard function that is U-shaped, the rate of failure being high for very new components because of flaws in the manufacturing process that show up very quickly, declining for components of intermediate age, and then increasing for older components as they wear out.

The empirical survival function and its logarithm can be expressed in terms of the ordered observations. For simplicity, suppose that there are no ties and that the ordered failure times are $T_{(1)} < T_{(2)} < \cdots < T_{(n)}$. Then if $t = T_{(i)}$, $F_n(t) = i/n$ and $S_n(t) = 1 - i/n$. Since $\log S_n(t)$ is then undefined for $t \geq T_{(n)}$, it is often defined as $S_n(t) = 1 - i/(n+1)$ for $T_{(i)} \leq t < T_{(i+1)}$.

E X A M P L E **B**   For the data of Example A, Figure 10.3 is a plot of the log of the empirical survival functions. We plotted $\log[1 - i/(n+1)]$ versus the ordered survival times $T_{(i)}$. From the slopes of these curves, we see that the hazard functions are initially fairly small. As the dosage level increases, the instantaneous mortality rates both increase more quickly and reach higher levels. The increased mortality rate sets in at an earlier age for the high-dosage group and seems greater (the slope is greater). (To see this, hold the figure at an angle so that you are "looking down" the curves.)   ∎

When interpreting plots such as that presented in Figure 10.3, we will find it useful to keep in mind the variability of the empirical log survival function. Using the method of propagation of error (Section 4.6), we have

$$\text{Var}\{\log[1 - F_n(t)]\} \approx \frac{\text{Var}[1 - F_n(t)]}{[1 - F(t)]^2}$$
$$= \frac{1}{n}\left(\frac{F(t)[1 - F(t)]}{[1 - F(t)]^2}\right)$$
$$= \frac{1}{n}\left(\frac{F(t)}{1 - F(t)}\right)$$

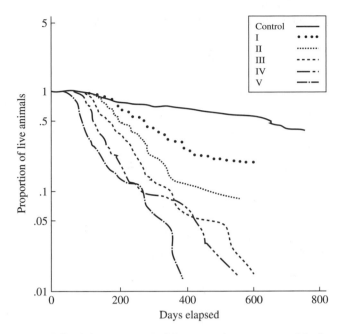

FIGURE **10.3** Log survival functions for guinea pig lifetimes. For purposes of visual clarity, the points have been joined by lines: The solid line corresponds to the control group, the dotted line to group I, the short-dash line to group II, the long-dash line to group III, the dot-and-long-dash line to group IV, and the short-and-long-dash line to group V.

From this expression, we see that for large values of t, the empirical log survival function is extremely unreliable, because $1 - F(t)$ is then very small. Thus, in practice, the last few data points are disregarded. (Note the large fluctuations of the log survival functions in Figure 10.3 for large times.)

10.2.3 Quantile-Quantile Plots

Quantile-quantile (Q-Q) plots are useful for comparing distribution functions. If X is a continuous random variable with a strictly increasing distribution function, F, the pth quantile of the distribution was defined in Section 2.2 to be that value of x such that

$$F(x) = p$$

or

$$x_p = F^{-1}(p)$$

In a Q-Q plot, the quantiles of one distribution are plotted against those of another. Suppose, for purposes of discussion, that one cdf (F) is a model for observations of a control group and another (G) is a model for observations of a group that has received some treatment. Let the observations of the control group be denoted by x with cdf F, and let the observations of the treatment group be denoted by y with cdf G. The simplest effect that the treatment could have would be to increase the expected response of every member of the treatment group by the same amount, say h units. That is, both the weakest and the strongest individuals would have their responses changed by h. Then $y_p = x_p + h$, and the Q-Q plot would be a straight line with slope 1 and intercept h. We will now show that this relationship between the quantiles implies that the cumulative distribution functions have the relationship $G(y) = F(y - h)$. This follows, because for every $0 \leq p \leq 1$,

$$p = G(y_p)$$
$$= F(x_p)$$
$$= F(y_p - h)$$

as in Figure 10.4.

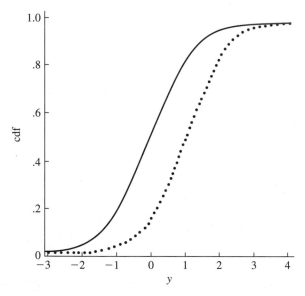

FIGURE 10.4   An additive treatment effect. The solid line is $F(y)$, and the dotted line is $G(y) = F(y - h)$.

Another possible effect of a treatment would be multiplicative: The response (such as lifetime or strength) is multiplied by a constant, c. The quantiles would then be related as $y_p = cx_p$, and the Q-Q plot would be a straight line with slope c and intercept 0. The cdf's would be related as $G(y) = F(y/c)$ (see Figure 10.5).

A simple summary of a treatment effect for the additive model would be of the form "the treatment increases lifetime by 2 mo." For the multiplicative model, one might say something like "the treatment increases lifetime by 25%."

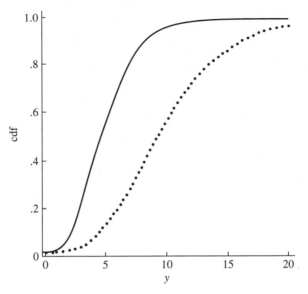

FIGURE **10.5**   A multiplicative treatment effect. The solid line is $F(y)$, and the dotted line is $G(y) = F(y/c)$.

The effect of a treatment can, of course, be much more complicated than either of these two simple models. For example, a treatment could benefit weaker individuals and be to the detriment of stronger individuals. An educational program that places very heavy emphasis on elementary or basic skills might be expected to have this sort of effect relative to a regular program.

Given a batch of numbers, or a sample from a probability distribution, quantiles are constructed from the order statistics. Given n observations and the order statistics $X_{(1)}, \ldots, X_{(n)}$, the $k/(n+1)$ quantile of data is assigned to $X_{(k)}$. (This convention is not unique; sometimes, for example, the quantile assigned to $X_{(k)}$ is defined as $(k - .5)/n$. For descriptive purposes, it makes little difference which definition we use.) In constructing probability plots in Chapter 9, we plotted sample quantiles defined as just described versus the quantiles of a theoretical distribution, such as the normal, and used these plots to informally assess goodness of fit.

To compare two batches of n numbers with order statistics $X_{(1)}, \ldots, X_{(n)}$ and $Y_{(1)}, \ldots, Y_{(n)}$, a Q-Q plot is simply constructed by plotting the points $(X_{(i)}, Y_{(i)})$. If the batches are of unequal size, an interpolation process can be used. A procedure for interpolating intermediate quantiles is described in the end-of-chapter problems.

E X A M P L E **A**   Cleveland et al. (1974) used Q-Q plots in a study of air pollution. They plotted the quantiles of distributions of the values of various variables on Sunday against the quantiles for weekdays (Figure 10.6). The Q-Q plot of the ozone maxima shows that the very highest quantiles occur on weekdays but that all the other quantiles are larger on Sundays. For carbon monoxide, nitrogen oxide, and aerosols, the differences in the

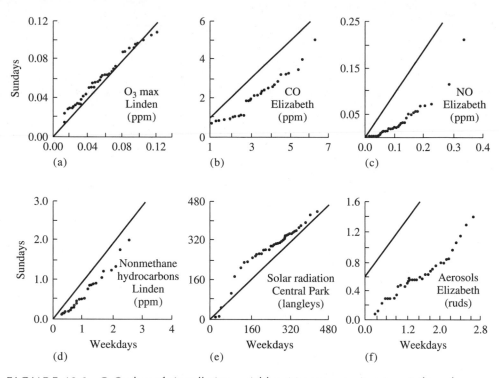

FIGURE 10.6   Q-Q plots of air pollution variables: (a) ozone maxima (ppm), (b) carbon monoxide concentration (ppm), (c) nitrogen oxide concentration (ppm), (d) nonmethane hydrocarbons (ppm), (e) solar radiation (langleys), (f) aerosols (ruds).

quantiles increase with increasing concentration. The very high and very low quantiles of solar radiation are about the same on Sundays and weekdays (presumably corresponding to very clear days and days with heavy cloud cover), but for intermediate quantiles, the Sunday quantiles are larger.   ■

EXAMPLE **B**   Figure 10.7 is a Q-Q plot for groups III and V of Bjerkdahl (see Example A in Section 10.2.2). It shows that the difference in the quantiles increases for the larger quantiles; this is consistent with the observations we made earlier. From his analysis of the data, Bjerkdahl concluded that the increases were proportionally the same for animals with little, average, or great resistance—that is, that the treatment effect is multiplicative in the sense defined earlier. If this were the case, the Q-Q plot would be a straight line. For times up to about 200 days, the animals in group III live approximately twice as long as those in group V, but beyond 100 days the difference is roughly constant. The Q-Q plot thus provides a simple and effective means of comparing the lifetimes in the two groups.   ■

Further discussion and examples of Q-Q plots can be found in Wilk and Gnanadesikan (1968).

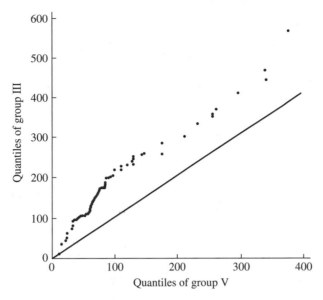

FIGURE **10.7**   Q-Q plots of groups III and V from Bjerkdahl (1960). For reference, the line $y = x$ has been added.

10.3 Histograms, Density Curves, and Stem-and-Leaf Plots

The histogram, a time-honored method of displaying data, has already been introduced. It displays the shape of the distribution of data values in the same sense that a density function displays probabilities. The range of the data is divided into intervals, or bins, and the number or proportion of the observations falling in each bin is plotted. If the bins are not of equal size, the resulting histogram can be misleading. A procedure that is often recommended is to plot the proportion of observations falling in the bin divided by the bin width; if this procedure is used, the area under the histogram is 1.

Figure 10.8 shows three histograms of the melting points of beeswax from Example A in Section 10.2.1 with increasingly larger bin width. If the bin width is too small, the histogram is too ragged; if the bin is too wide, the shape is oversmoothed and obscured. The choice of bin width is usually made subjectively in an attempt to strike a balance between a histogram that is too ragged and one that oversmooths. Rudemo (1982) discusses automatic methods for choosing the bin width.

Histograms are frequently used to display data for which there is no assumption of any stochastic model—for example, populations of U.S. cities. If the data are modeled as a random sample from some continuous distribution, the histogram may be viewed as an estimate of the probability density. Regarded in this light, it suffers from not being smooth.

A smooth probability density estimate can be constructed in the following way. Let $w(x)$ be a nonnegative, symmetric weight function, centered at zero and

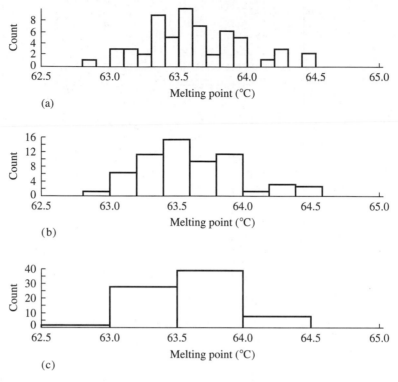

FIGURE **10.8**   Histograms of melting points of beeswax: (a) bin width $= .1$, (b) bin width $= .2$, (c) bin width $= .5$.

integrating to 1. For example, $w(x)$ can be the standard normal density. The function

$$w_h(x) = \frac{1}{h} w\left(\frac{x}{h}\right)$$

is a rescaled version of w. As h approaches zero, w_h becomes more concentrated and peaked about zero. As h approaches infinity, w_h becomes more spread out and flatter. If $w(x)$ is the standard normal density, then $w_h(x)$ is the normal density with standard deviation h. If $X_1, \ldots, X_n$ is a sample from a probability density function, f, an estimate of f is

$$f_h(x) = \frac{1}{n} \sum_{i=1}^{n} w_h(x - X_i)$$

This estimate, called a **kernel probability density estimate,** consists of the superposition of "hills" centered on the observations. In the case where $w(x)$ is the standard normal density, $w_h(x - X_i)$ is the normal density with mean X_i and standard deviation h.

The parameter h, the **bandwidth** of the estimating function, controls its smoothness and corresponds to the bin width of the histogram. If h is too small, the estimate

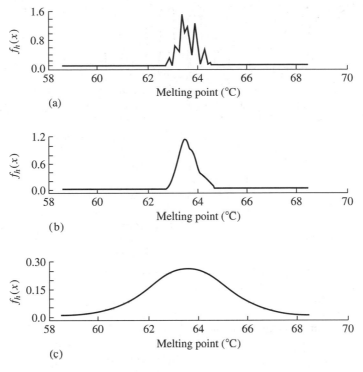

FIGURE 10.9   Probability density estimates from melting point data. The kernel w is the standard normal density with standard deviation (a) .025, (b) .125, and (c) 1.25. Note that the vertical scales are different.

is too rough; if it is too large, the shape of f is smeared out too much. Figure 10.9 shows estimates of the probability density of the melting points of beeswax (from Example A in Section 10.2.1) for various values of h. Making a reasonable choice of the bandwidth is important, just as is choosing the bin width for a histogram. From Figure 10.9, we see that too small a bandwidth yields a ragged curve and too large a bandwidth obscures the shape and spreads the probability mass out too much. Scott (1992) contains extensive discussion of probability density estimation, including methods for automatic, data-driven bandwidth choice and estimation of densities in more than one dimension.

One disadvantage of a histogram or a probability density estimate is that information is lost; neither allows the reconstruction of the original data. Furthermore, a histogram does not allow one to calculate a statistic such as a median; one can tell from a histogram only in which bin the median lies and not the median's actual value.

Stem-and-leaf plots (Tukey 1977) convey information about shape while retaining the numerical information. It is easiest to define this type of plot by an example, a stem-and-leaf plot of the beeswax melting-point data (the decimal point is one place to the left of the colon):

		STEM	LEAF
1	1	628	:5
1	0	629	:
4	3	630	:358
7	3	631	:033
9	2	632	:77
18	9	633	:001446669
23	5	634	:01335
	10	635	:0000113668
26	7	636	:0013689
19	2	637	:88
17	6	638	:334668
11	5	639	:22223
6	0	640	:
6	1	641	:2
5	3	642	:147
2	0	643	:
2	2	644	:02

The first three digits of the melting points have been selected to form the stem and are listed in the third column. The leaves on each stem are the fourth digit of all numbers with that stem. For example, the first stem is 628, and its leaf indicates the presence of the number 62.85 in the data. The third stem is 630, and its leaves indicate the presence of the numbers 63.03, 63.05, and 63.08. This stem-and-leaf plot was constructed by a computer, but they are very easy to make by hand. The second column of numbers gives the number of leaves on each stem. The first column of numbers facilitates finding order statistics, such as quartiles and the median; starting at the top of the plot and continuing down to the stem containing the median, the cumulative numbers of observations out to the smallest observation are listed. The numbering process is then extended symmetrically from the stem containing the median to the largest observation of the data.

Straightforward stem-and-leaf plots do not work well for data that range over several orders of magnitude. In such a situation, it is better to make a stem-and-leaf plot of the logarithms of the data.

10.4 Measures of Location

Sections 10.2 and 10.3 were concerned with data analogues of the cumulative distribution and density functions and with related curves, which convey visual information about the shape of the distribution of the data. Here and in Section 10.5, we discuss simple numerical summaries of data that are useful when there is not enough data to justify constructing a histogram or an ecdf, or when a more concise summary is desired.

A **measure of location** is a measure of the center of a batch of numbers. If the numbers result from different measurements of the same quantity, a measure of location is often used in the hope that it is more accurate than any single measurement. In other situations, a measure of location is used as a simple summary of the numbers—for example, "the average grade on the exam was 72." In this section, we will discuss several common measures of location and their relative advantages and disadvantages.

10.4.1   The Arithmetic Mean

The most commonly used measure of location is the arithmetic mean,

$$\bar{x} = \frac{1}{n} \sum_{i=1}^{n} x_i$$

For illustration, we consider a set of 26 measurements of the heat of sublimation of platinum from an experiment done by Hampson and Walker (1961). The data are listed here:

Heats of Sublimation of Platinum (kcal/mol)							
136.3	136.6	135.8	135.4	134.7	135.0	134.1	143.3
147.8	148.8	134.8	135.2	134.9	146.5	141.2	135.4
134.8	135.8	135.0	133.7	134.4	134.9	134.8	134.5
134.3	135.2						

The 26 measurements are all attempts to measure the "true" heat of sublimation, and we see that there is variability among them. Intuitively, it may seem that a measure of location or center for this batch of numbers would give a more accurate estimate of the heat of sublimation than any one of the numbers alone.

A common statistical model for the variability of a measurement process is the following:

$$X_i = \mu + \beta + \varepsilon_i$$

(See Section 4.2.1.) Here, X_i is the value of the ith measurement, μ is the true value of the heat of sublimation, β represents bias in the measurement procedure, and ε_i is the random error. The ε_i are usually assumed to be independent and identically distributed random variables with mean 0 and variance σ^2. The efficacy of measures of location is often judged by comparing their performances (mean squared error, for example) with this model. Note that with this model the data alone tell us nothing about β, the bias in the measurement procedure, which in some cases may be as or more important than the random variability.

The observations are listed across rows in the order in which the experiments were done. When observations are acquired sequentially, it is often informative to plot them in order, as in Figure 10.10. From this plot, we see that the first few observations were somewhat high. The most striking aspect of the plot is the presence of five extreme observations that occurred in groups of three and two. Such observations, which are quite far from the bulk of the data, are called **outliers.** Outliers occur all too frequently,

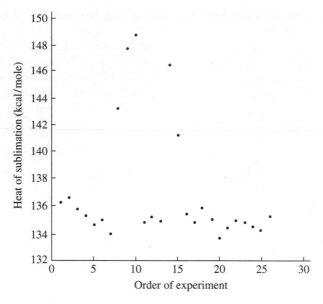

FIGURE **10.10** Plot showing time sequence of measurements of heat of sublimation of platinum.

even in carefully conducted studies. The outliers in this case might have been caused by improperly calibrated equipment, for example. Outliers can also be caused by recording and transcription errors or by equipment malfunctions. It is important to detect outliers, since they may have an undue influence on subsequent calculations. Graphical presentation is an effective means of detection. Careful reexamination of the data and the circumstances under which they were obtained can sometimes uncover the causes behind the outliers. Although outliers are often unexplainable aberrations, an examination of them and their causes can sometimes deepen an investigator's understanding of the phenomenon under study.

Figure 10.10 also makes us doubt that the model for measurement error given above is appropriate for this set of data. The fact that the outliers occur in groups of two and three, rather than being randomly scattered, makes the independence model somewhat implausible.

A stem-and-leaf plot provides another summary of this data (the decimal point is at the colon):

```
      1   1   133:7
      4   3   134:134
     11   7   134:5788899
          6   135:002244
      9   2   135:88
      7   1   136:3
      6   1   136:6
High: 141.2    143.3    146.5    147.8    148.8
```

On this stem-and-leaf plot, the outlying observations have been isolated and flagged as high.

In their analysis, Hampson and Walker set aside the seven largest observations and the smallest observation and found the average of the remaining observations to be 134.9. Calculated from all the observations, the arithmetic mean is 137.05. Note from the stem-and-leaf plot and from Figure 10.10 that this number is larger than the bulk of the data and is clearly not a good descriptive measure of the "center" of this batch of numbers. We would not be satisfied with it as an estimate of the true heat of sublimation.

If the data are modeled as a sample from a probability law, as with the measurement error model described above, an approximate $100(1 - \alpha)\%$ confidence interval for the population mean can be obtained from the central limit theorem as in Chapter 7. The interval is of the form

$$\bar{x} \pm z(\alpha/2)s_{\bar{x}}$$

Blindly applying this formula to the platinum data, with $\alpha = .05$, we obtain the interval 137.05 ± 1.71, or $(135.3, 138.8)$. Note where this interval falls on the stem-and-leaf plot!

Although the example presented here may be somewhat extreme, it illustrates the sensitivity of the sample mean to outlying observations. In fact, by changing a single number, the arithmetic mean of a batch of numbers can be made arbitrarily large or small. Thus, if used blindly, without careful attention to the data, the arithmetic mean can produce misleading results. When the data are automatically acquired, stored as files on disks or tapes, and not visually examined, this danger increases. For this reason, measures of location that are **robust,** or insensitive to outliers, are important.

10.4.2 The Median

If the sample size is an odd number, the **median** is defined to be the middle value of the ordered observations; if the sample size is even, the median is the average of the two middle values. Clearly, moving the extreme observations does not affect the sample median at all, so the median is quite robust. The median of the platinum data is 135.1, which, as can be seen from the stem-and-leaf plot, is more reasonable than the mean as a measure of the center.

When the data are a sample from a continuous probability law, the sample median can be viewed as an estimate of the population median, η, for which a simple confidence interval can be formed. We will now demonstrate that this interval is of the form

$$(X_{(k)}, X_{(n-k+1)})$$

The coverage probability of this interval is

$$P(X_{(k)} \leq \eta \leq X_{(n-k+1)}) = 1 - P(\eta < X_{(k)} \text{ or } \eta > X_{(n-k+1)})$$
$$= 1 - P(\eta < X_{(k)}) - P(\eta > X_{(n-k+1)})$$

since the events are mutually exclusive. To evaluate these terms, we first note that

$$P(\eta > X_{(n-k+1)}) = \sum_{j=0}^{k-1} P(j \text{ observations are greater than } \eta)$$

$$P(\eta < X_{(k)}) = \sum_{j=0}^{k-1} P(j \text{ observations are less than } \eta)$$

Since, by definition, the median satisfies

$$P(X_i > \eta) = P(X_i < \eta) = \tfrac{1}{2}$$

and since the n observations $X_1, \ldots, X_n$ are independent and identically distributed, the distribution of the number of observations greater than the median is binomial with n trials and probability $\frac{1}{2}$ of success on each trial. Thus,

$$P(\text{exactly } j \text{ observations are greater than } \eta) = \frac{1}{2^n} \binom{n}{j}$$

and

$$P(\eta > X_{(n-k+1)}) = \frac{1}{2^n} \sum_{j=0}^{k-1} \binom{n}{j}$$

From symmetry, we then have that the coverage probability of the interval in question is

$$1 - \frac{1}{2^{n-1}} \sum_{j=0}^{k-1} \binom{n}{j}$$

These probabilities can be found from tables of the cumulative binomial distribution since

$$\frac{1}{2^n} \sum_{j=0}^{k-1} \binom{n}{j} = P(Y \leq k - 1)$$

where Y is a binomial random variable with n trials and probability of success equal to $\frac{1}{2}$.

E X A M P L E **A**   As a concrete example, with $n = 26$, we have the following cumulative binomial probabilities:

k	$P(Y \leq k)$
5	.0012
6	.0047
7	.0145
8	.0378
9	.0843

If we choose $k = 8$,

$$P(Y < k) = .0145$$

and since $P(Y < k) = P(Y > n - k + 1)$, $P(Y > 19) = .0145$. Since $2 \times .0145 = .029$, the interval $(X_{(8)}, X_{(19)})$ is a 97% confidence interval. Note that this confidence interval is exact, not approximate, and does not depend on the form of the underlying cdf but only on the assumption that the cdf is continuous and that the observations are independent.

For the platinum data, this confidence interval is (134.8, 135.8). Compare this interval to the interval based on the sample mean. (But as we noted, there is reason to doubt the independence assumption for the platinum data, so these calculations should be viewed as an illustrative numerical exercise.)   ■

10.4.3 The Trimmed Mean

Another simple and robust measure of location is the **trimmed mean.** The $100\alpha\%$ trimmed mean is easy to calculate: Order the data, discard the lowest $100\alpha\%$ and the highest $100\alpha\%$, and take the arithmetic mean of the remaining data. It is generally recommended that the value chosen for α be from .1 to .2. Formally, we may write the trimmed mean as

$$\bar{x}_\alpha = \frac{x_{([n\alpha]+1)} + \cdots + x_{(n-[n\alpha])}}{n - 2[n\alpha]}$$

where $[n\alpha]$ denotes the greatest integer less than or equal to $n\alpha$. Note that the median can be regarded as a 50% trimmed mean.

The 20% trimmed mean for the platinum data listed in Section 10.4.1 is formed by discarding the highest and lowest five observations ($.2 \times 26 = 5.2$) and averaging the rest. The result is 135.29; for the same data, the median was 135.1 and the mean was 137.05.

10.4.4 M Estimates

The sample mean is the mle of μ, the location parameter, when the underlying distribution is normal. Equivalently, the sample mean minimizes the negative log likelihood, or

$$\sum_{i=1}^{n} \left(\frac{X_i - \mu}{\sigma} \right)^2$$

This is the simplest case of a **least squares estimate.** (We will discuss least squares estimates in more detail in the context of curve fitting.) Outliers have a great effect on this estimate, since the deviation of μ from X_i is measured by the square of their difference. In contrast, the median is the minimizer of (see Problem 34 of the end-of-chapter problems)

$$\sum_{i=1}^{n} \left| \frac{X_i - \mu}{\sigma} \right|$$

Here, large deviations are not weighted as heavily, and it is this property that causes the median to be robust.

Huber (1981) proposed a class of estimates, **M estimates,** which are the minimizers of

$$\sum_{i=1}^{n} \Psi \left(\frac{X_i - \mu}{\sigma} \right)$$

where the weight function Ψ is a compromise between the weight functions for the mean and the median. A wide variety of weight functions have been proposed. Huber discusses weight functions that are quadratic near zero and are linear beyond a cutoff point, k. Thus, $k = \infty$ corresponds to the mean and $k = 0$ to the median. A common choice is $k = 1.5$. With this choice, the influence of observations more than 1.5σ away from the center is reduced. In practice, a robust estimate of σ, such as those discussed in Section 10.5, must be used.

The computation of an M estimate is a nonlinear minimization problem and must be done iteratively (using the Newton-Raphson method, for example). If M is a convex function, the minimizer will be unique. Fairly simple computer programs that do this are common in statistical packages. The M estimate ($k = 1.5$) for the platinum data we have been considering is 135.38, close to the median (135.1) and the trimmed mean (135.29) but quite different from the mean (137.05).

10.4.5 Comparison of Location Estimates

We introduced several location estimates (and there are many others). Which one is best? There is no simple answer to this question. It is always important to bear in mind what is being estimated by the location estimates and to what purpose the estimate is being put. If the underlying distribution is symmetric, the trimmed mean, the sample mean, the sample median, and an M estimate all estimate the center of symmetry. If the underlying distribution is not symmetric, however, the four statistics estimate four different population parameters: the population mean, the population median, the population trimmed mean, and a functional of the cdf determined by the weight function Ψ. Moreover, there is no single estimate that is best for all symmetric distributions. Life isn't that simple. Simulations have been done to compare estimates for a variety of distributions. Andrews et al. (1972) report the results of a large number of simulations from symmetric distributions. Their results show that the 10% or 20% trimmed mean is overall quite an effective estimate: Its variance is never much larger than the variance of the ordinary mean (even in the Gaussian case for which the mean is optimal) and can be quite a lot smaller when the underlying distribution is heavy-tailed relative to the Gaussian. The median, although quite robust, has a substantially larger variance in the Gaussian case than does the trimmed mean. The trimmed mean and the median have a certain appealing simplicity and are easy to explain to someone who has little formal statistical training. M estimates performed quite well in the simulations of the Andrews et al. study, and they do generalize more naturally to other problems such as curve fitting. But they are somewhat more difficult to compute and have less immediate intuitive appeal. For the purpose of simply summarizing data, it is often useful to compute more than one measure of location and compare the results.

10.4.6 Estimating Variability of Location Estimates by the Bootstrap

If we view the observations $x_1, x_2, \ldots, x_n$ as realizations of independent random variables with common distribution function F, it is appropriate to investigate the variability and sampling distribution of a location estimate calculated from a sample of size n. Suppose we denote the location estimate as $\hat{\theta}$; it is important to keep in mind that $\hat{\theta}$ is a function of the random variables $X_1, X_2, \ldots, X_n$ and hence has a probability distribution, its sampling distribution, which is determined by n and F. We would like to know this sampling distribution, but we are faced with two problems: (1) We don't know F, and (2) even if we knew F, $\hat{\theta}$ may be such a complicated function of $X_1, X_2, \ldots, X_n$ that finding its distribution would exceed our analytic abilities.

First, we address the second problem. Suppose then, for the moment, that we knew F. How could we find the probability distribution of $\hat{\theta}$ without going through incredibly complicated analytic calculations? The computer comes to our rescue—we can do it by simulation. We generate many, many samples, say B in number, of size n from F; from each sample we calculate the value of $\hat{\theta}$. The empirical distribution of the resulting values $\theta_1^*, \theta_2^*, \ldots, \theta_B^*$ is an approximation to the distribution function of $\hat{\theta}$, which is good if B is very large. If we want to know the standard deviation of $\hat{\theta}$, we can find a good approximation to it by calculating the standard deviation of the collection of values $\theta_1^*, \theta_2^*, \ldots, \theta_B^*$. We can make these approximations arbitrarily accurate by taking B to be arbitrarily large.

All this would be well and good if we knew F, but we don't. So what do we do? The bootstrap solution is to view the empirical cdf F_n as an approximation to F and sample from F_n. That is, F_n would be used in place of F in the previous paragraph. How do we go about sampling from F_n? F_n is a discrete probability distribution that gives probability $1/n$ to each observed value $x_1, x_2, \ldots, x_n$. A sample of size n from F_n is thus a sample of size n drawn *with replacement* from the collection $x_1, x_2, \ldots, x_n$. We thus draw B samples of size n with replacement from the observed data, producing $\theta_1^*, \theta_2^*, \ldots, \theta_B^*$. The standard deviation of $\hat{\theta}$ is then estimated by

$$s_{\hat{\theta}} = \sqrt{\frac{1}{B}\sum_{i=1}^{B}(\theta_i^* - \bar{\theta}^*)^2}$$

where $\bar{\theta}^*$ is the mean of $\theta_1^*, \theta_2^*, \ldots, \theta_B^*$.

EXAMPLE **A**   We illustrate this idea on the platinum data by using the bootstrap to approximate the sampling distribution of the 20% trimmed mean and its standard error. To this end, 1000 samples of size $n = 26$ were drawn randomly with replacement from the collection of 26 values. A histogram of the 1000 trimmed means is displayed in Figure 10.11. The standard deviation of the 1000 values was .64, which is the estimated standard error of the 20% trimmed mean. The histogram is interesting—note the skewed tail to the right. We see that some of the trimmed means were far from the bulk of the data. This happened because some of the samples drawn with replacement included several

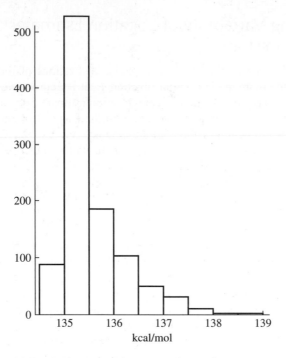

FIGURE **10.11**   Histogram of 1000 bootstrap 20% trimmed means.

replicates of the five outliers (see Figure 10.10). The computer calculation is telling us that if we sample from F_n, the 20% trimmed mean is not as robust as we might like; this is an extremely heavy-tailed distribution and a sample of 26 may contain a large number of outliers.

As in Chapter 8, we can use the bootstrap distribution to form an approximate 90% confidence interval. We proceed as in Examples D and E of Section 8.5.3, which you may want to review at this time. Denote the trimmed mean of the sample by $\hat{\theta} = 135.29$, and denote the 1000 ordered bootstrap trimmed means by $\theta_{(1)}^* \leq \theta_{(2)}^* \leq \cdots \leq \theta_{(1000)}^*$. Then the .05 quantile of the bootstrap distribution is $\underline{\theta} = \theta_{(50)}^* = 134.00$, and the .95 quantile is $\bar{\theta} = \theta_{(950)}^* = 136.93$. Following the notation of the examples of Section 8.5.3, the approximate 90% confidence interval is $(\hat{\theta} - \bar{\delta}, \hat{\theta} - \underline{\delta})$, where

$$\hat{\theta} - \bar{\delta} = \hat{\theta} - (\bar{\theta} - \hat{\theta})$$
$$= 2\hat{\theta} - \bar{\theta}$$
$$= 133.65$$

and

$$\hat{\theta} - \underline{\delta} = \hat{\theta} - (\underline{\theta} - \hat{\theta})$$
$$= 2\hat{\theta} - \underline{\theta}$$
$$= 135.58$$

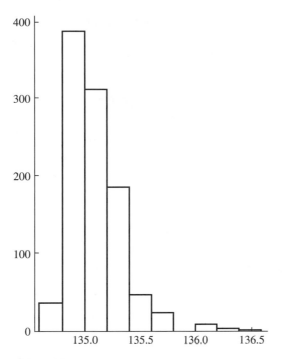

FIGURE **10.12**   Histogram of 1000 bootstrap medians.

Figure 10.12 is a histogram of 1000 bootstrapped medians. It is less dispersed than the histogram of trimmed means; the standard deviation of the medians is .24, considerably less than that of the trimmed mean. The bootstrap simulation is telling us that when sampling from a distribution like this, the median is more robust than the 20% trimmed mean. ■

How accurate are these bootstrap estimates? It is difficult to answer this question in a useful, explicit manner. Essentially, the accuracy depends on two factors: (1) the accuracy of F_n as an estimate of F, and (2) the dependence of the distribution of the statistic $\hat{\theta}$ on F. For example, if the distribution of $\hat{\theta}$ changes little if F changes, then F_n need not be a very good estimate of F, whereas if the distribution of $\hat{\theta}$ is extremely sensitive to F, F_n will have to be a good estimate of F, and hence the sample size will have to be large, in order for the bootstrap approximation to be accurate.

10.5 Measures of Dispersion

A measure of dispersion, or scale, gives a numerical indication of the "scatteredness" of a batch of numbers. Simple summaries of data often consist of a measure of location and a measure of dispersion. The most commonly used measure is the sample standard deviation, s, which is the square root of the sample variance,

$$s^2 = \frac{1}{n-1} \sum_{i=1}^{n} (X_i - \overline{X})^2$$

Using $n - 1$ as the divisor rather than the more obvious divisor n is based on the rationale that s^2 is an unbiased estimate of the population variance if the observations are independent and identically distributed with variance σ^2. (But s is not an unbiased estimate of σ because the square root is a nonlinear function.) If n is of moderate to large size, it makes little difference whether n or $n - 1$ is used.

If the observations are a sample from a normal distribution with variance σ^2,

$$\frac{(n-1)s^2}{\sigma^2} \sim \chi_{n-1}^2$$

This distributional result may be used to construct confidence intervals for σ^2 in the normal case (compare with Example A in Section 8.5.3), but the result is not robust against deviations from normality.

Like the sample mean, the sample standard deviation is sensitive to outlying observations. Two simple robust measures of dispersion are the **interquartile range (IQR)**, or the difference between the two sample quartiles; (the 25th and 75th percentiles) and the **median absolute deviation from the median (MAD)**. If the data are $x_1, \ldots, x_n$ with median $\tilde{x}$, the MAD is defined to be the median of the numbers $|x_i - \tilde{x}|$. These two measures of dispersion, the IQR and the MAD, can be converted into estimates of σ for a normal distribution by dividing them by 1.35 and .675, respectively. David (1981) discusses a method for finding a confidence interval for the population interquartile range, using reasoning similar to that used in Section 10.4.2 for developing a confidence interval for the population median.

Let us compare all three measures of dispersion for the platinum data:

$$s = 4.45$$
$$\frac{IQR}{1.35} = 1.26$$
$$\frac{MAD}{.675} = .934$$

The two robust estimates are similar. From the stem-and-leaf plot of the platinum values presented earlier, we can see that both the IQR and the MAD give measures of the spread of the central portion of the data, whereas the standard deviation is heavily influenced by the outliers.

10.6 Boxplots

A **boxplot** is a graphical display invented by Tukey that shows a measure of location (the median), a measure of dispersion (the interquartile range), and the presence of possible outliers and also gives an indication of the symmetry or skewness of the distribution. Figure 10.13 is a boxplot of the platinum data.

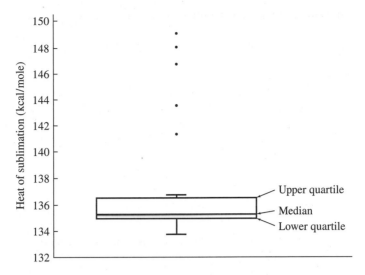

F I G U R E **10.13**   Boxplot of the platinum data.

We outline the construction of a boxplot:

1. Horizontal lines are drawn at the median and at the upper and lower quartiles and are joined by vertical lines to produce the box.
2. A vertical line is drawn up from the upper quartile to the most extreme data point that is within a distance of 1.5 (IQR) of the upper quartile. A similarly defined vertical line is drawn down from the lower quartile. Short horizontal lines are added to mark the ends of these vertical lines.
3. Each data point beyond the ends of the vertical lines is marked with an asterisk or dot (* or ·).

Boxplots are not uniformly standardized, but the basic structure is as outlined above, perhaps with additional embellishments or small variations. A boxplot thus gives an indication of the center of the data (the median), the spread of the data (the interquartile range), and the presence of outliers, and indicates the symmetry or asymmetry of the distribution of data values (the location of the median relative to the quartiles). In Figure 10.13, the five outliers of the platinum data are clearly displayed, and we see an indication that the central part of the distribution is somewhat skewed toward high values.

E X A M P L E **A**   Figure 10.14 is taken from Chambers et al. (1983). The data plotted are daily maximum concentrations in parts per billion of sulfur dioxide in Bayonne, N.J., from November 1969 to October 1972 grouped by month. There are thus 36 batches, each of size about 30. The investigators concluded:

> The boxplots . . . show many properties of the data rather strikingly. There is a general reduction in sulphur dioxide concentration through time due to the gradual conversion to low sulphur fuels in the region. The decline is most dramatic for the highest quantiles. Also, there are higher concentrations

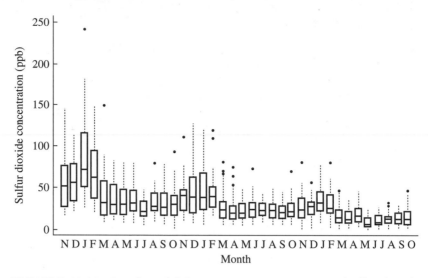

F I G U R E **10.14** Boxplots of daily maximum concentrations of sulfur dioxide.

during the winter months due to the use of heating oil. In addition, the boxplots show that the distributions are skewed toward high values and that the spread of the distributions ... is larger when the general level of concentration is higher.

The boxplot is clearly a very effective method of presenting and summarizing these data. As they are in this example, boxplots are generally useful for comparing batches of numbers, a purpose to which they will be put in the next two chapters. ■

10.7 Exploring Relationships with Scatterplots

Many interesting questions in statistics involve trying to understand the relationships among variables. The scatterplot is a basic method for displaying the empirical relationship between two variables based on a collection of pairs (x_i, y_i): one merely plots the points in the xy plane. This basic display can be augmented in various ways, as we will illustrate with some examples.

E X A M P L E **A**   Allison and Cicchetti (1976) examined the relationships of possible correlates of sleep behavior in mammals. Figure 10.15 is a scatterplot of total sleep versus brain weight. Other than that two mammals with very large brains slept very little, no relationships are apparent in the plot. There is in fact a relationship, but it is obscured in the plot because brain weights vary over orders of magnitude—the brain of the lesser short-tailed shrew weighs 0.14 grams, and at the other extreme the brain of the African elephant weighs 5,712 grams. It is thus much more informative to plot sleep versus the logarithm of brain weight, and annotating the plot helps further—as shown in Figure 10.16. It is now clear that mammals with heavier brains tend to sleep less.

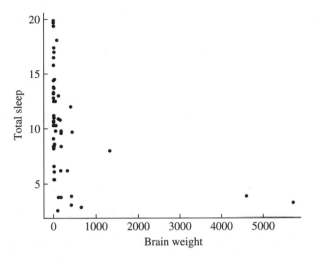

FIGURE **10.15** Sleep versus brain weight for a collection of mammals.

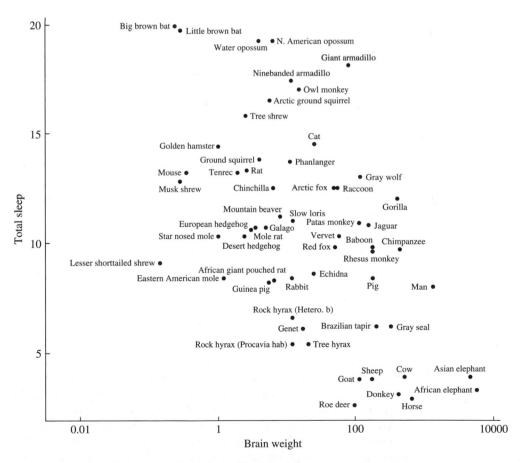

FIGURE **10.16** Sleep versus logarithm of brain weight.

Data on these and other variables (how much do elephants dream?) can be found at `http://lib.stat.cmu.edu/datasets/sleep`.   ∎

Correlation coefficients are often used as a simple numerical summary of the strength of a relationship. The Pearson correlation coefficient corresponding to the pairs (x_i, y_i) is

$$r = \frac{\sum (x_i - \bar{x})(y_i - \bar{y})}{\sqrt{\sum (x_i - \bar{x})^2 \sum (y_i - \bar{y})^2}}$$

This statistic measures the strength of a *linear* relationship. The correlation of brain weight and sleep is -0.36, and the correlation between the logarithm of brain weight and sleep is -0.56. These are different because a nonlinear transformation has been applied and the correlation coefficient measures the strength of a linear relationship. An alternative to the Pearson correlation coefficient is the rank correlation coefficient: the brain weights are replaced by their ordered ranks $(1, 2, \ldots)$, the sleeping times are replaced by their ranks, and then the Pearson correlation coefficient of the pairs of ranks is computed. The rank correlation turns out to be -0.39 in our example. Some advantages of the rank correlation coefficient are that it is insensitive to outliers and is invariant under any monotone transformation (thus the rank correlation does not depend on whether brain weight or log brain weight is used).

Arrays of scatterplots are useful for examining the relationships among more than two variables, as illustrated in the following example.

E X A M P L E **B**   Inductive loop detectors are wire loops embedded in the pavement of roadways. They operate by detecting the change in inductance caused by the metal in vehicles that pass over them. During successive intervals of time, a detector reports the number of passing vehicles, and the percentage of time that it was covered by a vehicle. The number of vehicles is called *flow*, the percentage of coverage is called the *occupancy*. Such detectors are widely used to measure freeway traffic but are subject to various kinds of malfunction. Faulty detectors must be identified by traffic management centers. One key to detecting malfunction is knowing that measurements in the several freeway lanes at a particular location should be highly related—the increases and decreases of traffic flow in one lane should tend to be mirrored in other lanes. Figure 10.17 shows an array of scatterplots of occupancy measured by detectors in four lanes at a particular location (Bickel et al. 2004). The detectors in lanes three and four were closely related to each other at all times and were correlated with measurements in lanes one and two some, but not all, of the time. Apparently the detectors in lanes 1 and 2 malfunctioned some of the time while this set of measurements was taken.   ∎

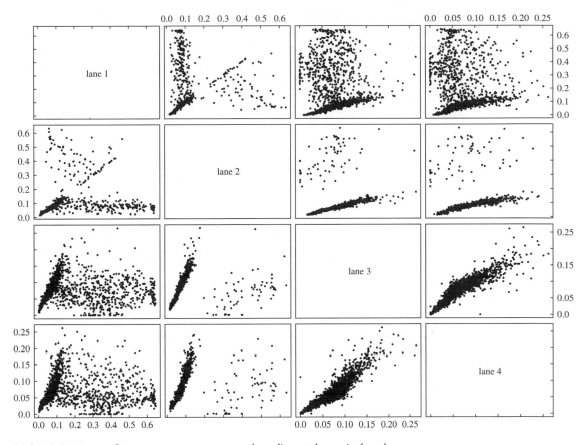

FIGURE **10.17**   Occupancy measurements by adjacent loops in four lanes.

10.8   Concluding Remarks

This chapter introduced several tools for summarizing data, some of which are graphical in nature. Under the assumption of a stochastic model for the data, some aspects of the sampling distributions of these summaries have been discussed. Summaries are very important in practice; an intelligent summary of data is often sufficient to fulfill the purposes for which the data were gathered, and more formal techniques such as confidence intervals or hypothesis tests sometimes add little to an investigator's understanding. Effective summaries can also point to "bad" data or to unexpected aspects of data that might have gone unnoticed if the data had been blindly crunched by a computer.

We saw the bootstrap appear again as a method for approximating a sampling distribution and functionals of it such as its standard deviation. The bootstrap, a relatively recent development in statistical methodology, relies on the availability of powerful and inexpensive computing resources. Our development of approximate confidence intervals based on the bootstrap followed that of Chapter 8, where we motivated the construction by using the bootstrap distribution of $\theta^* - \hat{\theta}$ to approximate the distribution of $\hat{\theta} - \theta_0$. We note that another popular method, known as the bootstrap percentile method, gives the interval $(\underline{\theta}, \bar{\theta})$ (see Example A of Section 8.5.3 for definition of the notation). The rationale for this is harder to understand. More accurate methods for constructing bootstrap confidence intervals have been proposed and are under study, but we will not pursue these developments.

10.9 Problems

1. Plot the ecdf of this batch of numbers: 1, 14, 10, 9, 11, 9.

2. Suppose that $X_1, X_2, \ldots, X_n$ are independent $U[0, 1]$ random variables.
    a. Sketch $F(x)$ and the standard deviation of $F_n(x)$.
    b. Generate many samples of size 16 on a computer; for each sample, plot $F_n(x)$ and $F_n(x) - F(x)$. Relate what you see to your answer to (a).

3. From Figure 10.1, roughly what are the upper and lower quartiles and the median of the distribution of melting points?

4. In Section 10.2.1, it was claimed that the random variables $I_{(-\infty, x]}(X_i)$ are independent. Why is this so?

5. Let $X_1, \ldots, X_n$ be a sample (i.i.d.) from a distribution function, F, and let F_n denote the ecdf. Show that

$$\text{Cov}\,[F_n(u), F_n(v)] = \frac{1}{n}[F(m) - F(u)F(v)]$$

where $m = \min(u, v)$. Conclude that $F_n(u)$ and $F_n(v)$ are positively correlated: If $F_n(u)$ overshoots $F(u)$, then $F_n(v)$ will tend to overshoot $F(v)$.

6. Various chemical tests were conducted on beeswax by White, Riethof, and Kushnir (1960). In particular, the percentage of hydrocarbons in each sample of wax was determined.
    a. Plot the ecdf, a histogram, and a normal probability plot of the percentages of hydrocarbons given in the following table. Find the .90, .75, .50, .25, and .10 quantiles. Does the distribution appear Gaussian?

14.27	14.80	12.28	17.09	15.10	12.92	15.56	15.38
15.15	13.98	14.90	15.91	14.52	15.63	13.83	13.66
13.98	14.47	14.65	14.73	15.18	14.49	14.56	15.03
15.40	14.68	13.33	14.41	14.19	15.21	14.75	14.41
14.04	13.68	15.31	14.32	13.64	14.77	14.30	14.62
14.10	15.47	13.73	13.65	15.02	14.01	14.92	15.47
13.75	14.87	15.28	14.43	13.96	14.57	15.49	15.13
14.23	14.44	14.57					

    b. The average percentage of hydrocarbons in microcrystalline wax (a synthetic commercial wax) is 85%. Suppose that beeswax was diluted with 1% microcrystalline wax. Could this be detected? What about a 3% or a 5% dilution? (Such questions were one of the main concerns of the beeswax study.)

7. Compare group I to group V in Figure 10.2. Roughly, what are the differences in lifetimes for the animals that are the 10% weakest, median, and 10% strongest?

8. Consider a sample of size 100 from an exponential distribution with parameter $\lambda = 1$.

   a. Sketch the approximate standard deviation of the empirical log survival function, $\log S_n(t)$, as a function of t.
   b. Generate several such samples of size 100 on a computer and for each sample plot the empirical log survival function. Relate the plots to your answer to (a).

9. Use the method of propagation of error to derive an approximation to the bias of the log survival function. Where is this bias large, and what is its sign?

10. Let $X_1, \ldots, X_n$ be a sample from cdf F and denote the order statistics by $X_{(1)}, X_{(2)}, \ldots, X_{(n)}$. We will assume that F is continuous, with density function f. From Theorem A in Section 3.7, the density function of $X_{(k)}$ is

$$f_k(x) = n \binom{n-1}{k-1} [F(x)]^{k-1} [1 - F(x)]^{n-k} f(x)$$

   a. Find the mean and variance of $X_{(k)}$ from a uniform distribution on $[0, 1]$. You will need to use the fact that the density of $X_{(k)}$ integrates to 1. Show that

$$\text{Mean} = \frac{k}{n+1}$$

$$\text{Variance} = \frac{1}{n+2} \left(\frac{k}{n+1} \right) \left(1 - \frac{k}{n+1} \right)$$

   b. Find the approximate mean and variance of $Y_{(k)}$, the kth-order statistic of a sample of size n from F. To do this, let

$$X_i = F(Y_i)$$

   or

$$Y_i = F^{-1}(X_i)$$

   The X_i are a sample from a $U[0, 1]$ distribution (why?). Use the propagation of error formula,

$$Y_{(k)} = F^{-1}(X_{(k)})$$

$$\approx F^{-1}\left(\frac{k}{n+1} \right) + \left(X_{(k)} - \frac{k}{n+1} \right) \frac{d}{dx} F^{-1}(x) \Big|_{k/(n+1)}$$

and argue that

$$EY_{(k)} \approx F^{-1}\left(\frac{k}{n+1}\right)$$

$$\text{Var } (Y_{(k)}) \approx \frac{k}{n+1}\left(1 - \frac{k}{n+1}\right)\frac{1}{(f\{F^{-1}[k/(n+1)]\})^2}\left(\frac{1}{n+2}\right)$$

c. Use the results of parts (a) and (b) to show that the variance of the pth sample quantile is approximately

$$\frac{1}{nf^2(x_p)}p(1-p)$$

where x_p is the pth quantile.

d. Use the result of part (c) to find the approximate variance of the median of a sample of size n from a $N(\mu, \sigma^2)$ distribution. Compare to the variance of the sample mean.

11. Calculate the hazard function for

$$F(t) = 1 - e^{-\alpha t^\beta}, \qquad t \geq 0$$

12. Let f denote the density function and h the hazard function of a nonnegative random variable. Show that

$$f(t) = h(t)e^{-\int_0^t h(s)ds}$$

that is, that the hazard function uniquely determines the density.

13. Give an example of a probability distribution with increasing failure rate.

14. Give an example of a probability distribution with decreasing failure rate.

15. A prisoner is told that he will be released at a time chosen uniformly at random within the next 24 hours. Let T denote the time that he is released. What is the hazard function for T? For what values of t is it smallest and largest? If he has been waiting for 5 hours, is it more likely that he will be released in the next few minutes than if he has been waiting for 1 hour?

16. Suppose that F is $N(0, 1)$ and G is $N(1, 1)$. Sketch a Q-Q plot. Repeat for G being $N(1, 4)$.

17. Suppose that F is an exponential distribution with parameter $\lambda = 1$ and that G is exponential with $\lambda = 2$. Sketch a Q-Q plot.

18. A certain chemotherapy treatment for cancer tends to lengthen the lifetimes of very seriously ill patients and decrease the lifetimes of the least ill patients. Suppose that an experiment is done that compares this treatment to a placebo. Draw a sketch showing the qualitative behavior of a Q-Q plot.

19. Consider the two cdfs:

$$F(x) = x, \qquad 0 \leq x \leq 1$$
$$G(x) = x^2, \qquad 0 \leq x \leq 1$$

Sketch a Q-Q plot of F versus G.

20. Sketch what you would expect the qualitative shape of the hazard function of human mortality to look like.

21. Make Q-Q plots for other pairs of treatment groups from Bjerkdahl's data (see Example A in Section 10.2.2). Does the model of a multiplicative effect appear reasonable?

22. By examining the survival function of group V of Bjerkdahl's data (see Example A in Section 10.2.2), make a rough sketch of the qualitative shape of a histogram. Then make a histogram, and compare it to your guess.

23. In the examples of Q-Q plots in the text, we only discussed the case in which quantiles of equal size batches are compared. From two batches of size n the $k/(n+1)$ quantiles are estimated as $X_{(k)}$ and $Y_{(k)}$, so one merely has to plot $X_{(k)}$ vs. $Y_{(k)}$. Write down a linear interpolation formula for the pth quantile where $k/(n+1) \leq p \leq (k+1)/(n+1)$. Now suppose that the batch sizes are not the same, being m and n, $m < n$ say. A Q-Q plot may be constructed by fixing the quantiles $k/(m+1)$ of the smaller data set and interpolating these quantiles for the larger data set.

Interpolate to find the upper and lower quartiles of the following batch of numbers: 1, 2, 3, 4, 5, 6.

24. Show that the probability plots discussed in Section 9.9 are Q-Q plots of the empirical distribution F_n versus a theoretical distribution F.

25. In Section 10.2.3, it was claimed that if $y_p = cx_p$, then $G(y) = F(y/c)$. Justify this claim.

26. Hampson and Walker also made measurements of the heats of sublimation of rhodium and iridium. Do the following calculations for each of the two given sets of data:

a. Make a histogram.
b. Make a stem-and-leaf plot.
c. Make a boxplot.
d. Plot the observations in the order of the experiment.
e. Does that statistical model of independent and identically distributed measurement errors seem reasonable?
f. Find the mean, 10% and 20% trimmed means, and median and compare them.
g. Find the standard error of the sample mean and a corresponding approximate 90% confidence interval.
h. Find a confidence interval based on the median that has as close to 90% coverage as possible.
i. Use the bootstrap to approximate the sampling distributions of the 10% and 20% trimmed means and their standard errors and compare.
j. Use the bootstrap to approximate the sampling distribution of the median and its standard error. Compare to the corresponding results for trimmed means above.
k. Find approximate 90% confidence intervals based on the trimmed means and compare to the intervals for the mean and median found previously.

Iridium (kcal/mol)								
136.6	145.2	151.5	162.7	159.1	159.8	160.8	173.9	160.1
160.4	161.1	160.6	160.2	159.5	160.3	159.2	159.3	159.6
160.0	160.2	160.1	160.0	159.7	159.5	159.5	159.6	159.5

Rhodium (kcal/mol)							
126.4	135.7	132.9	131.5	131.1	131.1	131.9	132.7
133.3	132.5	133.0	133.0	132.4	131.6	132.6	132.2
131.3	131.2	132.1	131.1	131.4	131.2	131.1	131.1
134.2	133.8	133.3	133.5	133.4	133.5	133.0	132.8
132.6	133.3	133.5	133.5	132.3	132.7	132.9	134.1

27. Demographers often refer to the hazard function as the "age specific mortality rate," or death rate. Until recently, most researchers in the field of gerontology thought that a death rate increasing with age was a universal fact in the biological world. There has been heavy debate over whether there is a genetically programmed upper limit to lifespan. Using a facility in which sterilized medflies are bred to be released to fight medfly infestations in California, James Carey and co-workers (Carey et al. 1992) bred more than a million medflies and recorded their pattern of mortality. The data file medflies, contains the number of medflies alive from an initial population of 1,203,646 as a function of age in days. Using these data, estimate and plot the age specific mortality rate. Does it increase with age?

28. For a sample of size $n = 3$ from a continuous probability distribution, what is $P(X_{(1)} < \eta < X_{(2)})$, where η is the median of the distribution? What is $P(X_{(1)} < \eta < X_{(3)})$?

29. Of the 26 measurements of the heat of sublimation of platinum, 5 are outliers (see Figure 10.10). Let N denote the number of these outliers that occur in a bootstrap sample (sample with replacement) of the 26 measurements.

    a. Explain why the distribution of N is binomial.
    b. Find $P(N \geq 10)$.
    c. In 1000 bootstrap samples, how many would you expect to contain 10 or more of these outliers?
    d. What is the probability that a bootstrap sample is composed entirely of these outliers?

30. In Example A of Section 10.4.6, a 90% bootstrap confidence interval based on the trimmed mean was found to be (133.65, 135.58). Compare these values to the list of data values given in Section 10.4.1 and observe that 133.65 is smaller than the smallest observation. Explain why the bootstrap confidence interval extends so far in this direction.

31. We have seen that the bootstrap entails sampling with replacement from the original observations.

a. If the original sample is of size n, how many samples with replacement are there?

b. Suppose for pedagogical purposes that $n = 3$ and we have the following observations: 1, 3, 4. List all the possible samples with replacement.

c. Now suppose that we want to find the bootstrap distribution of the sample mean. For each of the preceding samples, calculate the mean and use these results to construct the bootstrap distribution of the sample mean.

d. Based on the bootstrap distribution, what is the standard error of the sample mean? Compare this to the usual estimated standard error, $s_{\bar{X}}$.

32. Explain how the bootstrap could be used to approximate the sampling distribution of the MAD.

33. Which of the following statistics can be made arbitrarily large by making one number out of a batch of 100 numbers arbitrarily large: the mean, the median, the 10% trimmed mean, the standard deviation, the MAD, the interquartile range?

34. Show that the median is an M estimate if $\Psi(x) = |x|$. For what symmetric density function is this the mle of the mean?

35. What proportion of the observations from a normal sample would you expect to be marked by an asterisk on a boxplot?

36. Explain why the IQR and the MAD are divided by 1.35 and .675, respectively, to estimate σ for a normal sample.

37. For the data of Problem 6:

a. Find the mean, median, and 10% and 20% trimmed means.

b. Find an approximate 90% confidence interval for the mean.

c. Find a confidence interval with coverage near 90% for the median.

d. Use the bootstrap to find approximate standard errors of the trimmed means.

e. Use the bootstrap to find approximate 90% confidence intervals for the trimmed means.

f. Find and compare the standard deviation of the measurements, the interquartile range, and the MAD.

g. Use the bootstrap to find the approximate sampling distribution and standard error of the upper quartile.

38. The Cauchy distribution has the density function

$$f(x) = \frac{1}{\pi}\left(\frac{1}{1+x^2}\right), \qquad -\infty < x < \infty$$

which is symmetric about zero. This distribution has very heavy tails, which cause the arithmetic mean to be a very poor estimate of location. Simulate the distribution of the arithmetic mean and of the median from a sample of size 25 from the Cauchy distribution by drawing 100 samples of size 25 and compare. From Example B in Section 3.6.1, if Z_1 and Z_2 are independent and $N(0, 1)$, then their quotient follows a Cauchy distribution. (This gives a simple way of generating Cauchy random variables.)

39. Simiu and Filliben (1975), in a statistical analysis of extreme winds, analyzed the data contained in the file `windspeed 10.1`. Construct boxplots to examine and compare the forms of the distributions across cities and across years.

40. Olson, Simpson, and Eden (1975) discuss the analysis of data obtained from a cloud seeding experiment. A cloud was deemed "seedable" if it satisfied certain criteria; for each seedable cloud a decision was made at random whether to actually seed. The nonseeded clouds are referred to as control clouds. The following table presents the rainfall from 26 seeded and 26 control clouds. Make Q-Q plots for rainfall versus rainfall and log rainfall versus log rainfall. What do these plots suggest about the effect, if any, of seeding?

Seeded Clouds							
129.6	31.4	2745.6	489.1	430.0	302.8	119.0	4.1
92.4	17.5	200.7	274.7	274.7	7.7	1656.0	978.0
198.6	703.4	1697.8	334.1	118.3	255.0	115.3	242.5
32.7	40.6						

Control Clouds							
26.1	26.3	87.0	95.0	372.4	0.01	17.3	24.4
11.5	321.2	68.5	81.2	47.3	28.6	830.1	345.5
1202.6	36.6	4.9	4.9	41.1	29.0	163.0	244.3
147.8	21.7						

Based on your results, how would you expect boxplots of precipitation from seeded and unseeded clouds to compare? How would you expect boxplots of log precipitation to compare? Make the boxplots and see whether your predictions are confirmed.

41. Construct a nonparametric confidence interval for a quantile x_p by using the same reasoning as in the derivation of a confidence interval for a median.

42. In a study of the natural variability of rainfall, the rainfall of summer storms was measured by a network of rain gauges in southern Illinois for the years 1960–1964 (Changnon and Huff, in LeCam and Neyman 1967). The average amount of rainfall (in inches) from each storm, by year, is contained in the files, `Illinois60,...,Illinois64`.

a. Is the form of the distribution of rainfall per storm skewed or symmetric?

b. What is the average rainfall per storm? What is the median rainfall per storm? Explain why these measures differ, using the results of part (a).

c. You may have read statements like "10% of the storms account for 90% of the rain." Construct a graph that shows such a relationship for these data.

d. Compare the years using boxplots.

e. Which years were wet and which were dry? Are the wet years wet because there were more storms, because individual storms produced more rain, or for both of these reasons?

43. Barlow, Toland, and Freeman (1984) studied the lifetimes of Kevlar 49/epoxy strands subjected to sustained stress. (The space shuttle uses Kevlar/epoxy spherical vessels in an environment of sustained pressure.) The files `kevlar70`, `kevlar80`, and `kevlar90` contain the times to failure (in hours) of strands tested at 70%, 80%, and 90% stress levels. What do these data indicate about the nature of the distribution of lifetimes and the effect of increasing stress?

44. Hopper and Seeman (1994) studied the relationship between bone density and smoking among 41 pairs of middle-aged female twins. In each pair, one twin was a lighter smoker and one a heavier smoker, as measured by pack-years, the number of packages of cigarettes consumed in a year. Bone mineral density was measured at the lumbar spine, the femoral (hip) neck, and the femoral shaft. As well as smoking, other variables, such as alcohol consumption and tea and coffee consumption, were recorded. The data are contained in the file `bonden` and documentation is in the file `bonedendoc`. Use graphical methods to compare bone densities of the heavy and light smoking twins. Do any of the other variables bear a relationship to bone density? After completing your analysis, you may wish to compare your conclusions to those in the paper.

45. The 2000 U.S. Presidential election was very close and hotly contested. George W. Bush was ultimately appointed to the Presidency by the U.S. Supreme Court. Among the issues was a confusing ballot in Palm Beach County, Florida, the so-called Butterfly Ballot, shown in the following figure.

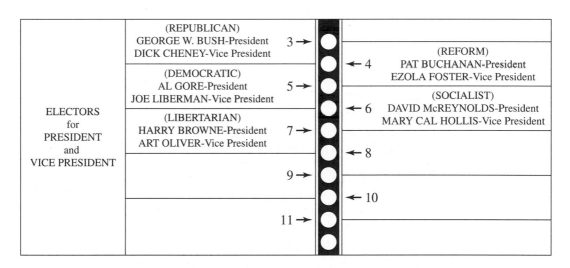

Notice that on this ballot, although the Democrats are listed in the second row on the left, a voter wishing to specify them would have to punch the third hole—punching the second hole would result in a vote for the Reform Party (Pat Buchanan). After the election, many distraught Democratic voters claimed that they had inadvertently voted for Buchanan, a right-wing candidate.

The file `PalmBeach` contains relevant data: vote counts by county in Florida for Buchanan and for four other presidential candidates in 2000, the total vote counts in 2000, the presidential vote counts for three presidential candidates in 1996, the vote count for Buchanan in the 1996 Republican primary, the registration in Buchanan's Reform Party, and the total registration in the county. Does this data support voters' claims that they were misled by the form of the ballot? Start by making two scatterplots: a plot of Buchanan's votes versus Bush's votes in 2000, and a plot of Buchanan's votes in 2000 versus his votes in the 1996 primary.

46. The file `bodytemp` contains normal body temperature readings (degrees Fahrenheit) and heart rates (beats per minute) of 65 males (coded by 1) and 65 females (coded by 2) from Shoemaker (1996).

   a. For both males and females, make scatterplots of heart rate versus body temperature. Comment on the relationship or lack thereof.
   b. Quantify the strengths of the relationships by calculating Pearson and rank correlation coefficients.
   c. Does the relationship for males appear to be the same as that for females? Examine this question graphically, by making a scatterplot showing both females and males and identifying females and males by different plotting symbols.

47. Old Faithful geyser in Yellowstone National Park, Wyoming, derives its name from the regularity of its eruptions. The file `oldfaithful` contains measurements on eight successive days of the durations of the eruptions (in minutes) and the subsequent time interval before the next eruption.

   a. Use histograms of durations and time intervals as well as other graphical methods to examine the fidelity of Old Faithful, and summarize your findings.
   b. Is there a relationship between the durations of eruptions and the time intervals between them?

48. In 1970, Congress instituted a lottery for the military draft to support the unpopular war in Vietnam. All 366 possible birth dates were placed in plastic capsules in a rotating drum and were selected one by one. Eligible males born on the first day drawn were first in line to be drafted followed by those born on the second day drawn, etc. The results were criticized by some who claimed that government incompetence at running a fair lottery resulted in a tendency of men born later in the year being more likely to be drafted. Indeed, later investigation revealed that the birthdates were placed in the drum by month and were not thoroughly mixed. The columns of the file `1970lottery` are month, month number, day of the year, and draft number.

   a. Plot draft number versus day number. Do you see any trend?
   b. Calculate the Pearson and rank correlation coefficients. What do they suggest?
   c. Is the correlation statistically significant? One way to assess this is via a permutation test. Randomly permute the draft numbers and find the correlation of this random permutation with the day numbers. Do this 100 times and see how many of the resulting correlation coefficients exceed the one observed in the data. If you are not satisfied with 100 times, do it 1,000 times.

d. Make parallel boxplots of the draft numbers by month. Do you see any pattern?

49. Olive oil from Spain, Tunisia, and other countries is imported into Italy and is then repackaged and exported with the label "Imported from Italy." Olive oils from different places have distinctive tastes. Can the oils from different regions and areas in Italy be distinguished based on their combinations of fatty acids? This question was considered by Forina et al. (1983). The data consists of the percentage composition of 8 fatty acids (palmitic, palmitoleic, stearic, oleic, linoleic, linolenic, arachidic, eicosenoic) found in the lipid fraction of 572 Italian olive oils. There are 9 collection areas, 4 from southern Italy (North and South Apulia, Calabria, Sicily), two from Sardinia (Inland and Coastal), and 3 from northern Italy (Umbria, East and West Liguria). The file `olive` contains the following variables for each of the 572 samples:

- Region: South, North, or Sardinia
- Area (subregions within the larger regions): North and South Apulia, Calabria, Sicily, Inland and Coastal Sardinia, Umbria, East and West Liguria
- Palmitic Acid Percentage
- Palmitoleic Acid Percentage
- Stearic Acid Percentage
- Oleic Acid Percentage
- Linoleic Acid Percentage
- Linolenic Acid Percentage
- Arachidic Acid Percentage
- Eicosenoic Acid Percentage

Examine this data with the aim of distinguishing between regions and areas by using fatty acid composition.

a. Make a table of the mean and median values of percentages for each area, grouping the areas within regions.

b. Complement the analysis by making parallel boxplots. Which variables look promising for separating the regions?

c. It is possible that the regions can be more clearly separated by considering pairs of variables. Use the variables that appear to be informative from the analysis up to this point to make scatterplots. How well can the regions be separated based on the scatterplots?

d. How well can the areas within regions be distinguished?

e. By interactively rotating point clouds, one can examine relationships among more than two variables at a time. Try this with the software `ggobi` available at `http://www.ggobi.org/`.

50. The file `flow-occ` contains data collected by loop detectors at a particular location of eastbound Interstate 80 in Sacramento, California, from March 14–20, 2003. (Source: `http://pems.eecs.berkeley.edu/`) For each of three lanes, the flow (the number of cars) and the occupancy (the percentage of time a car was over the loop) were recorded in successive five minute intervals. (See Example B of Section 10.7 for background information.) There were 1740 such five-minute intervals. Lane 1 is the farthest left lane, lane 2 is in the center, and lane 3 is the farthest right.

a. For each station, plot flow and occupancy versus time. Explain the patterns you see. Can you deduce from the plots what the days of the week were?

b. Compare the flows in the three lanes by making parallel boxplots. Which lane typically serves the most traffic?

c. Examine the relationships of the flows in the three lanes by making scatterplots. Can you explain the patterns you see? Are statements of the form, "The flow in lane 2 is typically about 50% higher than in lane 3," accurate descriptions of the relationships?

d. Occupancy can be viewed as a measure of congestion. Find the mean and median occupancy in each of the three lanes. Do you think that the distributions of occupancy are symmetric or skewed? Why?

e. Make histograms of the occupancies, varying the number of bins. What number of bins seems to give good representations for the shapes of the distributions? Are there any unusual features, and if so, how might they be explained?

f. Make plots to support or refute the statement, "When one lane is congested, the others are, too."

g. Flow can be regarded as a measure of the throughput of the system. How does this throughput depend on congestion? Consider the following conjecture: "When very few cars are on the road, flow is small and so is congestion. Adding a few more cars may increase congestion but not enough so that velocity is decreased, so flow will also increase. Beyond some point, increasing occupancy (congestion) will decrease velocity, but since there will then be more cars in total, flow will still continue to increase." Does this seem plausible to you? Plot flow versus occupancy for each of the three lanes. Does this conjecture appear to be true? Can you explain what you see? Is the relationship of flow to occupancy the same in all lanes?

h. This and the following exercises require the use of dynamic graphics, e.g., http://www.ggobi.org/. Make time series plots of all the variables. Consider lane 1. Make a one-dimensional display of occupancy and vary the smoothness until you can see some distinct modes. Use brushing to determine when in the time series plots those modes occured. Do the same for flow and then examine some other lanes.

i. Choose a lane and make one-dimensional displays for flow and occupancy and a scatterplot of flow versus occupancy. Use brushing to simultaneously identify regions in the three plots. Does what you see make sense?

j. From scatterplots of flow versus occupancy, examine when different regions of this scatterplot occur in time. In particular, identify when in the time series plots the flow breaks down because a critical point is reached.

k. You have now seen that all these variables, flow and occupancy in each of the three lanes, are closely related, but because scatterplots are two-dimensional, you have been able to examine only those relationships between pairs of variables. In these scatterplots, the points tend to lie along curves. What happens in higher dimensions?

i. Examine the relationship of the three flows. In three dimensions, do the points tend to lie along a curve (a one-dimensional object), or do they tend to concentrate on a two-dimensional manifold, or are they scattered over three dimensions?

ii. Examine the relationships of the three occupancies. In three dimensions, do the points tend to lie along a curve (a one-dimensional object), or do they tend to concentrate on a two-dimensional manifold, or are they scattered over three dimensions?

iii. How do the points lie in six dimensions (three flows and three occupancies)? When do different regions occur in time?

l. A taxi driver claims that when traffic breaks down, the fast lane breaks down first so he moves immediately to the right lane. Can you see any such phenomena in the data?

CHAPTER 11

Comparing Two Samples

11.1 Introduction

This chapter is concerned with methods for comparing samples from distributions that may be different and especially with methods for making inferences about how the distributions differ. In many applications, the samples are drawn under different conditions, and inferences must be made about possible effects of these conditions. We will be primarily concerned with effects that tend to increase or decrease the average level of response.

For example, in the end-of-chapter problems, we will consider some experiments performed to determine to what degree, if any, cloud seeding increases precipitation. In cloud-seeding experiments, some storms are selected for seeding, other storms are left unseeded, and the amount of precipitation from each storm is measured. This amount varies widely from storm to storm, and in the face of this natural variability, it is difficult to tell whether seeding has a systematic effect. The average precipitation from the seeded storms might be slightly higher than that from the unseeded storms, but a skeptic might not be convinced that the difference was due to anything but chance. We will develop statistical methods to deal with this type of problem based on a stochastic model that treats the amounts of precipitation as random variables. We will also see how a process of randomization allows us to make inferences about treatment effects even in the case where the observations are not modeled as samples from populations or probability laws.

This chapter will be concerned with analyzing measurements that are continuous in nature (such as temperature); Chapter 13 will take up the analysis of qualitative data. This chapter will conclude with some general discussion of the design and interpretation of experimental studies.

11.2 Comparing Two Independent Samples

In many experiments, the two samples may be regarded as being independent of each other. In a medical study, for example, a sample of subjects may be assigned to a particular treatment, and another independent sample may be assigned to a control (or placebo) treatment. This is often accomplished by randomly assigning individuals to the placebo and treatment groups. In later sections, we will discuss methods that are appropriate when there is some pairing, or dependence, between the samples, such as might occur if each person receiving the treatment were paired with an individual of similar weight in the control group.

Many experiments are such that if they were repeated, the measurements would not be exactly the same. To deal with this problem, a statistical model is often employed: The observations from the control group are modeled as independent random variables with a common distribution, F, and the observations from the treatment group are modeled as being independent of each other and of the controls and as having their own common distribution function, G. Analyzing the data thus entails making inferences about the comparison of F and G. In many experiments, the primary effect of the treatment is to change the overall level of the responses, so that analysis focuses on the difference of means or other location parameters of F and G. When only a small amount of data is available, it may not be practical to do much more than this.

11.2.1 Methods Based on the Normal Distribution

In this section, we will assume that a sample, $X_1, \ldots, X_n$, is drawn from a normal distribution that has mean μ_X and variance σ^2, and that an independent sample, $Y_1, \ldots, Y_m$, is drawn from another normal distribution that has mean μ_Y and the same variance, σ^2. If we think of the X's as having received a treatment and the Y's as being the control group, the effect of the treatment is characterized by the difference $\mu_X - \mu_Y$. A natural estimate of $\mu_X - \mu_Y$ is $\overline{X} - \overline{Y}$; in fact, this is the mle. Since $\overline{X} - \overline{Y}$ may be expressed as a linear combination of independent normally distributed random variables, it is normally distributed:

$$\overline{X} - \overline{Y} \sim N\left[\mu_X - \mu_Y, \sigma^2\left(\frac{1}{n} + \frac{1}{m}\right)\right]$$

If σ^2 were known, a confidence interval for $\mu_X - \mu_Y$ could be based on

$$Z = \frac{(\overline{X} - \overline{Y}) - (\mu_X - \mu_Y)}{\sigma\sqrt{\frac{1}{n} + \frac{1}{m}}}$$

which follows a standard normal distribution. The confidence interval would be of the form

$$(\overline{X} - \overline{Y}) \pm z(\alpha/2)\sigma\sqrt{\frac{1}{n} + \frac{1}{m}}$$

This confidence interval is of the same form as those introduced in Chapters 7 and 8—a statistic ($\overline{X} - \overline{Y}$ in this case) plus or minus a multiple of its standard deviation.

Generally, σ^2 will not be known and must be estimated from the data by calculating the **pooled sample variance,**

$$s_p^2 = \frac{(n-1)s_X^2 + (m-1)s_Y^2}{m+n-2}$$

where $s_X^2 = (n-1)^{-1}\sum_{i=1}^{n}(X_i - \overline{X})^2$ and similarly for s_Y^2. Note that s_p^2 is a weighted average of the sample variances of the X's and Y's, with the weights proportional to the degrees of freedom. This weighting is appropriate since if one sample is much larger than the other, the estimate of σ^2 from that sample is more reliable and should receive greater weight. The following theorem gives the distribution of a statistic that will be used for forming confidence intervals and performing hypothesis tests.

THEOREM A

Suppose that $X_1, \ldots, X_n$ are independent and normally distributed random variables with mean μ_X and variance σ^2, and that $Y_1, \ldots, Y_m$ are independent and normally distributed random variables with mean μ_Y and variance σ^2, and that the Y_i are independent of the X_i. The statistic

$$t = \frac{(\overline{X} - \overline{Y}) - (\mu_X - \mu_Y)}{s_p\sqrt{\dfrac{1}{n} + \dfrac{1}{m}}}$$

follows a t distribution with $m+n-2$ degrees of freedom.

Proof

According to the definition of the t distribution in Section 6.2, we have to show that the statistic is the quotient of a standard normal random variable and the square root of an independent chi-square random variable divided by its $n+m-2$ degrees of freedom. First, we note from Theorem B in Section 6.3 that $(n-1)s_X^2/\sigma^2$ and $(m-1)s_Y^2/\sigma^2$ are distributed as chi-square random variables with $n-1$ and $m-1$ degrees of freedom, respectively, and are independent since the X_i and Y_i are. Their sum is thus chi-square with $m+n-2$ df. Now, we express the statistic as the ratio U/V, where

$$U = \frac{(\overline{X} - \overline{Y}) - (\mu_X - \mu_Y)}{\sigma\sqrt{\dfrac{1}{n} + \dfrac{1}{m}}}$$

$$V = \sqrt{\left[\frac{(n-1)s_X^2}{\sigma^2} + \frac{(m-1)s_Y^2}{\sigma^2}\right]\frac{1}{m+n-2}}$$

U follows a standard normal distribution and from the preceding argument V has the distribution of the square root of a chi-square random variable divided by its degrees of freedom. The independence of U and V follows from Corollary A in Section 6.3. ∎

It is convenient and suggestive to use the following notation for the estimated standard deviation (or standard error) of $\overline{X} - \overline{Y}$:

$$s_{\overline{X}-\overline{Y}} = s_p\sqrt{\frac{1}{n} + \frac{1}{m}}$$

A confidence interval for $\mu_X - \mu_Y$ follows as a corollary to Theorem A.

COROLLARY A

Under the assumptions of Theorem A, a $100(1 - \alpha)\%$ confidence interval for $\mu_X - \mu_Y$ is

$$(\overline{X} - \overline{Y}) \pm t_{m+n-2}(\alpha/2)s_{\overline{X}-\overline{Y}} \qquad \blacksquare$$

E X A M P L E   A   Two methods, A and B, were used in a determination of the latent heat of fusion of ice (Natrella 1963). The investigators wanted to find out by how much the methods differed. The following table gives the change in total heat from ice at $-.72°C$ to water $0°C$ in calories per gram of mass:

Method A	Method B
79.98	80.02
80.04	79.94
80.02	79.98
80.04	79.97
80.03	79.97
80.03	80.03
80.04	79.95
79.97	79.97
80.05	
80.03	
80.02	
80.00	
80.02	

It is fairly obvious from the table and from boxplots (Figure 11.1) that there is a difference between the two methods (we will test this more formally later). If we assume the conditions of Theorem A, we can form a 95% confidence interval to estimate the magnitude of the average difference between the two methods. From the table, we calculate

$$\overline{X}_A = 80.02 \qquad S_a = .024$$
$$\overline{X}_B = 79.98 \qquad S_b = .031$$
$$s_p^2 = \frac{12 \times S_a^2 + 7 \times S_b^2}{19} = .0007178$$
$$s_p = .027$$

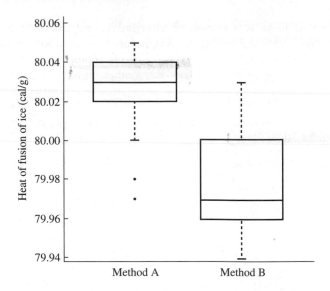

Our estimate of the average difference of the two methods is $\overline{X}_A - \overline{X}_B = .04$ and its estimated standard error is

$$s_{\overline{X}_A - \overline{X}_B} = s_p \sqrt{\frac{1}{13} + \frac{1}{8}}$$
$$= .012$$

From Table 4 of Appendix B, the .975 quantile of the t distribution with 19 df is 2.093, so $t_{19}(.025) = 2.093$ and the 95% confidence interval is $(\overline{X}_A - \overline{X}_B) \pm t_{19}(.025)s_{\overline{X}_A - \overline{X}_B}$, or (.015, .065). The estimated standard error and the confidence interval quantify the uncertainty in the point estimate $\overline{X}_A - \overline{X}_B = .04$. ∎

We will now discuss hypothesis testing for the two-sample problem. Although the hypotheses under consideration are different from those of Chapter 9, the general conceptual framework is the same (you should review that framework at this time). In the current case, the null hypothesis to be tested is

$$H_0: \mu_X = \mu_Y$$

This asserts that there is no difference between the distributions of the X's and Y's. If one group is a treatment group and the other a control, for example, this hypothesis asserts that there is no treatment effect. In order to conclude that there is a treatment effect, the null hypothesis must be rejected.

There are three common alternative hypotheses for the two-sample case:

$$H_1: \mu_X \neq \mu_Y$$
$$H_2: \mu_X > \mu_Y$$
$$H_3: \mu_X < \mu_Y$$

The first of these is a **two-sided alternative,** and the other two are **one-sided alternatives.** The first hypothesis is appropriate if deviations could in principle go in either direction, and one of the latter two is appropriate if it is believed that any deviation must be in one direction or the other. In practice, such a priori information is not usually available, and it is more prudent to conduct two-sided tests, as in Example A.

The test statistic that will be used to make a decision whether or not to reject the null hypothesis is

$$t = \frac{\overline{X} - \overline{Y}}{s_{\overline{X} - \overline{Y}}}$$

The t-statistic equals the multiple of its estimated standard deviation that $\overline{X} - \overline{Y}$ differs from zero. It plays the same role in the comparison of two samples as is played by the chi-square statistic in testing goodness of fit. Just as we rejected for large values of the chi-square statistic, we will reject in this case for extreme values of t. The distribution of t under H_0, its null distribution, is, from Theorem A, the t distribution with $m + n - 2$ degrees of freedom. Knowing this null distribution allows us to determine a rejection region for a test at level α, just as knowing that the null distribution of the chi-square statistic was chi-square with the appropriate degrees of freedom allowed the determination of a rejection region for testing goodness of fit. The rejection regions for the three alternatives just listed are

$$\text{For } H_1, |t| > t_{n+m-2}(\alpha/2)$$
$$\text{For } H_2, t > t_{n+m-2}(\alpha)$$
$$\text{For } H_3, t < -t_{n+m-2}(\alpha)$$

Note how the rejection regions are tailored to the particular alternatives and how knowing the null distribution of t allows us to determine the rejection region for any value of α.

E X A M P L E **B**   Let us continue Example A. To test $H_0: \mu_A = \mu_B$ versus a two-sided alternative, we form and calculate the following test statistic:

$$t = \frac{\overline{X}_A - \overline{X}_B}{s_p \sqrt{\frac{1}{n} + \frac{1}{m}}}$$
$$= 3.33$$

From Table 4 in Appendix B, $t_{19}(.005) = 2.861 < 3.33$. The two-sided test would thus reject at the level $\alpha = .01$. If there were no difference in the two conditions, differences as large or larger than that observed would occur only with probability less than .01—that is, the p-value is less than .01. There is little doubt that there is a difference between the two methods. ■

In Chapter 9, we developed a general duality between hypothesis tests and confidence intervals. In the case of the testing and confidence interval methods considered in this section, the t test rejects if and only if the confidence interval does not include zero (see Problem 10 at the end of this chapter).

We will now demonstrate that the test of H_0 versus H_1 is equivalent to a likelihood ratio test. (The rather long argument is sketched here and should be read with paper and pencil in hand.) Ω is the set of all possible parameter values:

$$\Omega = \{-\infty < \mu_X < \infty, -\infty < \mu_Y < \infty, 0 < \sigma < \infty\}$$

The unknown parameters are $\theta = (\mu_X, \mu_Y, \sigma)$. Under H_0, $\theta \in \omega_0$, where $\omega_0 = \{\mu_X = \mu_Y, 0 < \sigma < \infty\}$. The likelihood of the two samples $X_1, \ldots, X_n$ and $Y_1, \ldots, Y_m$ is

$$\text{lik}\left(\mu_X, \mu_Y, \sigma^2\right) = \prod_{i=1}^{n} \frac{1}{\sqrt{2\pi\sigma^2}} e^{-(1/2)[(X_i - \mu_X)^2/\sigma^2]} \prod_{j=1}^{m} \frac{1}{\sqrt{2\pi\sigma^2}} e^{-(1/2)[(Y_j - \mu_Y)^2/\sigma^2]}$$

and the log likelihood is

$$l\left(\mu_X, \mu_Y, \sigma^2\right) = -\frac{(m+n)}{2} \log 2\pi - \frac{(m+n)}{2} \log \sigma^2$$
$$-\frac{1}{2\sigma^2}\left[\sum_{i=1}^{n}(X_i - \mu_X)^2 + \sum_{j=1}^{m}(Y_j - \mu_Y)^2\right]$$

We must maximize the likelihood under ω_0 and under Ω and then calculate the ratio of the two maximized likelihoods, or the difference of their logarithms.

Under ω_0, we have a sample of size $m + n$ from a normal distribution with unknown mean μ_0 and unknown variance σ_0^2. The mle's of μ_0 and σ_0^2 are thus

$$\hat{\mu}_0 = \frac{1}{m+n}\left(\sum_{i=1}^{n} X_i + \sum_{j=1}^{m} Y_j\right)$$

$$\hat{\sigma}_0^2 = \frac{1}{m+n}\left[\sum_{i=1}^{n}(X_i - \hat{\mu}_0)^2 + \sum_{j=1}^{m}(Y_j - \hat{\mu}_0)^2\right]$$

The corresponding value of the maximized log likelihood is, after some cancellation,

$$l\left(\hat{\mu}_0, \hat{\sigma}_0^2\right) = -\frac{m+n}{2} \log 2\pi - \frac{m+n}{2} \log \hat{\sigma}_0^2 - \frac{m+n}{2}$$

To find the mle's $\hat{\mu}_X$, $\hat{\mu}_Y$, and $\hat{\sigma}_1^2$ under Ω, we first differentiate the log likelihood and obtain the equations

$$\sum_{i=1}^{n}(X_i - \hat{\mu}_X) = 0$$

$$\sum_{j=1}^{m}(Y_j - \hat{\mu}_Y) = 0$$

$$-\frac{m+n}{2\hat{\sigma}_1^2} + \frac{1}{2\hat{\sigma}_1^4}\left[\sum_{i=1}^{n}(X_i - \hat{\mu}_X)^2 + \sum_{j=1}^{m}(Y_j - \hat{\mu}_Y)^2\right] = 0$$

The mle's are, therefore,

$$\hat{\mu}_X = \overline{X}$$
$$\hat{\mu}_Y = \overline{Y}$$
$$\hat{\sigma}_1^2 = \frac{1}{m+n} \left[\sum_{i=1}^{n} (X_i - \hat{\mu}_X)^2 + \sum_{j=1}^{m} (Y_j - \hat{\mu}_Y)^2 \right]$$

When these are substituted into the log likelihood, we obtain

$$l\left(\hat{\mu}_X, \hat{\mu}_Y, \hat{\sigma}_1^2\right) = -\frac{m+n}{2} \log 2\pi - \frac{m+n}{2} \log \hat{\sigma}_1^2 - \frac{m+n}{2}$$

The log of the likelihood ratio is thus

$$\frac{m+n}{2} \log \left(\frac{\hat{\sigma}_1^2}{\hat{\sigma}_0^2} \right)$$

and the likelihood ratio test rejects for large values of

$$\frac{\hat{\sigma}_0^2}{\hat{\sigma}_1^2} = \frac{\sum_{i=1}^{n} (X_i - \hat{\mu}_0)^2 + \sum_{j=1}^{m} (Y_j - \hat{\mu}_0)^2}{\sum_{i=1}^{n} (X_i - \overline{X})^2 + \sum_{j=1}^{m} (Y_j - \overline{Y})^2}$$

We now find an alternative expression for the numerator of this ratio, by using the identities

$$\sum_{i=1}^{n} (X_i - \hat{\mu}_0)^2 = \sum_{i=1}^{n} (X_i - \overline{X})^2 + n(\overline{X} - \hat{\mu}_0)^2$$

$$\sum_{j=1}^{m} (Y_j - \hat{\mu}_0)^2 = \sum_{j=1}^{m} (Y_j - \overline{Y})^2 + m(\overline{Y} - \hat{\mu}_0)^2$$

We obtain

$$\hat{\mu}_0 = \frac{1}{m+n} (n\overline{X} + m\overline{Y})$$
$$= \frac{n}{m+n} \overline{X} + \frac{m}{m+n} \overline{Y}$$

Therefore,

$$\overline{X} - \hat{\mu}_0 = \frac{m(\overline{X} - \overline{Y})}{m+n}$$

$$\overline{Y} - \hat{\mu}_0 = \frac{n(\overline{Y} - \overline{X})}{m+n}$$

The alternative expression for the numerator of the ratio is thus

$$\sum_{i=1}^{n} (X_i - \overline{X})^2 + \sum_{j=1}^{m} (Y_j - \overline{Y})^2 + \frac{mn}{m+n} (\overline{X} - \overline{Y})^2$$

and the test rejects for large values of

$$1 + \frac{mn}{m+n} \left(\frac{(\overline{X} - \overline{Y})^2}{\sum\limits_{i=1}^{n}(X_i - \overline{X})^2 + \sum\limits_{j=1}^{m}(Y_j - \overline{Y})^2} \right)$$

or, equivalently, for large values of

$$\frac{|\overline{X} - \overline{Y}|}{\sqrt{\sum\limits_{i=1}^{n}(X_i - \overline{X})^2 + \sum\limits_{j=1}^{m}(Y_j - \overline{Y})^2}}$$

which is the t statistic apart from constants that do not depend on the data. Thus, the likelihood ratio test is equivalent to the t test, as claimed.

We have used the assumption that the two populations have the same variance. If the two variances are not assumed to be equal, a natural estimate of $\mathrm{Var}(\overline{X} - \overline{Y})$ is

$$\frac{s_X^2}{n} + \frac{s_Y^2}{m}$$

If this estimate is used in the denominator of the t statistic, the distribution of that statistic is no longer the t distribution. But it has been shown that its distribution can be closely approximated by the t distribution with degrees of freedom calculated in the following way and then rounded to the nearest integer:

$$\mathrm{df} = \frac{[(s_X^2/n) + (s_Y^2/m)]^2}{\dfrac{(s_X^2/n)^2}{n-1} + \dfrac{(s_Y^2/m)^2}{m-1}}$$

E X A M P L E **C**   Let us rework Example B, but without the assumption that the variances are equal. Using the preceding formula, we find the degrees of freedom to be 12 rather than 19. The t statistic is 3.12. Since the .995 quantile of the t distribution with 12 df is 3.055 (Table 4 of Appendix B), the test still rejects at level $\alpha = .01$.   ∎

If the underlying distributions are not normal and the sample sizes are large, the use of the t distribution or the normal distribution is justified by the central limit theorem, and the probability levels of confidence intervals and hypothesis tests are approximately valid. In such a case, however, there is little difference between the t and normal distributions. If the sample sizes are small, however, and the distributions are not normal, conclusions based on the assumption of normality may not be valid. Unfortunately, if the sample sizes are small, the assumption of normality cannot be tested effectively unless the deviation is quite gross, as we saw in Chapter 9.

11.2.1.1 An Example—A Study of Iron Retention   An experiment was performed to determine whether two forms of iron (Fe^{2+} and Fe^{3+}) are retained differently. (If one form of iron were retained especially well, it would be the better dietary supplement.) The investigators divided 108 mice randomly into 6 groups of 18 each; 3 groups were given Fe^{2+} in three different concentrations, 10.2, 1.2, and

.3 millimolar, and 3 groups were given Fe^{3+} at the same three concentrations. The mice were given the iron orally; the iron was radioactively labeled so that a counter could be used to measure the initial amount given. At a later time, another count was taken for each mouse, and the percentage of iron retained was calculated. The data for the two forms of iron are listed in the following table. We will look at the data for the concentration 1.2 millimolar. (In Chapter 12, we will discuss methods for analyzing all the groups simultaneously.)

Fe^{3+}			Fe^{2+}		
10.2	1.2	.3	10.2	1.2	.3
.71	2.20	2.25	2.20	4.04	2.71
1.66	2.93	3.93	2.69	4.16	5.43
2.01	3.08	5.08	3.54	4.42	6.38
2.16	3.49	5.82	3.75	4.93	6.38
2.42	4.11	5.84	3.83	5.49	8.32
2.42	4.95	6.89	4.08	5.77	9.04
2.56	5.16	8.50	4.27	5.86	9.56
2.60	5.54	8.56	4.53	6.28	10.01
3.31	5.68	9.44	5.32	6.97	10.08
3.64	6.25	10.52	6.18	7.06	10.62
3.74	7.25	13.46	6.22	7.78	13.80
3.74	7.90	13.57	6.33	9.23	15.99
4.39	8.85	14.76	6.97	9.34	17.90
4.50	11.96	16.41	6.97	9.91	18.25
5.07	15.54	16.96	7.52	13.46	19.32
5.26	15.89	17.56	8.36	18.4	19.87
8.15	18.3	22.82	11.65	23.89	21.60
8.24	18.59	29.13	12.45	26.39	22.25

As a summary of the data, boxplots (Figure 11.2) show that the data are quite skewed to the right. This is not uncommon with percentages or other variables that are bounded below by zero. Three observations from the Fe^{2+} group are flagged as possible outliers. The median of the Fe^{2+} group is slightly larger than the median of the Fe^{3+} groups, but the two distributions overlap substantially.

Another view of these data is provided by normal probability plots (Figure 11.3). These plots also indicate the skewness of the distributions. We should obviously doubt the validity of using normal distribution theory (for example, the t test) for this problem even though the combined sample size is fairly large (36).

The mean and standard deviation of the Fe^{2+} group are 9.63 and 6.69; for the Fe^{3+} group, the mean is 8.20 and the standard deviation is 5.45. To test the hypothesis that the two means are equal, we can use a t test without assuming that the population standard deviations are equal. The approximate degrees of freedom, calculated as described at the end of Section 11.2.1, are 32. The t statistic is .702, which corresponds to a p-value of .49 for a two-sided test; if the two populations had the same mean, values of the t statistic this large or larger would occur 49% of the time. There is thus insufficient evidence to reject the null hypothesis. A 95% confidence interval for the

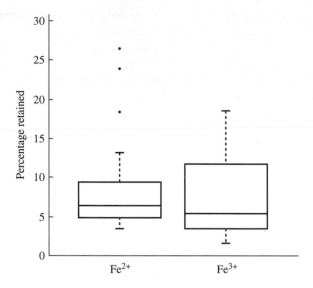

FIGURE **11.2** Boxplots of the percentages of iron retained for the two forms.

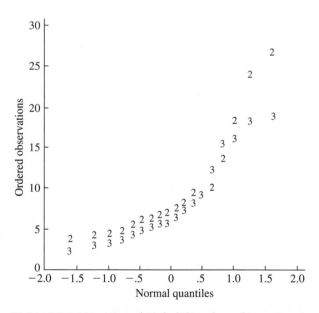

FIGURE **11.3** Normal probability plots of iron retention data.

difference of the two population means is $(-2.7, 5.6)$. But the t test assumes that the underlying populations are normally distributed, and we have seen there is reason to doubt this assumption.

It is sometimes advocated that skewed data be transformed to a more symmetric shape before normal theory is applied. Transformations such as taking the log or the square root can be effective in symmetrizing skewed distributions because they spread out small values and compress large ones. Figures 11.4 and 11.5 show boxplots and normal probability plots for the natural logs of the iron retention data we have been considering. The transformation was fairly successful in symmetrizing these distributions, and the probability plots are more linear than those in Figure 11.3, although some curvature is still evident.

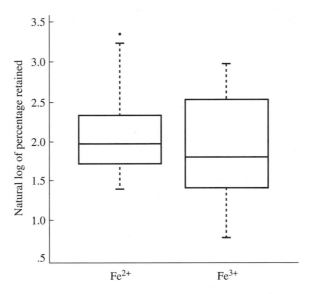

F I G U R E **11.4**   Boxplots of natural logs of percentages of iron retained.

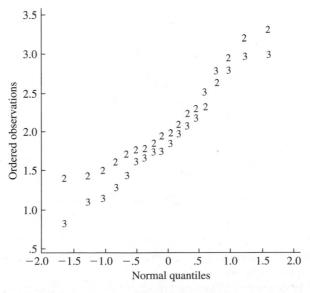

F I G U R E **11.5**   Normal probability plots of natural logs of iron retention data.

The following model is natural for the log transformation:

$$X_i = \mu_X(1 + \varepsilon_i), \qquad i = 1, \ldots, n$$
$$Y_j = \mu_Y(1 + \delta_j), \qquad j = 1, \ldots, m$$
$$\log X_i = \log \mu_X + \log(1 + \varepsilon_i)$$
$$\log Y_j = \log \mu_Y + \log(1 + \delta_j)$$

Here the ε_i and δ_j are independent random variables with mean zero. This model implies that if the variances of the errors are σ^2, then

$$E(X_i) = \mu_X$$
$$E(Y_j) = \mu_Y$$
$$\sigma_X = \mu_X \sigma$$
$$\sigma_Y = \mu_Y \sigma$$

or that

$$\frac{\sigma_X}{\mu_X} = \frac{\sigma_Y}{\mu_Y}$$

If the ε_i and δ_j have the same distribution, $\text{Var}(\log X) = \text{Var}(\log Y)$. The ratio of the standard deviation of a distribution to the mean is called the **coefficient of variation (CV)**; it expresses the standard deviation as a fraction of the mean. Coefficients of variation are sometimes expressed as percentages. For the iron retention data we have been considering, the CV's are .69 and .67 for the Fe^{2+} and Fe^{3+} groups; these values are quite close. These data are quite "noisy"—the standard deviation is nearly 70% of the mean for both groups.

For the transformed iron retention data, the means and standard deviations are given in the following table:

	Fe^{2+}	Fe^{3+}
Mean	2.09	1.90
Standard Deviation	.659	.574

For the transformed data, the t statistic is .917, which gives a p-value of .37. Again, there is no reason to reject the null hypothesis. A 95% confidence interval is $(-.61, .23)$. Using the preceding model, this is a confidence interval for

$$\log \mu_X - \log \mu_Y = \log\left(\frac{\mu_X}{\mu_Y}\right)$$

The interval is

$$-.61 \leq \log\left(\frac{\mu_X}{\mu_Y}\right) \leq .23$$

or

$$.54 \leq \frac{\mu_X}{\mu_Y} \leq 1.26$$

Other transformations, such as raising all values to some power, are sometimes used. Attitudes toward the use of transformations vary: Some view them as a very useful tool in statistics and data analysis, and others regard them as questionable manipulation of the data.

11.2.2 Power

Calculations of power are an important part of planning experiments in order to determine how large sample sizes should be. The power of a test is the probability of rejecting the null hypothesis when it is false. The power of the two-sample t test depends on four factors:

1. The real difference, $\Delta = |\mu_X - \mu_Y|$. The larger this difference, the greater the power.
2. The significance level α at which the test is done. The larger the significance level, the more powerful the test.
3. The population standard deviation σ, which is the amplitude of the "noise" that hides the "signal." The smaller the standard deviation, the larger the power.
4. The sample sizes n and m. The larger the sample sizes, the greater the power.

Before continuing, you should try to understand intuitively why these statements are true. We will express them quantitatively below.

The necessary sample sizes can be determined from the significance level of the test, the standard deviation, and the desired power against an alternative hypothesis,

$$H_1: \mu_X - \mu_Y = \Delta$$

To calculate the power of a t test exactly, special tables of the noncentral t distribution are required. But if the sample sizes are reasonably large, one can perform approximate power calculations based on the normal distribution, as we will now demonstrate.

Suppose that σ, α, and Δ are given and that the samples are both of size n. Then

$$\mathrm{Var}(\overline{X} - \overline{Y}) = \sigma^2 \left(\frac{1}{n} + \frac{1}{n} \right)$$

$$= \frac{2\sigma^2}{n}$$

The test at level α of $H_0: \mu_X = \mu_Y$ against the alternative $H_1: \mu_X \neq \mu_Y$ is based on the test statistic

$$Z = \frac{\overline{X} - \overline{Y}}{\sigma \sqrt{2/n}}$$

The rejection region for this test is $|Z| > z(\alpha/2)$, or

$$|\overline{X} - \overline{Y}| > z(\alpha/2)\sigma \sqrt{\frac{2}{n}}$$

The power of the test if $\mu_X - \mu_Y = \Delta$ is the probability that the test statistic falls in the rejection region, or

$$P\left[|\overline{X} - \overline{Y}| > z(\alpha/2)\sigma\sqrt{\frac{2}{n}}\right]$$

$$= P\left[\overline{X} - \overline{Y} > z(\alpha/2)\sigma\sqrt{\frac{2}{n}}\right] + P\left[\overline{X} - \overline{Y} < -z(\alpha/2)\sigma\sqrt{\frac{2}{n}}\right]$$

since the two events are mutually exclusive. Both probabilities on the right-hand side are calculated by standardizing. For the first one, we have

$$P\left[\overline{X} - \overline{Y} > z(\alpha/2)\sigma\sqrt{\frac{2}{n}}\right] = P\left[\frac{(\overline{X} - \overline{Y}) - \Delta}{\sigma\sqrt{2/n}} > \frac{z(\alpha/2)\sigma\sqrt{2/n} - \Delta}{\sigma\sqrt{2/n}}\right]$$

$$= 1 - \Phi\left[z(\alpha/2) - \frac{\Delta}{\sigma}\sqrt{\frac{n}{2}}\right]$$

where Φ is the standard normal cdf. Similarly, the second probability is

$$\Phi\left[-z(\alpha/2) - \frac{\Delta}{\sigma}\sqrt{\frac{n}{2}}\right]$$

Thus, the probability that the test statistic falls in the rejection region is equal to

$$1 - \Phi\left[z(\alpha/2) - \frac{\Delta}{\sigma}\sqrt{\frac{n}{2}}\right] + \Phi\left[-z(\alpha/2) - \frac{\Delta}{\sigma}\sqrt{\frac{n}{2}}\right]$$

Typically, as Δ moves away from zero, one of these terms will be negligible with respect to the other. For example, if Δ is greater than zero, the first term will be dominant. For fixed n, this expression can be evaluated as a function of Δ; or for fixed Δ, it can be evaluated as a function of n.

E X A M P L E   A   As an example, let us consider a situation similar to an idealized form of the iron retention experiment. Assume that we have samples of size 18 from two normal distributions whose standard deviations are both 5, and we calculate the power for various values of Δ when the null hypothesis is tested at a significance level of .05. The results of the calculations are displayed in Figure 11.6. We see from the plot that if the mean difference in retention is only 1%, the probability of rejecting the null hypothesis is quite small, only 9%. A mean difference of 5% in retention rate gives a more satisfactory power of 85%.

Suppose that we wanted to be able to detect a difference of $\Delta = 1$ with probability .9. What sample size would be necessary? Using only the dominant term in the expression for the power, the sample size should be such that

$$\Phi\left(1.96 - \frac{\Delta}{\sigma}\sqrt{\frac{n}{2}}\right) = .1$$

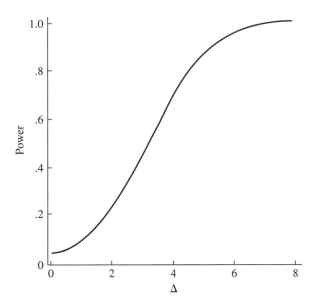

FIGURE **11.6** Plot of power versus Δ.

From the tables for the normal distribution, $.1 = \Phi(-1.28)$, so

$$1.96 - \frac{\Delta}{\sigma}\sqrt{\frac{n}{2}} = -1.28$$

Solving for n, we find that the necessary sample size would be 525! This is clearly unfeasible; if in fact the experimenters wanted to detect such a difference, some modification of the experimental technique to reduce σ would be necessary. ∎

11.2.3 A Nonparametric Method—The Mann-Whitney Test

Nonparametric methods do not assume that the data follow any particular distributional form. Many of them are based on replacement of the data by ranks. With this replacement, the results are invariant under any monotonic transformation; in comparison, we saw that the p-value of a t test may change if the log of the measurements is analyzed rather than the measurements on the original scale. Replacing the data by ranks also has the effect of moderating the influence of outliers.

For purposes of discussion, we will develop the **Mann-Whitney test** (also sometimes called the Wilcoxon rank sum test) in a specific context. Suppose that we have $m + n$ experimental units to assign to a treatment group and a control group. The assignment is made at random: n units are randomly chosen and assigned to the control, and the remaining m units are assigned to the treatment. We are interested in testing the null hypothesis that the treatment has no effect. If the null hypothesis is true, then any difference in the outcomes under the two conditions is due to the randomization.

procedure

A test statistic is calculated in the following way. First, we group all $m + n$ observations together and rank them in order of increasing size (we will assume for simplicity that there are no ties, although the argument holds even in the presence of ties). We next calculate the sum of the ranks of those observations that came from the control group. If this sum is too small or too large, we will reject the null hypothesis.

It is easiest to see how the procedure works by considering a very small example. Suppose that a treatment and a control are to be compared: Of four subjects, two are randomly assigned to the treatment and the other two to the control, and the following responses are observed (the ranks of the observations are shown in parentheses):

Treatment	Control
1 (1)	6 (4)
3 (2)	4 (3)

The sum of the ranks of the control group is $R = 7$, and the sum of the ranks of the treatment group is 3. Does this discrepancy provide convincing evidence of a systematic difference between treatment and control, or could it be just due to chance? To answer this question, we calculate the probability of such a discrepancy if the treatment had no effect at all, so that the difference was entirely due to the particular randomization—this is the null hypothesis. The key idea of the Mann-Whitney test is that we can explicitly calculate the distribution of R under the null hypothesis, since under this hypothesis every assignment of ranks to observations is equally likely and we can enumerate all $4! = 24$ such assignments. In particular, each of the $\binom{4}{2} = 6$ assignments of ranks to the control group shown in the following table is equally likely:

Ranks	R
{1, 2}	3
{1, 3}	4
{1, 4}	5
{2, 3}	5
{2, 4}	6
{3, 4}	7

From this table, we see that under the null hypothesis, the distribution of R (its null distribution) is:

r	3	4	5	6	7
$P(R = r)$	$\frac{1}{6}$	$\frac{1}{6}$	$\frac{1}{3}$	$\frac{1}{6}$	$\frac{1}{6}$

In particular, $P(R = 7) = \frac{1}{6}$, so this discrepancy would occur one time out of six purely on the basis of chance.

The small example of the previous paragraph has been laid out for pedagogical reasons, the point being that we could in principle go through similar calculations for any sample sizes m and n. Suppose that there are n observations in the treatment group and m in the control group. If the null hypothesis holds, every assignment of ranks to the $m + n$ observations is equally likely, and hence each of the $\binom{m+n}{m}$ possible assignments of ranks to the control group is equally likely. For each of these assignments, we can calculate the sum of the ranks and thus determine the null distribution of the test statistic—the sum of the ranks of the control group.

It is important to note that we have not made any assumption that the observations from the control and treatment groups are samples from a probability distribution. Probability has entered in only as a result of the random assignment of experimental units to treatment and control groups (this is similar to the way that probability enters into survey sampling). We should also note that, although we chose the sum of control ranks as the test statistic, any other test statistic could have been used and its null distribution computed in the same fashion. The rank sum is easy to compute and is sensitive to a treatment effect that tends to make responses larger or smaller. Also, its null distribution has to be computed only once and tabled; if we worked with the actual numerical values, the null distribution would depend on those particular values.

Tables of the null distribution of the rank sum are widely available and vary in format. Note that because the sum of the two rank sums is the sum of the integers from 1 to $m + n$, which is $[(m + n)(m + n + 1)/2]$, knowing one rank sum tells us the other. Some tables are given in terms of the rank sum of the smaller of the two groups, and some are in terms of the smaller of the two rank sums (the advantage of the latter scheme is that only one tail of the distribution has to be tabled). Table 8 of Appendix B makes use of additional symmetries. Let n_1 be the smaller sample size and let R be the sum of the ranks from that sample. Let $R' = n_1(m + n + 1) - R$ and $R^* = \min(R, R')$. The table gives critical values for R^*. (Fortunately, such fussy tables are largely obsolete with the increasing use of computers.)

When it is more appropriate to model the control values, $X_1, \ldots, X_n$, as a sample from some probability distribution F and the experimental values, $Y_1, \ldots, Y_m$, as a sample from some distribution G, the Mann-Whitney test is a test of the null hypothesis $H_0: F = G$. The reasoning is exactly the same: Under H_0, any assignment of ranks to the pooled $m + n$ observations is equally likely, etc.

We have assumed here that there are no ties among the observations. If there are only a small number of ties, tied observations are assigned average ranks (the average of the ranks for which they are tied); the significance levels are not greatly affected.

E X A M P L E **A**   Let us illustrate the Mann-Whitney test by referring to the data on latent heats of fusion of ice considered earlier (Example A in Section 11.2.1). The sample sizes are fairly small (13 and 8), so in the absence of any prior knowledge concerning the adequacy of the assumption of a normal distribution, it would seem safer to use a nonparametric method. The following table exhibits the ranks given to the measurements for each method (refer to Example A in Section 11.2.1 for the original data):

Method A	Method B
7.5	11.5
19.0	1.0
11.5	7.5
19.0	4.5
15.5	4.5
15.5	15.5
19.0	2.0
4.5	4.5
21.0	
15.5	
11.5	
9.0	
11.5	

Note how the ties were handled. For example, the four observations with the value 79.97 tied for ranks 3, 4, 5, and 6 were each assigned the rank of $4.5 = (3 + 4 + 5 + 6)/4$.

Table 8 of Appendix B is used as follows. The sum of the ranks of the smaller sample is $R = 51$.

$$R' = 8(8 + 13 + 1) - R$$
$$= 125$$

Thus, $R^* = 51$. From the table, 53 is the critical value for a two-tailed test with $\alpha = .01$, and 60 is the critical value for $\alpha = .05$. The Mann-Whitney test thus rejects at the .01 significance level. ∎

Let T_Y denote the sum of the ranks of $Y_1, Y_2, \ldots, Y_m$. Using results from Chapter 7, we can easily find $E(T_Y)$ and $\mathrm{Var}(T_Y)$ under the null hypothesis $F = G$.

THEOREM A

If $F = G$,

$$E(T_Y) = \frac{m(m + n + 1)}{2}$$

$$\mathrm{Var}(T_Y) = \frac{mn(m + n + 1)}{12}$$

Proof

Under the null hypothesis, T_Y is the sum of a random sample of size m drawn without replacement from a population consisting of the integers $\{1, 2, \dots, m + n\}$. T_Y thus equals m times the average of such a sample. From Theorems A and B of Section 7.3.1,

$$E(T_Y) = m\mu$$

$$\text{Var}(T_Y) = m\sigma^2 \left(\frac{N - m}{N - 1} \right)$$

where $N = m + n$ is the size of the population, and μ and σ^2 are the population mean and variance. Now, using the identities

$$\sum_{k=1}^{N} k = \frac{N(N + 1)}{2} \quad > \frac{(m+n)(m+n+1)}{2}$$

$$\sum_{k=1}^{N} k^2 = \frac{N(N + 1)(2N + 1)}{6}$$

we find that for the population $\{1, 2, \dots, m + n\}$

$$\mu = \frac{N + 1}{2}$$

$$\sigma^2 = \frac{N^2 - 1}{12}$$

The result then follows after algebraic simplification. ∎

Unlike the t test, the Mann-Whitney test does not depend on an assumption of normality. Since the actual numerical values are replaced by their ranks, the test is insensitive to outliers, whereas the t test is sensitive. It has been shown that even when the assumption of normality holds, the Mann-Whitney test is nearly as powerful as the t test and it is thus generally preferable, especially for small sample sizes.

The Mann-Whitney test can also be derived starting from a different point of view. Suppose that the X's are a sample from F and the Y's a sample from G, and consider estimating, as a measure of the effect of the treatment,

$$\pi = P(X < Y)$$

where X and Y are independently distributed with distribution functions F and G, respectively. The value π is the probability that an observation from the distribution F is smaller than an independent observation from the distribution G.

If, for example, F and G represent lifetimes of components that have been manufactured according to two different conditions, π is the probability that a component of one type will last longer than a component of the other type. An estimate of π can be obtained by comparing all n values of X to all m values of Y and calculating the proportion of the comparisons for which X was less than Y:

$$\hat{\pi} = \frac{1}{mn} \sum_{i=1}^{n} \sum_{j=1}^{m} Z_{ij}$$

where

$$Z_{ij} = \begin{cases} 1, & \text{if } X_i < Y_j \\ 0, & \text{otherwise} \end{cases}$$

To see the relationship of $\hat{\pi}$ to the rank sum introduced earlier, we will find it convenient to work with

$$V_{ij} = \begin{cases} 1, & \text{if } X_{(i)} < Y_{(j)} \\ 0, & \text{otherwise} \end{cases}$$

ranked

Clearly,

$$\sum_{i=1}^{n} \sum_{j=1}^{m} Z_{ij} = \sum_{i=1}^{n} \sum_{j=1}^{m} V_{ij}$$

since the V_{ij} are just a reordering of the Z_{ij}. Also,

$$\sum_{i=1}^{n} \sum_{j=1}^{m} V_{ij} = (\text{number of } X\text{'s that are less than } Y_{(1)})$$
$$+ (\text{number of } X\text{'s that are less than } Y_{(2)})$$
$$+ \cdots + (\text{number of } X\text{'s that are less than } Y_{(m)})$$

If the rank of $Y_{(k)}$ in the combined sample is denoted by R_{yk}, then the number of X's less than $Y_{(1)}$ is $R_{y1} - 1$, the number of X's less than $Y_{(2)}$ is $R_{y2} - 2$, etc. Therefore,

$$\sum_{i=1}^{n} \sum_{j=1}^{m} V_{ij} = (R_{y1} - 1) + (R_{y2} - 2) + \cdots + (R_{ym} - m)$$

$$= \sum_{i=1}^{m} R_{yi} - \sum_{i=1}^{m} i$$

$$= \sum_{i=1}^{m} R_{yi} - \frac{m(m+1)}{2}$$

$$= T_y - \frac{m(m+1)}{2}$$

Thus, $\hat{\pi}$ may be expressed in terms of the rank sum of the Y's (or in terms of the rank sum of the X's, since the two rank sums add up to a constant).

From Theorem A, we have

COROLLARY A

Let $U_Y = \sum_{i=1}^{n} \sum_{j=1}^{m} Z_{ij}$. Under the null hypothesis $H_0: F = G$,

$$E(U_Y) = \frac{mn}{2}$$

$$\mathrm{Var}(U_Y) = \frac{mn(m+n+1)}{12}$$ ∎

For m and n both greater than 10, the null distribution of U_Y is quite well approximated by a normal distribution,

$$\frac{U_Y - E(U_Y)}{\sqrt{\mathrm{Var}(U_Y)}} \sim N(0, 1)$$

(Note that this does not follow immediately from the ordinary central limit theorem; although U_Y is a sum of random variables, they are not independent.) Similarly, the distribution of the rank sum of the X's or Y's may be approximated by a normal distribution, since these rank sums differ from U_Y only by constants.

E X A M P L E **B**   Referring to Example A, let us use a normal approximation to the distribution of the rank sum from method B. For $n = 13$ and $m = 8$, we have from Theorem A that under the null hypothesis,

$$E(T) = \frac{8(8 + 13 + 1)}{2} = 88$$

$$\sigma_T = \sqrt{\frac{8 \times 13(8 + 13 + 1)}{12}} = 13.8$$

T is the sum of the ranks from method B, or 51, and the normalized test statistic is

$$\frac{T - E(T)}{\sigma_T} = -2.68$$

From the tables of the normal distribution, this corresponds to a p-value of .007 for a two-sided test, so the null hypothesis is rejected at level $\alpha = .01$, just as it was when we used the exact distribution. For this set of data, we have seen that the t test with the assumption of equal variances, the t test without that assumption, the exact Mann-Whitney test, and the approximate Mann-Whitney test all reject at level $\alpha = .01$. ∎

The Mann-Whitney test can be inverted to form confidence intervals. Let us consider a "shift" model: $G(x) = F(x - \Delta)$. This model says that the effect of the treatment (the Y's) is to add a constant Δ to what the response would have been with no treatment (the X's). (This is a very simple model, and we have already seen cases for which it is not appropriate.) We now derive a confidence interval for Δ. To test $H_0: F = G$, we used the statistic U_Y equal to the number of the $X_i - Y_j$ that are less than zero. To test the hypothesis that the shift parameter is Δ, we can similarly use

$$U_Y(\Delta) = \#[X_i - (Y_j - \Delta) < 0] = \#(Y_j - X_i > \Delta)$$

It can be shown that the null distribution of $U_Y(\Delta)$ is symmetric about $mn/2$:

$$P\left(U_Y(\Delta) = \frac{mn}{2} + k\right) = P\left(U_Y(\Delta) = \frac{mn}{2} - k\right)$$

for all integers k. Suppose that $k = k(\alpha)$ is such that $P(k \leq U_Y(\Delta) \leq mn - k) = 1 - \alpha$; the level α test then accepts for such $U_Y(\Delta)$. By the duality of confidence intervals and hypothesis tests, a $100(1 - \alpha)\%$ confidence interval for Δ is thus

$$C = \{\Delta \mid k \leq U_Y(\Delta) \leq mn - k\}$$

C consists of the set of values Δ for which the null hypothesis would not be rejected.

We can find an explicit form for this confidence interval. Let $D_{(1)}, D_{(2)}, \ldots, D_{(mn)}$ denote the ordered mn differences $Y_j - X_i$. We will show that

$$C = [D_{(k)}, D_{(mn-k+1)})$$

To see this, first suppose that $\Delta = D_{(k)}$. Then

$$U_Y(\Delta) = \#(X_i - Y_j + \Delta < 0)$$
$$= \#(Y_j - X_i > \Delta)$$
$$= mn - k$$

Similarly, if $\Delta = D_{(mn-k+1)}$,

$$U_Y(\Delta) = \#(Y_j - X_i > \Delta)$$
$$= k - 1$$

(You might find it helpful to consider the case $m = 3, n = 2, k = 2$.)

EXAMPLE C   We return to the data on iron retention (Section 11.2.1.1). The earlier analysis using the t test rested on the assumption that the populations were normally distributed, which, in fact, seemed rather dubious. The Mann-Whitney test does not make this assumption. The sum of the ranks of the Fe^{2+} group is used as a test statistic (we could have as easily used the U statistic). The rank sum is 362. Using the normal approximation to the null distribution of the rank sum, we get a p-value of .36. Again, there is insufficient evidence to reject the null hypothesis that there is no differential retention. The 95% confidence interval for the shift between the two distributions is $(-1.6, 3.7)$, which overlaps zero substantially. Note that this interval is shorter than the interval based on the t distribution; the latter was inflated by the contributions of the large observations to the sample variance. ∎

We close this section with an illustration of the use of the bootstrap in a two-sample problem. As before, suppose that $X_1, X_2, \ldots, X_n$ and $Y_1, Y_2, \ldots, Y_m$ are two independent samples from distributions F and G, respectively, and that $\pi = P(X < Y)$ is estimated by $\hat{\pi}$. How can the standard error of $\hat{\pi}$ be estimated and how can an approximate confidence interval for π be constructed? (Note that the calculations of Theorem A are not directly relevant, since they are done under the assumption that $F = G$.)

The problem can be approached in the following way: First suppose for the moment that F and G were known. Then the sampling distribution of $\hat{\pi}$ and its standard error could be estimated by simulation. A sample of size n would be generated from F, an independent sample of size m would be generated from G, and the resulting value of $\hat{\pi}$ would be computed. This procedure would be repeated many times, say B times, producing $\hat{\pi}_1, \hat{\pi}_2, \ldots, \hat{\pi}_B$. A histogram of these values would be an indication of the sampling distribution of $\hat{\pi}$ and their standard deviation would be an estimate of the standard error of $\hat{\pi}$.

Of course, this procedure cannot be implemented, because F and G are not known. But as in the previous chapter, an approximation can be obtained by using the empirical distributions F_n and G_n in their places. This means that a bootstrap value of $\hat{\pi}$ is generated by randomly selecting n values from $X_1, X_2, \ldots, X_n$ with replacement, m values from $Y_1, Y_2, \ldots, Y_m$ with replacement and calculating the resulting value of $\hat{\pi}$. In this way, a bootstrap sample $\hat{\pi}_1, \hat{\pi}_2, \ldots, \hat{\pi}_B$ is generated.

11.2.4 Bayesian Approach

We consider a Bayesian approach to the model, which stipulates that the X_i are i.i.d. normal with mean μ_X and precision ξ; and the Y_j are i.i.d. normal with mean μ_Y, precision ξ, and independent of the X_i. In general, a prior joint distribution assigned to (μ_X, μ_Y, ξ) would be multiplied by the likelihood and normalized to integrate to 1 to produce a three-dimensional joint posterior distribution for (μ_X, μ_Y, ξ). The marginal joint distribution of (μ_X, μ_Y) could be obtained by integrating out ξ. The marginal distribution of $\mu_X - \mu_Y$ could then be obtained by another integration as in Section 3.6.1. Several integrations would thus have to be done, either analytically or numerically. Special Monte Carlo methods have been devised for high dimensional Bayesian problems, but we will not consider them here.

An approximate result can be obtained using improper priors. We take (μ_X, μ_Y, ξ) to be independent. The means μ_X and μ_Y are given improper priors that are constant on $(-\infty, \infty)$, and ξ is given the improper prior $f_\Xi(\xi) = \xi^{-1}$. The posterior is thus proportional to the likelihood multiplied by ξ^{-1}:

$$f_{\text{post}}(\mu_X, \mu_Y, \xi) \propto \xi^{\frac{n+m}{2}-1} \exp\left(-\frac{\xi}{2}\left[\sum_{i=1}^{n}(x_i - \mu_X)^2 + \sum_{j=1}^{m}(y_j - \mu_Y)^2\right]\right)$$

Next, using $\sum_{i=1}^{n}(x_i - \mu_X)^2 = (n-1)s_x^2 + n(\mu_X - \bar{x})^2$ and the analogous expression for the y_j, we have

$$f_{\text{post}}(\mu_X, \mu_Y, \xi) \propto \xi^{\frac{n+m}{2}-1} \exp\left(-\frac{\xi}{2}\left[(n-1)s_x^2 + (m-1)s_y^2\right]\right)$$
$$\times \exp\left(-\frac{n\xi}{2}(\mu_X - \bar{x})^2\right) \exp\left(-\frac{m\xi}{2}(\mu_Y - \bar{y})^2\right)$$

From the form of this expression as a function of μ_X and μ_Y, we see that for fixed ξ, μ_X and μ_Y are independent normally distributed with means $\bar{x}$ and $\bar{y}$ and precisions $n\xi$ and $m\xi$. Their difference, $\mu_X - \mu_Y$, is thus normally distributed with mean $\bar{x} - \bar{y}$ and variance $\xi^{-1}(n^{-1} + m^{-1})$.

With further analysis similar to that of Section 8.6, it can be shown that the marginal posterior distribution of $\Delta = \mu_X - \mu_Y$ can be related to the t distribution:

$$\frac{\Delta - (\bar{x} - \bar{y})}{s_p\sqrt{n^{-1} + m^{-1}}} \sim t_{n+m-2}$$

Although formally similar to Theorem A of Section 11.2.1, the interpretation is different: $\bar{x} - \bar{y}$ and s_p are random in Theorem A but are fixed here, and $\Delta = \mu_X - \mu_Y$ is random here but fixed in Theorem A. The Bayesian formalism makes probability statements about Δ given the observed data.

The posterior probability that $\Delta > 0$ can thus be found using the t distribution. Let T denote a random variable with a t_{m+n-2} distribution. Then, denoting the observations by X and Y

$$P(\Delta > 0 \mid X, Y) = P\left(\frac{\Delta - (\bar{x} - \bar{y})}{s_p\sqrt{n^{-1} + m^{-1}}} \geq \frac{-(\bar{x} - \bar{y})}{s_p\sqrt{n^{-1} + m^{-1}}} \mid X, Y\right)$$
$$= P\left(T \geq \frac{\bar{y} - \bar{x}}{s_p\sqrt{n^{-1} + m^{-1}}}\right)$$

Letting X denote the measurements of method A, and Y denote the measurements of method B in Example A of Section 11.2.1, we find that for that example,

$$P(\Delta > 0 | X, Y) = t_{19}(-3.33) = .998$$

This posterior probability is very close to 1.0, and there is thus little doubt that the mean of method A is larger than the mean of method B.

The confidence interval calculated in Section 11.2.1 is formally similar but has a different interpretation under the Bayesian model, which concludes that

$$P(.015 \leq \Delta \leq .065 | X, Y) = .95$$

by integration of the posterior t distribution over a region containing 95% of the probability.

11.3 Comparing Paired Samples

In Section 11.2, we considered the problem of analyzing two independent samples. In many experiments, the samples are paired. In a medical experiment, for example, subjects might be matched by age or weight or severity of condition, and then one member of each pair randomly assigned to the treatment group and the other to the control group. In a biological experiment, the paired subjects might be littermates. In some applications, the pair consists of a "before" and an "after" measurement on the same object. Since pairing causes the samples to be dependent, the analysis of Section 11.2 does not apply.

Pairing can be an effective experimental technique, as we will now demonstrate by comparing a paired design and an unpaired design. First, we consider the paired design. Let us denote the pairs as (X_i, Y_i), where $i = 1, \ldots, n$, and assume the X's and Y's have means μ_X and μ_Y and variances σ_X^2 and σ_Y^2. We will assume that different pairs are independently distributed and that $\text{Cov}(X_i, Y_i) = \sigma_{XY}$. We will work with the differences $D_i = X_i - Y_i$, which are independent with

$$E(D_i) = \mu_X - \mu_Y$$

$$\text{Var}(D_i) = \sigma_X^2 + \sigma_Y^2 - 2\sigma_{XY}$$

$$= \sigma_X^2 + \sigma_Y^2 - 2\rho\sigma_X\sigma_Y$$

when ρ is the correlation of members of a pair. A natural estimate of $\mu_X - \mu_Y$ is $\overline{D} = \overline{X} - \overline{Y}$, the average difference. From the properties of D_i, it follows that

$$E(\overline{D}) = \mu_X - \mu_Y$$

$$\text{Var}(\overline{D}) = \frac{1}{n} \left(\sigma_X^2 + \sigma_Y^2 - 2\rho\sigma_X\sigma_Y \right)$$

Suppose, on the other hand, that an experiment had been done by taking a sample of n X's and an independent sample of n Y's. Then $\mu_X - \mu_Y$ would be estimated by $\overline{X} - \overline{Y}$ and

$$E(\overline{X} - \overline{Y}) = \mu_X - \mu_Y$$

$$\text{Var}(\overline{X} - \overline{Y}) = \frac{1}{n} \left(\sigma_X^2 + \sigma_Y^2 \right)$$

Comparing the variances of the two estimates, we see that the variance of $\overline{D}$ is smaller if the correlation is positive—that is, if the X's and Y's are positively correlated. In this circumstance, pairing is the more effective experimental design. In the simple case in which $\sigma_X = \sigma_Y = \sigma$, the two variances may be more simply expressed as

$$\text{Var}(\overline{D}) = \frac{2\sigma^2(1 - \rho)}{n}$$

in the paired case and as

$$\text{Var}(\overline{X} - \overline{Y}) = \frac{2\sigma^2}{n}$$

in the unpaired case, and the relative efficiency is

$$\frac{\text{Var}(\overline{D})}{\text{Var}(\overline{X} - \overline{Y})} = 1 - \rho$$

If the correlation coefficient is .5, for example, a paired design with n pairs of subjects yields the same precision as an unpaired design with $2n$ subjects per treatment. This additional precision results in shorter confidence intervals and more powerful tests if the degrees of freedom for estimating σ^2 are sufficiently large.

We next present methods based on the normal distribution for analyzing data from paired designs and then a nonparametric, rank-based method.

11.3.1 Methods Based on the Normal Distribution

In this section, we assume that the differences are a sample from a normal distribution with

$$E(D_i) = \mu_X - \mu_Y = \mu_D$$
$$\text{Var}(D_i) = \sigma_D^2$$

Generally, σ_D will be unknown, and inferences will be based on

$$t = \frac{\overline{D} - \mu_D}{s_{\overline{D}}}$$

which follows a t distribution with $n - 1$ degrees of freedom. Following familiar reasoning, a $100(1 - \alpha)\%$ confidence interval for μ_D is

$$\overline{D} \pm t_{n-1}(\alpha/2)s_{\overline{D}}$$

A two-sided test of the null hypothesis H_0: $\mu_D = 0$ (the natural null hypothesis for testing no treatment effect) at level α has the rejection region

$$|\overline{D}| > t_{n-1}(\alpha/2)s_{\overline{D}}$$

If the sample size n is large, the approximate validity of the confidence interval and hypothesis test follows from the central limit theorem. If the sample size is small and the true distribution of the differences is far from normal, the stated probability levels may be considerably in error.

E X A M P L E A   To study the effect of cigarette smoking on platelet aggregation, Levine (1973) drew blood samples from 11 individuals before and after they smoked a cigarette and measured the extent to which the blood platelets aggregated. Platelets are involved in the formation of blood clots, and it is known that smokers suffer more often from disorders involving blood clots than do nonsmokers. The data are shown in the following table, which gives the maximum percentage of all the platelets that aggregated after being exposed to a stimulus.

Before	After	Difference
25	27	2
25	29	4
27	37	10
44	56	12
30	46	16
67	82	15
53	57	4
53	80	27
52	61	9
60	59	−1
28	43	15

From the column of differences, $\overline{D} = 10.27$ and $s_{\overline{D}} = 2.40$. The uncertainty in $\overline{D}$ is quantified in $s_{\overline{D}}$ or in a confidence interval. Since $t_{10}(.05) = 1.812$, a 90% confidence interval is $\overline{D} \pm 1.812 s_{\overline{D}}$, or (5.9, 14.6). We can also formally test the null hypothesis that means before and after are the same. The t statistic is $10.27/2.40 = 4.28$, and since $t_{10}(.005) = 3.169$, the p-value of a two-sided test is less than .01. There is little doubt that smoking increases platelet aggregation.

The experiment was actually more complex than we have indicated. Some subjects also smoked cigarettes made of lettuce leaves and "smoked" unlit cigarettes. (You should reflect on why these additional experiments were done.)

Figure 11.7 is a plot of the after values versus the before values. They are correlated, with a correlation coefficient of .90. Pairing was a natural and effective experimental design in this case. ∎

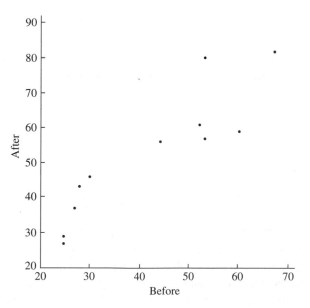

FIGURE **11.7** Plot of platelet aggregation after smoking versus aggregation before smoking.

11.3.2 A Nonparametric Method—The Signed Rank Test

A nonparametric test based on ranks can be constructed for paired samples. We illustrate the calculation with a very small example. Suppose there are four pairs, corresponding to "before" and "after" measurements listed in the following table:

Before	After	Difference	|Difference|	Rank	Signed Rank
25	27	2	2	2	2
29	25	−4	4	3	−3
60	59	−1	1	1	−1
27	37	10	10	4	4

The test statistic is calculated by the following steps:

1. Calculate the differences, D_i, and the absolute values of the differences and rank the latter.
2. Restore the signs of the differences to the ranks, obtaining signed ranks.
3. Calculate W_+, the sum of those ranks that have positive signs. For the table, this sum is $W_+ = 2 + 4 = 6$.

The idea behind the **signed rank test** (sometimes called the Wilcoxon signed rank test) is intuitively simple. If there is no difference between the two paired conditions, we expect about half the D_i to be positive and half negative, and W_+ will not be too small or too large. If one condition tends to produce larger values than the other, W_+ will tend to be more extreme. We therefore can use W_+ as a test statistic and reject for extreme values.

Before continuing, we need to specify more precisely the null hypothesis we are testing with the signed rank test: H_0 states that the distribution of the D_i is symmetric about zero. This will be true if the members of pairs of experimental units are assigned randomly to treatment and control conditions, and the treatment has no effect at all.

As usual, in order to define a rejection region for a test at level α, we need to know the sampling distribution of W_+ if the null hypothesis is true. The rejection region will be located in the tails of this null distribution in such a way that the test has level α. The null distribution may be calculated in the following way. If H_0 is true, it makes no difference which member of the pair corresponds to treatment and which to control. The difference $X_i - Y_i = D_i$ has the same distribution as the difference $Y_i - X_i = -D_i$, so the distribution of D_i is symmetric about zero. The kth largest value of D is thus equally likely to be positive or negative, and any particular assignment of signs to the integers $1, \ldots, n$ (the ranks) is equally likely. There are 2^n such assignments, and for each we can calculate W_+. We obtain a list of 2^n values (not all distinct) of W_+, each of which occurs with probability $1/2^n$. The probability of each distinct value of W_+ may thus be calculated, giving the desired null distribution.

The preceding argument has assumed that the D_i are a sample from some continuous probability distribution. If we do not wish to regard the X_i and Y_i as random variables and if the assignments to treatment and control have been made at random, the hypothesis that there is no treatment effect may be tested in exactly the same manner, except that inferences are based on the distribution induced by the randomization, as was done for the Mann-Whitney test.

The null distribution of W_+ is calculated by many computer packages, and tables are also available.

The signed rank test is a nonparametric version of the paired sample t test. Unlike the t test, it does not depend on an assumption of normality. Since differences are replaced by ranks, it is insensitive to outliers, whereas the t test is sensitive. It has been shown that even when the assumption of normality holds, the signed rank test is nearly as powerful as the t test. The nonparametric method is thus generally preferable, especially for small sample sizes.

EXAMPLE A The signed rank test can be applied to the data on platelet aggregation considered previously (Example A in Section 11.3.1). In this case, it is easier to work with W_- rather than W_+, since W_- is clearly 1. From Table 9 of Appendix B, the two-sided test is significant at $\alpha = .01$. ∎

If the sample size is greater than 20, a normal approximation to the null distribution can be used. To find this, we calculate the mean and variance of W_+.

THEOREM A

Under the null hypothesis that the D_i are independent and symmetrically distributed about zero,

$$E(W_+) = \frac{n(n+1)}{4}$$

$$\text{Var}(W_+) = \frac{n(n+1)(2n+1)}{24}$$

Proof

To facilitate the calculation, we represent W_+ in the following way:

$$W_+ = \sum_{k=1}^{n} kI_k$$

where

$$I_k = \begin{cases} 1, & \text{if the } k\text{th largest } |D_i| \text{ has } D_i > 0 \\ 0, & \text{otherwise} \end{cases}$$

Under H_0, the I_k are independent Bernoulli random variables with $p = \frac{1}{2}$, so

$$E(I_k) = \frac{1}{2}$$

$$\text{Var}(I_k) = \frac{1}{4}$$

We thus have

$$E(W_+) = \frac{1}{2} \sum_{k=1}^{n} k = \frac{n(n+1)}{4}$$

$$\mathrm{Var}(W_+) = \frac{1}{4} \sum_{k=1}^{n} k^2 = \frac{n(n+1)(2n+1)}{24}$$

as was to be shown.   ∎

If some of the differences are equal to zero, the most common technique is to discard those observations. If there are ties, each $|D_i|$ is assigned the average value of the ranks for which it is tied. If there are not too many ties, the significance level of the test is not greatly affected. If there are a large number of ties, modifications must be made. For further information on these matters, see Hollander and Wolfe (1973) or Lehmann (1975).

11.3.3   An Example—Measuring Mercury Levels in Fish

Kacprzak and Chvojka (1976) compared two methods of measuring mercury levels in fish. A new method, which they called "selective reduction," was compared to an established method, referred to as "the permanganate method." One advantage of selective reduction is that it allows simultaneous measurement of both inorganic mercury and methyl mercury. The mercury in each of 25 juvenile black marlin was measured by both techniques. The 25 measurements for each method (in ppm of mercury) and the differences are given in the following table.

Fish	Selective Reduction	Permanganate	Difference	Signed Rank
1	.32	.39	.07	+15.5
2	.40	.47	.07	+15.5
3	.11	.11	.00	
4	.47	.43	−.04	−11
5	.32	.42	.10	+19
6	.35	.30	−.05	−13.5
7	.32	.43	.11	+20
8	.63	.98	.35	+23
9	.50	.86	.36	+24
10	.60	.79	.19	+22
11	.38	.33	−.05	−13.5
12	.46	.45	−.01	−2.5

(*Continued*)

Fish	Selective Reduction	Permanganate	Difference	Signed Rank
13	.20	.22	.02	+6.5
14	.31	.30	−.01	−2.5
15	.62	.60	−.02	−6.5
16	.52	.53	.01	+2.5
17	.77	.85	.08	+17.5
18	.23	.21	−.02	−6.5
19	.30	.33	.03	+9.0
20	.70	.57	−.13	−21
21	.41	.43	.02	+6.5
22	.53	.49	−.04	−11
23	.19	.20	.01	+2.5
24	.31	.35	.04	+11
25	.48	.40	−.08	−17.5

In analyzing such data, it is often informative to check whether the differences depend in some way on the level or size of the quantity being measured. The differences versus the permanganate values are plotted in Figure 11.8. This plot is quite interesting. It appears that the differences are small for low permanganate values and larger for higher permanganate values. It is striking that the differences are all positive and large for the highest four values. The investigators do not comment on these phenomena. It is not uncommon for the size of fluctuations to increase as the value being measured increases; the percent error may remain nearly constant but the actual error does not. For this reason, data of this nature are often analyzed on a log scale.

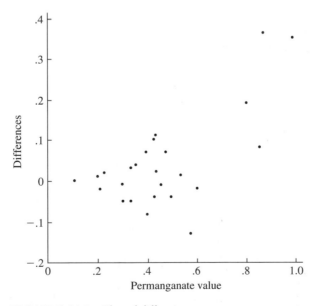

FIGURE **11.8** Plot of differences versus permanganate values.

Because the observations are paired (two measurements on each fish), we will use the paired t test for a parametric test. The sample size is large enough that the test should be robust against nonnormality. The mean difference is .04, and the standard deviation of the differences is .116. The t statistic is 1.724; with 24 degrees of freedom, this corresponds to a p-value of .094 for a two-sided test. Although this p-value is fairly small, the evidence against H_0: $\mu_D = 0$ is not overwhelming. The test does not reject at the significance level .05.

The signed ranks are shown in the last column of the table above. Note that the single zero difference was set aside, and also note how the tied ranks were handled. The test statistic W_+ is 194.5. Under H_0, its mean and variance are

$$E(W_+) = \frac{24 \times 25}{4} = 150$$

$$\text{Var}(W_+) = \frac{24 \times 25 \times 49}{24} = 1225$$

Since n is greater than 20, we use the normalized test statistic, or

$$Z = \frac{W_+ - E(W_+)}{\sqrt{\text{Var}(W_+)}} = 1.27$$

The p-value for a two-sided test from the normal approximation is .20, which is not strong evidence against the null hypothesis. It is possible to correct for the presence of ties, but in this case the correction only amounts to changing the standard deviation of W_+ from 35 to 34.95.

Neither the parametric nor the nonparametric test gives conclusive evidence that there is any systematic difference between the two methods of measurement. The informal graphical analysis does suggest, however, that there may be a difference for high concentrations of mercury.

11.4 Experimental Design

This section covers some basic principles of the interpretation and design of experimental studies and illustrates them with case studies.

11.4.1 Mammary Artery Ligation

A person with coronary artery disease suffers from chest pain during exercise because the constricted arteries cannot deliver enough oxygen to the heart. The treatment of ligating the mammary arteries enjoyed a brief vogue; the basic idea was that ligating these arteries forced more blood to flow into the heart. This procedure had the advantage of being quite simple surgically, and it was widely publicized in an article in *Reader's Digest* (Ratcliffe 1957). Two years later, the results of a more careful study (Cobb et al. 1959) were published. In this study, a control group and an experimental group were established in the following way. When a prospective patient entered surgery, the surgeon made the necessary preliminary incisions prior to tying off the mammary artery. At that point, the surgeon opened a sealed envelope that contained instructions about whether to complete the operation by tying off the artery. Neither the patient nor his attending physician knew whether the operation had actually been carried out. The study showed essentially no difference after the operation between the control group (no ligation) and the experimental group (ligation), although there was some suggestion that the control group had done better.

The Ratcliffe and Cobb studies differ in that in the earlier one there was no control group and thus no benchmark by which to gauge improvement. The reported improvement of the patients in this earlier study could have been due to the placebo effect, which we discuss next. The design of the later study protected against possible unconscious biases by randomly assigning the control and experimental groups and by concealing from the patients and their physicians the actual nature of the treatment. Such a design is called a double-blind, randomized controlled experiment.

11.4.2   The Placebo Effect

The **placebo effect** refers to the effect produced by any treatment, including dummy pills (placebos), when the subject believes that he or she has been given an effective treatment. The possibility of a placebo effect makes the use of a blind design necessary in many experimental investigations.

The placebo effect may not be due entirely to psychological factors, as was shown in an interesting experiment by Levine, Gordon, and Fields (1978). A group of subjects had teeth extracted. During the extraction, they were given nitrous oxide and local anesthesia. In the recovery room, they rated the amount of pain they were experiencing on a numerical scale. Two hours after surgery, the subjects were given a placebo and were again asked to rate their pain. An hour later, some of the subjects were given a placebo and some were given naloxone, a morphine antagonist. It is known that there are specific receptors to morphine in the brain and that the body can also release endorphins that bind to these sites. Naloxone blocks the morphine receptors. In the study, it was found that when those subjects who responded positively to the placebo received naloxone, they experienced an increase in pain that made their pain levels comparable to those of the patients who did not respond to the placebo. The implication is that those who responded to the placebo had produced endorphins, the actions of which were subsequently blocked by the naloxone.

An instance of the placebo effect was demonstrated by a psychologist, Claude Steele (2002), who gave a math exam to a group of male and female undergraduates at Stanford University. One group (treatment) was told that the exam was gender-neutral, and the other group (controls) was not so informed. The men outperformed the women in the control group. In the treatment group, men and women performed equally well. Men in the treatment group did worse than men in the control group. (*Economist* Feb 21, 2002).

11.4.3   The Lanarkshire Milk Experiment

The importance of the randomized assignment of individuals (or other experimental units) to treatment and control groups is illustrated by a famous study known as the Lanarkshire milk experiment. In the spring of 1930, an experiment was carried out in Lanarkshire, Scotland, to determine the effect of providing free milk to schoolchildren. In each participating school, some children (treatment group) were given free milk and others (controls) were not. The assignment of children to control or treatment was initially done at random; however, teachers were allowed to use their judgment in switching children between treatment and control to obtain a better balance of undernourished and well-nourished individuals in the groups.

A paper by Gosset (1931), who published under the name Student (as in Student's *t* test), is a very interesting critique of the experiment. An examination of the data revealed that at the start of the experiment the controls were heavier and taller. Student conjectured that the teachers, perhaps unconsciously, had adjusted the initial randomization in a manner that placed more of the undernourished children in the treatment group. A further complication was caused by weighing the children with their clothes on. The experimental data were weight gains measured in late spring relative to early spring or late winter. The more well-to-do children probably tended to be better nourished and may have had heavier winter clothing than the poor children. Thus, the well-to-do children's weight gains were vitiated as a result of differences in clothing, which may have influenced comparisons between the treatment and control groups.

11.4.4   The Portacaval Shunt

Cirrhosis of the liver, to which alcoholics are prone, is a condition in which resistance to blood flow causes blood pressure in the liver to build up to dangerously high levels. Vessels may rupture, which may cause death. Surgeons have attempted to relieve this condition by connecting the portal artery, which feeds the liver, to the vena cava, one of the main veins returning to the heart, thus reducing blood flow through the liver. This procedure, called the Portacaval shunt, had been used for more than 20 years when Grace, Muench, and Chalmers (1966) published an examination of 51 studies of the method. They examined the design of each study (presence or absence of a control group and presence or absence of randomization) and the investigators' conclusions (categorized as markedly enthusiastic, moderately enthusiastic, or not enthusiastic). The results are summarized in the following table, which speaks for itself:

	Enthusiasm		
Design	Marked	Moderate	None
No controls	24	7	1
Nonrandomized controls	10	3	2
Randomized controls	0	1	3

The differences between the experiments that used controls and those that did not is not entirely surprising, because the placebo effect was probably operating. The importance of randomized assignment to treatment and control groups is illustrated by comparing the conclusions for the randomized and nonrandomized controlled experiments. Randomization can help to ensure against subtle unconscious biases that may creep into an experiment. For example, a physician might tend to recommend surgery for patients who are somewhat more robust than the average. Articulate patients might be more likely to have an influence on the decision as to which group they are assigned to.

11.4.5 FD&C Red No. 40

This discussion follows Lagakos and Mosteller (1981). During the middle and late 1970s, experiments were conducted to determine possible carcinogenic effects of a widely used food coloring, FD&C Red No. 40. One of the experiments involved 500 male and 500 female mice. Both genders were divided into five groups: two control groups, a low-dose group, a medium-dose group, and a high-dose group. The mice were bred in the following way: Males and females were paired and before and during mating were given their prescribed dose of Red No. 40. The regime was continued during gestation and weaning of the young. From litters that had at least three pups of each sex, three of each sex were selected randomly and continued on their parents' dosage throughout their lives. After 109–111 weeks, all the mice still living were killed. The presence or absence of reticuloendothelial tumors was of particular interest. Although there were significant differences between some of the treatment groups, the results were rather confusing. For example, there was a significant difference between the incidence rates for the two male control groups, and among the males the medium-dose group had the lowest incidence.

Several experts were asked to examine the results of this and other experiments. Among them were Lagakos and Mosteller, who requested information on how the cages that housed the mice were arranged. There were three racks of cages, each containing five rows of seven cages in the front and five rows of seven cages in the back. Five mice were housed in each cage. The mice were assigned to the cages in a systematic way: The first male control group was in the top of the front of rack 1; the first female control group was in the bottom of the front of rack 1; and so on, ending with the high-dose females in the bottom of the back of rack 3 (Figure 11.9). Lagakos and Mosteller showed that there were effects due to cage position that could not be explained by gender or by dosage group. A random assignment of cage positions would have eliminated this confounding. Lagakos and Mosteller also suggested some experimental designs to systematically control for cage position.

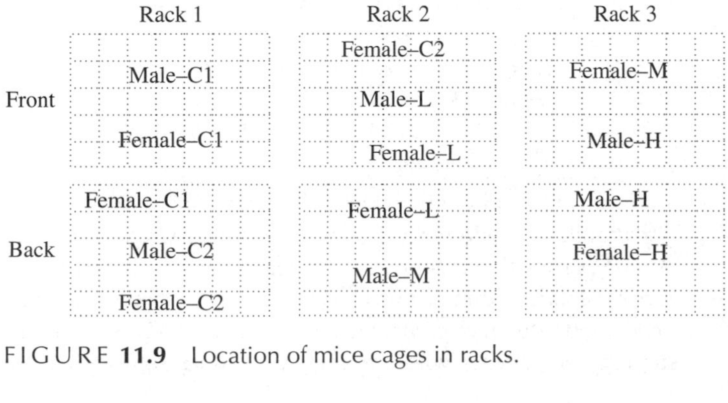

FIGURE **11.9** Location of mice cages in racks.

It was also possible that a litter effect might be complicating the analysis, since littermates received the same treatment and littermates of the same sex were housed in the same or contiguous cages. In the presence of a litter effect, mice from the same litter might show less variability than that present among mice from different litters. This reduces the effective sample size—in the extreme case in which littermates react identically, the effective sample size is the number of litters, not the total number of mice. One way around this problem would have been to use only one mouse from each litter.

The presence of a possible selection bias is another problem. Because mice were included in the experiment only if they came from a litter with at least three males and three females, offspring of possibly less healthy parents were excluded. This could be a serious problem since exposure to Red No. 40 might affect the parents' health and the birth process. If, for example, among the high-dose mice, only the most hardy produced large enough litters, their offspring might be hardier than the controls' offspring.

11.4.6   Further Remarks on Randomization

As well as guarding against possible biases on the part of the experimenter, the process of randomization tends to balance any factors that may be influential but are not explicitly controlled in the experiment. Time is often such a factor; background variables such as temperature, equipment calibration, line voltage, and chemical composition can change slowly with time. In experiments that are run over some period of time, therefore, it is important to randomize the assignments to treatment and control over time. Time is not the only factor that should be randomized, however. In agricultural experiments, the positions of test plots in a field are often randomly assigned. In biological experiments with test animals, the locations of the animals' cages may have an effect, as illustrated in the preceding section.

Although rarer than in other areas, randomized experiments have been carried out in the social sciences as well (*Economist* Feb 28, 2002). Randomized trials have been used to evaluate such programs as driver training, as well as the criminal justice system and reduced classroom size. In evaluations of "whole-language" approaches to reading (in which children are taught to read by evaluating contextual clues rather than breaking down words), 52 randomized studies carried out by the National Reading Panel in 2000 showed that effective reading instruction requires phonics. Randomized studies of "scared straight" programs, in which juvenile delinquents are introduced to prison inmates, suggested that the likelihood of subsequent arrests is actually increased by such programs.

Generally, if it is anticipated that a variable will have a significant effect, that variable should be included as one of the controlled factors in the experimental design. The matched-pairs design of this chapter can be used to control for a single factor. To control for more than one factor, factorial designs, which are briefly introduced in the next chapter, may be used.

11.4.7 Observational Studies, Confounding, and Bias in Graduate Admissions

It is not always possible to conduct controlled experiments or use randomization. In evaluating some medical therapies, for example, a randomized, controlled experiment would be unethical if one therapy was strongly believed to be superior. For many problems of psychological interest (effects of parental modes of discipline, for example), it is impossible to conduct controlled experiments. In such situations, recourse is often made to observational studies. Hospital records may be examined to compare the outcomes of different therapies, or psychological records of children raised in different ways may be analyzed. Although such studies may be valuable, the results are seldom unequivocal. Because there is no randomization, it is always possible that the groups under comparison differ in respects other than their "treatments."

As an example, let us consider a study of gender bias in admissions to graduate school at the University of California at Berkeley (Bickel and O'Connell 1975). In the fall of 1973, 8442 men applied for admission to graduate studies at Berkeley, and 44% were admitted; 4321 women applied, and 35% were admitted. If the men and women were similar in every respect other than sex, this would be strong evidence of sex bias. This was not a controlled, randomized experiment, however; sex was not randomly assigned to the applicants. As will be seen, the male and female applicants differed in other respects, which influenced admission.

The following table shows admission rates for the six most popular majors on the Berkeley campus.

Major	Men		Women	
	Number of Applicants	Percentage Admitted	Number of Applicants	Percentage Admitted
A	825	62	108	82
B	560	63	25	68
C	325	37	593	34
D	417	33	375	35
E	191	28	393	34
F	373	6	341	7

If the percentages admitted are compared, women do not seem to be unfavorably treated. But when the combined admission rates for all six majors are calculated, it is found that 44% of the men and only 30% of the women were admitted, which seems paradoxical. The resolution of the paradox lies in the observation that the women tended to apply to majors that had low admission rates (C through F) and the men to majors that had relatively high admission rates (A and B). This factor was not controlled for, because the study was observational in nature; it was also "confounded" with the factor of interest, sex; randomization, had it been possible, would have tended to balance out the confounded factor.

Confounding also plays an important role in studies of the effect of coffee drinking. Several studies have claimed to show a significant association of coffee consumption with coronary disease. Clearly, randomized, controlled trials are not possible here—a randomly selected individual cannot be told that he or she is in the treatment group and must drink 10 cups of coffee a day for the next five years. Also, it is known that heavy coffee drinkers also tend to smoke more than average, so smoking is confounded with coffee drinking. Hennekens et al. (1976) review several studies in this area.

11.4.8 Fishing Expeditions

Another problem that sometimes flaws observational studies, and controlled experiments as well, is that they engage in "fishing expeditions." For example, consider a hypothetical study of the effects of birth control pills. In such a case, it would be impossible to assign women to a treatment or a placebo at random, but a nonrandomized study might be conducted by carefully matching controls to treatments on such factors as age and medical history. The two groups might be followed up on for some time, with many variables being recorded for each subject such as blood pressure, psychological measures, and incidences of various medical problems. After termination of the study, the two groups might be compared on each of these variables, and it might be found, say, that there was a "significant difference" in the incidence of melanoma. The problem with this "significant finding" is the following. Suppose that 100 independent two-sample t tests are conducted at the .05 level and that, in fact, all the null hypotheses are true. We would expect that five of the tests would produce a "significant" result. Although each of the tests has probability .05 of type I error, as a collection they do not simultaneously have $\alpha = .05$. The combined significance level is the probability that at least one of the null hypotheses is rejected:

$$\alpha = P\{\text{at least one } H_0 \text{ rejected}\}$$
$$= 1 - P\{\text{no } H_0 \text{ rejected}\}$$
$$= 1 - .95^{100} = .994$$

Thus, with very high probability, at least one "significant" result will be found, even if all the null hypotheses are true.

There are no simple cures for this problem. One possibility is to regard the results of a fishing expedition as merely providing suggestions for further experiments. Alternatively, and in the same spirit, the data could be split randomly into two halves, one half for fishing in and the other half to be locked safely away, unexamined. "Significant" results from the first half could then be tested on the second half. A third alternative is to conduct each individual hypothesis test at a small significance level. To see how this works, suppose that all null hypotheses are true and that each of n null hypotheses is tested at level α. Let R_i denote the event that the ith null hypothesis is rejected, and let α^* denote the overall probability of a type I error. Then

$$\alpha^* = P\{R_1 \text{ or } R_2 \text{ or } \cdots \text{ or } R_n\}$$
$$\leq P\{R_1\} + P\{R_2\} + \cdots + P\{R_n\}$$
$$= n\alpha$$

Thus, if each of the n null hypotheses is tested at level α/n, the overall significance level is less than or equal to α. This is often called the **Bonferroni method.**

11.5 Concluding Remarks

This chapter was concerned with the problem of comparing two samples. Within this context, the fundamental statistical concepts of estimation and hypothesis testing, which were introduced in earlier chapters, were extended and utilized. The chapter also showed how informal descriptive and data analytic techniques are used in supplementing more formal analysis of data. Chapter 12 will extend the techniques of this chapter to deal with multisample problems. Chapter 13 is concerned with similar problems that arise in the analysis of qualitative data.

We considered two types of experiments, those with two independent samples and those with matched pairs. For the case of independent samples, we developed the t test, based on an assumption of normality, as well as a modification of the t test that takes into account possibly unequal variances. The Mann-Whitney test, based on ranks, was presented as a nonparametric method, that is, a method that is not based on an assumption of a particular distribution. Similarly, for the matched-pairs design, we developed a parametric t test and a nonparametric test, the signed rank test.

We discussed methods based on an assumption of normality and rank methods, which do not make this assumption. It turns out, rather surprisingly, that even if the normality assumption holds, the rank methods are quite powerful relative to the t test. Lehmann (1975) shows that the efficiency of the rank tests relative to that of the t test—that is, the ratio of sample sizes required to attain the same power—is typically around .95 if the distributions are normal. Thus, a rank test using a sample of size 100 is as powerful as a t test based on 95 observations. Collecting the extra 5 pieces of data is a small price to pay for a safeguard against nonnormality.

The bootstrap appeared again in this chapter. Indeed, uses of this recently developed technique are finding applications in a great variety of statistical problems. In contrast with earlier chapters, where bootstrap samples were generated from one distribution, here we have bootstrapped from two empirical distributions.

The chapter concluded with a discussion of experimental design, which emphasized the importance of incorporating controls and randomization in investigations. Possible problems associated with observational studies were discussed. Finally, the difficulties encountered in making many comparisons from a single data set were pointed out; such problems of multiplicity will come up again in Chapter 12.

11.6 Problems

1. A computer was used to generate four random numbers from a normal distribution with a set mean and variance: 1.1650, .6268, .0751, .3516. Five more random normal numbers with the same variance but perhaps a different mean were then generated (the mean may or may not actually be different): .3035, 2.6961, 1.0591, 2.7971, 1.2641.

   a. What do you think the means of the random normal number generators were? What do you think the difference of the means was?
   b. What do you think the variance of the random number generator was?
   c. What is the estimated standard error of your estimate of the difference of the means?

d. Form a 90% confidence interval for the difference of the means of the random number generators.

e. In this situation, is it more appropriate to use a one-sided test or a two-sided test of the equality of the means?

f. What is the p-value of a two-sided test of the null hypothesis of equal means?

g. Would the hypothesis that the means were the same versus a two-sided alternative be rejected at the significance level $\alpha = .1$?

h. Suppose you know that the variance of the normal distribution was $\sigma^2 = 1$. How would your answers to the preceding questions change?

2. The difference of the means of two normal distributions with equal variance is to be estimated by sampling an equal number of observations from each distribution. If it were possible, would it be better to halve the standard deviations of the populations or double the sample sizes?

3. In Section 11.2.1, we considered two methods of estimating $\text{Var}(\overline{X} - \overline{Y})$. Under the assumption that the two population variances were equal, we estimated this quantity by

$$s_p^2 \left(\frac{1}{n} + \frac{1}{m} \right)$$

and without this assumption by

$$\frac{s_X^2}{n} + \frac{s_Y^2}{m}$$

Show that these two estimates are identical if $m = n$.

4. Respond to the following:

Using the t distribution is absolutely ridiculous—another example of deliberate mystification! It's valid when the populations are normal and have equal variance. If the sample sizes were so small that the t distribution were practically different from the normal distribution, you would be unable to check these assumptions.

5. Respond to the following:

Here is another example of deliberate mystification—the idea of formulating and testing a null hypothesis. Let's take Example A of Section 11.2.1. It seems to me that it is inconceivable that the expected values of *any* two methods of measurement could be *exactly* equal. It is certain that there will be subtle differences at the very least. What is the sense, then, in testing H_0: $\mu_X = \mu_Y$?

6. Respond to the following:

I have two batches of numbers and I have a corresponding $\bar{x}$ and $\bar{y}$. Why should I test whether they are equal when I can just see whether they are or not?

7. In the development of Section 11.2.1, where are the following assumptions used? (1) $X_1, X_2, \ldots, X_n$ are independent random variables; (2) $Y_1, Y_2, \ldots, Y_n$ are independent random variables; (3) the X's and Y's are independent.

8. An experiment to determine the efficacy of a drug for reducing high blood pressure is performed using four subjects in the following way: two of the subjects are chosen at random for the control group and two for the treatment group. During the course of treatment with the drug, the blood pressure of each of the subjects in the treatment group is measured for ten consecutive days as is the blood pressure of each of the subjects in the control group.

   a. In order to test whether the treatment has an effect, do you think it is appropriate to use the two-sample t test with $n = m = 20$?

   b. Do you think it is appropriate to use the Mann-Whitney test with $n = m = 20$?

9. Referring to the data in Section 11.2.1.1, compare iron retention at concentrations of 10.2 and .3 millimolar using graphical procedures and parametric and nonparametric tests. Write a brief summary of your conclusions.

10. Verify that the two-sample t test at level α of $H_0: \mu_X = \mu_Y$ versus $H_A: \mu_X \neq \mu_Y$ rejects if and only if the confidence interval for $\mu_X - \mu_Y$ does not contain zero.

11. Explain how to modify the t test of Section 11.2.1 to test $H_0: \mu_X = \mu_Y + \Delta$ versus $H_A: \mu_X \neq \mu_Y + \Delta$ where Δ is specified.

12. An equivalence between hypothesis tests and confidence intervals was demonstrated in Chapter 9. In Chapter 10, a nonparametric confidence interval for the median, η, was derived. Explain how to use this confidence interval to test the hypothesis $H_0: \eta = \eta_0$. In the case where $\eta_0 = 0$, show that using this approach on a sample of differences from a paired experiment is equivalent to the **sign test.** The sign test counts the number of positive differences and uses the fact that in the case that the null hypothesis is true, the distribution of the number of positive differences is binomial with $(n, .5)$. Apply the sign test to the data from the measurement of mercury levels, listed in Section 11.3.3.

13. Let $X_1, \ldots, X_{25}$ be i.i.d. $N(.3, 1)$. Consider testing the null hypothesis $H_0: \mu = 0$ versus $H_A: \mu > 0$ at significance level $\alpha = .05$. Compare the power of the sign test and the power of the test based on normal theory assuming that σ is known.

14. Suppose that $X_1, \ldots, X_n$ are i.i.d. $N(\mu, \sigma^2)$. To test the null hypothesis $H_0: \mu = \mu_0$, the t test is often used:

$$t = \frac{\overline{X} - \mu_0}{s_{\overline{X}}}$$

   Under H_0, t follows a t distribution with $n - 1$ df. Show that the likelihood ratio test of this H_0 is equivalent to the t test.

15. Suppose that n measurements are to be taken under a treatment condition and another n measurements are to be taken independently under a control condition. It is thought that the standard deviation of a single observation is about 10 under both conditions. How large should n be so that a 95% confidence interval for $\mu_X - \mu_Y$ has a width of 2? Use the normal distribution rather than the t distribution, since n will turn out to be rather large.

16. Referring to Problem 15, how large should n be so that the test of $H_0: \mu_X = \mu_Y$ against the one-sided alternative $H_A: \mu_X > \mu_Y$ has a power of .5 if $\mu_X - \mu_Y = 2$ and $\alpha = .10$?

17. Consider conducting a two-sided test of the null hypothesis H_0: $\mu_X = \mu_Y$ as described in Problem 16. Sketch power curves for (a) $\alpha = .05$, $n = 20$; (b) $\alpha = .10$, $n = 20$; (c) $\alpha = .05$, $n = 40$; (d) $\alpha = .10$, $n = 40$. Compare the curves.

18. Two independent samples are to be compared to see if there is a difference in the population means. If a total of m subjects are available for the experiment, how should this total be allocated between the two samples in order to (a) provide the shortest confidence interval for $\mu_X - \mu_Y$ and (b) make the test of H_0: $\mu_X = \mu_Y$ as powerful as possible? Assume that the observations in the two samples are normally distributed with the same variance.

19. An experiment is planned to compare the mean of a control group to the mean of an independent sample of a group given a treatment. Suppose that there are to be 25 samples in each group. Suppose that the observations are approximately normally distributed and that the standard deviation of a single measurement in either group is $\sigma = 5$.

   a. What will the standard error of $\overline{Y} - \overline{X}$ be?
   b. With a significance level $\alpha = .05$, what is the rejection region of the test of the null hypothesis H_0: $\mu_Y = \mu_X$ versus the alternative H_A: $\mu_Y > \mu_X$?
   c. What is the power of the test if $\mu_Y = \mu_X + 1$?
   d. Suppose that the p-value of the test turns out to be 0.07. Would the test reject at significance level $\alpha = .10$?
   e. What is the rejection region if the alternative is H_A: $\mu_Y \neq \mu_X$? What is the power if $\mu_Y = \mu_X + 1$?

20. Consider Example A of Section 11.3.1 using a Bayesian model. As in the example, use a normal model for the differences and also use an improper prior for the expected difference and the precision (as in the case of unknown mean and variance in Section 8.6). Find the posterior probability that the expected difference is positive. Find a 90% posterior credibility interval for the expected difference.

21. A study was done to compare the performances of engine bearings made of different compounds (McCool 1979). Ten bearings of each type were tested. The following table gives the times until failure (in units of millions of cycles):

Type I	Type II
3.03	3.19
5.53	4.26
5.60	4.47
9.30	4.53
9.92	4.67
12.51	4.69
12.95	12.78
15.21	6.79
16.04	9.37
16.84	12.75

   a. Use normal theory to test the hypothesis that there is no difference between the two types of bearings.
   b. Test the same hypothesis using a nonparametric method.
   c. Which of the methods—that of part (a) or that of part (b)—do you think is better in this case?
   d. Estimate π, the probability that a type I bearing will outlast a type II bearing.
   e. Use the bootstrap to estimate the sampling distribution of $\hat{\pi}$ and its standard error.
   f. Use the bootstrap to find an approximate 90% confidence interval for π.

22. An experiment was done to compare two methods of measuring the calcium content of animal feeds. The standard method uses calcium oxalate precipitation followed by titration and is quite time-consuming. A new method using flame photometry is faster. Measurements of the percent calcium content made by each method of 118 routine feed samples (Heckman 1960) are contained in the file calcium. Analyze the data to see if there is any systematic difference between the two methods. Use both parametric and nonparametric tests and graphical methods.

23. Let $X_1, \ldots, X_n$ be i.i.d. with cdf F, and let $Y_1, \ldots, Y_m$ be i.i.d. with cdf G. The hypothesis to be tested is that $F = G$. Suppose for simplicity that $m + n$ is even so that in the combined sample of X's and Y's, $(m + n)/2$ observations are less than the median and $(m + n)/2$ are greater.

   a. As a test statistic, consider T, the number of X's less than the median of the combined sample. Show that T follows a hypergeometric distribution under the null hypothesis:

$$P(T = t) = \frac{\binom{(m+n)/2}{t}\binom{(m+n)/2}{n-t}}{\binom{m+n}{n}}$$

   Explain how to form a rejection region for this test.
   b. Show how to find a confidence interval for the difference between the median of F and the median of G under the shift model, $G(x) = F(x - \Delta)$. (*Hint:* Use the order statistics.)
   c. Apply the results (a) and (b) to the data of Problem 21.

24. Find the exact null distribution of the Mann-Whitney statistic, U_Y, in the case where $m = 3$ and $n = 2$.

25. Referring to Example A in Section 11.2.1, (a) if the smallest observation for method B (79.94) is made arbitrarily small, will the t test still reject? (b) If the largest observation for method B (80.03) is made arbitrarily large, will the t test still reject? (c) Answer the same questions for the Mann-Whitney test.

26. Let $X_1, \ldots, X_n$ be a sample from an $N(0, 1)$ distribution and let $Y_1, \ldots, Y_n$ be an independent sample from an $N(1, 1)$ distribution.

   a. Determine the expected rank sum of the X's.
   b. Determine the variance of the rank sum of the X's.

27. Find the exact null distribution of W_+ in the case where $n = 4$.

28. For $n = 10, 20,$ and 30, find the .05 and .01 critical values for a two-sided signed rank test from the tables and then by using the normal approximation. Compare the values.

29. (Permutation Test for Means) Here is another view on hypothesis testing that we will illustrate with Example A of Section 11.2.1. We ask whether the measurements produced by methods A and B are identical or exchangeable in the following sense. There are $13 + 8 = 21$ measurements in all and there are $\binom{21}{8}$, or about 2×10^5, ways that 8 of these could be assigned to method B. Is the particular assignment we have observed unusual among these in the sense that the means of the two samples are unusually different?

   a. It's not inconceivable, but it may be asking too much for you to generate all $\binom{21}{8}$ partitions. So just choose a random sample of these partitions, say of size 1000, and make a histogram of the resulting values of $\overline{X}_A - \overline{X}_B$. Where on this distribution does the value of $\overline{X}_A - \overline{X}_B$ that was actually observed fall? Compare to the result of Example B of Section 11.2.1.
   b. In what way is this procedure similar to the Mann-Whitney test?

30. Use the bootstrap to estimate the standard error of and a confidence interval for $\overline{X}_A - \overline{X}_B$ and compare to the result of Example A of Section 11.2.1.

31. In Section 11.2.3, if $F = G$, what are $E(\hat{\pi})$ and $\text{Var}(\hat{\pi})$? Would there be any advantage in using equal sample sizes $m = n$ in estimating π or does it make no difference?

32. If $X \sim N(\mu_X, \sigma_X^2)$ and Y is independent $N(\mu_Y, \sigma_Y^2)$, what is $\pi = P(X < Y)$ in terms of $\mu_X, \mu_Y, \sigma_X,$ and σ_Y?

33. To compare two variances in the normal case, let $X_1, \ldots, X_n$ be i.i.d. $N(\mu_X, \sigma_X^2)$, and let $Y_1, \ldots, Y_m$ be i.i.d. $N(\mu_Y, \sigma_Y^2)$, where the X's and Y's are independent samples. Argue that under $H_0: \sigma_X = \sigma_Y$,

$$\frac{s_X^2}{s_Y^2} \sim F_{n-1, \, m-1}$$

   a. Construct rejection regions for one- and two-sided tests of H_0.
   b. Construct a confidence interval for the ratio σ_X^2/σ_Y^2.
   c. Apply the results of parts (a) and (b) to Example A in Section 11.2.1. (*Caution:* This test and confidence interval are not robust against violations of the assumption of normality.)

34. This problem contrasts the power functions of paired and unpaired designs. Graph and compare the power curves for testing $H_0: \mu_X = \mu_Y$ for the following two designs.

   a. Paired: $\text{Cov}(X_i, Y_i) = 50, \sigma_X = \sigma_Y = 10, i = 1, \ldots, 25$.
   b. Unpaired: $X_1, \ldots, X_{25}$ and $Y_1, \ldots, Y_{25}$ are independent with variance as in part (a).

35. An experiment was done to measure the effects of ozone, a component of smog. A group of 22 seventy-day-old rats were kept in an environment containing ozone for 7 days, and their weight gains were recorded. Another group of 23 rats of a similar age were kept in an ozone-free environment for a similar time, and their weight gains were recorded. The data (in grams) are given below. Analyze the data to determine the effect of ozone. Write a summary of your conclusions. [This problem is from Doksum and Sievers (1976) who provide an interesting analysis.]

Controls			Ozone		
41.0	38.4	24.9	10.1	6.1	20.4
25.9	21.9	18.3	7.3	14.3	15.5
13.1	27.3	28.5	−9.9	6.8	28.2
−16.9	17.4	21.8	17.9	−12.9	14.0
15.4	27.4	19.2	6.6	12.1	15.7
22.4	17.7	26.0	39.9	−15.9	54.6
29.4	21.4	22.7	−14.7	44.1	−9.0
26.0	26.6		−9.0		

36. Lin, Sutton, and Qurashi (1979) compared microbiological and hydroxylamine methods for the analysis of ampicillin dosages. In one series of experiments, pairs of tablets were analyzed by the two methods. The data in the following table give the percentages of claimed amount of ampicillin found by the two methods in several pairs of tablets. What are $\overline{X} - \overline{Y}$ and $s_{\overline{X}-\overline{Y}}$? If the pairing had been erroneously ignored and it had been assumed that the two samples were independent, what would have been the estimate of the standard deviation of $\overline{X} - \overline{Y}$? Analyze the data to determine if there is a systematic difference between the two methods.

Microbiological Method	Hydroxylamine Method
97.2	97.2
105.8	97.8
99.5	96.2
100.0	101.8
93.8	88.0
79.2	74.0
72.0	75.0
72.0	67.5
69.5	65.8
20.5	21.2
95.2	94.8
90.8	95.8
96.2	98.0
96.2	99.0
91.0	100.2

37. Stanley and Walton (1961) ran a controlled clinical trial to investigate the effect of the drug stelazine on chronic schizophrenics. The trials were conducted on chronic schizophrenics in two closed wards. In each of the wards, the patients were divided into two groups matched for age, length of time in the hospital, and score on a behavior rating sheet. One member of each pair was given stelazine, and the other a placebo. Only the hospital pharmacist knew which member of each pair received the actual drug. The following table gives the behavioral rating scores for the patients at the beginning of the trial and after 3 mo. High scores are good.

Ward A			
Stelazine		Placebo	
Before	After	Before	After
2.3	3.1	2.4	2.0
2.0	2.1	2.2	2.6
1.9	2.45	2.1	2.0
3.1	3.7	2.9	2.0
2.2	2.54	2.2	2.4
2.3	3.72	2.4	3.18
2.8	4.54	2.7	3.0
1.9	1.61	1.9	2.54
1.1	1.63	1.3	1.72

Ward B			
Stelazine		Placebo	
Before	After	Before	After
1.9	1.45	1.9	1.91
2.3	2.45	2.4	2.54
2.0	1.81	2.0	1.45
1.6	1.72	1.5	1.45
1.6	1.63	1.5	1.54
2.6	2.45	2.7	1.54
1.7	2.18	1.7	1.54

a. For each of the wards, test whether stelazine is associated with improvement in the patients' scores.

b. Test if there is any difference in improvement between the wards. [These data are also presented in Lehmann (1975), who discusses methods of combining the data from the wards.]

38. Bailey, Cox, and Springer (1978) used high-pressure liquid chromatography to measure the amounts of various intermediates and by-products in food dyes. The following table gives the percentages added and found for two substances in the dye FD&C Yellow No. 5. Is there any evidence that the amounts found differ systematically from the amounts added?

Sulfanilic Acid		Pyrazolone-T	
Percentage Added	Percentage Found	Percentage Added	Percentage Found
.048	.060	.035	.031
.096	.091	.087	.084
.20	.16	.19	.16
.19	.16	.19	.17
.096	.091	.16	.15
.18	.19	.032	.040
.080	.070	.060	.076
.24	.23	.13	.11
0	0	.080	.082
.040	.042	0	0
.060	.056		

39. An experiment was done to test a method for reducing faults on telephone lines (Welch 1987). Fourteen matched pairs of areas were used. The following table shows the fault rates for the control areas and for the test areas:

Test	Control
676	88
206	570
230	605
256	617
280	653
433	2913
337	924
466	286
497	1098
512	982
794	2346
428	321
452	615
512	519

  a. Plot the differences versus the control rate and summarize what you see.
  b. Calculate the mean difference, its standard deviation, and a confidence interval.
  c. Calculate the median difference and a confidence interval and compare to the previous result.
  d. Do you think it is more appropriate to use a t test or a nonparametric method to test whether the apparent difference between test and control could be due to chance? Why? Carry out both tests and compare.

40. Biological effects of magnetic fields are a matter of current concern and research. In an early study of the effects of a strong magnetic field on the development of mice (Barnothy 1964), 10 cages, each containing three 30-day-old albino female mice, were subjected for a period of 12 days to a field with an average strength of 80 Oe/cm. Thirty other mice housed in 10 similar cages were not placed in a magnetic field and served as controls. The following table shows the weight gains, in grams, for each of the cages.

a. Display the data graphically with parallel dotplots. (Draw two parallel number lines and put dots on one corresponding to the weight gains of the controls and on the other at points corresponding to the gains of the treatment group.)

b. Find a 95% confidence interval for the difference of the mean weight gains.

c. Use a t test to assess the statistical significance of the observed difference. What is the p-value of the test?

d. Repeat using a nonparametric test.

e. What is the difference of the median weight gains?

f. Use the bootstrap to estimate the standard error of the difference of median weight gains.

g. Form a confidence interval for the difference of median weight gains based on the bootstrap approximation to the sampling distribution.

Field Present	Field Absent
22.8	23.5
10.2	31.0
20.8	19.5
27.0	26.2
19.2	26.5
9.0	25.2
14.2	24.5
19.8	23.8
14.5	27.8
14.8	22.0

41. The *Hodges-Lehmann shift estimate* is defined to be $\hat{\Delta} = \text{median}(X_i - Y_j)$, where $X_1, X_2, \ldots, X_n$ are independent observations from a distribution F and $Y_1, Y_2, \ldots, Y_m$ are independent observations from a distribution G and are independent of the X_i.

a. Show that if F and G are normal distributions, then $E(\hat{\Delta}) = \mu_X - \mu_Y$.

b. Why is $\hat{\Delta}$ robust to outliers?

c. What is $\hat{\Delta}$ for the previous problem and how does it compare to the differences of the means and of the medians?

d. Use the bootstrap to approximate the sampling distribution and the standard error of $\hat{\Delta}$.

e. From the bootstrap approximation to the sampling distribution, form an approximate 90% confidence interval for $\hat{\Delta}$.

42. Use the data of Problem 40 of Chapter 10.

  a. Estimate π, the probability that more rain will fall from a randomly selected seeded cloud than from a randomly selected unseeded cloud.

  b. Use the bootstrap to estimate the standard error of $\hat{\pi}$.

  c. Use the bootstrap to form an approximate confidence interval for π.

43. Suppose that $X_1, X_2, \ldots, X_n$ and $Y_1, Y_2, \ldots, Y_m$ are two independent samples. As a measure of the difference in location of the two samples, the difference of the 20% trimmed means is used. Explain how the bootstrap could be used to estimate the standard error of this difference.

44. Interest in the role of vitamin C in mental illness in general and schizophrenia in particular was spurred by a paper of Linus Pauling in 1968. This exercise takes its data from a study of plasma levels and urinary vitamin C excretion in schizophrenic patients (Subotičanec et al. 1986). Twenty schizophrenic patients and 15 controls with a diagnosis of neurosis of different origin who had been patients at the same hospital for a minimum of 2 months were selected for the study. Before the experiment, all the subjects were on the same basic hospital diet. A sample of 2 ml of venous blood for vitamin C determination was drawn from each subject before breakfast and after the subjects had emptied their bladders. Each subject was then given 1 g ascorbic acid dissolved in water. No foods containing ascorbic acid were available during the test. For the next 6 h all urine was collected from the subjects for assay of vitamin C. A second blood sample was also drawn 2 h after the dose of vitamin C.

The following two tables show the plasma concentrations (mg/dl).

Schizophrenics		Nonschizophrenics	
0 h	2 h	0 h	2 h
.55	1.22	1.27	2.00
.60	1.54	.09	.41
.21	.97	1.64	2.37
.09	.45	.23	.41
1.01	1.54	.18	.79
.24	.75	.12	.94
.37	1.12	.85	1.72
1.01	1.31	.69	1.75
.26	.92	.78	1.60
.30	1.27	.63	1.80
.26	1.08	.50	2.08
.10	1.19	.62	1.58
.42	.64	.19	.86
.11	.30	.66	1.92
.14	.24	.91	1.54
.20	.89		
.09	.24		
.32	1.68		
.24	.99		
.25	.67		

a. Graphically compare the two groups at the two times and for the difference in concentration at the two times.

b. Use the t test to assess the strength of the evidence for differences between the two groups at 0 h, at 2 h, and the difference 2 h − 0 h.

c. Use the Mann-Whitney test to test the hypotheses of (b).

The following tables show the amounts of urinary vitamin C, both total and milligrams per kilogram of body weight, for the two groups:

Schizophrenics		Nonschizophrenics	
Total	mg/kg	Total	mg/kg
16.6	.19	289.4	3.96
33.3	.44	0.0	0.00
34.1	.39	620.4	7.95
0.0	.00	0.0	0.00
119.8	1.75	8.5	.10
.1	.01	5.5	.09
25.3	.27	43.2	.91
359.3	5.99	91.7	1.00
6.6	.10	200.9	3.46
.4	.01	113.8	2.01
62.8	.68	102.2	1.50
.2	.01	108.2	1.98
13.0	.15	36.9	.49
0.0	0.00	122.0	1.72
0.0	0.00	101.9	1.52
5.9	.10		
.1	.01		
6.0	.07		
32.1	.42		
0.0	0.00		

d. Use descriptive statistics and graphical presentations to compare the two groups with respect to total excretion and mg/kg body weight. Do the data look normally distributed?

e. Use a t test to compare the two groups on both variables. Is the normality assumption reasonable?

f. Use the Mann-Whitney test to compare the two groups. How do the results compare with those obtained in part (e)?

The lower levels of plasma vitamin C in the schizophrenics before administration of ascorbic acid could be attributed to several factors. Interindividual differences in the intake of meals cannot be excluded, despite the fact that all patients were offered the same food. A more interesting possibility is that the differences are the result of poorer resorption or of higher ascorbic acid utilization in schizophrenics. In order to answer this question, another experiment was run on 15 schizophrenics and 15 controls. All subjects were given 70 mg of ascorbic acid daily for 4 weeks before the ascorbic acid loading test. The following table shows the concentration of plasma vitamin C (mg/dl) and the 6-h urinary excretion (mg) after administration of 1 g ascorbic acid.

Schizophrenics		Controls	
Plasma	Urine	Plasma	Urine
.72	86.20	1.02	190.14
1.11	21.55	.86	149.76
.96	182.07	.78	285.27
1.23	88.28	1.38	244.93
.76	76.58	.95	184.45
.75	18.81	1.00	135.34
1.26	50.02	.47	157.74
.64	107.74	.60	125.65
.67	.09	1.15	164.98
1.05	113.23	.86	99.65
1.28	34.38	.61	86.29
.54	8.44	1.01	142.23
.77	109.03	.77	144.60
1.11	144.44	.77	265.40
.51	172.09	.94	28.26

g. Use graphical methods and descriptive statistics to compare the two groups with respect to plasma concentrations and urinary excretion.

h. Use the t test to compare the two groups on the two variables. Does the normality assumption look reasonable?

i. Compare the two groups using the Mann-Whitney test.

45. This and the next two problems are based on discussions and data in Le Cam and Neyman (1967), which is devoted to the analysis of weather modification experiments. The examples illustrate some ways in which principles of experimental design have been used in this field. During the summers of 1957 through 1960, a series of randomized cloud-seeding experiments were carried out in the mountains of Arizona. Of each pair of successive days, one day was randomly selected for seeding to be done. The seeding was done during a two-hour to four-hour period starting at midday, and rainfall during the afternoon was measured by a network of 29 gauges. The data for the four years are given in the following table (in inches). Observations in this table are listed in chronological order.

a. Analyze the data for each year and for the years pooled together to see if there appears to be any effect due to seeding. You should use graphical descriptive methods to get a qualitative impression of the results and hypothesis tests to assess the significance of the results.

b. Why should the day on which seeding is to be done be chosen at random rather than just alternating seeded and unseeded days? Why should the days be paired at all, rather than just deciding randomly which days to seed?

1957		1958		1959		1960	
Seeded	Unseeded	Seeded	Unseeded	Seeded	Unseeded	Seeded	Unseeded
0	.154	.152	.013	.015	0	0	.010
.154	0	0	0	0	0	0	0
.003	.008	0	.445	0	.086	.042	.057
.084	.033	.002	0	.021	.006	0	0
.002	.035	.007	.079	0	.115	0	.093
.157	.007	.013	.006	.004	.090	0	.183
.010	.140	.161	.008	.010	0	.152	0
0	.022	0	.001	0	0	0	0
.002	0	.274	.001	.055	0	0	0
.078	.074	.001	.025	.004	.076	0	0
.101	.002	.122	.046	.053	.090	0	0
.169	.318	.101	.007	0	0	0	0
.139	.096	.012	.019	0	.078	.008	0
.172	0	.002	0	.090	.121	.040	.060
0	0	.066	0	.028	1.027	.003	.102
0	.050	.040	.012	0	.104	.011	.041
				.032	.023		
				.133	.172		
				.083	.002		
					0	0	

46. The National Weather Bureau's ACN cloud-seeding project was carried out in the states of Oregon and Washington. Cloud seeding was accomplished by dispersing dry ice from an aircraft; only clouds that were deemed "ripe" for seeding were candidates for seeding. On each occasion, a decision was made at random whether to seed, the probability of seeding being $\frac{2}{3}$. This resulted in 22 seeded and 13 control cases. Three types of targets were considered, two of which are dealt with in this problem. Type I targets were large geographical areas downwind from the seeding; type II targets were sections of type I targets located so as to have, theoretically, the greatest sensitivity to cloud seeding. The following table gives the average target rainfalls (in inches) for the seeded and control cases, listed in chronological order. Is there evidence that seeding has an effect on either type of target? In what ways is the design of this experiment different from that of the one in Problem 45?

Control Cases		Seeded Cases	
Type I	Type II	Type I	Type II
.0080	.0000	.1218	.0200
.0046	.0000	.0403	.0163
.0549	.0053	.1166	.1560
.1313	.0920	.2375	.2885
.0587	.0220	.1256	.1483
.1723	.1133	.1400	.1019
.3812	.2880	.2439	.1867
.1720	.0000	.0072	.0233
.1182	.1058	.0707	.1067
.1383	.2050	.1036	.1011
.0106	.0100	.1632	.2407
.2126	.2450	.0788	.0666
.1435	.1529	.0365	.0133
		.2409	.2897
		.0408	.0425
		.2204	.2191
		.1847	.0789
		.3332	.3570
		.0676	.0760
		.1097	.0913
		.0952	.0400
		.2095	.1467

47. During 1963 and 1964, an experiment was carried out in France; its design differed somewhat from those of the previous two problems. A 1500-km target area was selected, and an adjacent area of about the same size was designated as the control area; 33 ground generators were used to produce silver iodide to seed the target area. Precipitation was measured by a network of gauges for each suitable "rainy period," which was defined as a sequence of periods of continuous precipitation between dry spells of a specified length. When a forecaster determined that the situation was favorable for seeding, he telephoned an order to a service agent, who then opened a sealed envelope that contained an order to actually seed or not. The envelopes had been prepared in advance, using a table of random numbers. The following table gives precipitation (in inches) in the target and control areas for the seeded and unseeded periods.

a. Analyze the data, which are listed in chronological order, to see if there is an effect of seeding.

b. The analysis done by the French investigators used the square root transformation in order to make normal theory more applicable. Do you think that taking the square root was an effective transformation for this purpose?

c. Reflect on the nature of this design. In particular, what advantage is there to using the control area? Why not just compare seeded and unseeded periods on the target area?

Seeded		Unseeded	
Target	Control	Target	Control
1.6	1.0	1.1	2.2
28.1	27.0	3.5	5.2
7.8	.3	2.6	0.0
4.0	6.0	2.6	2.0
9.6	12.6	9.8	4.9
0.2	0.5	5.6	8.5
18.7	8.7	.1	3.5
16.5	21.5	0.0	1.1
4.6	13.9	17.7	11.0
9.3	6.7	19.4	19.8
3.5	4.5	8.9	5.3
0.1	0.7	10.6	8.9
11.5	8.7	10.2	4.5
0.0	0.0	16.0	13.0
9.3	10.7	9.7	21.1
5.5	4.7	21.4	15.9
70.2	29.1	6.1	19.5
0.7	1.9	24.3	16.3
38.6	34.7	20.9	6.3
11.3	10.2	60.2	47.0
3.3	2.7	15.2	10.8
8.9	2.8	2.7	4.8
11.1	4.3	0.3	0.0
64.3	38.7	12.2	5.7
16.6	11.1	2.2	5.1
7.3	6.5	23.3	30.6
3.2	3.0	9.9	3.7
23.9	13.6		
0.6	0.1		

48. Proteinuria, the presence of excess protein in urine, is a symptom of renal (kidney) distress among diabetics. Taguma et al. (1985) studied the effects of captopril for treating proteinuria in diabetics. Urinary protein was measured for 12 patients before and after eight weeks of captopril therapy. The amounts of urinary protein (in g/24 hrs) before and after therapy are shown in the following table. What can you conclude about the effect of captopril? Consider using parametric or nonparametric methods and analyzing the data on the original scale or on a log scale.

Before	After
24.6	10.1
17.0	5.7
16.0	5.6
10.4	3.4
8.2	6.5
7.9	0.7
8.2	6.5
7.9	0.7
5.8	6.1
5.4	4.7
5.1	2.0
4.7	2.9

49. Egyptian researchers, Kamal et al. (1991), took a sample of 126 police officers subject to inhalation of vehicle exhaust in downtown Cairo and found an average blood level concentration of lead equal to 29.2 μg/dl with a standard deviation of 7.5 μg/dl. A sample of 50 policemen from a suburb, Abbasia, had an average concentration of 18.2 μg/dl and a standard deviation of 5.8 μg/dl. Form a confidence interval for the population difference and test the null hypothesis that there is no difference in the populations.

50. The file `bodytemp` contains normal body temperature readings (degrees Fahrenheit) and heart rates (beats per minute) of 65 males (coded by 1) and 65 females (coded by 2) from Shoemaker (1996).

a. Using normal theory, form a 95% confidence interval for the difference of mean body temperatures between males and females. Is the use of the normal approximation reasonable?

b. Using normal theory, form a 95% confidence interval for the difference of mean heart rates between males and females. Is the use of the normal approximation reasonable?

c. Use both parametric and nonparametric tests to compare the body temperatures and heart rates. What do you conclude?

51. A common symptom of otitis-media (inflamation of the middle ear) in young children is the prolonged presence of fluid in the middle ear, called *middle-ear effusion*. It is hypothesized that breast-fed babies tend to have less prolonged effusions than do bottle-fed babies. Rosner (2006) presents the results of a study of 24 pairs of infants who were matched according to sex, socioeconomic status, and type of medication taken. One member of each pair was bottle-fed and the other was breast-fed. The file `ears` gives the durations (in days) of middle-ear effusions after the first episode of otitis-media.

a. Examine the data using graphical methods and summarize your conclusions.

b. In order to test the hypothesis of no difference, do you think it is more appropriate to use a parametric or a nonparametric test? Carry out a test. What do you conclude?

52. The media often present short reports of the results of experiments. To the critical reader or listener, such reports often raise more questions than they answer. Comment on possible pitfalls in the interpretation of each of the following.

   a. It is reported that patients whose hospital rooms have a window recover faster than those whose rooms do not.

   b. Nonsmoking wives whose husbands smoke have a cancer rate twice that of wives whose husbands do not smoke.

   c. A 2-year study in North Carolina found that 75% of all industrial accidents in the state happened to workers who had skipped breakfast.

   d. A school integration program involved busing children from minority schools to majority (primarily white) schools. Participation in the program was voluntary. It was found that the students who were bused scored lower on standardized tests than did their peers who chose not to be bused.

   e. When a group of students were asked to match pictures of newborns with pictures of their mothers, they were correct 36% of the time.

   f. A survey found that those who drank a moderate amount of beer were healthier than those who totally abstained from alcohol.

   g. A 15-year study of more than 45,000 Swedish soldiers revealed that heavy users of marijuana were six times more likely than nonusers to develop schizophrenia.

   h. A University of Wisconsin study showed that within 10 years of the wedding, 38% of those who had lived together before marriage had split up, compared to 27% of those who had married without a "trial period."

   i. A study of nearly 4,000 elderly North Carolinians has found that those who attended religious services every week were 46% less likely to die over a six-year period than people who attended less often or not at all, according to researchers at Duke University Medical Center.

53. Explain why in Levine's experiment (Example A in Section 11.3.1) subjects also smoked cigarettes made of lettuce leaves and unlit cigarettes.

54. This example is taken from an interesting article by Joiner (1981) and from data in Ryan, Joiner, and Ryan (1976). The National Institute of Standards and Technology supplies standard materials of many varieties to manufacturers and other parties, who use these materials to calibrate their own testing equipment. Great pains are taken to make these reference materials as homogeneous as possible. In an experiment, a long homogeneous steel rod was cut into 4-inch lengths, 20 of which were randomly selected and tested for oxygen content. Two measurements were made on each piece. The 40 measurements were made over a period of 5 days, with eight measurements per day. In order to avoid possible bias from time-related trends, the sequence of measurements was randomized. The file `steelrods` contains the measurements. There is an unexpected systematic source of variability in these data. Can you find it by making an appropriate plot? Would this effect have been detectable if the measurements had not been randomized over time?

CHAPTER 12

The Analysis of Variance

12.1 Introduction

Chapter 11 was concerned with the analysis of data arising from experimental designs with two samples. Experiments frequently involve more than two samples; they may compare several treatments, such as different drugs, and perhaps other factors, such as sex, at the same time. This chapter is an introduction to the statistical analysis of such experiments. The methods we will discuss are called *analysis of variance*. Contrary to what this phrase seems to imply, we will be primarily concerned with the comparison of the means of the data, not their variances. We will consider the two most elementary multisample designs: the one-way and two-way layouts. Methods based on the normal distribution and nonparametric methods will be developed.

12.2 The One-Way Layout

A **one-way layout** is an experimental design in which independent measurements are made under each of several treatments. The techniques we will introduce are thus generalizations of the techniques for comparing two independent samples that were covered in Chapter 11.

In this section, we will use as an example data from Kirchhoefer (1979), who studied the measurement of chlorpheniramine maleate in tablets. Measurements of composites that had nominal dosages equal to 4 mg were made by seven laboratories, each laboratory making 10 measurements. Data is shown in the following table. There are two possible sources of variability in the data: variability within labs and variability between labs.

Lab 1	Lab 2	Lab 3	Lab 4	Lab 5	Lab 6	Lab 7
4.13	3.86	4.00	3.88	4.02	4.02	4.00
4.07	3.85	4.02	3.88	3.95	3.86	4.02
4.04	4.08	4.01	3.91	4.02	3.96	4.03
4.07	4.11	4.01	3.95	3.89	3.97	4.04
4.05	4.08	4.04	3.92	3.91	4.00	4.10
4.04	4.01	3.99	3.97	4.01	3.82	3.81
4.02	4.02	4.03	3.92	3.89	3.98	3.91
4.06	4.04	3.97	3.90	3.89	3.99	3.96
4.10	3.97	3.98	3.97	3.99	4.02	4.05
4.04	3.95	3.98	3.90	4.00	3.93	4.06

Figure 12.1, a boxplot of these data, shows some variation in the medians among the seven labs, as well as some variation in the interquartile ranges. It appears from the figure that there may be some systematic differences between the labs and that there is less variability in some labs than in others. We will discuss the following question: Are the differences in the means of the measurements from the various labs significant, or might they be due to chance?

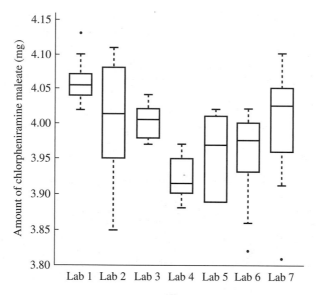

FIGURE 12.1 Boxplots of determinations of amounts of chlorpheniramine maleate in tablets by seven laboratories.

12.2.1 Normal Theory; the *F* Test

We first discuss the analysis of variance and the *F* test in the case of *I* groups, each containing *J* samples. The *I* groups will be referred to generically as *treatments,* or *levels.* (In the preceding example, $I = 7$ and $J = 10$. We will discuss the case of unequal sample sizes later.)

We first define some notation and introduce the basic model. Let

$$Y_{ij} = \text{the } j\text{th observation of the } i\text{th treatment}$$

Our model is that the observations are corrupted by random errors and that the error in one observation is independent of the errors in the other observations. The statistical model is

$$Y_{ij} = \mu + \alpha_i + \varepsilon_{ij}$$

Here μ is the overall mean level, α_i is the differential effect of the ith treatment, and ε_{ij} is the random error in the jth observation under the ith treatment. The errors are assumed to be independent, normally distributed with mean zero and variance σ^2. The α_i are normalized:

$$\sum_{i=1}^{I} \alpha_i = 0$$

The expected response to the ith treatment is $E(Y_{ij}) = \mu + \alpha_i$. Thus, if $\alpha_i = 0$, for $i = 1, \ldots, I$, all treatments have the same expected response, and, in general, $\alpha_i - \alpha_j$ is the difference between the expected values under treatments i and j. We will derive a test for the null hypothesis, which is that all the means are equal.

The analysis of variance is based on the following identity:

$$\sum_{i=1}^{I} \sum_{j=1}^{J} (Y_{ij} - \overline{Y}_{..})^2 = \sum_{i=1}^{I} \sum_{j=1}^{J} (Y_{ij} - \overline{Y}_{i.})^2 + J \sum_{i=1}^{I} (\overline{Y}_{i.} - \overline{Y}_{..})^2$$

where

$$\overline{Y}_{i.} = \frac{1}{J} \sum_{j=1}^{J} Y_{ij}$$

is the average of the observations under the ith treatment and

$$\overline{Y}_{..} = \frac{1}{IJ} \sum_{i=1}^{I} \sum_{j=1}^{J} Y_{ij}$$

is the overall average. The terms appearing in the first identity above are called *sums of squares,* and the identity may be symbolically expressed as

$$SS_{TOT} = SS_W + SS_B$$

In words, this means that the total sum of squares equals the sum of squares *within* groups plus the sum of squares *between* groups. The terminology reflects that SS_W is a measure of the variation of the data within the treatment groups and that SS_B is a measure of the variation of the treatment means among or between treatments.

To establish the identity, we express the left-hand side as

$$\sum_{i=1}^{I} \sum_{j=1}^{J} (Y_{ij} - \overline{Y}_{..})^2 = \sum_{i=1}^{I} \sum_{j=1}^{J} [(Y_{ij} - \overline{Y}_{i.}) + (\overline{Y}_{i.} - \overline{Y}_{..})]^2$$

$$= \sum_{i=1}^{I} \sum_{j=1}^{J} (Y_{ij} - \overline{Y}_{i.})^2 + \sum_{i=1}^{I} \sum_{j=1}^{J} (\overline{Y}_{i.} - \overline{Y}_{..})^2$$

$$+ 2 \sum_{i=1}^{I} \sum_{j=1}^{J} (Y_{ij} - \overline{Y}_{i.})(\overline{Y}_{i.} - \overline{Y}_{..})$$

$$= \sum_{i=1}^{I} \sum_{j=1}^{J} (Y_{ij} - \overline{Y}_{i.})^2 + \sum_{i=1}^{I} \sum_{j=1}^{J} (\overline{Y}_{i.} - \overline{Y}_{..})^2$$

$$+ 2 \sum_{i=1}^{I} \left[(\overline{Y}_{i.} - \overline{Y}_{..}) \sum_{j=1}^{J} (Y_{ij} - \overline{Y}_{i.}) \right]$$

The last term of the final expression vanishes because the sum of deviations from a mean is zero.

As we will see, the basic idea underlying analysis of variance is the comparison of the sizes of various sums of squares. We can calculate the expected values of the sums of squares defined previously using the following lemma.

LEMMA A

Let X_i, where $i = 1, \ldots, n$, be independent random variables with $E(X_i) = \mu_i$ and $Var(X_i) = \sigma^2$. Then

$$E(X_i - \overline{X})^2 = (\mu_i - \overline{\mu})^2 + \frac{n-1}{n} \sigma^2$$

where

$$\bar{\mu} = \frac{1}{n} \sum_{i=1}^{n} \mu_i$$

Proof

We use the fact that $E(U^2) = [E(U)]^2 + Var(U)$ for any random variable U with finite variance. The first term on the right-hand side of the equation in the lemma follows immediately. For the second term, we have to calculate $Var(X_i - \overline{X})$:

$$Var(X_i - \overline{X}) = Var(X_i) + Var(\overline{X}) - 2Cov(X_i, \overline{X})$$

and

$$Var(X_i) = \sigma^2$$
$$Var(\overline{X}) = \frac{1}{n} \sigma^2$$

$$Cov(X_i, \overline{X}) = Cov\left(X_i, \frac{1}{n} \sum_{j=1}^{n} X_j \right) = \frac{1}{n} \sigma^2$$

(Here we have used $Cov(X_i, X_j) = 0$ if $i \neq j$, since the X's are independent.) Putting these results together proves the lemma. ∎

Lemma A may be applied to the sums of squares discussed before, yielding the following theorem.

THEOREM **A**

Under the assumptions for the model stated at the beginning of this section,

$$E(SS_W) = \sum_{i=1}^{I} \sum_{j=1}^{J} E(Y_{ij} - \overline{Y}_{i.})^2$$

$$= \sum_{i=1}^{I} \sum_{j=1}^{J} \frac{J-1}{J} \sigma^2$$

$$= I(J-1)\sigma^2$$

Here we have used Lemma A, with the role of X_i being played by Y_{ij} and that of $\overline{X}$ being played by $\overline{Y}_{i.}$. The second line then follows since $E(Y_{ij}) = E(\overline{Y}_i) = \mu + \alpha_i$. To find $E(SS_B)$, we again use the lemma with $\overline{Y}_{i.}$ and $\overline{Y}_{..}$ in place of X_i and $\overline{X}$:

$$E(SS_B) = J \sum_{i=1}^{I} E(\overline{Y}_{i.} - \overline{Y}_{..})^2$$

$$= J \sum_{i=1}^{I} \left[\alpha_i^2 + \frac{(I-1)\sigma^2}{IJ} \right]$$

$$= J \sum_{i=1}^{I} \alpha_i^2 + (I-1)\sigma^2 \qquad \blacksquare$$

SS_W may be used to estimate σ^2; the estimate is

$$s_p^2 = \frac{SS_W}{I(J-1)}$$

which is unbiased. The subscript p stands for *pooled*. Estimates of σ^2 from the I treatments are pooled together, since SS_W can be written as

$$SS_W = \sum_{i=1}^{I} (J-1)s_i^2$$

where s_i^2 is the sample variance in the ith group.

If all the α_i are equal to zero, then the expectation of $SS_B/(I-1)$ is also σ^2. Thus, in this case, $SS_W/[I(J-1)]$ and $SS_B/(I-1)$ should be about equal. If some of the α_i are nonzero, SS_B will be inflated. We next develop a method of comparing the two sums of squares to find a test statistic for testing the null hypothesis that all the α_i are equal. Under the assumption that the errors are normally distributed, the probability distributions of the sums of squares can be calculated.

THEOREM **B**

If the errors are independent and normally distributed with means 0 and variances σ^2, then SS_W/σ^2 follows a chi-square distribution with $I(J-1)$ degrees of freedom. If, additionally, the α_i are all equal to zero, then SS_B/σ^2 follows a chi-square distribution with $I-1$ degrees of freedom and is independent of SS_W.

Proof

We first consider SS_W. From Theorem B of Section 6.3,

$$\frac{1}{\sigma^2} \sum_{j=1}^{J} (Y_{ij} - \overline{Y}_{i.})^2$$

follows a chi-square distribution with $J-1$ degrees of freedom. There are I such sums in SS_W, and they are independent of each other since the observations are independent. The sum of I independent chi-square random variables that each have $J-1$ degrees of freedom follows a chi-square distribution with $I(J-1)$ degrees of freedom. Theorem B of Section 6.3 can also be applied to SS_B, noting that $\text{Var}(\overline{Y}_{i.}) = \sigma^2/J$.

We next prove that the two sums of squares are independent of each other. SS_W is a function of the vector **U**, which has elements $Y_{ij} - \overline{Y}_{i.}$, where $i = 1, \ldots, I$ and $j = 1, \ldots, J$. SS_B is a function of the vector **V**, whose elements are $\overline{Y}_{i.}$, where $i = 1, \ldots, I$, since $\overline{Y}_{..}$ can be obtained from the $\overline{Y}_{i.}$. Thus, it is sufficient to show that these two vectors are independent of each other. First, if $i \neq i'$, $Y_{ij} - \overline{Y}_{i.}$ and $\overline{Y}_{i'.}$ are independent since they are functions of different observations. Second, $Y_{ij} - \overline{Y}_{i.}$ and $\overline{Y}_{i.}$ are independent by Theorem A of Section 6.3. This completes the proof of the theorem. ∎

The statistic

$$F = \frac{SS_B/(I-1)}{SS_W/[I(J-1)]}$$

is used to test the following null hypothesis:

$$H_0: \alpha_1 = \alpha_2 = \cdots = \alpha_I = 0$$

By Theorem A, the denominator of the F statistic has expected value equal to σ^2, and the expectation of the numerator is $J(I-1)^{-1} \sum_{i=1}^{I} \alpha_i^2 + \sigma^2$. Thus, if the null hypothesis is true, the F statistic should be close to 1, whereas if it is false, the statistic should be larger. If the null hypothesis is false, the numerator reflects variation between the different groups as well as variation within groups, whereas the denominator reflects only variation within groups. The hypothesis is thus rejected for large values of F. As usual, in order to apply this test, we must know the null distribution of the test statistic.

THEOREM **C**

Under the assumption that the errors are normally distributed, the null distribution of F is the F distribution with $(I-1)$ and $I(J-1)$ degrees of freedom.

Proof

The theorem follows from Theorem B and from the definition of the F distribution (Section 6.2), since, under H_0, F is the ratio of two independent chi-square random variables divided by their degrees of freedom. ■

Percentage points of the F distribution are widely tabled. It can show that, under the normality assumption, the F test is equivalent to the likelihood ratio test.

EXAMPLE **A**   We can illustrate the use of the F statistic by applying it to the tablet data from Section 12.2. In doing so, we adopt an explicit statistical model for the variability seen in Figure 12.1. According to this model, there is an unknown mean level associated with each laboratory, and the deviations from this mean level of the 10 measurements within a laboratory are independent, normally distributed, random variables. With the aid of this model, we will see whether it is plausible that the unknown laboratory means are all equal, so that the variability between labs displayed in Figure 12.1 is entirely due to chance.

The sums of squares defined previously are calculated and presented in a table called the **analysis of variance table:**

Source	df	SS	MS	F
Labs	6	.125	.021	5.66
Error	63	.231	.0037	
Total	69	.356		

In the table, SS_W is the sum of squares due to error, and SS_B is the sum of squares due to labs. MS stands for *mean square* and equals the sum of squares divided by the degrees of freedom. The column headed F gives the F statistic for testing the null hypothesis that there is no systematic difference among the seven labs. The F statistic has 6 and 63 df and a value of 5.66. This particular combination of degrees of freedom is not included in Table 5 of Appendix B, but upon examining the entries with 6 and 60 df, it is clear that the p-value is less than .01. We may thus conclude that the means of the measurements from the various labs are significantly different.

Figure 12.2 is a normal probability plot of the residuals from the analysis of variance model (the residuals are formed by simply subtracting from the measurements of each lab the mean value for that lab). There is some indication of deviation

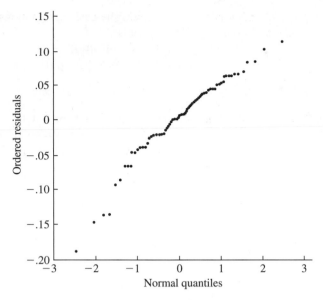

F I G U R E **12.2** Normal probability plot of residuals from one-way analyses of variance of tablet data.

from normality in the lower tail of the distribution, but the data do not appear grossly nonnormal. ∎

We now outline the procedure for the case in which the numbers of observations under the various treatments are not necessarily equal. The only difficulties with this case are algebraic; conceptually, the analysis is the same as for the case of equal sample sizes. Suppose that there are J_i observations under treatment i, for $i = 1, \ldots, I$. The basic identity still holds; that is, we have

$$\sum_{i=1}^{I} \sum_{j=1}^{J_i} (Y_{ij} - \overline{Y}_{..})^2 = \sum_{i=1}^{I} \sum_{j=1}^{J_i} (Y_{ij} - \overline{Y}_{i.})^2 + \sum_{i=1}^{I} J_i (\overline{Y}_i - \overline{Y}_{..})^2$$

By reasoning similar to that used here for the simple case, it can be shown that

$$E(SS_W) = \sigma^2 \sum_{i=1}^{I} (J_i - 1)$$

$$E(SS_B) = (I - 1)\sigma^2 + \sum_{i=1}^{I} J_i \alpha_i^2$$

The degrees of freedom for these sums of squares are $\sum_{i=1}^{I} J_i - I$ and $I - 1$, respectively. It may be argued, as in the proof of Theorem B, that the normalized sums of squares follow chi-square distributions and that the ratio of mean squares follows an F distribution under the null hypothesis of no treatment differences.

To conclude this section, let us review the basic assumptions of the model and comment on their importance. The model is

$$Y_{ij} = \mu + \alpha_i + \varepsilon_{ij}$$

We assume the following:

1. The ε_{ij} are normally distributed. The F test, like the t test, remains approximately valid for moderate to large samples from moderately nonnormal distributions.
2. The error variance, σ^2, is constant. In many applications, the error variances may be different in different groups. For example, Figure 12.1 suggests that some labs may be more precise in their measurements than others. Fortunately, if there are an equal number of observations in each group, the F test is not strongly affected.
3. The ε_{ij} are independent. This assumption is very important, both for normal theory and for the nonparametric analysis we will present later.

12.2.2 The Problem of Multiple Comparisons

The application of the F test in Example A in Section 12.2.1 has an anticlimactic character. We concluded that the means of measurements from different labs are not all equal, but the test gives no information about how they differ, in particular about which pairs are significantly different. In many applications, the null hypothesis is a "straw man" that is not seriously entertained. Real interest may be focused on comparing pairs or groups of treatments and estimating the treatment means and their differences. A naive approach would be to compare all pairs of treatment means using t tests. The difficulty with such a procedure was pointed out in the section on experimental design in Chapter 11: Although each individual comparison would have a type I error rate of α, the collection of all comparisons considered simultaneously would not. In this section, we discuss two solutions to this problem—Tukey's method and the Bonferroni method. More discussion can be found in Miller (1981).

12.2.2.1 Tukey's Method   Tukey's method is used to construct confidence intervals for the differences of all pairs of means in such a way that the intervals simultaneously have a set coverage probability. The duality of confidence intervals and tests can then be used to determine which particular pairs are significantly different.

If the sample sizes are all equal and the errors are normally distributed with a constant variance, the centered sample means, $\overline{Y}_{i.} - \mu_i$, are independent and normally distributed with means 0 and variances σ^2/J, which may be estimated by s_p^2/J. Tukey's method is based on the probability distribution of the random variable

$$\max_{i_1, i_2} \frac{|(\overline{Y}_{i_1.} - \mu_{i_1}) - (\overline{Y}_{i_2.} - \mu_{i_2})|}{s_p/\sqrt{J}}$$

where the maximum is taken over all pairs i_1 and i_2. This distribution is called the **studentized range distribution** with parameters I (the number of samples being compared) and $I(J-1)$ (the degrees of freedom in s_p). The upper 100α percentage point of the distribution is denoted by $q_{I,I(J-1)}(\alpha)$. Now,

$$P\left[|(\overline{Y}_{i_1.} - \mu_{i_1}) - (\overline{Y}_{i_2.} - \mu_{i_2})| \le q_{I,I(J-1)}(\alpha)\frac{s_p}{\sqrt{J}}, \text{ for all } i_1 \text{ and } i_2\right]$$

$$= P\left[\max_{i_1,i_2}|(\overline{Y}_{i_1.} - \mu_{i_1}) - (\overline{Y}_{i_2.} - \mu_{i_2})| \le q_{I,I(J-1)}(\alpha)\frac{s_p}{\sqrt{J}}\right]$$

By definition, this latter probability equals $1 - \alpha$. The idea is that all the differences are less than some number if and only if the largest difference is. The above probability statement can be converted directly into a set of confidence intervals that hold simultaneously for all differences $\mu_{i_1} - \mu_{i_2}$ with confidence level $100(1 - \alpha)\%$. The intervals are

$$(\overline{Y}_{i_1.} - \overline{Y}_{i_2.}) \pm q_{I,I(J-1)}(\alpha)\frac{s_p}{\sqrt{J}}$$

By the duality of confidence intervals and hypothesis tests, if the $100(1 - \alpha)\%$ confidence interval for $(\overline{Y}_{i_1.} - \overline{Y}_{i_2.})$ does not include zero—that is, if

$$|\overline{Y}_{i_1.} - \overline{Y}_{i_2.}| > q_{I,I(J-1)}(\alpha)\frac{s_p}{\sqrt{J}}$$

the null hypothesis that there is no difference between u_{i_1} and u_{i_2} may be rejected at level α. Also, all such hypothesis tests considered collectively have level α.

E X A M P L E A   We can illustrate Tukey's method by applying it to the tablet data of Section 12.2. We list the labs in decreasing order of the mean of their measurements:

Lab	Mean
1	4.062
3	4.003
7	3.998
2	3.997
5	3.957
6	3.955
4	3.920

s_p is the square root of the mean square for error in the analysis of variance table of Example A of Section 12.2.1: $s_p = .06$. The degrees of freedom of the appropriate studentized range distribution are 7 and 63. Using 7 and 60 df in Table 6 of Appendix B as an approximation, $q_{7,60}(.05) = 4.31$, two of the means in the preceding table are significantly different at the .05 level if they differ by more than

$$q_{7,63}(.05)\frac{s_p}{\sqrt{J}} = .082$$

The mean from lab 1 is thus significantly different from those from labs 4, 5, and 6; The mean of lab 3 is significantly greater than that of lab 4. No other comparisons are significant at the .05 level.

At the 95% confidence level, the other differences in mean level that are seen in Figure 12.1 cannot be judged to be significantly different from zero. Although differences between these labs must certainly exist, we cannot reliably establish the signs of the differences.

It is interesting to note that a price is paid here for performing multiple comparisons simultaneously. If separate t tests had been conducted using the pooled sample variance, labs would have been declared significantly different if their means had differed by more than

$$t_{63}(.025)s_p\sqrt{\frac{2}{J}} = .053$$
∎

12.2.2.2 The Bonferroni Method
The Bonferroni method was briefly introduced in Section 11.4.8. The idea is very simple. If k null hypotheses are to be tested, a desired overall type I error rate of at most α can be guaranteed by testing each null hypothesis at level α/k. Equivalently, if k confidence intervals are each formed to have confidence level $100(1 - \alpha/k)\%$, they all hold simultaneously with confidence level at least $100(1 - \alpha)\%$.

The method is simple and versatile and, although crude, gives surprisingly good results if k is not too large.

E X A M P L E A To apply the Bonferroni method to the data on tablets, we note that there are $k = \binom{7}{2} = 21$ pairwise comparisons among the seven labs. A set of simultaneous 95% confidence intervals for the pairwise comparisons is

$$(\overline{Y}_{i_1.} - \overline{Y}_{i_2.}) \pm s_p \frac{t_{63}(.025/21)}{\sqrt{5}}$$

Special tables for such values of the t distribution have been prepared; from Table 7 of Appendix B, we find

$$t_{60}\left(\frac{.025}{20}\right) = 3.16$$

which we will use as an approximation to $t_{63}(.025/21)$, giving confidence intervals

$$(\overline{Y}_{i_1.} - \overline{Y}_{i_2.}) \pm .085$$

that we will use as an approximation. Given the crude nature of the Bonferroni method, these are surprisingly close to the intervals produced by Tukey's method, which have a half-width of .082. Here, too, we conclude that lab 1 produced significantly higher measurements than those of labs 4, 5, and 6.
∎

A significant advantage of the Bonferroni method over Tukey's method is that it does not require equal sample sizes in each treatment.

12.2.3   A Nonparametric Method—The Kruskal-Wallis Test

The Kruskal-Wallis test is a generalization of the Mann-Whitney test that is conceptually quite simple. The observations are assumed to be independent, but no particular distributional form, such as the normal, is assumed. The observations are pooled together and ranked. Let

$$R_{ij} = \text{the rank of } Y_{ij} \text{ in the combined sample}$$

Let

$$\overline{R}_{i.} = \frac{1}{J_i} \sum_{j=1}^{J_i} R_{ij}$$

be the average rank in the ith group. Let

$$\overline{R}_{..} = \frac{1}{N} \sum_{i=1}^{I} \sum_{j=1}^{J_i} R_{ij}$$

$$= \frac{N+1}{2}$$

where N is the total number of observations. As in the analysis of variance, let

$$SS_B = \sum_{i=1}^{I} J_i (\overline{R}_{i.} - \overline{R}_{..})^2$$

be a measure of the dispersion of the $\overline{R}_{i.}$. SS_B may be used to test the null hypothesis that the probability distributions generating the observations under the various treatments are identical. The larger SS_B is, the stronger is the evidence against the null hypothesis. The exact null distribution of this statistic for various combinations of I and J_i can be enumerated, as for the Mann-Whitney test. The null distribution is commonly available in computer packages. Tables are given in Lehmann (1975) and in references therein. For $I = 3$ and $J_i \geq 5$ or $I > 3$ and $J_i \geq 4$, a chi-square approximation to a normalized version of SS_B is fairly accurate. Under the null hypothesis that the probability distributions of the I groups are identical, the statistic

$$K = \frac{12}{N(N+1)} SS_B$$

is approximately distributed as a chi-square random variable with $I - 1$ degrees of freedom. The value of K can be found by running the ranks through an analysis of variance program and multiplying SS_B by $12/[N(N+1)]$. It can be shown that K can also be expressed as

$$K = \frac{12}{N(N+1)} \left(\sum_{i=1}^{I} J_i \overline{R}_{i.}^2 \right) - 3(N+1)$$

which is easier to compute by hand.

EXAMPLE A   For the data on the tablets, $K = 29.51$. Referring to Table 3 of Appendix B with 6 df, we see that the p-value is less than .005. The nonparametric analysis, too, indicates that there is a systematic difference among the labs.   ■

Multiple comparison procedures for nonparametric methods are discussed in detail in Miller (1981). The Bonferroni method requires no special discussion; it can be applied to all comparisons tested by Mann-Whitney tests.

Like the Mann-Whitney test, the Kruskal-Wallis test makes no assumption of normality and thus has a wider range of applicability than does the F test. It is especially useful in small-sample situations. Also, because data are replaced by their ranks, outliers will have less influence on this nonparametric test than on the F test. In some applications, the data consist of ranks—for example, in a wine tasting, judges usually rank the wines—which makes the use of the Kruskal-Wallis test natural.

12.3 The Two-Way Layout

A **two-way layout** is an experimental design involving two factors, each at two or more levels. The levels of one factor might be various drugs, for example, and the levels of the other factor might be genders. If there are I levels of one factor and J of the other, there are $I \times J$ combinations. We will assume that K independent observations are taken for each of these combinations. (The last section of this chapter will outline the advantages of such an experimental design.)

The next section defines the parameters that we might want to estimate from a two-way layout. Later sections present statistical methods based on normal theory and nonparametric methods.

12.3.1 Additive Parametrization

To develop and illustrate the ideas in this section, we will use a portion of the data contained in a study of electric range energy consumption (Fechter and Porter 1978). The following table shows the mean number of kilowatt-hours used by three electric ranges in cooking on each of three menu days (means are over several cooks).

Menu Day	Range 1	Range 2	Range 3
1	3.97	4.24	4.44
2	2.39	2.61	2.82
3	2.76	2.75	3.01

We wish to describe the variation in the numbers in this table in terms of the effects of different ranges and different menu days. Denoting the number in the ith row and jth column by Y_{ij}, we first calculate a grand average

$$\hat{\mu} = \overline{Y}_{..} = \frac{1}{9} \sum_{i=1}^{3} \sum_{j=1}^{3} Y_{ij} = 3.22$$

This gives a measure of typical energy consumption per menu day.

The menu day means, averaged over the ranges, are

$$\overline{Y}_{1.} = 4.22$$
$$\overline{Y}_{2.} = 2.61$$
$$\overline{Y}_{3.} = 2.84$$

We will define the differential effect of a menu day as the difference between the mean for that day and the overall mean; we will denote these differential effects by $\hat{\alpha}_i$, where $i = 1, 2,$ or 3.

$$\hat{\alpha}_1 = \overline{Y}_{1.} - \overline{Y}_{..} = 1.00$$
$$\hat{\alpha}_2 = \overline{Y}_{2.} - \overline{Y}_{..} = -.61$$
$$\hat{\alpha}_3 = \overline{Y}_{3.} - \overline{Y}_{..} = -.38$$

(Note that, except for rounding error, the α_i would sum to zero.) In words, on menu day 1, 1 kwh more than the average is consumed, and so on.

The range means, averaged over the menu days, are

$$\overline{Y}_{.1} = 3.04$$
$$\overline{Y}_{.2} = 3.20$$
$$\overline{Y}_{.3} = 3.42$$

The differential effects of the ranges are

$$\hat{\beta}_1 = \overline{Y}_{.1} - \overline{Y}_{..} = -.18$$
$$\hat{\beta}_2 = \overline{Y}_{.2} - \overline{Y}_{..} = -.02$$
$$\hat{\beta}_3 = \overline{Y}_{.3} - \overline{Y}_{..} = .20$$

The effects of the ranges are smaller than the effects of the menu days.

The preceding description of the values in the table incorporates an overall average level plus differential effects of ranges and menu days. This is a simple **additive model.**

$$\hat{Y}_{ij} = \hat{\mu} + \hat{\alpha}_i + \hat{\beta}_j$$

Here we use $\hat{Y}_{ij}$ to denote the fitted or predicted values of Y_{ij} from the additive model. According to this additive model, the differences between the three ranges are the same on all menu days. For example, for $i = 1, 2, 3,$

$$\hat{Y}_{i1} - \hat{Y}_{i2} = (\hat{\mu} + \hat{\alpha}_i + \hat{\beta}_1) - (\hat{\mu} + \hat{\alpha}_i + \hat{\beta}_2)$$
$$= \hat{\beta}_1 - \hat{\beta}_2$$

Figure 12.3 shows that this is not quite the case. If the differences were exactly the same on all menu days, the three lines would be exactly parallel. The differences between menu days 1 and 2 appear nearly the same—the lines are nearly parallel. But on menu day 3, the difference between ranges 2 and 3 increased and the difference between

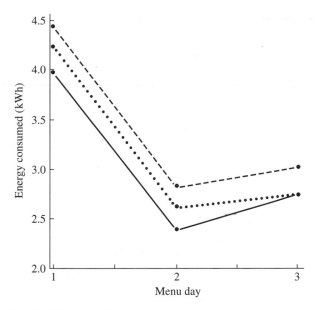

FIGURE **12.3** Plot of energy consumption versus menu day for three electric ranges. The dashed line corresponds to range 3, the dotted line to range 2, and the solid line to range 1.

ranges 1 and 2 decreased. This phenomenon is called an **interaction** between menu days and ranges—it is as if there were something about menu day 3 that especially affected adversely the energy consumption of range 1 relative to range 2.

The differences of the observed values and the fitted values, $Y_{ij} - \hat{Y}_{ij}$, are the residuals from the additive model and are shown in the following table:

Menu Day	Range 1	Range 2	Range 3
1	−.07	.04	.02
2	−.04	.02	.01
3	.10	−.07	−.03

The residuals are small relative to the main effects, with the possible exception of those for menu day 3.

Interactions can be incorporated into the model to make it fit the data exactly. The residual in cell ij is

$$Y_{ij} - \hat{\mu} - \hat{\alpha}_i - \hat{\beta}_j = Y_{ij} - Y_{..} - (\overline{Y}_{i.} - \overline{Y}_{..}) - (\overline{Y}_{.j} - \overline{Y}_{..})$$
$$= Y_{ij} - \overline{Y}_{i.} - \overline{Y}_{.j} + \overline{Y}_{..}$$
$$= \hat{\delta}_{ij}$$

Note that

$$\sum_{i=1}^{3} \hat{\delta}_{ij} = \sum_{j=1}^{3} \hat{\delta}_{ij} = 0$$

For example,

$$\sum_{i=1}^{3} \hat{\delta}_{ij} = \sum_{i=1}^{3} (Y_{ij} - \overline{Y}_{i.} - \overline{Y}_{.j} + \overline{Y}_{..})$$

$$= 3\overline{Y}_{.j} - 3\overline{Y}_{..} - 3\overline{Y}_{.j} + 3\overline{Y}_{..}$$

$$= 0$$

In the preceding table of residuals, the row and column sums are not exactly zero because of rounding errors. The model

$$Y_{ij} = \hat{\mu} + \hat{\alpha}_i + \hat{\beta}_j + \hat{\delta}_{ij}$$

thus fits the data exactly; it is merely another way of expressing the numbers listed in the table.

An additive model is simple and easy to interpret, especially in the absence of interactions. Transformations of the data are sometimes used to improve the adequacy of an additive model. The logarithmic transformation, for example, converts a multiplicative model into an additive one. Transformations are also used to stabilize the variance (to make the variance independent of the mean) and to make normal theory more applicable. There is no guarantee, of course, that a given transformation will accomplish all these aims.

The discussion in this section has centered on the parametrization and interpretation of the additive model as used in the analysis of variance. We have not taken into account the possibility of random errors and their effects on the inferences about the parameters, but will do so in the next section.

12.3.2   Normal Theory for the Two-Way Layout

In this section, we will assume that there are $K > 1$ observations per cell in a two-way layout. A design with an equal number of observations per cell is called **balanced.** Let Y_{ijk} denote the kth observation in cell ij; the statistical model is

$$Y_{ijk} = \mu + \alpha_i + \beta_j + \delta_{ij} + \varepsilon_{ijk}$$

We will assume that the random errors, ε_{ijk}, are independent and normally distributed with mean zero and common variance σ^2. Thus, $E(Y_{ijk}) = \mu + \alpha_i + \beta_j + \delta_{ij}$. The parameters satisfy the following constraints:

$$\sum_{i=1}^{I} \alpha_i = 0$$

$$\sum_{j=1}^{J} \beta_j = 0$$

$$\sum_{i=1}^{I} \delta_{ij} = \sum_{j=1}^{J} \delta_{ij} = 0$$

We now find the mle's of the unknown parameters. Since the observations in cell ij are normally distributed with mean $\mu + \alpha_i + \beta_j + \delta_{ij}$ and variance σ^2, and since all the observations are independent, the log likelihood is

$$l = -\frac{IJK}{2} \log(2\pi\sigma^2) - \frac{1}{2\sigma^2} \sum_{i=1}^{I} \sum_{j=1}^{J} \sum_{k=1}^{K} (Y_{ijk} - \mu - \alpha_i - \beta_j - \delta_{ij})^2$$

Maximizing the likelihood subject to the constraints given above yields the following estimates (see Problem 17 at the end of this chapter):

$$\hat{\mu} = \overline{Y}_{...}$$
$$\hat{\alpha}_i = \overline{Y}_{i..} - \overline{Y}_{...}, \qquad i = 1, \ldots, I$$
$$\hat{\beta}_j = \overline{Y}_{.j.} - \overline{Y}_{...}, \qquad j = 1, \ldots, J$$
$$\hat{\delta}_{ij} = \overline{Y}_{ij.} - \overline{Y}_{i..} - \overline{Y}_{.j.} + \overline{Y}_{...}$$

as is expected from the discussion in Section 12.3.1.

Like one-way analysis of variance, two-way analysis of variance is conducted by comparing various sums of squares. The sums of squares are as follows:

$$SS_A = JK \sum_{i=1}^{I} (\overline{Y}_{i..} - \overline{Y}_{...})^2$$

$$SS_B = IK \sum_{j=1}^{J} (\overline{Y}_{.j.} - \overline{Y}_{...})^2$$

$$SS_{AB} = K \sum_{i=1}^{I} \sum_{j=1}^{J} (\overline{Y}_{ij.} - \overline{Y}_{i..} - \overline{Y}_{.j.} + \overline{Y}_{...})^2$$

$$SS_E = \sum_{i=1}^{I} \sum_{j=1}^{J} \sum_{k=1}^{K} (Y_{ijk} - \overline{Y}_{ij.})^2$$

$$SS_{TOT} = \sum_{i=1}^{I} \sum_{j=1}^{J} \sum_{k=1}^{K} (Y_{ijk} - \overline{Y}_{...})^2$$

The sums of squares satisfy this algebraic identity:

$$SS_{TOT} = SS_A + SS_B + SS_{AB} + SS_E$$

This identity may be proved by writing

$$Y_{ijk} - \overline{Y}_{...} = (Y_{ijk} - \overline{Y}_{ij.}) + (\overline{Y}_{i..} - \overline{Y}_{...}) + (\overline{Y}_{.j.} - \overline{Y}_{...})$$
$$+ (\overline{Y}_{ij.} - \overline{Y}_{i..} - \overline{Y}_{.j.} + \overline{Y}_{...})$$

and then squaring both sides, summing, and verifying that the cross products vanish.

The following theorem gives the expectations of these sums of squares.

THEOREM **A**

Under the assumption that the errors are independent with mean zero and variance σ^2,

$$E(SS_A) = (I-1)\sigma^2 + JK \sum_{i=1}^{I} \alpha_i^2$$

$$E(SS_B) = (J-1)\sigma^2 + IK \sum_{j=1}^{J} \beta_j^2$$

$$E(SS_{AB}) = (I-1)(J-1)\sigma^2 + K \sum_{i=1}^{I} \sum_{j=1}^{J} \delta_{ij}^2$$

$$E(SS_E) = IJ(K-1)\sigma^2$$

Proof

The results for SS_A, SS_B, and SS_E follow from Lemma A of Section 12.2.1. Applying the lemma to SS_{TOT}, we have

$$E(SS_{TOT}) = E \sum_{i=1}^{I} \sum_{j=1}^{J} \sum_{k=1}^{K} (Y_{ijk} - \overline{Y}_{...})^2$$

$$= (IJK-1)\sigma^2 + \sum_{i=1}^{I} \sum_{j=1}^{J} \sum_{k=1}^{K} (\alpha_i + \beta_j + \delta_{ij})^2$$

$$= (IJK-1)\sigma^2 + JK \sum_{i=1}^{I} \alpha_i^2 + IK \sum_{j=1}^{J} \beta_j^2 + K \sum_{i=1}^{I} \sum_{j=1}^{J} \delta_{ij}^2$$

In the last step, we used the constraints on the parameters. For example, the cross product involving α_i and β_j is

$$\sum_{i=1}^{I} \sum_{j=1}^{J} \sum_{k=1}^{K} \alpha_i \beta_j = K \left(\sum_{i=1}^{I} \alpha_i \right) \left(\sum_{j=1}^{J} \beta_j \right)$$
$$= 0$$

The desired expression for $E(SS_{AB})$ now follows, since

$$E(SS_{TOT}) = E(SS_A) + E(SS_B) + E(SS_{AB}) + E(SS_E) \qquad ■$$

The distributions of these sums of squares are given by the following theorem.

THEOREM **B**

Assume that the errors are independent and normally distributed with means zero and variances σ^2. Then

a. SS_E/σ^2 follows a chi-square distribution with $IJ(K-1)$ degrees of freedom.
b. Under the null hypothesis

$$H_A: \alpha_i = 0, i = 1, \ldots, I$$

SS_A/σ^2 follows a chi-square distribution with $I-1$ degrees of freedom.
c. Under the null hypothesis

$$H_B: \beta_j = 0, \; j = 1, \ldots, J$$

SS_B/σ^2 follows a chi-square distribution with $J-1$ degrees of freedom.
d. Under the null hypothesis

$$H_{AB}: \delta_{ij} = 0, \; i = 1, \ldots, I, \; j = 1, \ldots, J$$

SS_{AB}/σ^2 follows a chi-square distribution with $(I-1)(J-1)$ degrees of freedom.
e. The sums of squares are independently distributed.

Proof

We will not give a full proof of this theorem. The results for SS_A, SS_B, and SS_E follow from arguments similar to those used in proving Theorem B of Section 12.2.1. The result for SS_{AB} requires some additional argument. ∎

F tests of the various null hypotheses are conducted by comparing the appropriate sums of squares to the sum of squares for error, as was done for the simpler case of the one-way layout. The mean squares are the sums of squares divided by their degrees of freedom and the F statistics are ratios of mean squares. When such a ratio is substantially larger than 1, the presence of an effect is suggested. Note, for example, that from Theorem A, $E(MS_A) = \sigma^2 + (JK/(I-1)) \sum_i \alpha_i^2$ and that $E(MS_E) = \sigma^2$. So if the ratio MS_A/MS_E is large, it suggests that some of the α_i are nonzero. The null distribution of this F statistic is the F distribution with $(I-1)$ and $IJ(K-1)$ degrees of freedom, and knowing this null distribution allows us to assess the significance of the ratio.

EXAMPLE **A**   As an example, we return to the experiment on iron retention discussed in Section 11.2.1.1. In the complete experiment, there were $I = 2$ forms of iron, $J = 3$ dosage levels, and $K = 18$ observations per cell. In Section 11.2.1.1, we discussed a logarithmic transformation of the data to make it more nearly normal and to stabilize the variance. Figure 12.4 shows boxplots of the data on the original scale; boxplots of the log data are given in Figure 12.5. The distribution of the log data is more symmetrical, and the interquartile ranges are less variable. Figure 12.6 is a plot of cell standard

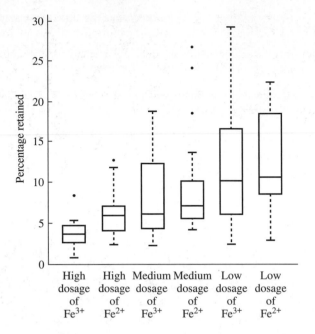

FIGURE 12.4 Boxplots of iron retention for two forms of iron at three dosage levels.

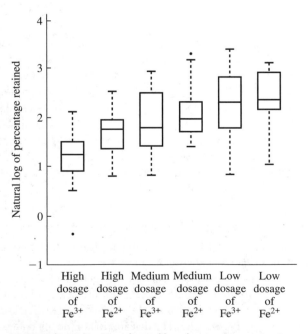

FIGURE 12.5 Boxplots of log data on iron retention.

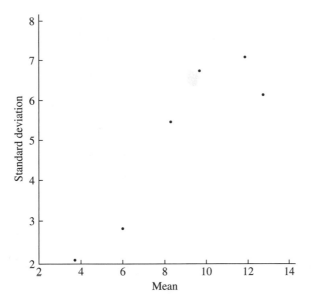

FIGURE **12.6**   Plot of cell standard deviations versus cell means for iron retention data.

deviations versus cell means for the untransformed data; it shows that the error variance increases with the mean. Figure 12.7 is a plot of cell standard deviations versus means for the log data; it shows that the transformation is successful in stabilizing the variance. Note that one of the assumptions of Theorem B is that the errors have equal variance.

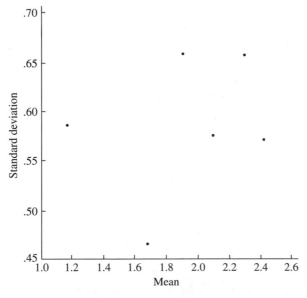

FIGURE **12.7**   Plot of cell standard deviations versus cell means for log data on iron retention.

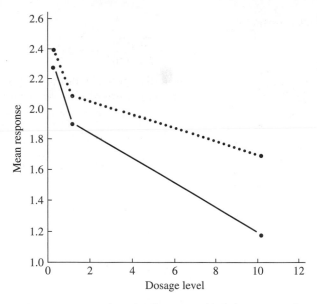

F I G U R E **12.8**   Plot of cell means of log data versus dosage level. The dashed line corresponds to Fe^{2+} and the solid line to Fe^{3+}.

Figure 12.8 is a plot of the cell means of the transformed data versus the dosage levels for the two forms of iron. It suggests that Fe^{2+} may be retained more than Fe^{3+}. If there is no interaction, the two curves should be parallel except for random variation. This appears to be roughly the case, although there is a hint that the difference in retention of the two forms of iron increases with dosage level. To check this, we will perform a quantitative test for interaction.

In the following analysis of variance table, SS_A is the sum of squares due to the form of iron, SS_B is the sum of squares due to dosage, and SS_{AB} is the sum of squares due to interaction. The F statistics were found by dividing the appropriate mean square by the mean square for error.

Analysis of Variance Table

Source	df	SS	MS	F
Iron form	1	2.074	2.074	5.99
Dosage	2	15.588	7.794	22.53
Interaction	2	.810	.405	1.17
Error	102	35.296	.346	
Total	107	53.768		

To test the effect of the form of iron, we test

$$H_A: \alpha_1 = \alpha_2 = 0$$

using the statistic

$$F = \frac{SS_{IRON}/1}{SS_E/102} = 5.99$$

From computer evaluation of the F distribution with 1 and 102 df, the p-value is less than .025. There is an effect due to the form of iron. An estimate of the difference $\alpha_1 - \alpha_2$ is

$$\overline{Y}_{1..} - \overline{Y}_{2..} = .28$$

and a confidence interval for the difference may be obtained by noting that $\overline{Y}_{1..}$ and $\overline{Y}_{2..}$ are uncorrelated since they are averages over different observations and that

$$\mathrm{Var}(\overline{Y}_{1..}) = \mathrm{Var}(\overline{Y}_{2..}) = \frac{\sigma^2}{JK}$$

Thus,

$$\mathrm{Var}(\overline{Y}_{1..} - \overline{Y}_{2..}) = \frac{2\sigma^2}{JK}$$

Estimating σ^2 by the mean square for error, $\mathrm{Var}(\overline{Y}_{1..} - \overline{Y}_{2..})$ is estimated by

$$s^2_{\overline{Y}_{1.}-\overline{Y}_{2.}} = \frac{2 \times .346}{54} = .0128$$

A confidence interval can be constructed using the t distribution with $IJ(K-1)$ degrees of freedom. The interval is of the form

$$(\overline{Y}_{1..} - \overline{Y}_{2..}) \pm t_{IJ(K-1)}(\alpha/2)s_{\overline{Y}_{1.}-\overline{Y}_{2.}}$$

There are 102 df; to form a 95% confidence interval we use $t_{120}(.025) = 1.98$ from Table 4 of Appendix B as an approximation, producing the interval $.28 \pm 1.98\sqrt{.0128}$, or $(.06, .5)$.

Recall that we are working on a log scale. The additive effect of .28 on the log scale corresponds to a multiplicative effect of $e^{.28} = 1.32$ on a linear scale and the interval $(.06, .50)$ transforms to $(e^{.06}, e^{.50})$, or $(1.06, 1.65)$. Thus, we estimate that Fe^{2+} increases retention by a factor of 1.32, and the uncertainty in this factor is expressed in the confidence interval $(1.06, 1.65)$.

The F statistic for testing the effect of dosage is significant, but this effect is expected and is not of major interest.

To test the hypothesis H_{AB} which states that there is no interaction, we consider the following F statistic:

$$F = \frac{SS_{AB}/(I-1)(J-1)}{SS_E/IJ(K-1)} = 1.17$$

From computer evaluation of the F distribution with 2 and 102 df, the p-value is .31, so there is insufficient evidence to reject this hypothesis. Thus, the deviation of the lines of Figure 12.8 from parallelism could easily be due to chance.

In conclusion, it appears that there is a difference of 6–65% in the ratio of percentage retained between the two forms of iron and that there is little evidence that this difference depends on dosage. ∎

12.3.3 Randomized Block Designs

Randomized block designs originated in agricultural experiments. To compare the effects of I different fertilizers, J relatively homogeneous plots of land, or blocks, are selected, and each is divided into I plots. Within each block, the assignment of fertilizers to plots is made at random. By comparing fertilizers within blocks, the variability between blocks, which would otherwise contribute "noise" to the results, is controlled. This design is a multisample generalization of a matched-pairs design.

A randomized block design might be used by a nutritionist who wants to compare the effects of three different diets on experimental animals. To control for genetic variation in the animals, the nutritionist might select three animals from each of several litters and randomly determine their assignments to the diets. Randomized block designs are used in many areas. If an experiment is to be carried out over a substantial period of time, the blocks may consist of stretches of time. In industrial experiments, the blocks are often batches of raw material.

Randomization helps ensure against unintentional bias and can form a basis for inference. In principle, the null distribution of a test statistic can be derived by permutation arguments, just as we derived the null distribution of the Mann-Whitney test statistic in Section 11.2.3. Parametric procedures often give a good approximation to the permutation distribution.

As a model for the responses in the randomized block design, we will use

$$Y_{ij} = \mu + \alpha_i + \beta_j + \varepsilon_{ij}$$

where α_i is the differential effect of the ith treatment, β_j is the differential effect of the jth block, and the ε_{ij} are independent random errors. This is the model of Section 12.3.2 but with the additional assumption of no interactions between blocks and treatments. Interest is focused on the α_i.

From Theorem A of Section 12.3.2, if there is no interaction,

$$E(MS_A) = \sigma^2 + \frac{J}{I-1} \sum_{i=1}^{I} \alpha_i^2$$

$$E(MS_B) = \sigma^2 + \frac{I}{J-1} \sum_{j=1}^{J} \beta_j^2$$

$$E(MS_{AB}) = \sigma^2$$

Thus, σ^2 can be estimated from MS_{AB}. Also, since these mean squares are independently distributed, F tests can be performed to test H_A or H_B. For example, to test

$$H_A: \alpha_i = 0, \ i = 1, \ldots, I$$

this statistic can be used:

$$F = \frac{MS_A}{MS_{AB}}$$

From Theorem B in Section 12.3.2, under H_A, the statistic follows an F distribution with $I-1$ and $(I-1)(J-1)$ degrees of freedom. H_B may be tested similarly but is not usually of interest. Note that if, contrary to the assumption, there is an interaction, then

$$E(MS_{AB}) = \sigma^2 + \frac{1}{(I-1)(J-1)} \sum_{i=1}^{I} \sum_{j=1}^{J} \delta_{ij}^2$$

MS_{AB} will tend to overestimate σ^2. This will cause the F statistic to be smaller than it should be and will result in a test that is conservative; that is, the actual probability of type I error will be smaller than desired.

EXAMPLE A   Let us consider an experimental study of drugs to relieve itching (Beecher 1959). Five drugs were compared to a placebo and no drug with 10 volunteer male subjects aged 20–30. (Note that this set of subjects limits the scope of inference; from a statistical point of view, one cannot extrapolate the results of the experiment to older women, for example. Any such extrapolation could be justified only on grounds of medical judgment.) Each volunteer underwent one treatment per day, and the time-order was randomized. Thus, individuals were "blocks." The subjects were given a drug (or placebo) intravenously, and then itching was induced on their forearms with cowage, an effective itch stimulus. The subjects recorded the duration of the itching. More details are in Beecher (1959). The following table gives the durations of the itching (in seconds):

Subject	No Drug	Placebo	Papa-verine	Morphine	Amino-phylline	Pento-barbital	Tripelen-namine
BG	174	263	105	199	141	108	141
JF	224	213	103	143	168	341	184
BS	260	231	145	113	78	159	125
SI	255	291	103	225	164	135	227
BW	165	168	144	176	127	239	194
TS	237	121	94	144	114	136	155
GM	191	137	35	87	96	140	121
SS	100	102	133	120	222	134	129
MU	115	89	83	100	165	185	79
OS	189	433	237	173	168	188	317
Average	191.0	204.8	118.2	148.0	144.3	176.5	167.2

Figure 12.9 shows boxplots of the responses to the six treatments and to the control (no drugs). Although the boxplot is probably not the ideal visual display of these data, since it takes no account of the blocking, Figure 12.9 does show some interesting aspects of the data. There is a suggestion that all the drugs had some effect and that papaverine was the most effective. There is a lot of scatter relative to the differences between the medians, and there are some outliers. It is interesting that the placebo responses have the greatest spread; this might be because some subjects responded to the placebo and some did not.

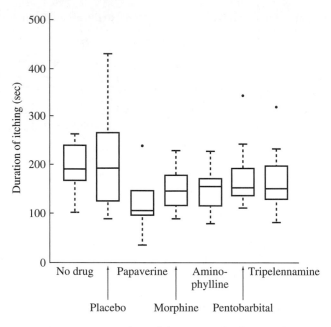

FIGURE **12.9** Boxplots of durations of itching under seven treatments.

We next construct an analysis of variance table for this experiment:

Source	df	SS	MS	F
Drugs	6	53013	8835	2.85
Subjects	9	103280	11476	3.71
Interaction	54	167130	3095	
Total	69	323422		

The F statistic for testing differences between drugs is 2.85 with 6 and 54 df, corresponding to a p-value less than .025. The null hypothesis that there is no difference between subjects is not experimentally interesting.

Figure 12.10 is a probability plot of the residuals from the two-way analysis of variance model. The residual in cell ij is

$$r_{ij} = Y_{ij} - \hat{\mu} - \hat{\alpha}_i - \hat{\beta}_j$$
$$= Y_{ij} - \overline{Y}_{i.} - \overline{Y}_{.j} + \overline{Y}_{..}$$

There is a slightly bowed character to the probability plot, indicating some skewness in the distribution of the residuals. But because the F test is robust against moderate deviations from normality, we should not be overly concerned.

Tukey's method may be applied to make multiple comparisons. Suppose that we want to compare the drug means, $\overline{Y}_{1.}, \ldots, \overline{Y}_{7.}$ ($I = 7$). These have expectations $\mu + \alpha_i$, where $i = 1, \ldots, I$, and each is an average over $J = 10$ independent observations. The error variance is estimated by MS_{AB} with 54 df. Simultaneous 95%

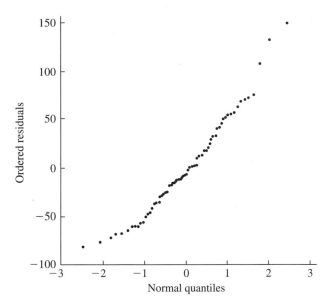

F I G U R E **12.10**   Normal probability plot of residuals from two-way analysis of variance of data on duration of itching.

confidence intervals for all differences between drug means have half-widths of

$$\frac{q_{7,54}(.05)s}{\sqrt{J}} = 4.31\sqrt{\frac{3095}{10}}$$
$$= 75.8$$

[Here we have used $q_{7,60}(.05)$ from Table 6 of Appendix B as an approximation to $q_{7,54}(.05)$.] Examining the table of means, we see that, at the 95% confidence level, we can conclude only that papaverine achieves a reduction of itching over the effect of a placebo.   ■

12.3.4 A Nonparametric Method—Friedman's Test

This section presents a nonparametric method for the randomized block design. Like other nonparametric methods we have discussed, Friedman's test relies on ranks and does not make an assumption of normality. The test is very simple. Within each of the J blocks, the observations are ranked. To test the hypothesis that there is no effect due to the factor corresponding to treatments (I), the following statistic is calculated:

$$SS_A = J \sum_{i=1}^{I} (\overline{R}_{i.} - \overline{R}_{..})^2$$

just as in the ordinary analysis of variance. Under the null hypothesis that there is no treatment effect and that the only effect is due to the randomization within blocks, the permutation distribution of the statistic can, in principle, be calculated.

For sample sizes such as that of the itching experiment, a chi-square approximation to this distribution is perfectly adequate. The null distribution of

$$Q = \frac{12J}{I(I+1)} \sum_{i=1}^{I} (\overline{R}_{i.} - \overline{R}_{..})^2$$

is approximately chi-square with $I - 1$ degrees of freedom.

EXAMPLE A   To carry out Friedman's test on the data from the experiment on itching, we first construct the following table by ranking durations of itching for each subject:

	No Drug	Placebo	Papaverine	Morphine	Amino-phylline	Pento-barbital	Tripelen-namine
BG	5	7	1	6	3.5	2	3.5
JF	6	5	1	2	3	7	4
BS	7	6	4	2	1	5	3
SI	6	7	1	4	3	2	5
BW	3	4	2	5	1	7	6
TS	7	3	1	5	2	4	6
GM	7	5	1	2	3	6	4
SS	1	2	5	3	7	6	4
MU	5	3	2	4	6	7	1
OS	4	7	5	2	1	3	6
Average	5.10	4.90	2.30	3.50	3.05	4.90	4.25

Note that we have handled ties in the usual way by assigning average ranks. From the preceding table, no drug, placebo, and pentobarbitol have the highest average ranks. From these average ranks, we find $\overline{R} = 4$, $\sum(\overline{R}_{i.} - \overline{R}_{..})^2 = 6.935$ and $Q = 14.86$. From Table 3 of Appendix B with 6 df, the p-value is less than .025. The nonparametric analysis also rejects the hypothesis that there is no drug effect.   ■

Procedures for using Friedman's test for multiple comparisons are discussed by Miller (1981). When these methods are applied to the data from the experiment on itching, the conclusions reached are identical to those reached by the parametric analysis.

12.4 Concluding Remarks

The most complicated experimental design considered in this chapter was the two-way layout; more generally, a **factorial design** incorporates several factors with one or more observations per cell. With such a design, the concept of interaction becomes more complicated—there are interactions of various orders. For instance, in a three-factor experiment, there are two-factor and three-factor interactions. It is both interesting and useful that the two-factor interactions in a three-way layout can be estimated using only one observation per cell.

To gain some insight into why factorial designs are effective, we can begin by considering a two-way layout, with each factor at five levels, no interaction, and one observation per cell. With this design, comparisons of two levels of any factor are based on 10 observations. A traditional alternative to this design is to do first an experiment comparing the levels of factor A and then another experiment comparing the levels of factor B. To obtain the same precision as is achieved by the two-way layout in this case, 25 observations in each experiment, or a total of 50 observations, would be needed. The factorial design achieves its economy by using the same observations to compare the levels of factor A as are used to compare the levels of factor B.

The advantages of factorial designs become greater as the number of factors increases. For example, in an experiment with four factors, with each factor at two levels (which might be the presence or absence of some chemical, for example) and one observation per cell, there are 16 observations that may be used to compare the levels of each factor. Furthermore, it can be shown that two- and three-factor interactions can be estimated. By comparison, if each of the four factors were investigated in a separate experiment, 64 observations would be required to attain the same precision.

As the number of factors increases, the number of observations necessary for a factorial experiment with only one observation per cell grows very rapidly. To decrease the cost of an experiment, certain cells, designated in a systematic way, can be left empty, and the main effects and some interactions can still be estimated. Such arrangements are called **fractional factorial designs.**

Similarly, with a randomized block design, the individual blocks may not be able to accommodate all the treatments. For example, in a chemical experiment that compares a large number of treatments, the blocks of the experiment, batches of raw material of uniform quality, may not be large enough. In such situations, **incomplete block designs** may be used to retain the advantages of blocking.

The basic theoretical assumptions underlying the analysis of variance are that the errors are independent and normally distributed with constant variance. Because we cannot fully check the validity of these assumptions in practice and can probably detect only gross violations, it is natural to ask how robust the procedures are with respect to violations of the assumptions. It is impossible to give a complete and conclusive answer to this question. Generally speaking, the independence assumption is probably the most important (and this is true for nonparametric procedures as well). The F test is robust against moderate departures from normality; if the design is balanced, the F test is also robust against unequal error variance.

For further reading, Box, Hunter, and Hunter (1978) is recommended.

12.5 Problems

1. Simulate observations like those of Figure 12.1 under the null hypothesis of no treatment effects. That is, simulate seven batches of ten normally distributed random numbers with mean 4 and variance .0037. Make parallel boxplots of these seven batches like those of Figure 12.1. Do this several times. Your figures

display the kind of variability that random fluctuations can cause; do you see any pairs of labs that appear quite different in either mean level or dispersion?

2. Verify that if $I = 2$, the estimate s_p^2 of Theorem A of Section 11.2.1 is the s_p^2 given in Section 12.2.1.

3. For a one-way analysis of variance with $I = 2$ treatment groups, show that the F statistic is t^2, where t is the usual t statistic for a two-sample case.

4. Prove the analogues of Theorems A and B in Section 12.2.1 for the case of unequal numbers of observations in the cells of a one-way layout.

5. Derive the likelihood ratio test for the null hypothesis of the one-way layout, and show that it is equivalent to the F test.

6. Prove this version of the Bonferroni inequality:

$$P\left(\bigcap_{i=1}^{n} A_i\right) \geq 1 - \sum_{i=1}^{n} P\left(A_i^c\right)$$

(Use Venn diagrams if you wish.) In the context of simultaneous confidence intervals, what is A_i and what is A_i^c?

7. Show that, as claimed in Theorem B of Section 12.2.1, $SS_B/\sigma^2 \sim \chi_{I-1}^2$.

8. Form simultaneous confidence intervals for the difference of the mean of lab 1 and those of labs 4, 5, and 6 in Example A of Section 12.2.2.1.

9. Compare the tables of the t distribution and the studentized range in Appendix B. For example, consider the column corresponding to $t_{.95}$; multiply the numbers in that column by $\sqrt{2}$ and observe that you get the numbers in the column $t = 2$ of the table of $q_{.90}$. Why is this?

10. Suppose that in a one-way layout there are 10 treatments and seven observations under each treatment. What is the ratio of the length of a simultaneous confidence interval for the difference of two means formed by Tukey's method to that of one formed by the Bonferroni method? How do both of these compare in length to an interval based on the t distribution that does not take account of multiple comparisons?

11. Consider a hypothetical two-way layout with four factors (A, B, C, D) each at three levels (I, III, III). Construct a table of cell means for which there is no interaction.

12. Consider a hypothetical two-way layout with three factors (A, B, C) each at two levels (I, II). Is it possible for there to be interactions but no main effects?

13. Show that for comparing two groups the Kruskal-Wallis test is equivalent to the Mann-Whitney test.

14. Show that for comparing two groups Friedman's test is equivalent to the sign test.

15. Show the equality of the two forms of K given in Section 12.2.3:

$$K = \frac{12}{N(N+1)} \sum_{i=1}^{I} J_i (\bar{R}_{i.} - \bar{R}_{..})^2$$

$$= \frac{12}{N(N+1)} \left(\sum_{i=1}^{I} J_i \bar{R}_{i.}^2 \right) - 3(N+1)$$

16. Prove the sums of squares identity for the two-way layout:

$$SS_{TOT} = SS_A + SS_B + SS_{AB} + SS_E$$

17. Find the mle's of the parameters α_i, β_j, δ_{ij}, and μ of the model for the two-way layout.

18. The table below gives the energy use of five gas ranges for seven menu days. (The units are equivalent kilowatt-hours; .239 kwh = 1 ft^3 of natural gas.) Estimate main effects and discuss interaction, paralleling the discussion of Section 12.3.

Menu Day	Range 1	Range 2	Range 3	Range 4	Range 5
1	8.25	8.26	6.55	8.21	6.69
2	5.12	4.81	3,87	4.81	3.99
3	5.32	4.37	3.76	4.67	4.37
4	8.00	6.50	5.38	6.51	5.60
5	6.97	6.26	5.03	6.40	5.60
6	7.65	5.84	5.23	6.24	5.73
7	7.86	7.31	5.87	6.64	6.03

19. Develop a parametrization for a balanced three-way layout. Define main effects and two-factor and three-factor interactions, and discuss their interpretation. What linear constraints do the parameters satisfy?

20. This problem introduces a **random effects model** for the one-way layout. Consider a balanced one-way layout in which the I groups being compared are regarded as being a sample from some larger population. The random effects model is

$$Y_{ij} = \mu + A_i + \varepsilon_{ij}$$

where the A_i are random and independent of each other with $E(A_i) = 0$ and $\text{Var}(A_i) = \sigma_A^2$. The ε_{ij} are independent of the A_i and of each other, and $E(\varepsilon_{ij}) = 0$ and $\text{Var}(\varepsilon_{ij}) = \sigma_\varepsilon^2$.

To fix these ideas, we can consider an example from Davies (1960). The variation of the strength (coloring power) of a dyestuff from one manufactured batch to another was studied. Strength was measured by dyeing a square of cloth with a standard concentration of dyestuff under carefully controlled conditions and visually comparing the result with a standard. The result was numerically scored by a technician. Large samples were taken from six batches of a dyestuff; each sample was well mixed, and from each six subsamples were taken. These

36 subsamples were submitted to the laboratory in random order over a period of several weeks for testing as described. The percentage strengths of the dyestuff are given in the following table.

Batch	Subsample 1	Subsample 2	Subsample 3	Subsample 4	Subsample 5	Subsample 6
I	94.5	93.0	91.0	89.0	96.5	88.0
II	89.0	90.0	92.5	88.5	91.5	91.5
III	88.5	93.5	93.5	88.0	92.5	91.5
IV	100.0	99.0	100.0	98.0	95.0	97.5
V	91.5	93.0	90.0	92.5	89.0	91.0
VI	98.5	100.0	98.0	100.0	96.5	98.0

There are two sources of variability in these numbers: batch-to-batch variability and measurement variability. It is hoped that variability between subsamples has been eliminated by the mixing. We will consider the random effects model,

$$Y_{ij} = \mu + A_i + \varepsilon_{ij}$$

Here, μ is the overall mean level, A_i is the random effect of the ith batch, and ε_{ij} is the measurement error on the jth subsample from the ith batch. We assume that the A_i are independent of each other and of the measurement errors, with $E(A_i) = 0$ and $\text{Var}(A_i) = \sigma_A^2$. The ε_{ij} are assumed to be independent of each other and to have mean 0 and variance σ_ε^2. Thus,

$$\text{Var}(Y_{ij}) = \sigma_A^2 + \sigma_\varepsilon^2$$

Large variability in the Y_{ij} could be caused by large variability among batches, large measurement error, or both. The former could be decreased by changing the manufacturing process to make the batches more homogeneous, and the latter by controlling the scoring process more carefully.

a. Show that for this model

$$E(MS_W) = \sigma_\varepsilon^2$$
$$E(MS_B) = \sigma_\varepsilon^2 + J\sigma_A^2$$

and that therefore σ_ε^2 and σ_A^2 can be estimated from the data. Calculate these estimates.

b. Suppose that the samples had not been mixed, but that duplicate measurements had been made on each subsample. Formulate a model that also incorporates variability between subsamples. How could the parameters of this model be estimated?

21. During each of four experiments on the use of carbon tetrachloride as a worm killer, ten rats were infested with larvae (Armitage 1983). Eight days later, five rats were treated with carbon tetrachloride; the other five were kept as controls. After two more days, all the rats were killed and the numbers of worms were counted. The table below gives the counts of worms for the four control groups. Significant differences, although not expected, might be attributable to changes in

experimental conditions. A finding of significant differences could result in more carefully controlled experimentation and thus greater precision in later work. Use both graphical techniques and the F test to test whether there are significant differences among the four groups. Use a nonparametric technique as well.

Group I	Group II	Group III	Group IV
279	378	172	381
338	275	335	346
334	412	335	340
198	265	282	471
303	286	250	318

22. Referring to Section 12.2, the file `tablets` gives the measurements on chlorpheniramine maleate tablets from another manufacturer. Are there systematic differences between the labs? If so, which pairs differ significantly? How do these data compare to those given for the other manufacturer in Section 12.2?

23. For a study of the release of luteinizing hormone (LH), male and female rats kept in constant light were compared to male and female rats in a regime of 14 h of light and 10 h of darkness. Various dosages of luteinizing releasing factor (LRF) were given: control (saline), 10, 50, 250, and 1250 ng. Levels of LH (in nanograms per milliliter of serum) were measured in blood samples at a later time. Analyze the data given in file LHfemale, LHmale to determine the effects of light regime and LRF on release of LH for both males and females. Use both graphical techniques and more formal analyses.

24. A collaborative study was conducted to study the precision and homogeneity of a method of determining the amount of niacin in cereal products (Campbell and Pelletier 1962). Homogenized samples of bread and bran flakes were enriched with 0, 2, 4, or 8 mg of niacin per 100 g. Portions of the samples were sent to 12 labs, which were asked to carry out the specified procedures on each of three separate days. The data (in milligrams per 100 g) are given in the file `niacin`. Conduct two-way analyses of variance for both the bread and bran data and discuss the results. (Two data points are missing. Substitute for them the corresponding cell means.)

25. This problem deals with an example from Youden (1962). An ingot of magnesium alloy was drawn into a square rod about 100 m long with a cross section of about 4.5 cm on a side. The rod was then cut into 100 bars, each a meter long. Five of these were selected at random, and a test piece 1.2 cm thick was cut from each. From each of these five specimens, 10 test points were selected in a particular geometric pattern. Two determinations of the magnesium content were made at each test point (the analyst ran all 50 points once and then made a set of repeat measurements). The overall purpose of the experiment was to test for homogeneity of magnesium content in the different bars and different locations. Analyze the data in the file `magnesium` (giving percentage of magnesium times 1000) to determine if there is significant variability between bars and between

locations. There are a couple of unexpected aspects of these data—can you find them?

26. The concentrations (in nanograms per milliliter) of plasma epinephrine were measured for 10 dogs under isofluorane, halothane, and cyclopropane anesthesia; the measurements are given in the following table (Perry et al. 1974). Is there a difference in treatment effects? Use a parametric and a nonparametric analysis.

	Dog 1	Dog 2	Dog 3	Dog 4	Dog 5	Dog 6	Dog 7	Dog 8	Dog 9	Dog 10
Isofluorane	.28	.51	1.00	.39	.29	.36	.32	.69	.17	.33
Halothane	.30	.39	.63	.68	.38	.21	.88	.39	.51	.32
Cyclopropane	1.07	1.35	.69	.28	1.24	1.53	.49	.56	1.02	.30

27. Three species of mice were tested for "aggressiveness." The species were A/J, C57, and F2 (a cross of the first two species). A mouse was placed in a 1-m² box, which was marked off into 49 equal squares. The mouse was let go on the center square, and the number of squares traversed in a 5-min period was counted. Analyze the file C57, AJ, F2, using the Bonferroni method, to determine if there is a significant difference among species.

28. Samples of each of three types of stopwatches were tested. The following table gives thousands of cycles (on-off-restart) survived until some part of the mechanism failed (Natrella 1963). Test whether there is a significant difference among the types, and if there is, determine which types are significantly different. Use both a parametric and a nonparametric technique.

Type I	Type II	Type III
1.7	13.6	13.4
1.9	19.8	20.9
6.1	25.2	25.1
12.5	46.2	29.7
16.5	46.2	46.9
25.1	61.1	
30.5		
42.1		
82.5		

29. The performance of a semiconductor depends upon the thickness of a layer of silicon dioxide. In an experiment (Czitrom and Reece, 1997), layer thicknesses were measured at three furnace locations for three types of wafers (virgin wafers, recycled in-house wafers, and recycled wafers from an external source). The data are contained in the file waferlayers. Conduct a two-way analysis of variance and test for significance of main effects and interactions. Construct a graph such as that shown in Figure 12.3. Does the comparison of layer thicknesses depend on furnace location?

30. Ten varieties of linseed were grown on six different plots (Adguna and Labuschagne, 2002). The file `linseed` contains the yields (kg per hectare). Can you conclude that the varieties have different yields? Use Tukey's method to compare the varieties.

31. Problem 39 of Chapter 10 involved a table of maximum windspeeds for 35 years at each of 21 cities. Would you expect an additive model (no interaction) to provide a good fit to the numbers in this table? Why or why not? Check it out.

32. It is known that increased reproductions leads to reduced longevity for female fruitflies. Patridge and Farquhar (1981) studied whether the same phenomenon held for male fruitflies. The data are also discussed in Hanley and Shapiro (1994). The experiment set up five treatment groups, each consisting of 25 randomly assigned male fruitflies. The males in one treatment were housed with eight virgin females per day. In another treatment, the males were housed with one virgin female day. There were three control groups: males housed with eight newly impregnated females, housed with one newly impregnated female, and housed alone. (Newly inseminated females will not usually mate within two days).

  The data are contained in the file `fruitfly`, with a row for each male in the following format:

Column 1: the number of females
Column 2: the type of female—0 denoting newly pregnant, 1 denoting virgin, and 9 when there were no females
Column 3: lifespan in days
Column 4: length of thorax (mm), which is fixed at birth
Column 5: percentage of time spent sleeping

a. Calculate summary statistics for lifespan in each group and compare. Display the data in parallel boxplots. Qualitatively, what do you conclude?
b. Do the same for percentage of time spent sleeping.
c. Make a scatterplot of lifespan versus thorax length. Is thorax length predictive of lifespan and did the randomization balance thorax length between the groups?
d. Use the F test to test for differences in longevity between the groups. Use both Tukey's method and the Bonferroni method to compare all pairs of means. Summarize your conclusions.
e. Repeat the analysis using the Kruskal-Wallis test and the Bonferroni method.
f. How does the availability of virgin females affect the sleep of male fruitflies?

33. How does diet affect longevity? Studies on animals have shown that restricting caloric intake can increase lifespan. Weindruch et al. did an experiment involving six treatment groups of female mice. The data, contained in the file `diet-and-longevity`, are also discussed in Ramsey and Shafer (2002). The groups were:

1. NP: mice ate as much as they wished of a standard diet.
2. N/N85: mice were fed normally before and after weaning. After weaning, their caloric intake was 85 kcal per week, which is the normal average level.

3. N/R50: mice were fed normally before weaning; and after weaning, their caloric intake was restricted to 50 kcal per week.
4. R/R50: mice were fed 50 kcal per week before and after weaning.
5. lopro: mice were fed normally before weaning, a restricted diet of 50 kcal per week after weaning, and the dietary protein content decreased as they got older.
6. N/R40: mice were fed normally before weaning and were given 40 kcal per week after weaning.

As well as making parallel boxplots and conducting an overall test for equality of means, scientific questions of interest involve some specific comparisons. For example, to determine whether reducing from a normal 85 kcal per week to 50 kcal per week, groups N/N85 and N/R50 would be compared. Which groups would you compare to answer the following questions:

   a. Do preweaning dietary restrictions have an effect?
   b. Does reduction in protein have an effect?
   c. Does reduction to 40 kcal per week have an effect?

Formulate the comparisons you wish to make and carry them out using an appropriate Bonferroni correction. What is the purpose of including the group NP?

34. The following table gives the survival times (in hours) for animals in an experiment whose design consisted of three poisons, four treatments, and four observations per cell.

   a. Conduct a two-way analysis of variance to test the effects of the two main factors and their interaction.
   b. Box and Cox (1964) analyzed the reciprocals of the data, pointing out that the reciprocal of a survival time can be interpreted as the rate of death. Conduct a two-way analysis of variance, and compare to the results of part (a). Comment on how well the standard two-way analysis of variance model fits and on the interaction in both analyses.

Poison					Treatment			
	A		B		C		D	
I	3.1	4.5	8.2	11.0	4.3	4.5	4.5	7.1
	4.6	4.3	8.8	7.2	6.3	7.6	6.6	6.2
II	3.6	2.9	9.2	6.1	4.4	3.5	5.6	10.0
	4.0	2.3	4.9	12.4	3.1	4.0	7.1	3.8
III	2.2	2.1	3.0	3.7	2.3	2.5	3.0	3.6
	1.8	2.3	3.8	2.9	2.4	2.2	3.1	3.3

35. The concentration of follicle stimulating hormone (FSH) can be measured through a bioassay. The basic idea is that when FSH is added to a certain culture, a proportional amount of estrogen is produced; hence, after calibration, the amount of FSH can be found by measuring estrogen production. However, determining FSH levels in serum samples is difficult because some factor(s) in the serum inhibit estrogen production and thus screw up the bioassay. An experiment was done to see if it would be effective to pretreat the serum with polyethyleneglycol (PEG) which, it was hoped, precipitates some of the inhibitory substance(s).

Three treatments were applied to prepared cultures: no serum, PEG-treated FSH free serum, and untreated FSH free serum. Each culture had one of eight doses of FSH: 4, 2, 1, .5, .25, .125, .06, or 0.0 mIU/μl. For each serum-dose combination, there were three cultures, and after incubation for three days, each culture was assayed for estrogen by radioimmunoassay. The table that follows gives the results (units are nanograms of estrogen per milliliter). Analyze these data with a view to determining to what degree PEG treatment is successful in removing the inhibitory substances from the serum. Write a brief report summarizing and documenting your conclusions.

Dose	No Serum	PEG Serum	Untreated Serum
.00	1,814.4	372.7	1,745.3
.00	3,043.2	350.1	2,470.0
.00	3,857.1	426.0	1,700.0
.06	2,447.9	628.3	1,919.2
.06	3,320.9	655.0	1,605.1
.06	3,387.6	700.0	2,796.0
.12	4,887.8	1,701.8	1,929.7
.12	5,171.2	2,589.4	1,537.3
.12	3,370.7	1,117.1	1,692.7
.25	10,255.6	4,114.6	1,149.1
.25	9,431.8	2,761.5	743.4
.25	10,961.2	1,975.8	948.5
.50	14,538.8	6,074.3	4,471.9
.50	14,214.3	12,273.9	2,772.1
.50	16,934.5	14,240.9	5,782.3
1.00	19,719.8	17,889.9	11,588.7
1.00	20,801.4	11,685.7	8,249.5
1.00	32,740.7	11,342.4	18,481.5
2.00	16,453.8	11,843.5	10,433.5
2.00	28,793.8	18,320.7	8,181.0
2.00	19,148.5	23,580.6	11,104.0
4.00	17,967.0	12,380.0	10,020.0
4.00	18,768.6	20,039.0	8,448.5
4.00	19,946.9	15,135.6	10,482.8

The Analysis of Categorical Data

13.1 Introduction

This chapter introduces the analysis of data that are in the form of counts in various categories. We will deal primarily with two-way tables, the rows and columns of which represent categories. Suppose that the rows of such a table represent various hair colors and its columns various eye colors and that each cell contains a count of the number of people who fall in that particular cross-classification. We might be interested in dependencies between the row and column classifications—that is, is hair color related to eye color?

We emphasize that the data considered in this chapter are counts, rather than continuous measurements as they were in Chapter 12. Thus, in this chapter, we will make heavy use of the multinomial and chi-square distributions.

13.2 Fisher's Exact Test

We will develop Fisher's exact test in the context of the following example. Rosen and Jerdee (1974) conducted several experiments, using as subjects male bank supervisors attending a management institute. As part of their training, the supervisors had to make decisions on items in an in-basket. The investigators embedded their experimental materials in the contents of the in-baskets. In one experiment, the supervisors were given a personnel file and had to decide whether to promote the employee or to hold the file and interview additional candidates. By random selection, 24 of the supervisors examined a file labeled as being that of a male employee and 24 examined a file labeled as being that of a female employee; the files were otherwise identical. The results are summarized in the following table:

	Male	Female
Promote	21	14
Hold File	3	10

From the results, it appears that there is a sex bias—21 of 24 males were promoted, but only 14 of 24 females were. But someone who was arguing against the presence of sex bias could claim that the results occurred by chance; that is, even if there were no bias and the supervisors were completely indifferent to sex, discrepancies like those observed could occur with fairly large probability by chance alone. To rephrase this argument, the claim is that 35 of the 48 supervisors chose to promote the employee and 13 chose not to, and that 21 of the 35 promotions were of a male employee merely because of the random assignment of supervisors to male and female files.

The strength of the argument against sex bias must be assessed by a calculation of probability. If it is likely that the randomization could result in such an imbalance, the argument is difficult to refute; however, if only a small proportion of all possible randomizations would give such an imbalance, the argument has less force. We take as the null hypothesis that there is no sex bias and that any differences observed are due to the randomization. We denote the counts in the table and on the margins as follows:

N_{11}	N_{12}	$n_{1.}$
N_{21}	N_{22}	$n_{2.}$
$n_{.1}$	$n_{.2}$	$n_{..}$

According to the null hypothesis, the margins of the table are fixed: There are 24 females, 24 males, 35 supervisors who choose to promote, and 13 who choose not to. Also, the process of randomization determines the counts in the interior of the table (denoted by capital letters since they are random) subject to the constraints of the margins. With these constraints, there is only 1 degree of freedom in the interior of the table; if any interior count is fixed, the others may be determined.

Consider the count N_{11}, the number of males who are promoted. Under the null hypothesis, the distribution of N_{11} is that of the number of successes in 24 draws without replacement from a population of 35 successes and 13 failures; that is, the distribution of N_{11} induced by the randomization is hypergeometric. The probability that $N_{11} = n_{11}$ is

$$p(n_{11}) = \frac{\binom{n_{1.}}{n_{11}} \binom{n_{2.}}{n_{21}}}{\binom{n_{..}}{n_{.1}}}$$

We will use N_{11} as the test statistic for testing the null hypothesis. The preceding hypergeometric probability distribution is the null distribution of N_{11} and is tabled here. A two-sided test rejects for extreme values of N_{11}.

n_{11}	$p(n_{11})$
11	.000
12	.000
13	.004
14	.021
15	.072
16	.162
17	.241
18	.241
19	.162
20	.072
21	.021
22	.004
23	.000
24	.000

From this table, a rejection region for a two-sided test with $\alpha = .05$ consists of the following values for N_{11}: 11, 12, 13, 14, 21, 22, 23, and 24. The observed value of N_{11} falls in this region, so the test would reject at level .05. An imbalance in promotions as or more extreme than that observed would occur only by chance with probability .05, so there is fairly strong evidence of gender bias.

13.3 The Chi-Square Test of Homogeneity

Suppose that we have independent observations from J multinomial distributions, each of which has I cells, and that we want to test whether the cell probabilities of the multinomials are equal—that is, to test the homogeneity of the multinomial distributions.

As an example, we will consider a quantitative study of an aspect of literary style. Several investigators have used probabilistic models of word counts as indices of literary style, and statistical techniques applied to such counts have been used in controversies about disputed authorship. An interesting account is given by Morton (1978), from whom we take the following example.

When Jane Austen died, she left the novel *Sanditon* only partially completed, but she left a summary of the remainder. A highly literate admirer finished the novel, attempting to emulate Austen's style, and the hybrid was published. Morton counted the occurrences of various words in several works: Chapters 1 and 3 of *Sense and Sensibility*, Chapters 1, 2, and 3 of *Emma*, Chapters 1 and 6 of *Sanditon* (written by Austen); and Chapters 12 and 24 of *Sanditon* (written by her admirer). The counts Morton obtained for six words are given in the following table:

Word	*Sense and Sensibility*	*Emma*	*Sanditon* I	*Sanditon* II
a	147	186	101	83
an	25	26	11	29
this	32	39	15	15
that	94	105	37	22
with	59	74	28	43
without	18	10	10	4
Total	375	440	202	196

We will compare the relative frequencies with which these words appear and will examine the consistency of Austen's usage of them from book to book and the degree to which her admirer was successful in imitating this aspect of her style. A stochastic model will be used for this purpose: The six counts for *Sense and Sensibility* will be modeled as a realization of a multinomial random variable with unknown cell probabilities and total count 375; the counts for the other works will be similarly modeled as independent multinomial random variables.

Thus, we must consider comparing J multinomial distributions each having I categories. If the probability of the ith category of the jth multinomial is denoted π_{ij}, the null hypothesis to be tested is

$$H_0: \pi_{i1} = \pi_{i2} = \cdots = \pi_{iJ}, \qquad i = 1, \ldots, I$$

We may view this as a goodness-of-fit test: Does the model prescribed by the null hypothesis fit the data? To test goodness of fit, we will compare observed values with expected values as in Chapter 9, using likelihood ratio statistics or Pearson's chi-square statistic. We will assume that the data consist of independent samples from each multinomial distribution, and we will denote the count in the ith category of the jth multinomial as n_{ij}.

Under H_0, each of the J multinomials has the same probability for the ith category, say π_i. The following theorem shows that the mle of π_i is simply $n_{i.}/n_{..}$, which is an obvious estimate. Here, $n_{i.}$ is the total count in the ith category, $n_{..}$ is the grand total count, $n_{.j}$ is the total count for the jth multinomial.

THEOREM A

Under H_0, the mle's of the parameters $\pi_1, \pi_2, \ldots, \pi_I$ are

$$\hat{\pi}_i = \frac{n_{i.}}{n_{..}}, \qquad i = 1, \ldots, I$$

where $n_{i.}$ is the total number of responses in the ith category and $n_{..}$ is the grand total number of responses.

Proof

Since the multinomial distributions are independent,

$$\text{lik}(\pi_1, \pi_2, \ldots, \pi_I) = \prod_{j=1}^{J} \binom{n_{.j}}{n_{1j}n_{2j}\cdots n_{Ij}} \pi_1^{n_{1j}} \pi_2^{n_{2j}} \cdots \pi_I^{n_{Ij}}$$

$$= \pi_1^{n_{1.}} \pi_2^{n_{2.}} \cdots \pi_I^{n_{I.}} \prod_{j=1}^{J} \binom{n_{.j}}{n_{1j}n_{2j}\cdots n_{Ij}}$$

Let us consider maximizing the log likelihood subject to the constraint $\sum_{i=1}^{I} \pi_i = 1$. Introducing a Lagrange multiplier, we have to maximize

$$l(\pi, \lambda) = \sum_{j=1}^{J} \log \binom{n_{.j}}{n_{1j}n_{2j}\cdots n_{Ij}} + \sum_{i=1}^{I} n_{i.} \log \pi_i + \lambda \left(\sum_{i=1}^{I} \pi_i - 1 \right)$$

Now,

$$\frac{\partial l}{\partial \pi_i} = \frac{n_{i.}}{\pi_i} + \lambda, \qquad i = 1, \ldots, I$$

or

$$\hat{\pi}_i = -\frac{n_{i.}}{\lambda}$$

Summing over both sides and applying the constraint, we find $\lambda = -n_{..}$ and $\hat{\pi}_i = n_{i.}/n_{..}$, as was to be proved. ∎

For the jth multinomial, the expected count in the ith category is the estimated probability of that cell times the total number of observations for the jth multinomial, or

$$E_{ij} = \frac{n_{.j}n_{i.}}{n_{..}}$$

Pearson's chi-square statistic is therefore

$$X^2 = \sum_{i=1}^{I} \sum_{j=1}^{J} \frac{(O_{ij} - E_{ij})^2}{E_{ij}}$$

$$= \sum_{i=1}^{I} \sum_{j=1}^{J} \frac{(n_{ij} - n_{i.}n_{.j}/n_{..})^2}{n_{i.}n_{.j}/n_{..}}$$

For large sample sizes, the approximate null distribution of this statistic is chi-square. (The usual recommendation concerning the sample size necessary for this approximation to be reasonable is that the expected counts should all be greater than 5.) The degrees of freedom are the number of independent counts minus the number of independent parameters estimated from the data. Each multinomial has $I - 1$ independent counts, since the totals are fixed, and $I - 1$ independent parameters have been estimated. The degrees of freedom are therefore

$$df = J(I - 1) - (I - 1) = (I - 1)(J - 1)$$

We now apply this method to the word counts from Austen's works. First, we consider Austen's consistency from one work to another. The following table gives the observed count and, below it, the expected count in each cell of the table.

Word	Sense and Sensibility	Emma	Sanditon I
a	147	186	101
	160.0	187.8	86.2
an	25	26	11
	22.9	26.8	12.3
this	32	39	15
	31.7	37.2	17.1
that	94	105	37
	87.0	102.1	46.9
with	59	74	28
	59.4	69.7	32.0
without	18	10	10
	14.0	16.4	7.5

The observed counts look fairly close to the expected counts, and the chi-square statistic is 12.27. The 10% point of the chi-square distribution with 10 degrees of freedom is 15.99, and the 25% point is 12.54. The data are thus consistent with the model that the word counts in the three works are realizations of multinomial random variables with the same underlying probabilities. The relative frequencies with which Austen used these words did not change from work to work.

To compare Austen and her imitator, we can pool all Austen's work together in light of the above findings. The following table shows the observed and expected frequencies for the imitator and Austen:

Word	Imitator	Austen
a	83 83.5	434 433.5
an	29 14.7	62 76.3
this	15 16.3	86 84.7
that	22 41.7	236 216.3
with	43 33.0	161 171.0
without	4 6.8	38 35.2

The chi-square statistic is 32.81 with 5 degrees of freedom, giving a *p*-value of less than .001. The imitator was not successful in imitating this aspect of Austen's style. To see which discrepancies are large, it is helpful to examine the contributions to the chi-square statistic cell by cell, as tabulated here:

Word	Imitator	Austen
a	0.00	0.00
an	13.90	2.68
this	0.11	0.02
that	9.30	1.79
with	3.06	0.59
without	1.14	0.22

Inspecting this and the preceding table, we see that the relative frequency with which Austen used the word *an* was much smaller than that with which her imitator used it, and that the relative frequency with which she used *that* was much larger.

13.4 The Chi-Square Test of Independence

This section develops a chi-square test that is very similar to the one of the preceding section but is aimed at answering a slightly different question. We will again use an example.

In a demographic study of women who were listed in *Who's Who*, Kiser and Schaefer (1949) compiled the following table for 1436 women who were married at least once:

Education	Married Once	Married More Than Once	Total
College	550	61	611
No College	681	144	825
Total	1231	205	1436

Is there a relationship between marital status and educational level? Of the women who had a college degree, $\frac{61}{611} = 10\%$ were married more than once; of those who had no college degree, $\frac{144}{825} = 17\%$ were married more than once. Alternatively, the question might be addressed by noting that of those women who were married more than once, $\frac{61}{205} = 30\%$ had a college degree, whereas of those married only once, $\frac{550}{1231} = 45\%$ had a college degree. For this sample of 1436 women, having a college degree is positively associated with being married only once, but it is impossible to make causal inferences from the data. Marital stability could be influenced by educational level, or both characteristics could be influenced by other factors, such as social class.

A critic of the study could in any case claim that the relationship between marital status and educational level is "statistically insignificant." Since the data are not a sample from any population, and since no randomization has been performed, the role of probability and statistics is not clear. One might respond to such a criticism by saying that the data speak for themselves and that there is no chance mechanism at work. The critic might then rephrase his argument: "If I were to take a sample of 1436 people cross-classified into two categories which were in fact unrelated in the population from which the sample was drawn, I might find associations as strong or stronger than those observed in this table. Why should I believe that there is any real association in your table?" Even though this argument may not seem compelling, statistical tests are often carried out in situations in which stochastic mechanisms are at best hypothetical.

We will discuss statistical analysis of a sample of size n cross-classified in a table with I rows and J columns. Such a configuration is called a **contingency table.** The joint distribution of the counts n_{ij}, where $i = 1, \ldots, I$ and $j = 1, \ldots, J$, is multinomial with cell probabilities denoted as π_{ij}. Let

$$\pi_{i.} = \sum_{j=1}^{J} \pi_{ij}$$

$$\pi_{.j} = \sum_{i=1}^{I} \pi_{ij}$$

denote the marginal probabilities that an observation will fall in the ith row and jth column, respectively. If the row and column classifications are independent of each other,

$$\pi_{ij} = \pi_{i.}\pi_{.j}$$

We thus consider testing the following null hypothesis:

$$H_0: \pi_{ij} = \pi_{i.}\pi_{.j}, \quad i = 1, \ldots, I, \quad j = 1, \ldots, J$$

versus the alternative that the π_{ij} are free. Under H_0, the mle of π_{ij} is

$$\hat{\pi}_{ij} = \hat{\pi}_{i.}\hat{\pi}_{.j}$$
$$= \frac{n_{i.}}{n} \times \frac{n_{.j}}{n}$$

(see Problem 10 at the end of this chapter). Under the alternative, the mle of π_{ij} is simply

$$\tilde{\pi}_{ij} = \frac{n_{ij}}{n}$$

These estimates can be used to form a likelihood ratio test or an asymptotically equivalent Pearson's chi-square test,

$$X^2 = \sum_{i=1}^{I}\sum_{j=1}^{J} \frac{(O_{ij} - E_{ij})^2}{E_{ij}}$$

Here the O_{ij} are the observed counts (n_{ij}). The expected counts, the E_{ij}, are the fitted counts:

$$E_{ij} = n\hat{\pi}_{ij} = \frac{n_{i.}n_{.j}}{n}$$

Pearson's chi-square statistic is, therefore,

$$X^2 = \sum_{i=1}^{I}\sum_{j=1}^{J} \frac{(n_{ij} - n_{i.}n_{.j}/n)^2}{n_{i.}n_{.j}/n}$$

The degrees of freedom for the chi-square statistic are calculated as in Section 9.5. Under Ω, the cell probabilities sum to 1 but are otherwise free and there are thus $IJ - 1$ independent parameters. Under the null hypothesis, the marginal probabilities, are estimated from the data and are specified by $(I - 1) + (J - 1)$ independent parameters. Thus,

$$df = IJ - 1 - (I - 1) - (J - 1) = (I - 1)(J - 1)$$

Returning to the data on 1436 women from the demographic study, we calculate expected values and construct the following table:

Education	Married Once	More Than Once
College	550	61
	523.8	87.2
No College	681	144
	707.2	117.8

The chi-square statistic is 16.01 with 1 degree of freedom, giving a p-value less than .001. We would reject the hypothesis of independence and conclude that there is a relationship between marital status and educational level.

The chi-square statistic used here to test independence is identical in form and degrees of freedom to that used in the preceding section to test homogeneity; however, the hypotheses are different and the sampling schemes are different. The test of homogeneity was derived under the assumption that the column (or row) margins were fixed, and the test of independence was derived under the assumption that only the grand total was fixed. Because the test statistics are computed in an identical fashion and have the same number of degrees of freedom, the distinction between them is often slurred over. Furthermore, the notions of homogeneity and independence are closely related and easily confused. Independence can be thought of as homogeneity of conditional distributions; for example, if education level and marital status are independent, then the conditional probabilities of marital status given educational level are homogeneous—P(Married Once | College) $=$ P(Married Once | No College).

13.5 Matched-Pairs Designs

Matched-pairs designs can be effective for experiments involving categorical data; as with experiments involving continuous data, pairing can control for extraneous sources of variability and can increase the power of a statistical test. Appropriate techniques, however, must be used in the analysis of the data. This section begins with an extended example illustrating these concepts.

EXAMPLE A   Vianna, Greenwald, and Davies (1971) collected data comparing the percentages of tonsillectomies for a group of patients suffering from Hodgkin's disease and a comparable control group:

	Tonsillectomy	No Tonsillectomy
Hodgkin's	67	34
Control	43	64

The table shows that 66% of the Hodgkin's sufferers had had a tonsillectomy, compared to 40% of the control group. The chi-square test for homogeneity gives a chi-square statistic of 14.26 with 1 degree of freedom, which is highly significant. The investigators conjectured that the tonsils act as a protective barrier in some fashion against Hodgkin's disease.

Johnson and Johnson (1972) selected 85 Hodgkin's patients who had a sibling of the same sex who was free of the disease and whose age was within 5 years of the patient's. These investigators presented the following table:

	Tonsillectomy	No Tonsillectomy
Hodgkin's	41	44
Control	33	52

They calculated a chi-square statistic of 1.53, which is not significant. Their findings thus appeared to be at odds with those of Vianna, Greenwald, and Davies.

Several letters to the editor of the journal that published Johnson and Johnson's results pointed out that those investigators had made an error in their analysis by ignoring the pairings. The assumption behind the chi-square test of homogeneity is that independent multinomial samples are compared, and Johnson and Johnson's samples were not independent, because siblings were paired. An appropriate analysis of Johnson and Johnson's data is suggested once we set up a table that exhibits the pairings:

		Sibling	
		No Tonsillectomy	Tonsillectomy
Patient	No Tonsillectomy	37	7
	Tonsillectomy	15	26

Viewed in this way, the data are a sample of size 85 from a multinomial distribution with four cells. We can represent the probabilities in the table as follows:

π_{11}	π_{12}	$\pi_{1.}$
π_{21}	π_{22}	$\pi_{2.}$
$\pi_{.1}$	$\pi_{.2}$	1

The appropriate null hypothesis states that the probabilities of tonsillectomy and no tonsillectomy are the same for patients and siblings—that is, $\pi_{1.} = \pi_{.1}$ and $\pi_{2.} = \pi_{.2}$, or

$$\pi_{11} + \pi_{12} = \pi_{11} + \pi_{21}$$
$$\pi_{12} + \pi_{22} = \pi_{21} + \pi_{22}$$

These equations simplify to $\pi_{12} = \pi_{21}$.

The relevant null hypothesis is thus

$$H_0\colon \pi_{12} = \pi_{21}$$

Under the null hypothesis, the off-diagonal probabilities are equal, and under the alternative they are not. The diagonal probabilities do not distinguish the null and alternative hypotheses. We will derive a test, called **McNemar's test,** of this hypothesis. Under H_0, the mle's of the cell probabilities are (see Problem 10 at the end of this chapter)

$$\hat{\pi}_{11} = \frac{n_{11}}{n}$$

$$\hat{\pi}_{22} = \frac{n_{22}}{n}$$

$$\hat{\pi}_{12} = \hat{\pi}_{21} = \frac{n_{12} + n_{21}}{2n}$$

The contributions to the chi-square statistic from the n_{11} and n_{22} cells are equal to zero; the remainder of the statistic is

$$X^2 = \frac{[n_{12} - (n_{12} + n_{21})/2]^2}{(n_{12} + n_{21})/2} + \frac{[n_{21} - (n_{12} + n_{21})/2]^2}{(n_{12} + n_{21})/2}$$

$$= \frac{(n_{12} - n_{21})^2}{n_{12} + n_{21}}$$

Let us count degrees of freedom: Under Ω there are three free parameters since there are four cell probabilities which are constrained to sum to 1. Under the null hypothesis, there is the additional constraint, $\pi_{12} = \pi_{21}$, and there are two free parameters. The chi-square statistic thus has 1 degree of freedom. For the data table exhibiting the pairings, $X^2 = 2.91$, with a corresponding p-value of .09. This casts doubt on the null hypothesis, contrary to Johnson and Johnson's original analysis. ∎

E X A M P L E **B** *Cell Phones and Driving*
Does the use of cell phones while driving cause accidents? This is a difficult question to study empirically. An observational study comparing accident rates of users and nonusers would be subject to numerous sources of confounding, such as age, gender, and time and place of driving. A randomized, controlled experiment in which drivers were randomly assigned to use or not use cell phones is infeasible, partly because it would be unethical to deliberately expose people to a potentially hazardous condition. Double blinding would clearly be impossible. Redelmeier and Tibshirani (1997) conducted a clever study, designed in the following way. They identified 699 drivers who owned cell phones and who had been involved in motor vehicle collisions. They then used billing records to determine whether each individual used a cell phone during the 10 minutes preceding the collision and also at the same time during the previous week. (For more details, see the cited paper.) Each person thus served as his own control, eliminating various sources of confounding. The results are laid out in the

following table:

Collision	Before Collision		
	On Phone	Not on Phone	Total
On Phone	13	157	170
Not on Phone	24	505	529
Total	37	662	699

From the table, on the day of the collision, 24% of the drivers had been on the phone as compared to 5% the day before the collision. McNemar's test can be applied to test the null hypothesis of no association:

$$X^2 = \frac{(157 - 24)^2}{157 + 24}$$
$$= 97.7$$

There is thus no doubt that the association is statistically significant. However, the authors pointed out that this result does not necessarily imply that cell phone use while driving *causes* more accidents—for example, it is possible that during times of emotional stress, drivers are more likely to use cell phones, and because of the emotional stress are also less attentive to their driving. ■

13.6 Odds Ratios

If an event A has probability $P(A)$ of occurring, the **odds** of A occurring are defined to be

$$\text{odds}(A) = \frac{P(A)}{1 - P(A)}$$

Since this implies that

$$P(A) = \frac{\text{odds}(A)}{1 + \text{odds}(A)}$$

odds of 2 (or 2 to 1), for example, correspond to $P(A) = 2/3$.

Now suppose that X denotes the event that an individual is exposed to a potentially harmful agent and that D denotes the event that the individual becomes diseased. We denote the complementary events as $\overline{X}$ and $\overline{D}$. The odds of an individual contracting the disease given that he is exposed are

$$\text{odds}(D|X) = \frac{P(D|X)}{1 - P(D|X)}$$

and the odds of contracting the disease given that he is not exposed are

$$\text{odds}(D|\overline{X}) = \frac{P(D|\overline{X})}{1 - P(D|\overline{X})}$$

The **odds ratio**

$$\Delta = \frac{\text{odds}(D|X)}{\text{odds}(D|\overline{X})}$$

is a measure of the influence of exposure on subsequent disease.

We will consider how the odds and odds ratio could be estimated by sampling from a population with joint and marginal probabilities defined as in the following table:

	$\overline{D}$	D	
$\overline{X}$	π_{00}	π_{01}	$\pi_{0.}$
X	π_{10}	π_{11}	$\pi_{1.}$
	$\pi_{.0}$	$\pi_{.1}$	1

With this notation,

$$P(D|X) = \frac{\pi_{11}}{\pi_{10} + \pi_{11}}$$

$$P(D|\overline{X}) = \frac{\pi_{01}}{\pi_{00} + \pi_{01}}$$

so that

$$\text{odds}(D|X) = \frac{\pi_{11}}{\pi_{10}}$$

$$\text{odds}(D|\overline{X}) = \frac{\pi_{01}}{\pi_{00}}$$

and the odds ratio is

$$\Delta = \frac{\pi_{11}\pi_{00}}{\pi_{01}\pi_{10}}$$

the product of the diagonal probabilities in the preceding table divided by the product of the off-diagonal probabilities.

Now we will consider three possible ways to sample this population to study the relationship of disease and exposure. First, we might consider drawing a random sample from the entire population; from such a sample we could estimate all the probabilities directly. However, if the disease is rare, the total sample size would have to be quite large to guarantee that a substantial number of diseased individuals was included.

A second method of sampling is called a **prospective study**—a fixed number of exposed and nonexposed individuals are sampled, and the incidences of disease in those two groups are compared. In this case the data allow us to estimate and compare $P(D|X)$ and $P(D|\overline{X})$ and, hence, the odds ratio. For example, $P(D|X)$ would be estimated by the proportion of exposed individuals who had the disease. However, note that the individual probabilities π_{ij} cannot be estimated from the data, because the marginal counts of exposed and unexposed individuals have been fixed arbitrarily by the sampling design.

A third method of sampling—a **retrospective study**—is one in which a fixed number of diseased and undiseased individuals are sampled and the incidences of exposure in the two groups are compared. The study of Vianna, Greenwald, and Davies (1971) discussed in the previous section was of this type. From such data, we can directly estimate $P(X|D)$ and $P(X|\overline{D})$ by the proportions of diseased and nondiseased individuals who were exposed. Because the marginal counts of diseased and nondiseased are fixed, we cannot estimate the joint probabilities or the important conditional probabilities $P(D|X)$ and $P(D|\overline{X})$. However, as will be shown, we can estimate the odds ratio Δ. Observe that

$$P(X|D) = \frac{\pi_{11}}{\pi_{01} + \pi_{11}}$$

$$1 - P(X|D) = \frac{\pi_{01}}{\pi_{01} + \pi_{11}}$$

$$\text{odds}(X|D) = \frac{\pi_{11}}{\pi_{01}}$$

Similarly,

$$\text{odds}(X|\overline{D}) = \frac{\pi_{10}}{\pi_{00}}$$

We thus see that the odds ratio, Δ, defined previously, can also be expressed as

$$\Delta = \frac{\text{odds}(X|D)}{\text{odds}(X|\overline{D})}$$

Specifically, suppose that the counts in such a study are denoted as in the following table:

	$\overline{D}$	D
$\overline{X}$	n_{00}	n_{01}
X	n_{10}	n_{11}
	$n_{.0}$	$n_{.1}$

Then the conditional probabilities and the odds ratios are estimated as

$$\hat{P}(X|D) = \frac{n_{11}}{n_{.1}}$$

$$1 - \hat{P}(X|D) = \frac{n_{01}}{n_{.1}}$$

$$\widehat{\text{odds}}(X|D) = \frac{n_{11}}{n_{01}}$$

Similarly,

$$\widehat{\text{odds}}(X|\overline{D}) = \frac{n_{10}}{n_{00}}$$

so that the estimate of the odds ratio is

$$\hat{\Delta} = \frac{n_{00}n_{11}}{n_{01}n_{10}}$$

the product of the diagonal counts divided by the product of the off-diagonal counts.

As an example, consider the table in the previous section that displays the data of Vianna, Greenwald, and Davies. The odds ratio is estimated to be

$$\hat{\Delta} = \frac{67 \times 64}{43 \times 34} = 2.93$$

According to this study, the odds of contracting Hodgkin's disease are increased by about a factor of three by undergoing a tonsillectomy.

As well as having a point estimate $\hat{\Delta} = 2.93$, it would be useful to attach an approximate standard error to the estimate to indicate its uncertainty. Since $\hat{\Delta}$ is a nonlinear function of the counts, it appears that an analytical derivation of its standard error would be difficult. Once again, however, the convenience of simulation (the bootstrap) comes to our aid. In order to approximate the distribution of $\hat{\Delta}$ by simulation, we need to generate random numbers according to a statistical model for the counts in the table of Vianna, Greenwald, and Davies. The model is that the count in the first row and first column, N_{11}, is binomially distributed with $n = 101$ and probability π_{11}. The count in the second row and second column, N_{22}, is independently binomially distributed with $n = 107$ and probability π_{22}. The distribution of the random variable

$$\hat{\Delta} = \frac{N_{11} N_{22}}{(101 - N_{11})(107 - N_{22})}$$

is thus determined by the two binomial distributions, and we could approximate it arbitrarily well by drawing a large number of samples from them.

Since the probabilities π_{11} and π_{22} are unknown, they are estimated from the observed counts by $\hat{\pi}_{11} = 67/101 = .663$ and $\hat{\pi}_{22} = 64/107 = .598$. One thousand realizations of binomial random variables N_{11} and N_{22} were generated on a computer and Figure 13.1 shows a histogram of the resulting 1000 values of $\hat{\Delta}$. The standard deviation of these 1000 values was .89, which can be used as an estimated standard error for our observed estimate $\hat{\Delta} = 2.93$.

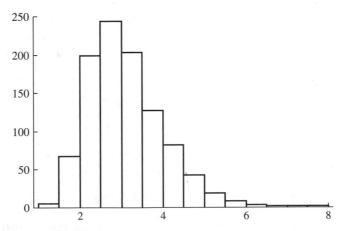

FIGURE **13.1** Histogram of 1000 bootstrapped estimates of the odds ratio, Δ.

13.7 Concluding Remarks

This chapter has introduced two-way classifications, which are the simplest form of contingency tables. For higher-order classifications, which frequently occur in practice, a greater variety of forms of dependence arise. For example, for a three-way table, the factors of which are denoted A, B, and C, we might consider testing whether, conditionally on C, A and B are independent.

Dependencies can be specified by means of **log linear models.** If the row and column classifications in a two-way table are independent, then

$$\pi_{ij} = \pi_i \pi_j$$

or

$$\log \pi_{ij} = \log \pi_i + \log \pi_j$$

We can denote $\log \pi_i$ by α_i and $\log \pi_j$ by β_j. Then, if there is dependence, $\log \pi_{ij}$ may be written as

$$\log \pi_{ij} = \alpha_i + \beta_j + \gamma_{ij}$$

This form mimics the additive analysis of variance models introduced in Chapter 12. The idea can readily be extended to higher-order tables. For example, a possible model for a three-way table is

$$\log \pi_{ijk} = \alpha_i + \beta_j + \gamma_k + \delta_{ij} + \epsilon_{ik} + \phi_{jk}$$

which allows second-order dependencies, but no third-order dependencies. The parameters of log-linear models may be estimated by mle's and likelihood ratio tests may be employed. Agresti (1996) treats these and other topics in the analysis of categorical data.

13.8 Problems

1. Adult-onset diabetes is known to be highly genetically determined. A study was done comparing frequencies of a particular allele in a sample of such diabetics and a sample of nondiabetics. The data are shown in the following table:

	Diabetic	Normal
Bb or *bb*	12	4
BB	39	49

    Are the relative frequencies of the alleles significantly different in the two groups?

2. Phillips and Smith (1990) conducted a study to investigate whether people could briefly postpone their deaths until after the occurrence of a significant occasion. The senior woman of the household plays a central ceremonial role in the Chinese

Harvest Moon Festival. Phillips and Smith compared the mortality patterns of old Jewish women and old Chinese women who died of natural causes for the weeks immediately preceding and following the festival, using records from California for the years 1960–1984. Compare the mortality patterns shown in the table. (Week -1 is the week preceding the festival, week 1 is the week following, etc.)

Week	Chinese	Jewish
-2	55	141
-1	33	145
1	70	139
2	49	161

3. Overfield and Klauber (1980) published the following data on the incidence of tuberculosis in relation to blood groups in a sample of Eskimos. Is there any association of the disease and blood group within the ABO system or within the MN system?

ABO system

Severity	O	A	AB	B
Moderate-Advanced	7	5	3	13
Minimal	27	32	8	18
Not Present	55	50	7	24

MN system

Severity	MM	MN	NN
Moderate-Advanced	21	6	1
Minimal	54	27	5
Not Present	74	51	11

4. In a famous sociological study called *Middletown*, Lynd and Lynd (1956) administered questionnaires to 784 white high school students. The students were asked which two of ten given attributes were most desirable in their fathers. The following table shows how the desirability of the attribute "being a college graduate" was rated by male and female students. Did the males and females value this attribute differently?

	Male	Female
Mentioned	86	55
Not Mentioned	283	360

5. Dowdall (1974) [also discussed in Haberman (1978)] studied the effect of ethnic background on role attitude of women of ages 15–64 in Rhode Island. Respondents were asked whether they thought it was all right for a woman to have a job instead of taking care of the house and children while her husband worked. The following table breaks down the responses by ethnic origin of the respondent. Is there a relationship between response and ethnic group? If so, describe it.

Ethnic Origin	Yes	No
Italian	78	47
Northern European	56	29
Other European	43	29
English	53	32
Irish	43	30
French Canadian	36	22
French	42	23
Portuguese	29	7

6. It is conventional wisdom in military squadrons that pilots tend to father more girls than boys. Snyder (1961) gathered data for military fighter pilots. The sex of the pilots' offspring were tabulated for three kinds of flight duty during the month of conception, as shown in the following table. Is there any significant difference between the three groups? In the United States in 1950, 105.37 males were born for every 100 females. Are the data consistent with this sex ratio?

Father's Activity	Female Offspring	Male Offspring
Flying Fighters	51	38
Flying Transports	14	16
Not Flying	38	46

7. Grades in an elementary statistics class were classified by the students' majors. Is there any relationship between grade and major?

	Major		
Grade	Psychology	Biology	Other
A	8	15	13
B	14	19	15
C	15	4	7
D-F	3	1	4

8. A randomized double-blind experiment compared the effectiveness of several drugs in ameliorating postoperative nausea. All patients were anesthetized with nitrous oxide and ether. The following table shows the incidence of nausea during the first four postoperative hours for each of several drugs and a placebo (Beecher 1959). Compare the drugs to each other and to the placebo.

	Number of Patients	Incidence of Nausea
Placebo	165	95
Chlorpromazine	152	52
Dimenhydrinate	85	52
Pentobarbital (100 mg)	67	35
Pentobarbital (150 mg)	85	37

9. This problem considers some more data on Jane Austen and her imitator (Morton 1978). The following table gives the relative frequency of the word *a* preceded by (PB) and not preceded by (NPB) the word *such*, the word *and* followed by (FB) or not followed by (NFB) *I*, and the word *the* preceded by and not preceded by *on*.

Words	Sense and Sensibility	Emma	Sanditon I	Sanditon II
a PB *such*	14	16	8	2
a NPB *such*	133	180	93	81
and FB *I*	12	14	12	1
and NFB *I*	241	285	139	153
the PB *on*	11	6	8	17
the NPB *on*	259	265	221	204

Was Austen consistent in these habits of style from one work to another? Did her imitator successfully copy this aspect of her style?

10. Verify that the mle's of the cell probabilities, π_{ij}, are as given in Section 13.4 for the test of independence and in Section 13.5 for McNemar's test.

11. (a) Derive the likelihood ratio test of homogeneity. (b) Calculate the likelihood ratio test statistic for the example of Section 13.3, and compare it to Pearson's chi-square statistic. (c) Derive the likelihood ratio test of independence. (d) Calculate the likelihood ratio test statistic for the example of Section 13.4, and compare it to Pearson's chi-square statistic.

12. Show that McNemar's test is nearly equivalent to scoring each response as a 0 or a 1 and calculating a paired-sample t test on the resulting data.

13. A sociologist is studying influences on family size. He finds pairs of sisters, both of whom are married, and determines for each sister whether she has 0, 1, or 2 or more children. He wants to compare older and younger sisters. Explain what the following hypotheses mean and how to test them.

   a. The number of children the younger sister has is independent of the number of children the older sister has.
   b. The distribution of family sizes is the same for older and younger sisters. Could one hypothesis be true and the other false? Explain.

14. Lazarsfeld, Berelson, and Gaudet (1948) present the following tables relating degree of interest in political elections to education and age:

No high school education

Degree of Interest	Under 45	Over 45
Great	71	217
Little	305	652

Some high school or more

Degree of Interest	Under 45	Over 45
Great	305	180
Little	869	259

Since there are three factors—education, age, and interest—these tables considered jointly are more complicated than the tables considered in this chapter.

   a. Examine the tables informally and analyze the dependence of interest in political elections on age and education. What do the numbers suggest?
   b. Extend the ideas of this chapter to test two hypotheses, H_1: given educational level, age and degree of interest are unrelated, and H_2: given age, educational level and degree of interest are unrelated.

15. Reread Section 11.4.5, which contains a discussion of methodological problems in a study of the effects of FD&C Red No. 40. The following tables give the

numbers of mice that developed RE tumors in each of several groups:

Incidence among males

	Control I	Control 2	Low Dose	Med. Dose	High Dose
Number with tumor	25	10	20	9	17
Total number	100	100	99	100	99

Incidence among females

	Control I	Control 2	Low Dose	Med. Dose	High Dose
Number with tumor	33	25	32	26	22
Total number	100	99	99	99	100

Use chi-square tests to compare the incidences in the different groups for males and for females. Which differences are significant? What would you conclude from this analysis had you not known about the possibility of cage position effects?

16. A market research team conducted a survey to investigate the relationship of personality to attitude toward small cars. A sample of 250 adults in a metropolitan area were asked to fill out a 16-item self-perception questionnaire, on the basis of which they were classified into three types: cautious conservative, middle-of-the-roader, and confident explorer. They were then asked to give their overall opinion of small cars: favorable, neutral, or unfavorable. Is there a relationship between personality type and attitude toward small cars? If so, what is the nature of the relationship?

	Personality Type		
Attitude	Cautious	Midroad	Explorer
Favorable	79	58	49
Neutral	10	8	9
Unfavorable	10	34	42

17. In a study of the relation of blood type to various diseases, the following data were gathered in London and Manchester (Woolf 1955):

London

	Control	Peptic Ulcer
Group A	4219	579
Group O	4578	911

Manchester

	Control	Peptic Ulcer
Group A	3775	246
Group O	4532	361

First, consider the two tables separately. Is there a relationship between blood type and propensity to peptic ulcer? If so, evaluate the strength of the relationship. Are the data from London and Manchester comparable?

18. Records of 317 patients at least 48 years old who were diagnosed as having endometrial carcinoma were obtained from two hospitals (Smith et al. 1975). Matched controls for each case were obtained from the two institutions; the controls had cervical cancer, ovarian cancer, or carcinoma of the vulva. Each control was matched by age at diagnosis (within four years) and year of diagnosis (within two years) to a corresponding case of endometrial carcinoma. This sort of design, called a **retrospective case-control study,** is frequently used in medical investigations where a randomized experiment is not possible. The following table gives the numbers of cases and controls who had taken estrogen for at least 6 mo prior to the diagnosis of cancer. Is there a significant relationship between estrogen use and endometrial cancer? Do you see any possible weak points in a retrospective case-control design?

		Controls	
		Estrogen Used	Not Used
Cases	Estrogen Used	39	113
	Not Used	15	150

19. A psychological experiment was done to investigate the effect of anxiety on a person's desire to be alone or in company (Schacter 1959; Lehmann 1975). A group of 30 subjects was randomly divided into two groups of sizes 13 and 17. The subjects were told that they would be subjected to some electric shocks, but one group was told that the shocks would be quite painful and the other group was told that they would be mild and painless. The former group was the "high-anxiety" group, and the latter was the "low-anxiety" group. Both groups were told that there would be a 10-min wait before the experiment began, and each subject was given the choice of waiting alone or with the other subjects. The following are the results:

	Wait Together	Wait Alone
High-Anxiety	12	5
Low-Anxiety	4	9

Use Fisher's exact test to test whether there is a significant difference between the high- and low-anxiety groups.

20. Define appropriate notation for the three sample designs considered in Section 13.6 (simple random sample, prospective study, and retrospective study).

    a. Show how to estimate the odds ratio Δ.
    b. Use the method of propagation of error to find approximately $\text{Var}(\log(\hat{\Delta}))$ ($\log(\hat{\Delta})$ is sometimes used in place of $\hat{\Delta}$).

21. For Problem 1, what is the relevant odds ratio and what is its estimate?

22. A study was done to identify factors affecting physicians' decisions to advise or not to advise patients to stop smoking (Cummings et al. 1987). The study was related to a training program to teach physicians ways to counsel patients to stop smoking and was carried out in a family practice outpatient center in Buffalo, New York. The study population consisted of the cigarette-smoking patients of residents in family medicine seen in the center between February and May 1984.

    a. We first consider whether certain patient characteristics are related to being advised or not being advised. The following table shows a breakdown by sex:

	Advised	Not Advised
Male	48	47
Female	80	136

    What proportion of the males were advised to quit and what proportion of the females were advised? What are the standard errors of these proportions? What is the standard error of their difference? Test whether the difference in the proportions is statistically significant.

    Next consider a breakdown by race: White and Other versus African-American:

	Advised	Not Advised
White	26	34
African-American	102	149

    What proportions of the African-Americans and Whites were asked to quit and what are the standard errors of these proportions? What is the standard error of the difference of the proportions? Is the difference statistically significant?

    Finally consider the relation of the number of cigarettes smoked daily to being advised or not:

	Advised	Not Advised
< 15	64	112
15–25	39	54
>25	25	16

For each of the three groups, what proportion was advised to quit smoking and what are the standard errors of the proportions? Is the difference in proportions statistically significant?

b. Next consider the relationship of certain physician characteristics to the decision whether to advise. First, the physician's sex:

	Advised	Not Advised
Male	78	94
Female	50	89

What proportions of the patients of male and female physicians were advised? What are the standard errors of the proportions and their difference? Is the difference statistically significant?

The following table shows the breakdown according to whether the physician smokes:

	Advised	Not Advised
Smoker	13	37
Nonsmoker	115	146

Of those patients who saw a smoking physician, what proportion were advised to quit, and of those who saw a nonsmoker, what proportion were so advised? What are the standard errors of the proportions and of their difference? Is the difference statistically significant?

Finally, this table gives a breakdown by age of physician:

	Advised	Not Advised
< 30	88	128
30–39	28	37
> 39	12	18

What are the proportions advised to quit in each of the three age categories and what are their standard errors? Are the differences statistically significant?

23. Does heavy exercise increase the risk of myocardial infarction? Mittleman et al. (1993) studied this question by examining the activities of 1228 patients who had suffered myocardial infarctions. It was determined whether each patient had participated in heavy exertion in the hour before the onset of the infarction and also whether each had participated in heavy exertion at the same time the previous day. Their results are displayed in the following table:

Previous Day	Day of Infarction		
	Exertion	No Exertion	Total
Exertion	4	9	13
No Exertion	50	1165	1215
Total	54	1174	1228

Does the study demonstrate that heavy exertion is associated with myocardial infarction? How does the design of this study relate to that of the cell phone study in Example B of Section 13.5?

24. Is it advantageous to wear the color red in a sporting contest? According to Hill and Barton (2005):

> Although other colours are also present in animal displays, it is specifically the presence and intensity of red coloration that correlates with male dominance and testosterone levels. In humans, anger is associated with a reddening of the skin due to increased blood flow, whereas fear is associated with increased pallor in similarly threatening situations. Hence, increased redness during aggressive interactions may reflect relative dominance. Because artificial stimuli can exploit innate responses to natural stimuli, we tested whether wearing red might influence the outcome of physical contests in humans.
>
> In the 2004 Olympic Games, contestants in four combat sports (boxing, tae kwon do, Greco-Roman wrestling, and freestyle wrestling) were randomly assigned red or blue outfits (or body protectors). If colour has no effect on the outcome of contests, the number of winners wearing red should be statistically indistinguishable from the number of winners wearing blue.

They thus tabulated the colors worn by the winners in these contests:

Sport	Red	Blue
Boxing	148	120
Freestyle Wrestling	27	24
Greco Roman Wrestling	25	23
Tae Kwon Do	45	35

Some supplementary information is given in the file `red-blue.txt`.

a. Let π_R denote the probability that the contestant wearing red wins. Test the null hypothesis that $\pi_R = \frac{1}{2}$ versus the alternative hypothesis that π_R is the same in each sport, but $\pi_R \neq \frac{1}{2}$.

b. Test the null hypothesis $\pi_R = \frac{1}{2}$ against the alternative hypothesis that allows π_R to be different in different sports, but not equal to $\frac{1}{2}$.

    c. Are either of these hypothesis tests equivalent to that which would test the null hypothesis $\pi_R = \frac{1}{2}$ versus the alternative hypothesis $\pi_R \neq \frac{1}{2}$, using as data the total numbers of wins summed over all the sports?

    d. Is there any evidence that wearing red is more favorable in some of the sports than others?

    e. From an analysis of the points scored by winners and losers, Hill and Barton concluded that color had the greatest effect in close contests. Data on the points of each match are contained in the file `red-blue.xls`. Analyze this data and see whether you agree with their conclusion.

25. The Physicians' Health Study was a randomized, double-blind, placebo-controlled trial designed to determine whether low-dose aspirin (325 mg every other day) decreases cardiovascular mortality. The experiment assigned 11,037 physicians at random to receive aspirin, and 11,034 to receive a placebo.

    a. The following table shows the incidence of cardiovascular events. What would you conclude about the effects of aspirin?

	Aspirin	Placebo
Myocardial Infarction		
Fatal	10	26
Nonfatal	129	213
Stroke		
Fatal	9	6
Nonfatal	110	92

    b. The following tables details cardiovascular mortality. What would you conclude about the effects of aspirin?

Cause	Aspirin	Placebo
Acute myocardial infarction	10	28
Other ischemic heart disease	24	25
Sudden death	22	12
Stroke	10	7
Other cardiovascular	15	11

26. Insulin pumps are used by diabetic patients to control blood glucose levels, but a side effect, diabetic ketoacidosis (DKA), may occur. Mecklenburg et al. (1984) gathered data on incidence of DKA before and after pump therapy, shown in the following table. Test whether the rate of DKA is the same before and after therapy.

After Therapy	Before Therapy	
	No DKA	DKA
No DKA	128	7
DKA	19	7

27. The data in the following table are taken from an article in the *New York Times* (April 20, 2001), "Victim's Race Affects Killer's Sentence." The data are from a study of all homicide cases in North Carolina for the period 1993–1997 in which it was possible that a murder conviction would result in the death penalty. Such data have played an important role in the debate about the death penalty in the U.S., the only wealthy western nation which imposes it. Qualitatively, what do you conclude from looking at the numbers? Discuss whether it is appropriate to use a chi-square test to test that the combination of the victim's race and the defendant's race was independent of whether the defendant received the death penalty for convicted murderers in North Carolina during the years 1993–1997.

Defendant's Race	Victim's Race	Death Penalty	No Death Penalty
Not white	White	33	251
White	White	33	508
Not white	Not white	29	587
White	Not white	4	76

28. In Section 13.3, a chi-square test of homogeneity was carried out on the frequencies of word counts in four works. The test used the actual counts (e.g, 147 occurrences of the word "a" in Sense and Sensibility). Suppose that instead of the counts, the relative frequencies (e.g., $147/375 = 0.39$) were presented in the table and the chi-square statistic was calculated using the relative frequencies rather than the counts. Would the value of the chi-square statistic be the same? What would happen if percentages were used?

29. Suppose that a company wishes to examine the relationship of gender to job satisfaction, grouping job satisfaction into four categories: very satisfied, somewhat satisfied, somewhat dissatisfied, and very dissatisfied. The company plans to ask the opinions of 100 employees. Should you, the company's statistician, carry out a chi-square test of independence or a test of homogeneity?

CHAPTER 14

Linear Least Squares

14.1 Introduction

In order to fit a straight line to a plot of points (x_i, y_i), where $i = 1, \ldots, n$, the slope and intercept of the line $y = \beta_0 + \beta_1 x$ must be found from the data in some manner. In order to fit a pth-order polynomial, $p + 1$ coefficients must be determined. Other functional forms besides linear and polynomial ones may be fit to data, and in order to do so parameters associated with those forms must be determined.

The most common, but by no means only, method for determining the parameters in curve-fitting problems is the method of least squares. The principle underlying this method is to minimize the sum of squared deviations of the predicted, or fitted, values (given by the curve) from the actual observations. For example, suppose that a straight line is to be fit to the points (y_i, x_i), where $i = 1, \ldots, n$; y is called the **dependent variable** and x is called the **independent variable,** and we want to predict y from x. (This usage of the terms *independent* and *dependent* is different from their probabilistic meaning.) Sometimes x and y are called the **predictor variable** and the **response variable,** respectively. Applying the method of least squares, we choose the slope and intercept of the straight line to minimize

$$S(\beta_0, \beta_1) = \sum_{i=1}^{n} (y_i - \beta_0 - \beta_1 x_i)^2$$

Note that β_0 and β_1 are chosen to minimize the sum of squared vertical deviations, or prediction errors (see Figure 14.1). The procedure is not symmetric in y and x.

Curves are often fit to data as part of the process of calibrating instruments. For example, Bailey, Cox, and Springer (1978) discuss a method for measuring the concentrations of food dyes and other substances by high-pressure chromatography. Measurements of the chromatographic peak areas corresponding to sulfanilic acid were taken for several known concentrations of FD&C Yellow No. 5. Figure 14.2

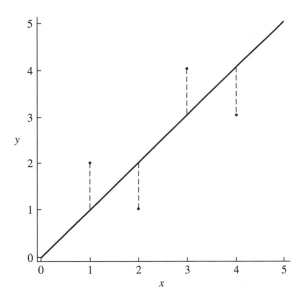

FIGURE **14.1** The least squares line minimizes the sum of squared vertical distances (dotted lines) from the points to the line.

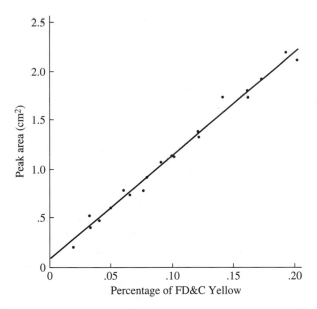

FIGURE **14.2** Data points and the least squares line for the relation of sulfanilic acid peak area to percentage of FD&C Yellow.

shows a plot of peak area versus percentage of FD&C Yellow. To casual examination, the plot looks fairly linear.

Once the equation of the line was established, it could be used in estimating concentrations of the dye from measurements of peak area.

To find β_0 and β_1, we calculate

$$\frac{\partial S}{\partial \beta_0} = -2 \sum_{i=1}^{n} (y_i - \beta_0 - \beta_1 x_i)$$

$$\frac{\partial S}{\partial \beta_1} = -2 \sum_{i=1}^{n} x_i (y_i - \beta_0 - \beta_1 x_i)$$

Setting these partial derivatives equal to zero, we have that the minimizers $\hat{\beta}_0$ and $\hat{\beta}_1$ satisfy

$$\sum_{i=1}^{n} y_i = n\hat{\beta}_0 + \hat{\beta}_1 \sum_{i=1}^{n} x_i$$

$$\sum_{i=1}^{n} x_i y_i = \hat{\beta}_0 \sum_{i=1}^{n} x_i + \hat{\beta}_1 \sum_{i=1}^{n} x_i^2$$

Solving for $\hat{\beta}_0$ and $\hat{\beta}_1$, we obtain

$$\hat{\beta}_0 = \frac{\left(\sum_{i=1}^{n} x_i^2\right)\left(\sum_{i=1}^{n} y_i\right) - \left(\sum_{i=1}^{n} x_i\right)\left(\sum_{i=1}^{n} x_i y_i\right)}{n \sum_{i=1}^{n} x_i^2 - \left(\sum_{i=1}^{n} x_i\right)^2}$$

$$\hat{\beta}_1 = \frac{n \sum_{i=1}^{n} x_i y_i - \left(\sum_{i=1}^{n} x_i\right)\left(\sum_{i=1}^{n} y_i\right)}{n \sum_{i=1}^{n} x_i^2 - \left(\sum_{i=1}^{n} x_i\right)^2}$$

Problem 10 at the end of the chapter asks you to derive the following useful equivalent expressions:

$$\hat{\beta}_0 = \bar{y} - \hat{\beta}_1 \bar{x}$$
$$\hat{\beta}_1 = \frac{\sum_{i=1}^{n} (x_i - \bar{x})(y_i - \bar{y})}{\sum_{i=1}^{n} (x_i - \bar{x})^2}$$

The fitted line, with the parameters determined from the expressions above to be $\hat{\beta}_0 = .073$ and $\hat{\beta}_1 = 10.8$, is drawn in Figure 14.2. Is this a "reasonable" fit? How much faith do we have in these values for $\hat{\beta}_0$ and $\hat{\beta}_1$, since there is apparently some "noise" in the data? We will answer these questions in later sections of this chapter.

Functional forms more complicated than straight lines are often fit to data. For example, to determine the proper placement in college mathematics courses for entering freshmen, data that have a bearing on predicting performance in first-year calculus may be available. Suppose that score on a placement exam, high school grade-point average in math courses, and quantitative college board scores are available; we can denote these values by x_1, x_2, and x_3, respectively. We might try to predict a student's grade in first-year calculus, y, by the form

$$y \approx \beta_0 + \beta_1 x_1 + \beta_2 x_2 + \beta_3 x_3$$

where the β_i could be estimated from data on the performance of students in previous years. It could be ascertained how reliable this prediction equation was, and if it were sufficiently reliable, it could be used in a counseling program for entering freshmen.

In biological and chemical work, it is common to fit functions of the following form to decay curves:

$$f(t) = Ae^{-\alpha t} + Be^{-\beta t}$$

Note that the function f is linear in the parameters A and B and nonlinear in the parameters α and β. From data (y_i, t_i), $i = 1, \ldots, n$, where, for example, y_i is the measured concentration of a substance at time t_i, the parameters are determined by the method of least squares as being the minimizers of

$$S(A, B, \alpha, \beta) = \sum_{i=1}^{n} (y_i - Ae^{-\alpha t_i} - Be^{-\beta t_i})^2$$

In fitting periodic phenomena, functions of the following form occur:

$$f(t) = A \cos \omega_1 t + B \sin \omega_1 t + C \cos \omega_2 t + D \sin \omega_2 t$$

This function is linear in the parameters $A, B, C,$ and D and nonlinear in the parameters ω_1 and ω_2.

When the function to be fit is linear in the unknown parameters, the minimization is relatively straightforward, because calculating partial derivatives and setting them equal to zero produces a set of simultaneous linear equations that can be solved in closed form. This important special case is known as **linear least squares.** If the function to be fit is not linear in the unknown parameters, a system of nonlinear equations must be solved to find the coefficients. Typically, the solution cannot be found in closed form, so an iterative procedure must be used.

For our purposes, the general formulation of the linear least squares problem is as follows: A function of the form

$$f(x_1, x_2, \ldots, x_{p-1}) = \beta_0 + \beta_1 x_1 + \beta_2 x_2 + \cdots + \beta_{p-1} x_{p-1}$$

involving p unknown parameters, $\beta_0, \beta_1, \beta_2, \ldots, \beta_{p-1}$, is to be fit to n data points,

$$y_1, x_{11}, x_{12}, \ldots, x_{1,p-1}$$
$$y_2, x_{21}, x_{22}, \ldots, x_{2,p-1}$$
$$\vdots$$
$$y_n, x_{n1}, x_{n2}, \ldots, x_{n,p-1}$$

The function $f(x)$ is called the **linear regression** of y on x. We will always assume that $p < n$, that is, that there are fewer unknown parameters than observations. Fitting a straight line clearly follows this format. A quadratic can be fit in this way by setting $x_1 = x$, and $x_2 = x^2$. If the frequencies in the trigonometric fitting problem referred to above are known, we can let $x_1 = \cos \omega_1 t$, $x_2 = \sin \omega_1 t$, $x_3 = \cos \omega_2 t$, and $x_4 = \sin \omega_2 t$ and identify the unknown amplitudes $A, B, C,$ and D as the β_i. If the frequencies are unknown and must be determined from the data, the problem is nonlinear.

Many functions that are not initially linear in the unknowns can be put into linear form by means of a suitable transformation. An example of this type of function that occurs frequently in chemistry and biochemistry is the Arrhenius equation,

$$\alpha = Ce^{-e_A/(KT)}$$

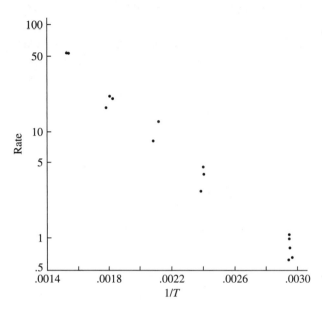

F I G U R E **14.3** A plot of log rate versus $1/T$ for a reaction involving atomic oxygen.

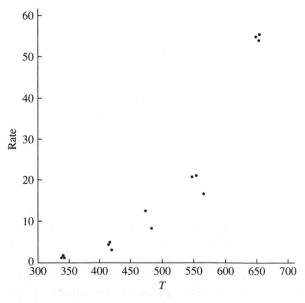

F I G U R E **14.4** A plot of rate versus temperature for a reaction involving atomic oxygen.

Here, α is the rate of a chemical reaction, C is an unknown constant called the frequency factor, e_A is the activation energy of the reaction, K is Boltzmann's constant, and T is absolute temperature. If a reaction is run at several temperatures and the rate is measured, the activation energy and the frequency factor can be estimated by fitting the equation to the data. The function as written above is linear in the parameter C and nonlinear in e_A, but

$$\log \alpha = \log C - e_A \frac{1}{KT}$$

is linear in $\log C$ and e_A. As an example, a plot of log rate versus $1/T$ for a reaction involving atomic oxygen, taken from Huie and Herron (1972), is shown in Figure 14.3. Figure 14.4 is a plot of rate versus temperature, which, in contrast, is quite nonlinear.

14.2 Simple Linear Regression

This section deals with the very common problem of fitting a straight line to data. Later sections of this chapter will generalize the results of this section. First, statistical properties of least squares estimates will be discussed and then methods of assessing goodness of fit, largely through the examination of residuals. Finally, the relation of regression to correlation is presented.

14.2.1 Statistical Properties of the Estimated Slope and Intercept

Up to now we have presented the method of least squares simply as a reasonable principle, without any explicit discussion of statistical models. Consequently, we have not addressed such pertinent questions as the reliability of the slope and intercept in the presence of "noise." In order to address this question, we must have a statistical model for the noise. The simplest model, which we will refer to as the standard statistical model, stipulates that the observed value of y is a linear function of x plus random noise:

$$y_i = \beta_0 + \beta_1 x_i + e_i, \qquad i = 1, \ldots, n$$

Here the e_i are independent random variables with $E(e_i) = 0$ and $\text{Var}(e_i) = \sigma^2$. The x_i are assumed to be fixed.

In Section 14.1, we derived formulas for the slope, $\hat{\beta}_1$, and the intercept, $\hat{\beta}_0$. Referring to those equations, we see that they are linear functions of the y_i, and thus linear functions of the e_i. $\hat{\beta}_0$ and $\hat{\beta}_1$ are estimates of β_0 and β_1. The standard statistical model thus makes computation of the means and variances of $\hat{\beta}_0$ and $\hat{\beta}_1$ straightforward.

THEOREM A

Under the assumptions of the standard statistical model, the least squares estimates are unbiased: $E(\hat{\beta}_j) = \beta_j$, for $j = 0, 1$.

Proof

From the assumptions, $E(y_i) = \beta_0 + \beta_1 x_i$. Thus, from the equation for $\hat{\beta}_0$ in Section 14.1,

$$E(\hat{\beta}_0) = \frac{\left(\sum_{i=1}^{n} x_i^2\right)\left(\sum_{i=1}^{n} E(y_i)\right) - \left(\sum_{i=1}^{n} x_i\right)\left(\sum_{i=1}^{n} x_i E(y_i)\right)}{n\sum_{i=1}^{n} x_i^2 - \left(\sum_{i=1}^{n} x_i\right)^2}$$

$$= \frac{\left(\sum_{i=1}^{n} x_i^2\right)\left(n\beta_0 + \beta_1 \sum_{i=1}^{n} x_i\right) - \left(\sum_{i=1}^{n} x_i\right)\left(\beta_0 \sum_{i=1}^{n} x_i + \beta_1 \sum_{i=1}^{n} x_i^2\right)}{n\sum_{i=1}^{n} x_i^2 - \left(\sum_{i=1}^{n} x_i\right)^2}$$

$$= \beta_0$$

The proof for β_1 is similar. ∎

Note that the proof of Theorem A does not depend on the assumptions that the e_i are independent and have the same variance, only on the assumptions that the errors are additive and $E(e_i) = 0$.

From the standard statistical model, $\text{Var}(y_i) = \sigma^2$ and $\text{Cov}(y_i, y_j) = 0$, where $i \neq j$. This makes the computation of the variances of the $\hat{\beta}_i$ straightforward.

THEOREM B

Under the assumptions of the standard statistical model,

$$\text{Var}(\hat{\beta}_0) = \frac{\sigma^2 \sum_{i=1}^{n} x_i^2}{n\sum_{i=1}^{n} x_i^2 - \left(\sum_{i=1}^{n} x_i\right)^2}$$

$$\text{Var}(\hat{\beta}_1) = \frac{n\sigma^2}{n\sum_{i=1}^{n} x_i^2 - \left(\sum_{i=1}^{n} x_i\right)^2}$$

$$\mathrm{Cov}(\hat{\beta}_0, \hat{\beta}_1) = \frac{-\sigma^2 \sum_{i=1}^{n} x_i}{n \sum_{i=1}^{n} x_i^2 - \left(\sum_{i=1}^{n} x_i\right)^2}$$

Proof

From a form for $\hat{\beta}_1$ given in Section 14.1

$$\hat{\beta}_1 = \frac{\sum_{i=1}^{n}(x_i - \bar{x})(y_i - \bar{y})}{\sum_{i=1}^{n}(x_i - \bar{x})^2} = \frac{\sum_{i=1}^{n}(x_i - \bar{x})y_i}{\sum_{i=1}^{n}(x_i - \bar{x})^2}$$

The identity for the numerator follows from expanding the product and using $\sum(x_i - \bar{x}) = 0$. We then have

$$\mathrm{Var}(\hat{\beta}_1) = \frac{\sigma^2}{\sum_{i=1}^{n}(x_i - \bar{x})^2}$$

which reduces to the desired expression. The other expressions may be derived similarly. Later we will give a more general proof. ■

From Theorem B, we see that the variances of the slope and intercept depend on the x_i and on the error variance, σ^2. The x_i are known; therefore, to estimate the variance of the slope and intercept, we need to estimate only σ^2. Since, in the standard statistical model, σ^2 is the expected squared deviation of the y_i from the line $\beta_0 + \beta_1 x_i$, it is natural to base an estimate of σ^2 on the average squared deviations of the data about the fitted line. We define the **residual sum of squares (RSS)** to be

$$\mathrm{RSS} = \sum_{i=1}^{n}(y_i - \hat{\beta}_0 - \hat{\beta}_1 x_i)^2$$

We will show in Section 14.4.3 that

$$s^2 = \frac{\mathrm{RSS}}{n - 2}$$

is an unbiased estimate of σ^2. The divisor $n - 2$ is used rather than n because two parameters have been estimated from the data, giving $n - 2$ degrees of freedom.

The variances of $\hat{\beta}_0$ and $\hat{\beta}_1$ as given in Theorem B are thus estimated by replacing σ^2 by s^2, yielding estimates that we will denote $s_{\hat{\beta}_0}^2$ and $s_{\hat{\beta}_1}^2$.

If the errors, e_i, are independent normal random variables, then the estimated slope and intercept, being linear combinations of independent normally distributed random variables, are normally distributed as well. More generally, if the e_i are independent and the x_i satisfy certain assumptions, a version of the central limit theorem implies that, for large n, the estimated slope and intercept are approximately normally distributed. The normality assumption, or its approximation, makes possible the construction of confidence intervals and hypothesis tests. It can then be shown that

$$\frac{\hat{\beta}_i - \beta_i}{s_{\hat{\beta}_i}} \sim t_{n-2}$$

which allows the t distribution to be used for confidence intervals and hypothesis tests.

EXAMPLE A   We apply these procedures to the 21 data points on chromatographic peak area. The following table presents some of the statistics from the fit (tables like this are produced by regression programs of software packages):

Coefficient	Estimate	Standard Error	t Value
β_0	.0729	.0297	2.45
β_1	10.77	.27	40.20

The estimated standard deviation of the errors is $s = .068$. The standard error of the intercept is $s_{\hat{\beta}_0} = .0297$. A 95% confidence interval for the intercept, β_0, based on the t distribution with 19 df is

$$\hat{\beta}_0 \pm t_{19}(.025)s_{\hat{\beta}_0}$$

or $(.011, .135)$. Similarly, a 95% confidence interval for the slope, β_1, is

$$\hat{\beta}_1 \pm t_{19}(.025)s_{\hat{\beta}_1}$$

or $(10.21, 11.33)$. To test the null hypothesis $H_0: \beta_0 = 0$, we would use the t statistic $\hat{\beta}_0/s_{\hat{\beta}_0} = 2.45$. The hypothesis would be rejected at significance level $\alpha = .05$, so there is strong evidence that the intercept is nonzero.   ∎

14.2.2   Assessing the Fit

As an aid in assessing the quality of the fit, we will make extensive use of the residuals, which are the differences between the observed and fitted values:

$$\hat{e}_i = y_i - \hat{\beta}_0 - \hat{\beta}_1 x_i$$

It is most useful to examine the residuals graphically. Plots of the residuals versus the x values may reveal systematic misfit or ways in which the data do not conform to the fitted model. Ideally, the residuals should show no relation to the x values, and the plot should look like a horizontal blur.

EXAMPLE A   Figure 14.5 is a plot of the residuals for the data on chromatographic peak area. There is no apparent deviation from randomness in the residuals, so this plot confirms

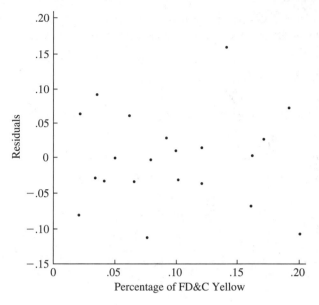

FIGURE **14.5**   A plot of residuals for the data on chromatographic peak area.

the impression from Figure 14.2 that it is reasonable to model the relation as linear.   ∎

We next consider an example in which the plot shows curvature.

EXAMPLE **B**   The data in the following table were gathered for an environmental impact study that examined the relationship between the depth of a stream and the rate of its flow (Ryan, Joiner, and Ryan 1976).

Depth	Flow Rate
.34	.636
.29	.319
.28	.734
.42	1.327
.29	.487
.41	.924
.76	7.350
.73	5.890
.46	1.979
.40	1.124

A plot of flow rate versus depth suggests that the relation is not linear (Figure 14.6). This is even more immediately apparent from the bowed shape of the plot

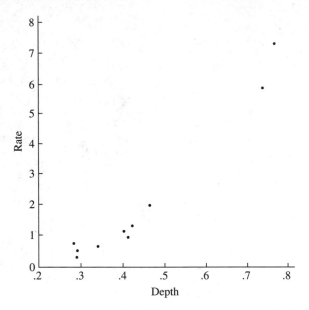

FIGURE **14.6**   A plot of flow rate versus stream depth.

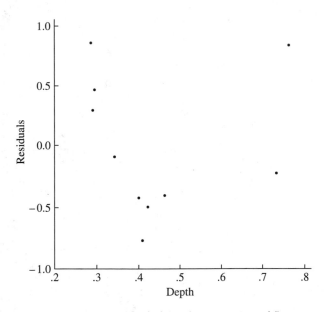

FIGURE **14.7**   Residuals from the regression of flow rate on depth.

of the residuals versus depth (Figure 14.7). In order to empirically linearize relationships, transformations are frequently employed. Figure 14.8 is a plot of log rate versus log depth, and Figure 14.9 shows the residuals for the corresponding fit. There is no sign of obvious misfit. (The possibility of expressing flow rate as a quadratic function of depth will be explored in a later example.)   ■

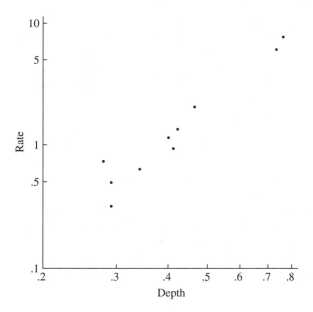

FIGURE **14.8**   Plot of log flow rate versus log depth.

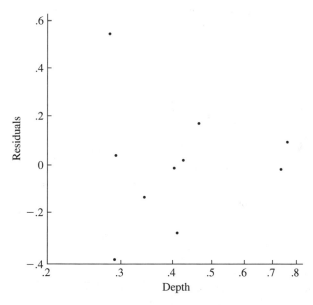

FIGURE **14.9**   Residuals from the regression of log flow rate on log depth.

We have seen that one of the assumptions of the standard statistical model is that the variance of the errors is constant and does not depend on x. Errors with this property are said to be **homoscedastic.** If the variance of the errors is not constant, the errors are said to be **heteroscedastic.** If in fact the error variance is not constant, standard errors and confidence intervals based on the assumption that s^2 is an estimate of σ^2 may be misleading.

EXAMPLE C   In Problem 65 at the end of Chapter 7, data on the population and number of breast cancer mortalities in 301 counties were presented. A scatterplot of the number of cases (y) versus population (x) is shown in Figure 14.10. This plot appears to be consistent with the simple model that the number of cases is proportional to the population size, or $y \approx \beta x$. (We will test whether or not the intercept is zero below.) Accordingly, we fit a model with zero intercept by least squares to the data, yielding $\hat{\beta} = 3.559 \times 10^{-3}$. (See Problem 15 at the end of this chapter for fitting a zero intercept model.) Figure 14.11 shows the residuals from the regression of the number of cases on population plotted versus population. Since it is very hard to see what is going on in the left-hand side of this plot, the residuals are plotted versus log population in Figure 14.12, from which it is quite clear that the error variance is not constant but grows with population size.

The residual plot in Figure 14.12 shows no curvature but indicates that the variance is not constant. For counted data, the variability often grows with the mean, and frequently a square root transformation is used in an attempt to stabilize the variance. We therefore fit a model of the form $\sqrt{y} \approx \gamma \sqrt{x}$. Figure 14.13 shows the plot of residuals for this fit. The residual variability is more nearly constant

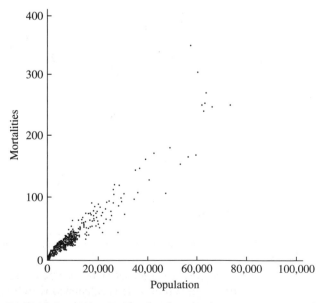

FIGURE **14.10**   Scatterplot showing breast cancer mortality versus population.

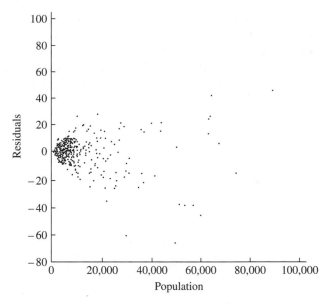

FIGURE **14.11**   Residuals from the regression of mortality on population.

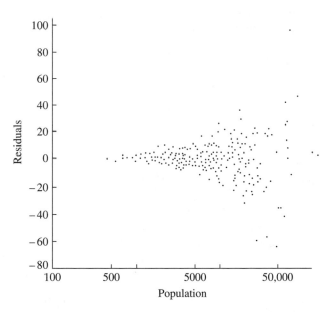

FIGURE **14.12**   A plot of residuals versus log population.

here; β is estimated by the square of the slope, $\hat{\gamma}$, which for this example gives $\tilde{\beta} = \hat{\gamma}^2 = 3.471 \times 10^{-3}$.

Finally, we note that the zero intercept model can be tested in the following way. A linear regression on a square root scale is calculated with both slope and intercept terms, and the intercept is found to be .066 with a standard error $s_{\hat{\gamma}_0} = 9.74 \times 10^{-2}$.

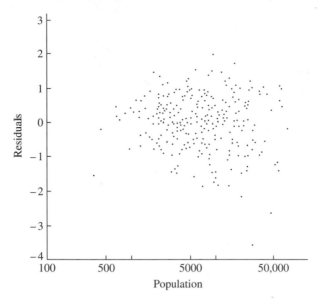

F I G U R E **14.13**   Residuals from the regression of the square root of mortality on the square root of population.

The t statistic for testing $H_0: \gamma_0 = 0$ is

$$ t = \frac{\hat{\gamma}_0}{s_{\hat{\gamma}_0}} = .68 $$

The null hypothesis cannot be rejected for these data.   ■

A normal probability plot of residuals may be useful in indicating gross departures from normality and the presence of outliers. Least squares estimates are not robust against outliers, which can have a large effect on the estimated coefficients, their standard errors, and s, especially if the corresponding x values are at the extremes of the data. It can happen, however, that an outlier with an extreme x value will pull the line toward itself and produce a small residual, as illustrated in Figure 14.14.

E X A M P L E **D**   Figures 14.15 and 14.16 are normal probability plots of the residuals from the fits of Example C. For Figure 14.15, the residuals are from the ordinary linear regression with zero intercept; for Figure 14.16, the residuals are from the zero intercept model with the square root transformation. Note that the distribution in Figure 14.16 is more nearly normal (although there is a hint of skewness) and that the distribution in Figure 14.15 is heavier-tailed than the normal distribution because of the presence of the large residuals from the heavily populated counties.   ■

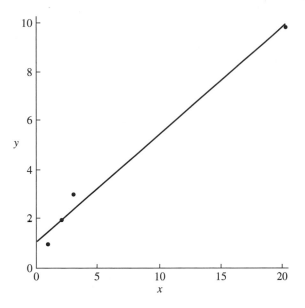

FIGURE **14.14**   An extreme *x* value exerts great leverage on the fitted line and produces a small residual at that point.

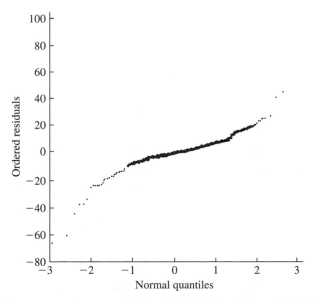

FIGURE **14.15**   Normal probability plot of the residuals from the regression of mortality on population.

It is often useful to plot residuals against variables that are not in the model but might be influential. If the data were collected over a period of time, a plot of the residuals versus time might reveal unexpected time dependencies.

We conclude this section with an extended example.

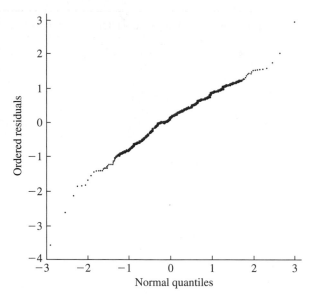

FIGURE **14.16** Normal probability plot of the residuals from the regression of the square root of mortality on the square root of population.

EXAMPLE **E** Houck (1970) studied the bismuth I–II transition pressure as a function of temperature. The data are listed in the following table (Residuals have been rescaled to have standard deviation equal to 1—this process is discussed further in Section 14.4.4).

Pressure (bar)	Temperature (°C)	Standardized Residual
25366	20.8	1.67
25356	20.9	1.48
25336	21.0	.97
25256	21.9	.40
25267	22.1	.22
25306	22.1	1.46
25237	22.4	−.35
25267	22.5	.74
25138	24.8	−.34
25148	24.8	−.02
25143	25.0	.08
24731	34.0	−1.20
24751	34.0	−.57
24771	34.1	.19
24424	42.7	.46
24444	42.7	1.11
24419	42.7	.30

(Continued)

Pressure (bar)	Temperature (°C)	Standardized Residual
24117	49.9	.15
24102	50.1	−.08
24092	50.1	−.42
25202	22.5	−1.33
25157	23.1	−1.97
25157	23.0	−2.10

From Figure 14.17, a plot of the tabulated data, it appears that the relationship is fairly linear. The least squares line is

$$\text{Pressure} = 26172(\pm 21) - 41.3(\pm .6) \times \text{temperature}$$

where the estimated standard errors of the parameters are given in parentheses. The residual standard deviation is $s = 32.5$ with 21 df. An approximate 95% confidence interval for the slope is

$$\hat{\beta}_1 \pm s_{\hat{\beta}_1} t_{21}(.025)$$

or (40.05, 42.55).

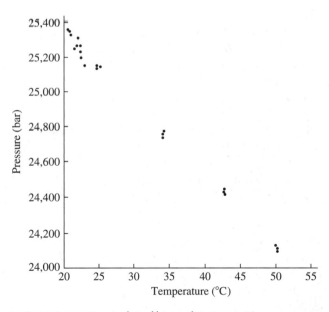

FIGURE **14.17**   A plot of bismuth I–II transition pressure versus temperature.

In order to check how well the model fits, we look at a plot of the standardized residuals versus temperature (Figure 14.18). The plot is rather odd. At first glance, it appears that the variability is greater at lower temperatures. (Bear in mind that the error variance was assumed to be constant in the derivation of the statistical properties of $\hat{\beta}$.) There is another possible explanation for the wedge-shaped

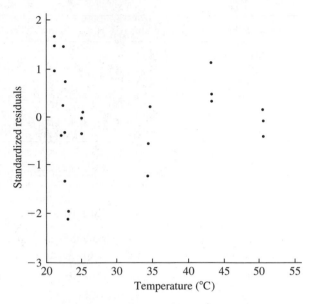

FIGURE **14.18** A plot of standardized residuals versus temperature.

appearance of the residual plot. The table reveals that the data were apparently collected in the following way: three measurements at about 21°C, five at about 22°C, three at about 25°C, three at about 34°C, three at about 43°C, three at about 50°C, and three at about 23°C. It is quite possible that the measurements were taken in the order in which they are listed. We can circle these groups of measurements on the residual plot and note the offsets among them. The last three measurements, taken at about 23°C, particularly stand out, and the three taken at 43°C appear out of line with those at 34°C and 50°C. A plausible explanation for this pattern is as follows: The experimental equipment was set up for a given temperature and several measurements were made; then the equipment was set for another temperature and more measurements were made, and so on; at each setting, errors were introduced that affected every measurement at that temperature. Calibration errors are a possibility.

The standard statistical model, which assumes that the errors at each point are independent, does not provide a faithful representation of such a phenomenon. The standard errors given above for $\hat{\beta}_0$ and $\hat{\beta}_1$ and the confidence interval for $\hat{\beta}_1$ are clearly suspect. (Recall, however, that the estimates $\hat{\beta}_0$ and $\hat{\beta}_1$ are unbiased even if the errors are dependent.) ∎

14.2.3 Correlation and Regression

There is a close relationship between correlation analysis and fitting straight lines by the least squares method. Let us introduce some notation:

$$s_{xx} = \frac{1}{n} \sum_{i=1}^{n} (x_i - \bar{x})^2$$

$$s_{yy} = \frac{1}{n} \sum_{i=1}^{n} (y_i - \bar{y})^2$$

$$s_{xy} = \frac{1}{n} \sum_{i=1}^{n} (x_i - \bar{x})(y_i - \bar{y})$$

The correlation coefficient between the x's and y's is

$$r = \frac{s_{xy}}{\sqrt{s_{xx} s_{yy}}}$$

The slope of the least squares line is (see Problem 10 at the end of this chapter)

$$\hat{\beta}_1 = \frac{s_{xy}}{s_{xx}}$$

and therefore

$$r = \hat{\beta}_1 \sqrt{\frac{s_{xx}}{s_{yy}}}$$

In particular, the correlation is zero if and only if the slope is zero.

To further investigate the relationship of correlation and regression, it is instructive to standardize the variables. If in the regression equation $\hat{y} = \hat{\beta}_0 + \hat{\beta}_1 x$ the coefficients are expressed as

$$\hat{\beta}_0 = \bar{y} - \hat{\beta}_1 \bar{x}$$

$$\hat{\beta}_1 = \frac{\sum_{i=1}^{n} (x_i - \bar{x})(y_i - \bar{y})}{\sum_{i=1}^{n} (x_i - \bar{x})^2}$$

and $\hat{\beta}_1$ is expressed in terms of r as above, then after some manipulation, we arrive at

$$\frac{\hat{y} - \bar{y}}{\sqrt{s_{yy}}} = r \frac{x - \bar{x}}{\sqrt{s_{xx}}}$$

(You should check this calculation.) The equation can be interpreted as follows: Suppose that $r > 0$ and that x, the predictor variable, is one standard deviation greater than its average; then the predicted value of y is r standard deviations bigger than its average, $r \le 1$. The predicted value thus deviates from its average by fewer standard deviations than does the predictor. In units of standard deviations, it is closer to its average than is the predictor.

The term *regression* stems from the work of Sir Francis Galton (1822–1911), a famous geneticist who studied the sizes of seeds and their offspring and the heights of fathers and their sons. In both cases, he found that the offspring of parents of larger than average size tended to be smaller than their parents and that the offspring of parents of smaller than average size tended to be larger than their parents. He called this phenomenon "regression towards mediocrity." This is exactly what the regression line predicts, as in the previous paragraph.

E X A M P L E **A**   Figure 14.19 (from Freedman, Pisani, and Purves, 1998) is a scatterplot of the heights of 1078 pairs of fathers and sons. The fathers' average height is 67.7 in. with a standard deviation of 2.74 in.; the sons' average and standard deviation are 68.7 in. and 2.81 in., respectively; the correlation coefficient is 0.501. The solid line, in the figure is the regression line, and the dashed one is the line $y = x + 1$ (since the sons are 1 in. taller than the fathers on average). Notice how the prediction son's height = father's height $+ 1$ under-predicts on the left and over-predicts on the right.

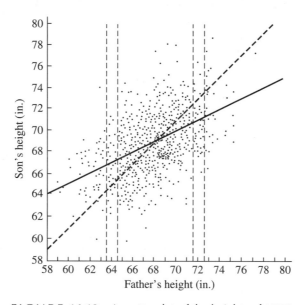

F I G U R E **14.19**   A scatterplot of the heights of 1078 sons versus the heights of their fathers.

In the vertical strip on the right, the fathers' heights are 72 in. to the nearest inch; the average height of the sons in that strip is 71 in.—one inch shorter than their fathers'. The regression line is

$$\frac{\hat{y} - 68.7}{2.81} = .5 \times \frac{x - 67.7}{2.74}$$

Evaluating this for $x = 72$ predicts the sons' height to be 70.9 in., which is very close to the empirical average in the strip.

In the vertical strip on the left, the fathers' heights are 64 in. to the nearest inch, and the average height of a son in that strip is 67 in.—three inches taller than their fathers'. The prediction from the regression line is 66.8 in.   ■

E X A M P L E **B**   Statistics from the sport of baseball have been extensively gathered and studied; the statistical analysis of baseball records is called "sabermetrics." (See Albert and Bennett, 2003.) Analysis has shown that one of the key statistics relating to a player's

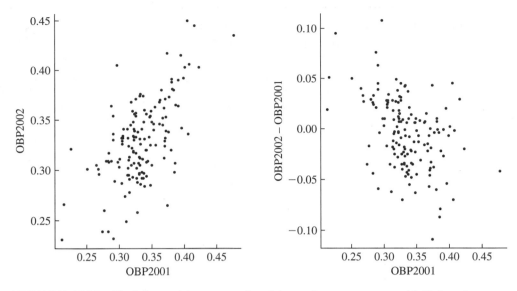

FIGURE **14.20**   The left panel is a scatterplot of the on-base-percentage of 148 American League players in 2002 versus their percentages in 2001. In the right panel, the change is plotted versus on-base-percentage in 2001.

offensive effectiveness is the percentage of the time he gets on base. The left panel of Figure 14.20 shows the on-base percentage for all American League players in 2001 and 2002 with at least 100 plate appearances each season. There is a strong correlation ($r = 0.62$) between the players' performances in the two consecutive seasons. The right panel of the figure shows the difference (2002−2001) plotted against the 2001 performance. Observe that the scatterplot exhibits a negative slope—players who did relatively poorly in 2001 tended to improve in 2002, whereas those who did relatively well in 2001 tended to worsen in 2002.   ∎

We have already encountered the phenomenon of regression in Example B in Section 4.4.1, where we saw that if X and Y follow a bivariate normal distribution with $\sigma_X = \sigma_Y = 1$, the conditional expectation of Y given X does not lie along the major axis of the elliptical contours of the joint density; rather, $E(Y|X) = \rho X$. Regression to the mean was also discussed in Example B of Section 4.4.2.

The regression effect must be taken into account in test-retest situations. Suppose, for example, that a group of preschool children are given an IQ test at age four and another test at age five. The results of the tests will certainly be correlated, and according to the analysis above, children who do poorly on the first test will tend to score higher on the second test. If, on the basis of the first test, low-scoring children were selected for supplemental educational assistance, their gains might be mistakenly attributed to the program. A comparable control group is needed in this situation to tighten up the experimental design.

14.3 The Matrix Approach to Linear Least Squares

With problems more complex than fitting a straight line, it is very useful to approach the linear least squares analysis via linear algebra. As well as providing a compact notation, the conceptual framework of linear algebra can generate theoretical and practical insights. Developments in numerical analysis have resulted in the availability of high-quality software packages; see, for example, LINPACK (www.netlib.org/linpack/)

Suppose that a model of the form

$$y = \beta_0 + \beta_1 x_1 + \cdots + \beta_{p-1} x_{p-1}$$

is to be fit to data, which we denote as

$$y_i, \ x_{i1}, \ x_{i2}, \ldots, x_{i, p-1}, \qquad i = 1, \ldots, n$$

The observations y_i, where $i = 1, \ldots, n$, will be represented by a vector $\mathbf{Y}$. The unknowns, $\beta_0, \ldots, \beta_{p-1}$, will be represented by a vector $\boldsymbol{\beta}$. Let $\mathbf{X}_{n \times p}$ be the matrix

$$\mathbf{X} = \begin{bmatrix} 1 & x_{11} & x_{12} & \cdots & x_{1,\,p-1} \\ 1 & x_{21} & x_{22} & \cdots & x_{2,\,p-1} \\ \vdots & \vdots & \vdots & \vdots & \vdots \\ 1 & x_{n1} & x_{n2} & \cdots & x_{n,\,p-1} \end{bmatrix}$$

For a given $\boldsymbol{\beta}$, the vector of fitted or predicted values, $\hat{\mathbf{Y}}$, can be written

$$\underset{n \times 1}{\hat{\mathbf{Y}}} = \underset{n \times p}{\mathbf{X}} \ \underset{p \times 1}{\boldsymbol{\beta}}$$

(Verify this by writing out explicitly the first row of the system of equations.) The least squares problem can then be phrased as follows: Find $\boldsymbol{\beta}$ to minimize

$$S(\boldsymbol{\beta}) = \sum_{i=1}^{n} (y_i - \beta_0 - \beta_1 x_{i1} - \cdots - \beta_{p-1} x_{i,\,p-1})^2$$
$$= ||\mathbf{Y} - \mathbf{X}\boldsymbol{\beta}||^2$$
$$= ||\mathbf{Y} - \hat{\mathbf{Y}}||^2$$

(If $\mathbf{u}$ is a vector, $||\mathbf{u}||^2 = \sum_{i=1}^{n} u_i^2$.)

E X A M P L E   A   Let us consider fitting a straight line, $y = \beta_0 + \beta_1 x$, to points (y_i, x_i), where $i = 1, \ldots, n$. In this case,

$$\mathbf{Y} = \begin{bmatrix} y_1 \\ y_2 \\ \vdots \\ y_n \end{bmatrix}$$

$$\boldsymbol{\beta} = \begin{bmatrix} \beta_0 \\ \beta_1 \end{bmatrix}$$

$$\mathbf{X} = \begin{bmatrix} 1 & x_1 \\ 1 & x_2 \\ \vdots & \vdots \\ 1 & x_n \end{bmatrix}$$

and

$$\mathbf{Y} - \mathbf{X}\boldsymbol{\beta} = \begin{bmatrix} y_1 - \beta_0 - \beta_1 x_1 \\ y_2 - \beta_0 - \beta_1 x_2 \\ \vdots \\ y_n - \beta_0 - \beta_1 x_n \end{bmatrix} \qquad \blacksquare$$

Returning to the general case, if we differentiate S with respect to each β_k and set the derivatives equal to zero, we see that the minimizers $\hat{\beta}_0, \ldots, \hat{\beta}_{p-1}$ satisfy the p linear equations

$$n\hat{\beta}_0 + \hat{\beta}_1 \sum_{i=1}^{n} x_{i1} + \cdots + \hat{\beta}_{p-1} \sum_{i=1}^{n} x_{i, p-1} = \sum_{i=1}^{n} y_i$$

$$\hat{\beta}_0 \sum_{i=1}^{n} x_{ik} + \hat{\beta}_1 \sum_{i=1}^{n} x_{i1} x_{ik} + \cdots + \hat{\beta}_{p-1} \sum_{i=1}^{n} x_{ik} x_{i, p-1} = \sum_{i=1}^{n} y_i x_{ik}, k = 1, \ldots, p-1$$

These p equations can be written in matrix form

$$\mathbf{X}^T \mathbf{X} \hat{\boldsymbol{\beta}} = \mathbf{X}^T \mathbf{Y}$$

and are called the **normal equations.** If $\mathbf{X}^T \mathbf{X}$ is nonsingular, the formal solution is

$$\hat{\boldsymbol{\beta}} = (\mathbf{X}^T \mathbf{X})^{-1} \mathbf{X}^T \mathbf{Y}$$

We stress that this is a formal solution; computationally, it is sometimes unwise even to form the normal equations because the multiplications involved in forming $\mathbf{X}^T \mathbf{X}$ can introduce undesirable round-off error. Alternative methods of finding the least squares solution $\hat{\boldsymbol{\beta}}$ are developed in Problems 8 and 9 at the end of this chapter.

The following lemma gives a criterion for the existence and uniqueness of solutions of the normal equations.

LEMMA A

$\mathbf{X}^T\mathbf{X}$ is nonsingular if and only if the rank of $\mathbf{X}$ equals p.

Proof

First suppose that $\mathbf{X}^T\mathbf{X}$ is singular. There exists a nonzero vector $\mathbf{u}$ such that $\mathbf{X}^T\mathbf{X}\mathbf{u} = 0$. Multiplying the left-hand side of this equation by $\mathbf{u}^T$, we have

$$0 = \mathbf{u}^T\mathbf{X}^T\mathbf{X}\mathbf{u}$$
$$= (\mathbf{X}\mathbf{u})^T(\mathbf{X}\mathbf{u})$$

so $\mathbf{X}\mathbf{u} = \mathbf{0}$, the columns of $\mathbf{X}$ are linearly dependent, and the rank of $\mathbf{X}$ is less than p.

Next, suppose that the rank of $\mathbf{X}$ is less than p so that there exists a nonzero vector $\mathbf{u}$ such that $\mathbf{X}\mathbf{u} = 0$. Then $\mathbf{X}^T\mathbf{X}\mathbf{u} = 0$, and hence $\mathbf{X}^T\mathbf{X}$ is singular. ∎

For example, suppose that a straight line is to be fitted to the points (y_i, x_i), $i = 1, 2, 3$. Then the design matrix is

$$\mathbf{X} = \begin{pmatrix} 1 & x_1 \\ 1 & x_2 \\ 1 & x_3 \end{pmatrix}$$

If $x_1 = x_2 = x_3$, the matrix is singular since the two columns are proportional to each other. In this case, we would be trying to fit a line to a single point. You should calculate $\mathbf{X}^T\mathbf{X}$ and check that it is singular.

The vector $\hat{\boldsymbol{\beta}} = (\mathbf{X}^T\mathbf{X})^{-1}\mathbf{X}^T\mathbf{Y}$ is the vector of fitted parameters, and the corresponding vector of fitted, or predicted, y values is $\hat{\mathbf{Y}} = \mathbf{X}\hat{\boldsymbol{\beta}}$. The residuals $\mathbf{Y} - \hat{\mathbf{Y}} = \mathbf{Y} - \mathbf{X}\hat{\boldsymbol{\beta}}$ are the differences between the observed and fitted values. We will make use of these residuals in examining goodness of fit.

E X A M P L E **B**   Returning to Example A on fitting a straight line, we have

$$\mathbf{X}^T\mathbf{X} = \begin{bmatrix} 1 & \cdots & 1 \\ x_1 & \cdots & x_n \end{bmatrix} \begin{bmatrix} 1 & x_1 \\ \vdots & \vdots \\ 1 & x_n \end{bmatrix}$$

$$= \begin{bmatrix} n & \sum_{i=1}^{n} x_i \\ \sum_{i=1}^{n} x_i & \sum_{i=1}^{n} x_i^2 \end{bmatrix}$$

$$(\mathbf{X}^T\mathbf{X})^{-1} = \frac{1}{n\sum_{i=1}^{n}x_i^2 - \left(\sum_{i=1}^{n}x_i\right)^2} \begin{bmatrix} \sum_{i=1}^{n}x_i^2 & -\sum_{i=1}^{n}x_i \\ -\sum_{i=1}^{n}x_i & n \end{bmatrix}$$

$$\mathbf{X}^T\mathbf{Y} = \begin{bmatrix} \sum_{i=1}^{n}y_i \\ \sum_{i=1}^{n}x_i y_i \end{bmatrix}$$

Thus,

$$\hat{\beta} = \begin{bmatrix} \hat{\beta}_0 \\ \hat{\beta}_1 \end{bmatrix}$$

$$= (\mathbf{X}^T\mathbf{X})^{-1}\mathbf{X}^T\mathbf{Y}$$

$$= \frac{1}{n\sum_{i=1}^{n}x_i^2 - \left(\sum_{i=1}^{n}x_i\right)^2} \begin{bmatrix} \sum_{i=1}^{n}x_i^2 & -\sum_{i=1}^{n}x_i \\ -\sum_{i=1}^{n}x_i & n \end{bmatrix} \begin{bmatrix} \sum_{i=1}^{n}y_i \\ \sum_{i=1}^{n}x_i y_i \end{bmatrix}$$

$$= \frac{1}{n\sum_{i=1}^{n}x_i^2 - \left(\sum_{i=1}^{n}x_i\right)^2} \begin{bmatrix} \left(\sum_{i=1}^{n}y_i\right)\left(\sum_{i=1}^{n}x_i^2\right) - \left(\sum_{i=1}^{n}x_i\right)\left(\sum_{i=1}^{n}x_i y_i\right) \\ n\sum_{i=1}^{n}x_i y_i - \left(\sum_{i=1}^{n}x_i\right)\left(\sum_{i=1}^{n}y_i\right) \end{bmatrix}$$

which agrees with the earlier calculation. ∎

14.4 Statistical Properties of Least Squares Estimates

In this section, we develop some statistical properties of the vector $\hat{\beta}$, which is found by the least squares method, under some assumptions on the vector of errors. In order to do this, we must use concepts and notation for the analysis of random vectors.

14.4.1 Vector-Valued Random Variables

In Section 14.3, we found expressions for least squares estimates in terms of matrices and vectors. We now develop methods and notation for dealing with random vectors, vectors whose components are random variables. These concepts will be applied to finding statistical properties of the least squares estimates.

We consider the random vector

$$\mathbf{Y} = \begin{bmatrix} Y_1 \\ Y_2 \\ \vdots \\ Y_n \end{bmatrix}$$

the elements of which are jointly distributed random variables with

$$E(Y_i) = \mu_i$$

and

$$\text{Cov}(Y_i, Y_j) = \sigma_{ij}$$

The **mean vector** is defined to be simply the vector of means, or

$$E(\mathbf{Y}) = \boldsymbol{\mu}_Y = \begin{bmatrix} \mu_1 \\ \mu_2 \\ \vdots \\ \mu_n \end{bmatrix}$$

The **covariance matrix** of **Y**, denoted $\boldsymbol{\Sigma}$, is defined to be an $n \times n$ matrix with the ij element σ_{ij}, which is the covariance of Y_i and Y_j. Note that $\boldsymbol{\Sigma}$ is a symmetric matrix.

Suppose that

$$\underset{m \times 1}{\mathbf{Z}} = \underset{m \times 1}{\mathbf{c}} + \underset{m \times n}{\mathbf{A}} \underset{n \times 1}{\mathbf{Y}}$$

is another random vector formed from a fixed vector, **c,** and a fixed linear transformation, **A,** of the random vector **Y.** The next two theorems show how the mean vector and covariance matrix of **Z** are determined from the mean vector and covariance matrix of **Y** and the matrix **A.** Each of the theorems is followed by two examples; the results in those examples could easily be derived without using matrix algebra, but they illustrate how the matrix formalisms work.

THEOREM A

If $\mathbf{Z} = \mathbf{c} + \mathbf{AY}$, where **Y** is a random vector and **A** is a fixed matrix and **c** is a fixed vector, then

$$E(\mathbf{Z}) = \mathbf{c} + \mathbf{A}E(\mathbf{Y})$$

Proof

The ith component of **Z** is

$$Z_i = c_i + \sum_{j=1}^{n} a_{ij} Y_j$$

By the linearity of the expectation,

$$E(Z_i) = c_i + \sum_{j=1}^{n} a_{ij} E(Y_j)$$

Writing these equations in matrix form completes the proof. ∎

E X A M P L E **A**   As a simple example, let us consider the case where $Z = \sum_{i=1}^{n} a_i Y_i$. In matrix notation, this can be written $Z = \mathbf{a}^T \mathbf{Y}$. According to Theorem A,

$$E(\mathbf{Z}) = \mathbf{a}^T \boldsymbol{\mu} = \sum_{i=1}^{n} a_i \mu_i$$

as we already knew. ∎

E X A M P L E **B**   As another example, let us consider a moving average. Suppose that $Z_i = Y_i + Y_{i+1}$, for $i = 1, \ldots, n - 1$. We can write this in matrix notation as $\mathbf{Z} = \mathbf{AY}$ where $\mathbf{A}$ is the matrix

$$\begin{bmatrix} 1 & 1 & 0 & 0 & \cdots & 0 & 0 \\ 0 & 1 & 1 & 0 & \cdots & 0 & 0 \\ \vdots & \vdots & \vdots & \vdots & \vdots & \vdots & \vdots \\ 0 & 0 & 0 & 0 & \cdots & 1 & 1 \end{bmatrix}$$

Using Theorem A to find $E(\mathbf{Z})$, it is easy to see that $\mathbf{A}\boldsymbol{\mu}$ has ith component $\mu_i + \mu_{i+1}$. ∎

THEOREM **B**

Under the assumptions of Theorem A, if the covariance matrix of $\mathbf{Y}$ is $\sum_{YY}$, then the covariance matrix of $\mathbf{Z}$ is

$$\Sigma_{ZZ} = \mathbf{A} \Sigma_{YY} \mathbf{A}^T$$

Proof

The constant $\mathbf{c}$ does not affect the covariance.

$$\text{Cov}(Z_i, Z_j) = \text{Cov}\left(\sum_{k=1}^{n} a_{ik}Y_k, \sum_{l=1}^{n} a_{jl}Y_l\right)$$

$$= \sum_{k=1}^{n}\sum_{l=1}^{n} a_{ik}a_{jl}\text{Cov}(Y_k, Y_l)$$

$$= \sum_{k=1}^{n}\sum_{l=1}^{n} a_{ik}\sigma_{kl}a_{jl}$$

The last expression is the ij element of the desired matrix.   ∎

EXAMPLE C   Continuing Example A, suppose that the Y_i are uncorrelated with constant variance σ^2. The covariance matrix of $\mathbf{Y}$ can then be expressed as $\Sigma_{YY} = \sigma^2\mathbf{I}$, where $\mathbf{I}$ is the identity matrix. The role of $\mathbf{A}$ in Theorem B is played by $\mathbf{a}^T$. Therefore, the covariance matrix of Z, which is a 1×1 matrix in this case, is

$$\Sigma_{ZZ} = \sigma^2\mathbf{a}^T\mathbf{a} = \sigma^2 \sum_{i=1}^{n} a_i^2$$ ∎

EXAMPLE D   Suppose that the Y_i of Example B have the covariance matrix $\sigma^2\mathbf{I}$. Then $\Sigma_{ZZ} = \sigma^2\mathbf{A}^T\mathbf{A}$, or

$$\sigma^2 \begin{bmatrix} 2 & 1 & 0 & 0 & \cdots & 0 \\ 1 & 2 & 1 & 0 & \cdots & 0 \\ 0 & 1 & 2 & 1 & \cdots & 0 \\ \vdots & \vdots & \vdots & \vdots & \vdots & \vdots \\ 0 & 0 & 0 & 0 & \cdots & 2 \end{bmatrix}$$ ∎

The proofs of both these theorems are straightforward, although the unfamiliarity of the notation may present a difficulty. But one of the advantages of using matrices and vectors when dealing with collections of random variables is that this notation is much more compact and easier to follow once one has mastered it, because all the subscripts have been suppressed.

Let $\mathbf{A}$ be a symmetric $n \times n$ matrix and $\mathbf{x}$ an n vector. The expression

$$\mathbf{x}^T\mathbf{A}\mathbf{x} = \sum_{i=1}^{n}\sum_{j=1}^{n} x_i a_{ij} x_j$$

is called a **quadratic form.** We will next calculate the expectation of a quadratic form in the case where $\mathbf{x}$ is a random vector.

THEOREM C

Let $\mathbf{X}$ be a random n vector with mean $\boldsymbol{\mu}$ and covariance $\boldsymbol{\Sigma}$, and let $\mathbf{A}$ be a fixed matrix. Then

$$E(\mathbf{X}^T \mathbf{A} \mathbf{X}) = \text{trace}(\mathbf{A}\boldsymbol{\Sigma}) + \boldsymbol{\mu}^T \mathbf{A} \boldsymbol{\mu}$$

Proof

The trace of a square matrix is defined to be the sum of its diagonal terms. Since

$$E(X_i X_j) = \sigma_{ij} + \mu_i \mu_j$$

we have that

$$E\left(\sum_{i=1}^{n} \sum_{j=1}^{n} X_i X_j a_{ij}\right) = \sum_{i=1}^{n} \sum_{j=1}^{n} \sigma_{ij} a_{ij} + \sum_{i=1}^{n} \sum_{j=1}^{n} \mu_i \mu_j a_{ij}$$

$$= \text{trace}(\mathbf{A}\boldsymbol{\Sigma}) + \boldsymbol{\mu}^T \mathbf{A} \boldsymbol{\mu} \qquad \blacksquare$$

E X A M P L E **E**   Consider $E[\sum_{i=1}^{n}(X_i - \overline{X})^2]$, where the X_i are uncorrelated random variables with common mean μ. We recognize that this is the squared length of a vector $\mathbf{AX}$ for some matrix $\mathbf{A}$. To figure out what $\mathbf{A}$ must be, we first note that $\overline{X}$ can be expressed as

$$\overline{X} = \frac{1}{n} \mathbf{1}^T \mathbf{X}$$

where $\mathbf{1}$ is a vector consisting of all ones. The vector consisting of entries all of which are $\overline{X}$ can thus be written as $(1/n)\mathbf{1}\mathbf{1}^T\mathbf{X}$, and $\mathbf{A}$ can be written as

$$\mathbf{A} = \mathbf{I} - \frac{1}{n}\mathbf{1}\mathbf{1}^T$$

Thus,

$$\sum_{i=1}^{n}(X_i - \overline{X})^2 = ||\mathbf{AX}||^2 = \mathbf{X}^T \mathbf{A}^T \mathbf{A} \mathbf{X}$$

The matrix $\mathbf{A}$ has some special properties. In particular, $\mathbf{A}$ is symmetric, and $\mathbf{A}^2 = \mathbf{A}$, as can be verified by simply multiplying $\mathbf{A}$ by $\mathbf{A}$, noting that $\mathbf{1}^T\mathbf{1} = n$. Thus,

$$\mathbf{X}^T \mathbf{A}^T \mathbf{A} \mathbf{X} = \mathbf{X}^T \mathbf{A} \mathbf{X}$$

and by Theorem C,

$$E(\mathbf{X}^T \mathbf{A} \mathbf{X}) = \sigma^2 \text{trace}(\mathbf{A}) + \boldsymbol{\mu}^T \mathbf{A} \boldsymbol{\mu}$$

Since $\boldsymbol{\mu}$ can be written as $\boldsymbol{\mu} = \mu\mathbf{1}$, it can be verified that $\mathbf{A}\boldsymbol{\mu} = 0$. Also, trace $\mathbf{A} = n - 1$, so the expectation above is $\sigma^2(n - 1)$. $\blacksquare$

If $\mathbf{Y}_{p \times 1}$ and $\mathbf{Z}_{m \times 1}$ are random vectors, the **cross-covariance matrix** of $\mathbf{Y}$ and $\mathbf{Z}$ is defined to be the $p \times m$ matrix Σ_{YZ} with the ij element $\sigma_{ij} = \text{Cov}(Y_i, Z_j)$.

The entries of the cross-covariance matrix quantify the strengths of linear relationships between the elements of $\mathbf{Y}$ and $\mathbf{Z}$. The covariance of Y_i and Z_j can be converted to a correlation coefficient by dividing by the product of the standard deviations of Y_i and Z_j.

THEOREM D

Let $\mathbf{X}$ be a random vector with covariance matrix Σ_{XX}. If

$$\mathbf{Y} = \underset{p \times n}{\mathbf{A}} \, \mathbf{X}$$

and

$$\mathbf{Z} = \underset{m \times n}{\mathbf{B}} \, \mathbf{X}$$

where $\mathbf{A}$ and $\mathbf{B}$ are fixed matrices, the cross-covariance matrix of $\mathbf{Y}$ and $\mathbf{Z}$ is

$$\Sigma_{YZ} = \mathbf{A}\Sigma_{XX}\mathbf{B}^T$$

Proof

The proof follows the lines of that of Theorem B (you should work it through for yourself). ∎

E X A M P L E **F**   Let $\mathbf{X}$ be a random n vector with $E(\mathbf{X}) = \mu\mathbf{1}$ and $\Sigma_{XX} = \sigma^2\mathbf{I}$. Let $Y = \overline{X}$, and let $\mathbf{Z}$ be the vector with ith element $X_i - \overline{X}$. We will find Σ_{ZY}, an $n \times 1$ matrix. In matrix form,

$$\mathbf{Z} = \left(\mathbf{I} - \frac{1}{n}\mathbf{1}\mathbf{1}^T\right)\mathbf{X}$$

$$\mathbf{Y} = \frac{1}{n}\mathbf{1}^T\mathbf{X}$$

From Theorem D,

$$\Sigma_{ZY} = \left(\mathbf{I} - \frac{1}{n}\mathbf{1}\mathbf{1}^T\right)(\sigma^2\mathbf{I})\left(\frac{1}{n}\mathbf{1}\right)$$

which becomes an $n \times 1$ matrix of zeros after multiplying out. Thus, the mean $\overline{X}$ is uncorrelated with each of $X_i - \overline{X}$, $i = 1, \ldots, n$. In the case that the elements of $\mathbf{X}$ are normal random variables, for which being uncorrelated implies independence, this result implies Theorem A of Section 6.3 and hence that $\overline{X}$ and S^2 are independent (Corollary A of Section 6.3). ∎

14.4.2 Mean and Covariance of Least Squares Estimates

Once a function has been fit to data by the least squares method, it may be necessary to consider the stability of the fit and of the estimated parameters, since if the measurements were to be taken again they would often be slightly different. To address the question of the variability of least squares estimates in the presence of noise, we will use the following model:

$$Y_i = \beta_0 + \sum_{j=1}^{p-1} \beta_j x_{ij} + e_i, \qquad i = 1, \ldots, n$$

where the e_i are random errors with

$$E(e_i) = 0$$
$$\mathrm{Var}(e_i) = \sigma^2$$
$$\mathrm{Cov}(e_i, e_j) = 0, \qquad i \neq j$$

In matrix notation, we have

$$\underset{n \times 1}{\mathbf{Y}} = \underset{n \times p}{\mathbf{X}} \underset{p \times 1}{\boldsymbol{\beta}} + \underset{n \times 1}{\mathbf{e}}$$

and

$$E(\mathbf{e}) = 0$$
$$\Sigma_{ee} = \sigma^2 \mathbf{I}$$

In words, the y measurements are equal to the true values of the function plus random, uncorrelated errors with constant variance. Note that in this model, the X's are fixed, not random. A useful theorem follows immediately from Theorem A of Section 14.4.1.

THEOREM A

Under the assumption that the errors have mean zero, the least squares estimates are unbiased.

Proof

The least squares estimate of $\boldsymbol{\beta}$ is

$$\hat{\boldsymbol{\beta}} = (\mathbf{X}^T \mathbf{X})^{-1} \mathbf{X}^T \mathbf{Y}$$
$$= (\mathbf{X}^T \mathbf{X})^{-1} \mathbf{X}^T (\mathbf{X}\boldsymbol{\beta} + \mathbf{e})$$
$$= \boldsymbol{\beta} + (\mathbf{X}^T \mathbf{X})^{-1} \mathbf{X}^T \mathbf{e}$$

From Theorem A of Section 14.4.1,

$$E\hat{\boldsymbol{\beta}} = \boldsymbol{\beta} + (\mathbf{X}^T \mathbf{X})^{-1} \mathbf{X}^T E(\mathbf{e})$$
$$= \boldsymbol{\beta} \qquad \blacksquare$$

It should be noted that the only assumption on the errors used in this proof of Theorem A is that they have mean zero. Thus, even if the errors are correlated and have a nonconstant variance, the least squares estimates are unbiased. The covariance matrix of $\hat{\beta}$ can also be calculated; the proof of the following theorem does depend on assumptions concerning the covariance of the errors.

THEOREM B

Under the assumption that the errors have mean zero and are uncorrelated with constant variance σ^2, the covariance matrix of the least squares estimate $\hat{\beta}$ is

$$\Sigma_{\hat{\beta}\hat{\beta}} = \sigma^2(\mathbf{X}^T\mathbf{X})^{-1}$$

Proof

From Theorem B of Section 14.4.1, the covariance matrix of $\hat{\beta}$ is

$$\Sigma_{\hat{\beta}\hat{\beta}} = (\mathbf{X}^T\mathbf{X})^{-1}\mathbf{X}^T\Sigma_{ee}\mathbf{X}(\mathbf{X}^T\mathbf{X})^{-1}$$
$$= \sigma^2(\mathbf{X}^T\mathbf{X})^{-1}$$

since the covariance matrix of $\mathbf{e}$ is $\sigma^2\mathbf{I}$, and $\mathbf{X}^T\mathbf{X}$ and therefore $(\mathbf{X}^T\mathbf{X})^{-1}$ as well are symmetric. ∎

These theorems generalize Theorems A and B of Section 14.2.1. Note how the use of matrix algebra simplifies the derivation.

EXAMPLE A   We return to the case of fitting a straight line. From the computation of $(\mathbf{X}^T\mathbf{X})^{-1}$ in Example B in Section 14.3, we have

$$\Sigma_{\hat{\beta}\hat{\beta}} = \frac{\sigma^2}{n\sum_{i=1}^{n}x_i^2 - \left(\sum_{i=1}^{n}x_i\right)^2}\begin{bmatrix} \sum_{i=1}^{n}x_i^2 & -\sum_{i=1}^{n}x_i \\ -\sum_{i=1}^{n}x_i & n \end{bmatrix}$$

Therefore,

$$\text{Var}(\hat{\beta}_0) = \frac{\sigma^2\sum_{i=1}^{n}x_i^2}{n\sum_{i=1}^{n}x_i^2 - \left(\sum_{i=1}^{n}x_i\right)^2}$$

$$\text{Var}(\hat{\beta}_1) = \frac{n\sigma^2}{n\sum_{i=1}^{n}x_i^2 - \left(\sum_{i=1}^{n}x_i\right)^2}$$

$$\text{Cov}(\hat{\beta}_0, \hat{\beta}_1) = \frac{-\sigma^2\sum_{i=1}^{n}x_i}{n\sum_{i=1}^{n}x_i^2 - \left(\sum_{i=1}^{n}x_i\right)^2}$$ ∎

14.4.3 Estimation of σ^2

In order to use the formulas for variances developed in the preceding section (to form confidence intervals, for example), σ^2 must be known or estimated. In this section, we develop an estimate of σ^2.

Because σ^2 is the expected squared value of an error, e_i, it is natural to use the sample average squared value of the residuals. The vector of residuals is

$$\hat{\mathbf{e}} = \mathbf{Y} - \hat{\mathbf{Y}}$$
$$= \mathbf{Y} - \mathbf{X}\hat{\boldsymbol{\beta}}$$
$$= \mathbf{Y} - \mathbf{X}(\mathbf{X}^T\mathbf{X})^{-1}\mathbf{X}^T\mathbf{Y}$$

or

$$\hat{\mathbf{e}} = \mathbf{Y} - \mathbf{P}\mathbf{Y}$$

where $\mathbf{P} = \mathbf{X}(\mathbf{X}^T\mathbf{X})^{-1}\mathbf{X}^T$ is an $n \times n$ matrix.

Two useful properties of $\mathbf{P}$ are given in the following lemma (you should be able to write out its proof).

LEMMA A

Let $\mathbf{P}$ be defined as before. Then
$$\mathbf{P} = \mathbf{P}^T = \mathbf{P}^2$$
$$(\mathbf{I} - \mathbf{P}) = (\mathbf{I} - \mathbf{P})^T = (\mathbf{I} - \mathbf{P})^2 \qquad \blacksquare$$

Since $\mathbf{P}$ has the properties given in this lemma, it is a projection matrix—that is, $\mathbf{P}$ projects on the subspace of $\mathbf{R}^n$ spanned by the columns of $\mathbf{X}$. Thus, we may think geometrically of the fitted values, $\hat{\mathbf{Y}}$, as being the projection of $\mathbf{Y}$ onto the subspace spanned by the columns of $\mathbf{X}$. However, we will not pursue the implications of this geometrical interpretation.

The sum of squared residuals is, using Lemma A,

$$\sum_{i=1}^{n}(Y_i - \hat{Y}_i)^2 = \|\mathbf{Y} - \mathbf{P}\mathbf{Y}\|^2$$
$$= \|(\mathbf{I} - \mathbf{P})\mathbf{Y}\|^2$$
$$= \mathbf{Y}^T(\mathbf{I} - \mathbf{P})^T(\mathbf{I} - \mathbf{P})\mathbf{Y}$$
$$= \mathbf{Y}^T(\mathbf{I} - \mathbf{P})\mathbf{Y}$$

From Theorem C of Section 14.4.1, we can compute the expected value of this quadratic form:

$$E[\mathbf{Y}^T(\mathbf{I} - \mathbf{P})\mathbf{Y}] = [E(\mathbf{Y})]^T(\mathbf{I} - \mathbf{P})[E(\mathbf{Y})] + \sigma^2\text{trace}(\mathbf{I} - \mathbf{P})$$

Now $E(\mathbf{Y}) = \mathbf{X}\beta$, so

$$(\mathbf{I} - \mathbf{P})E(\mathbf{Y}) = [\mathbf{I} - \mathbf{X}(\mathbf{X}^T\mathbf{X})^{-1}\mathbf{X}^T]\mathbf{X}\beta$$
$$= \mathbf{0}$$

Furthermore,

$$\text{trace}(\mathbf{I} - \mathbf{P}) = \text{trace}(\mathbf{I}) - \text{trace}(\mathbf{P})$$

and, using the cyclic property of the trace—that is, $\text{trace}(\mathbf{AB}) = \text{trace}(\mathbf{BA})$—we have

$$\text{trace }(\mathbf{P}) = \text{trace }[\mathbf{X}(\mathbf{X}^T\mathbf{X})^{-1}\mathbf{X}^T]$$
$$= \text{trace }[\mathbf{X}^T\mathbf{X}(\mathbf{X}^T\mathbf{X})^{-1}]$$
$$= \text{trace }\left(\underset{p \times p}{\mathbf{I}}\right) = p$$

Since $\text{trace}(\mathbf{I}_{n \times n}) = n$, we have shown that

$$E(\|\mathbf{Y} - \hat{\mathbf{Y}}\|^2) = (n - p)\sigma^2$$

and have proved the following theorem.

THEOREM A

Under the assumption that the errors are uncorrelated with constant variance σ^2, an unbiased estimate of σ^2 is

$$s^2 = \frac{\|\mathbf{Y} - \hat{\mathbf{Y}}\|^2}{n - p}$$ ■

The sum of the squared residuals, $\|\mathbf{Y} - \hat{\mathbf{Y}}\|^2$, is often denoted by RSS, for residual sum of squares.

14.4.4 Residuals and Standardized Residuals

Information concerning whether or not a model fits is contained in the vector of residuals,

$$\hat{\mathbf{e}} = \mathbf{Y} - \hat{\mathbf{Y}} = (\mathbf{I} - \mathbf{P})\mathbf{Y}$$

As we did for the case of fitting a straight line, we will use the residuals to check on the adequacy of the fit of a presumed functional form and on the assumptions underlying the statistical analysis (such as that the errors are uncorrelated with constant variance).

The covariance matrix of the residuals is

$$\Sigma_{\hat{e}\hat{e}} = (\mathbf{I} - \mathbf{P})(\sigma^2\mathbf{I})(\mathbf{I} - \mathbf{P})^T$$
$$= \sigma^2(\mathbf{I} - \mathbf{P})$$

where we have used Lemma A of Section 14.4.3. We see that the residuals are correlated with one another and that different residuals have different variances. In order to make the residuals comparable to one another, they are often standardized. Also, standardization puts the residuals on the familiar scale corresponding to a normal distribution with mean 0 and variance 1 and thus makes their magnitudes easier to interpret. The ith standardized residual is

$$\frac{Y_i - \hat{Y}_i}{s\sqrt{1 - p_{ii}}}$$

where p_{ii} is the ith diagonal element of $\mathbf{P}$.

A further property of the residuals is given by the following theorem.

THEOREM A

If the errors have the covariance matrix $\sigma^2 \mathbf{I}$, the residuals are uncorrelated with the fitted values.

Proof

The residuals are

$$\hat{\mathbf{e}} = (\mathbf{I} - \mathbf{P})\mathbf{Y}$$

and the fitted values are

$$\hat{\mathbf{Y}} = \mathbf{P}\mathbf{Y}$$

From Theorem D of Section 14.4.1, the cross-covariance matrix of $\hat{\mathbf{e}}$ and $\hat{\mathbf{Y}}$ is

$$\begin{aligned}
\Sigma_{\hat{e}\hat{y}} &= (\mathbf{I} - \mathbf{P})(\sigma^2 \mathbf{I})\mathbf{P}^T \\
&= \sigma^2(\mathbf{P}^T - \mathbf{P}\mathbf{P}^T) \\
&= 0
\end{aligned}$$

This result follows from Lemma A of Section 14.4.3. ∎

In Section 14.2.2, we considered plotting residuals versus fitted values (see Figure 14.9). According to this theorem, there should be no linear relationship in such a plot.

14.4.5 Inference about β

In this section, we continue the discussion of the statistical properties of the least squares estimate $\hat{\boldsymbol{\beta}}$. In addition to the assumptions made previously, we will assume that the errors, e_i, are independent and normally distributed. Because the components of $\hat{\boldsymbol{\beta}}$ are in this case linear combinations of independent normally distributed random variables, they are also normally distributed.

In particular, each component $\hat{\beta}_i$ of $\hat{\boldsymbol{\beta}}$ is normally distributed with mean β_i and variance $\sigma^2 c_{ii}$, where $\mathbf{C} = (\mathbf{X}^T\mathbf{X})^{-1}$. The standard error of $\hat{\beta}_i$ may thus be estimated as

$$s_{\hat{\beta}_i} = s\sqrt{c_{ii}}$$

This result will be used to construct confidence intervals and hypothesis tests that will be exact under the assumption of normality and approximate otherwise (because $\hat{\beta}_i$ may be expressed as a linear combination of the independent random variables e_i, a version of the central limit theorem with certain assumptions on $\mathbf{X}$ implies the approximate result).

Under the normality assumption, it can be shown that

$$\frac{\hat{\beta}_i - \beta_i}{s_{\hat{\beta}_i}} \sim t_{n-p}$$

although we will not derive this result. It follows that a $100(1 - \alpha)\%$ confidence interval for β_i is

$$\hat{\beta}_i \pm t_{n-p}(\alpha/2)s_{\hat{\beta}_i}$$

To test the null hypothesis $H_0: \beta_i = \beta_{i0}$, where β_{i0} is a fixed number, we can use the test statistic

$$t = \frac{\hat{\beta}_i - \beta_{i0}}{s_{\hat{\beta}_i}}$$

Under H_0, this statistic follows a t distribution with $n - p$ degrees of freedom. The most commonly tested null hypothesis is $H_0: \beta_i = 0$, which states that x_i has no predictive value.

We will illustrate these concepts in the context of polynomial regression.

EXAMPLE A   *Peak Area*

Let us return to Example A in Section 14.2.2 concerning the regression of peak area on percentage of FD&C Yellow No. 5. We have seen from the residual plot in Figure 14.5 that a straight line appears to give a reasonable fit. Consider enlarging the model so that it is quadratic:

$$y = \beta_0 + \beta_1 x + \beta_2 x^2$$

where y is peak area and x is percentage of Yellow No. 5. The following table gives the statistics of the fit:

Coefficient	Estimate	Standard Error	t Value
β_0	.058	.054	1.07
β_1	11.17	1.20	9.33
β_2	−1.90	5.53	−.35

To test the hypothesis $H_0: \beta_2 = 0$, we would use −.35 as the value of the t statistic, which would not reject H_0. Thus, this test, like the residual analysis, gives no evidence that a quadratic term is needed.   ∎

EXAMPLE B   In Example B in Section 14.2.2, we saw that a residual plot for the linear regression of stream flow rate on depth indicated the inadequacy of that model. The statistics for a quadratic model are given in the following table:

Coefficient	Estimate	Standard Error	t Value
β_0	1.68	1.06	1.59
β_1	-10.86	4.52	-2.40
β_2	23.54	4.27	5.51

Here, the linear and quadratic terms are both statistically significant, and a residual plot, Figure 14.21, shows no signs of systematic misfit.

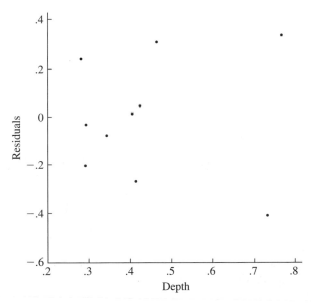

FIGURE **14.21**   Residual plot from the quadratic regression of flow rate on stream depth.

We have seen that the estimated covariance matrix of $\hat{\beta}$ is

$$\hat{\Sigma}_{\hat{\beta}\hat{\beta}} = s^2(\mathbf{X}^T\mathbf{X})^{-1}$$

The corresponding correlation matrix for the coefficients is

$$\begin{bmatrix} 1.00 & -.99 & .97 \\ -.99 & 1.00 & -.99 \\ .97 & -.99 & 1.00 \end{bmatrix}$$

(Note that the correlation matrix does not depend on s and is therefore completely determined by $\mathbf{X}$.) The correlation matrix shows that fluctuations in the components of $\hat{\beta}$ are strongly interrelated. The linear coefficient, $\hat{\beta}_1$, is negatively correlated with both the constant and quadratic coefficients, which in turn are positively correlated with each other. This partly explains why the values of the constant and linear coefficients change so much (they become $\hat{\beta}_0 = -3.98$ and $\hat{\beta}_1 = 13.83$) when the quadratic term is absent from the model.   ■

The estimated covariance matrix of $\hat{\boldsymbol{\beta}}$ is useful for other purposes. Suppose that $\mathbf{x}_0$ is a vector of predictor variables and that we wish to estimate the regression function at $\mathbf{x}_0$. The obvious estimate is

$$\hat{\mu}_0 = \mathbf{x}_0^T \hat{\boldsymbol{\beta}}$$

The variance of this estimate is

$$\text{Var}(\hat{\mu}_0) = \mathbf{x}_0^T \Sigma_{\hat{\beta}\hat{\beta}} \mathbf{x}_0$$

$$= \sigma^2 \mathbf{x}_0^T (\mathbf{X}^T \mathbf{X})^{-1} \mathbf{x}_0$$

This variance can be estimated by substituting s^2 for σ^2, yielding a confidence interval for μ_0,

$$\hat{\mu}_0 \pm t_{n-p}(\alpha/2)s_{\hat{\mu}_0}$$

Note that $\text{Var}(\hat{\mu}_0)$ depends on $\mathbf{x}_0$. This dependency is explored further in Problem 13 of the end-of-chapter problems.

14.5 Multiple Linear Regression—An Example

This section gives a brief introduction to the subject of multiple regression. We will consider the statistical model

$$y_i = \beta_0 + \beta_1 x_{i1} + \beta_2 x_{i2} + \cdots + \beta_{p-1} x_{i,p-1} + e_i \qquad i = 1, \ldots, n$$

As before, $\beta_0, \beta_1, \ldots, \beta_{p-1}$ are unknown parameters and the e_i are independent random variables with mean zero and variance σ^2. The β_i have a simple interpretation: β_k is the change in the expected value of y if x_k is increased by one unit and the other x's are held fixed. Usually, the x's are measurements on different variables, but polynomial regression can be incorporated into this model by letting $x_{i2} = x_{i1}^2$, $x_{i3} = x_{i1}^3$, and so on.

We will develop and illustrate several concepts by means of an example (Weindling 1977). Other examples are included in the end-of-chapter problems. Heart catheterization is sometimes performed on children with congenital heart defects. A Teflon tube (catheter) 3 mm in diameter is passed into a major vein or artery at the femoral region and pushed up into the heart to obtain information about the heart's physiology and functional ability. The length of the catheter is typically determined by a physician's educated guess. In a small study involving 12 children, the exact catheter length required was determined by using a fluoroscope to check that the tip of the catheter had reached the pulmonary artery. The patients' heights and weights were recorded. The objective was to see how accurately catheter length could be determined by these two variables. The data are given in the following table:

Height (in.)	Weight (lb)	Distance to Pulmonary Artery (cm)
42.8	40.0	37.0
63.5	93.5	49.5
37.5	35.5	34.5
39.5	30.0	36.0
45.5	52.0	43.0
38.5	17.0	28.0
43.0	38.5	37.0
22.5	8.5	20.0
37.0	33.0	33.5
23.5	9.5	30.5
33.0	21.0	38.5
58.0	79.0	47.0

Because this is a very small sample, any conclusions must be regarded as tentative.

Figure 14.22 presents scatterplots of all pairs of variables, providing a useful visual presentation of their relationships. We will refer to these plots as we proceed through the analysis.

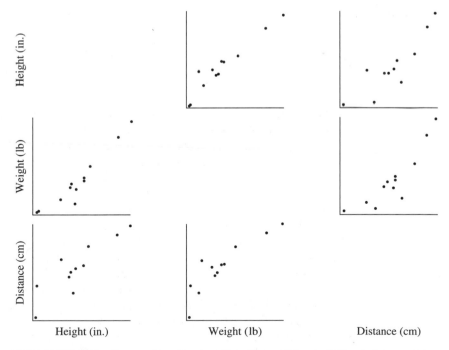

FIGURE **14.22**   Scatterplots showing all pairings of the variables height, weight, and catheter length.

We first consider predicting the length by height alone and by weight alone. The results of simple linear regressions are tabulated below:

	Height	Weight
$\hat{\beta}_0$	12.1 ($\pm$4.3)	25.6 ($\pm$2.0)
$\hat{\beta}_1$	.60 ($\pm$.10)	.28 ($\pm$.04)
s	4.0	3.8
r^2	.78	.80

The standard errors of $\hat{\beta}_0$ and $\hat{\beta}_1$ are given in parentheses. To test the null hypothesis $H_0: \beta_1 = 0$, the appropriate test statistic is $t = \beta_1/s_{\hat{\beta}_1}$. (These null hypotheses are of no real interest in this problem, but we show the tests for pedagogical purposes.) Clearly, this null hypothesis would be rejected in this case. The predictions from both models are similar; the standard deviations of the residuals about the fitted lines are 4.0 and 3.8, respectively, and the squared correlation coefficients are .78 and .80.

The panels of Figure 14.23 are plots of the standardized residuals from each of the simple linear regressions versus the respective independent variable. The plot of residuals versus weight shows some hint of curvature, which is also apparent in the bottom middle scatterplot in Figure 14.22. The largest standardized residual from this fit comes from the lightest and shortest child (see eighth row of data table).

We next consider the multiple regression of length on height and weight together, since perhaps better predictions may be obtained by using both variables rather than

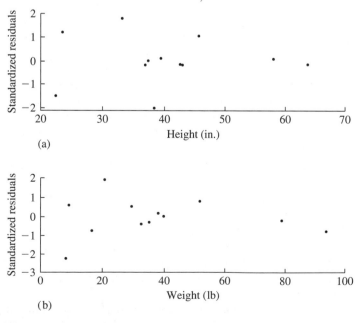

(a)

(b)

FIGURE 14.23   Standardized residuals from simple linear regressions of catheter length plotted against the independent variables (a) height and (b) weight.

either one alone. The method of least squares produces the following relationship:

$$\text{Length} = 21(\pm 8.8) + .20(\pm .36) \times \text{height} + .19(\pm .17) \times \text{weight}$$

where the standard errors of the coefficients are shown in parentheses. The standard deviation of the residuals is 3.9.

The **squared multiple correlation coefficient,** or **coefficient of determination,** is sometimes used as a crude measure of the strength of a relationship that has been fit by least squares. This coefficient is simply defined as the squared correlation of the dependent variable and the fitted values. It can be shown that the squared multiple correlation coefficient, denoted by R^2, can be expressed as

$$R^2 = \frac{s_y^2 - s_{\hat{e}}^2}{s_y^2}$$

Since this is the ratio of the difference between the variance of the dependent variable and the variance of the residuals from the fit to the variance of the dependent variable, it can be interpreted as the proportion of the variability of the dependent variable that can be explained by the independent variables. For the catheter example, $R^2 = .81$.

Consider the coefficients and their standard error shown in the table above. It may seem surprising that the standard errors of the coefficients of height and weight are large relative to the coefficients themselves. Applying t tests would not lead to rejection of either of the hypotheses $H_1: \beta_1 = 0$ or $H_2: \beta_2 = 0$. Yet in the simple linear regressions carried out above, the coefficients were highly significant. A partial explanation of this is that the coefficients in the simple regressions and the coefficients in the multiple regression have different interpretations. In the multiple regression, β_1 is the change in the expected value of the catheter length if height is increased by one unit and weight is held constant. It is the slope along the height axis of the plane that describes the relation of length to height and weight; the large standard error indicates that this slope is not well resolved. To see why, consider the scatterplot of height versus weight in Figure 14.22. The method of least squares fits a plane to the catheter length values that correspond to the pairs of height and weight values in this plot. It should be intuitively clear from the figure that the slope of the fitted plane is relatively well resolved along the line about which the data points fall but poorly resolved along lines on which either height or weight is constant. Imagine how the fitted plane might move if values of length corresponding to pairs of height and weight values were perturbed. Variables that are strongly linearly related, such as height and weight in this example, are said to be highly **collinear.** If the values of height and weight had fallen exactly on a straight line, we would not have been able to determine a plane at all; in fact, **X** would not have had full column rank.

The plot of height versus weight should also serve as a caution concerning making predictions from such a study. Obviously, we would not want to make a prediction for any pair of height and weight values quite dissimilar to those used in making the original fit. Any empirical relationship developed in the region in which the observed data fall might break down if it were extrapolated to a region in which no data had been observed.

Little or no reduction in s has been obtained by fitting simultaneously to height and weight rather than fitting to either height or weight alone. (In fact, fitting to weight alone gives a smaller value of s than does fitting to height and weight together. This

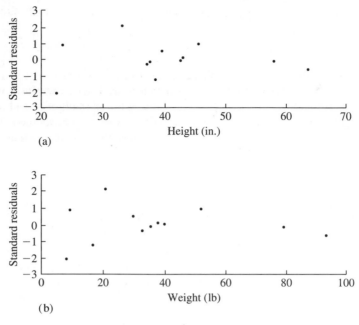

FIGURE **14.24**   Standardized residuals from the multiple regression of catheter length on height and weight plotted against the independent variables (a) height and (b) weight.

may seem paradoxical, but recall that there are 10 degrees of freedom in the former case and 9 in the latter, and that s is the square root of the residual sum of squares divided by the degrees of freedom.) Again, this is partially explained by Figure 14.22, which shows that weight can be predicted quite well from height. Thus, it should not seem surprising that adding weight to the equation for predicting from height produces little gain.

Finally, the panels of Figure 14.24 show the residuals from the multiple regression plotted versus height and weight. The plots are very similar to those in Figure 14.23.

This simple example illustrates that the interpretation of regression coefficients is problematical, since the coefficient of a given variable depends on what other variables are included in the regression—that coefficient can change dramatically, and can even change in sign, as other variables are included or dropped from the model. Tukey and Mosteller (1977) give an example of the use of multiple regression in a study of influences on student achievement. The variables are

$$y = \text{verbal achievement score of 6th graders}$$
$$x_1 = \text{staff salaries per pupil}$$
$$x_2 = \text{percentage of white collar fathers}$$
$$x_3 = \text{socioeconomic status}$$
$$x_4 = \text{teachers' average verbal scores}$$
$$x_5 = \text{mothers' average education (1 unit = 2 school years)}$$

A multiple regression fit results in

$$y = 19.9 - 1.79x_1 + .0432x_2 + 0.556x_3 + 1.11x_4 - 1.79x_5$$

Is the policy implication that it is best to pay teachers small salaries and not educate mothers? Clearly, many of the predictors are highly correlated with each other and are also correlated with variables that are not in the model, and literal interpretation of a coefficient as being the effect if that variable is increased by one unit and the others are held fixed is fallacious. Also note that this is an observational study, not a controlled experiment.

14.6 Conditional Inference, Unconditional Inference, and the Bootstrap

The results in this chapter on the statistical properties of least squares estimates have been derived under the assumptions of a linear model relating independent variables $\mathbf{X}$ to dependent variables $\mathbf{Y}$ of the form

$$\mathbf{Y} = \mathbf{X}\boldsymbol{\beta} + \mathbf{e}$$

In this formulation, the independent variables have been assumed to be *fixed* with randomness arising only through the errors $\mathbf{e}$. This model seems appropriate for some experimental setups, such as that of Section 14.1, where fixed percentages of dies, $\mathbf{X}$, were used and peak areas on a chromatograph, $\mathbf{Y}$, were measured. However, consider Example B of Section 14.2.2, where the flow rate of a stream was related to its depth. The data consisted of measurements from 10 streams and it would seem to be rather forced to model the depths of those streams as being fixed and the flow rates as being random. In this section, we pursue the consequences of a model in which both $\mathbf{X}$ and $\mathbf{Y}$ are random, and we discuss the use of the bootstrap to quantify the uncertainty in parameter estimates under such a model.

First we need to develop some notation. The design matrix will be denoted as a random matrix $\boldsymbol{\Xi}$ and a particular realization of this random matrix will be denoted, as before, by $\mathbf{X}$. The rows of $\boldsymbol{\Xi}$ will be denoted by $\boldsymbol{\xi}_1, \boldsymbol{\xi}_2, \ldots, \boldsymbol{\xi}_n$ and the rows of a realization $\mathbf{X}$ by $\mathbf{x}_1, \mathbf{x}_2, \ldots, \mathbf{x}_n$. In place of the model $Y_i = \mathbf{x}_i\boldsymbol{\beta} + e_i$, where $\mathbf{x}_i$ is fixed and e_i is random with mean 0 and variance σ^2, we will use the model $E(Y|\boldsymbol{\xi} = \mathbf{x}) = \mathbf{x}\boldsymbol{\beta}$ and $\text{Var}(Y|\boldsymbol{\xi} = \mathbf{x}) = \sigma^2$. In the fixed $\mathbf{X}$ model, the e_i were independent of each other. In the random $\mathbf{X}$ model, Y and $\boldsymbol{\xi}$ have a joint distribution (for which the conditional distribution of Y given $\boldsymbol{\xi}$ has mean and variance as specified before) and the data are modeled as n independent random vectors, $(Y_1, \boldsymbol{\xi}_1), (Y_2, \boldsymbol{\xi}_2), \ldots, (Y_n, \boldsymbol{\xi}_n)$ drawn from that joint distribution. The previous model is seen to be a *conditional* version of the new model—the analysis is conditional on the observed values $\mathbf{x}_1, \mathbf{x}_2, \ldots, \mathbf{x}_n$.

We will now deduce some of the consequences for least squares parameter estimation under the new, unconditional, model. First, we have seen that in the old model, the least squares estimate of $\boldsymbol{\beta}$ is unbiased (Theorem A of Section 14.4.2); viewed within the context of the new model we would express this result as $E(\hat{\boldsymbol{\beta}}|\boldsymbol{\Xi} = \mathbf{X}) = \boldsymbol{\beta}$.

We can use Theorem A of Section 4.4.1 to find $E(\hat{\boldsymbol{\beta}})$ under the new model:

$$E(\hat{\boldsymbol{\beta}}) = E(E(\hat{\boldsymbol{\beta}}|\boldsymbol{\Xi}))$$
$$= E(\boldsymbol{\beta})$$
$$= \boldsymbol{\beta}$$

where the outer expectation is with respect to the distribution of $\boldsymbol{\Xi}$. The least squares estimate is thus unbiased under the new model as well.

Next we consider the variance of the least squares estimate. From Theorem B of Section 14.4.2, $\text{Var}(\hat{\beta}_i|\boldsymbol{\Xi} = \mathbf{X}) = \sigma^2(\mathbf{X}^T\mathbf{X})_{ii}^{-1}$. This is the conditional variance. To find the unconditional variance we can use Theorem B of Section 4.4.1, according to which

$$\text{Var}(\hat{\beta}_i) = \text{Var}\left(E(\hat{\beta}_i|\boldsymbol{\Xi})\right) + E\left(\text{Var}(\hat{\beta}_i|\boldsymbol{\Xi})\right)$$
$$= \text{Var}(\beta_i) + E\left(\sigma^2(\boldsymbol{\Xi}^T\boldsymbol{\Xi})_{ii}^{-1}\right)$$
$$= \sigma^2 E(\boldsymbol{\Xi}^T\boldsymbol{\Xi})_{ii}^{-1}$$

This is a highly nonlinear function of the random vectors $\boldsymbol{\xi}_1, \boldsymbol{\xi}_2, \ldots, \boldsymbol{\xi}_n$ and would generally be difficult to evaluate analytically.

Thus for the new, unconditional model, the least squares estimates are still unbiased, but their variances (and covariances) are different. Surprisingly, it turns out that the confidence intervals we have developed still hold at their nominal levels of coverage. Let $C(\mathbf{X})$ denote the $100(1 - \alpha)\%$ confidence interval for β_j that we developed under the old model. Using I_A to denote the indicator variable of the event A, we can express the fact that this is a $100(1 - \alpha)\%$ confidence interval as

$$E\left(I_{\{\beta_j \in C(\mathbf{X})\}}|\boldsymbol{\Xi} = \mathbf{X}\right) = 1 - \alpha$$

that is, the conditional probability of coverage is $1 - \alpha$. Because the conditional probability of coverage is the same for every value of $\boldsymbol{\Xi}$, the unconditional probability of coverage is also $1 - \alpha$:

$$EI_{\{\beta_j \in C(\boldsymbol{\Xi})\}} = E\left(E(I_{\{\beta_j \in C(\boldsymbol{\Xi})\}}|\boldsymbol{\Xi})\right)$$
$$= E(1 - \alpha)$$
$$= 1 - \alpha$$

This very useful result says that for forming confidence intervals we can use the old fixed-$\mathbf{X}$ model and that the intervals we thus form have the correct coverage in the new random-$\mathbf{X}$ model as well.

We complete this section by discussing how the bootstrap can be used to estimate the variability of a parameter estimate under the new model according to which the parameter estimate, say $\hat{\theta}$, is based on n i.i.d. random vectors $(Y_1, \xi_1), (Y_2, \xi_2), \ldots, (Y_n, \xi_n)$. Depending on the context, there are a variety of parameters θ that might be of interest. For example, θ could be one of the regression coefficients, β_i; θ could be $E(Y|\xi = \mathbf{x}_0)$, the expected response at a fixed level $\mathbf{x}_0$ of the independent variables (see Problem 13); in simple linear regression, θ could be that value x_0 such that $E(Y|\xi = x_0) = \mu_0$ for some fixed μ_0; in simple linear regression, θ could be the correlation coefficient of Y and ξ. Now if we knew the probability distribution of the random vector (Y, ξ), we could simulate the sampling distribution of the parameter estimate in the following way: On the computer draw B (a large number) of n-tuples, $(Y_1, \xi_1), (Y_2, \xi_2), \ldots, (Y_B, \xi_B)$, from that distribution and for each draw compute the parameter estimate $\hat{\theta}$. This would yield $\hat{\theta}_1, \hat{\theta}_2, \ldots, \hat{\theta}_B$ and the empirical distribution of this collection would be an approximation to the sampling distribution of $\hat{\theta}$. In particular, the standard deviation of this collection would be an approximation to the standard error of $\hat{\theta}$.

This procedure is, of course, predicated on knowing the distribution of the random vector (Y, ξ), which is unlikely to be the case in practice. The bootstrap principle says to approximate this unknown distribution by the observed empirical distribution of $(Y_1, \mathbf{x}_1), (Y_2, \mathbf{x}_2), \ldots, (Y_n, \mathbf{x}_n)$—that is, draw B samples of size n with replacement from $(Y_1, \mathbf{x}_1), (Y_2, \mathbf{x}_2), \ldots, (Y_n, \mathbf{x}_n)$. For example, to approximate the sampling distribution of a correlation coefficient, r, computed from n pairs $(Y_1, X_1), (Y_2, X_2), \ldots, (Y_n, X_n)$ one would draw B samples each of size n with replacement from these pairs and from each sample one would compute the correlation coefficient, yielding $r_1^*, r_2^*, \ldots, r_B^*$. The standard deviation of these would then be used as an estimate of the standard error of r.

14.7 Local Linear Smoothing

We motivate the material in this section with an example. Recapitulating the material in Example B of Section 10.7, recall that an inductive loop detector is a wire loop embedded in the pavement of a roadway. From the output of the detector, the number of passing vehicles (flow), and the percentage of time that the detector was covered by a vehicle (occupancy), is reported to a traffic management center. If a detector is faulty or not operating at all, it may be desirable to estimate its flow and occupancy from flow and occupancy in other lanes. Such estimates might be used in summaries of traffic patterns, for example.

Figure 14.25 is a plot of occupancy in lane 3 of a particular freeway location versus occupancy in lane 1. Lane 1 is the leftmost lane and lane 3 is the rightmost. The two are clearly strongly related, but the relationship is not linear. The dashed line is the line occupancy 3 = occupancy 1, and the solid line is the regression line. The data depart systematically from both of these relationships. It is interesting that at low occupancies, the values of lane 3 tend to be larger than those of lane 1. As occupancies increase, those in lane 1 are larger, except for very high occupancy, in which case they are about equal. These very high occupancies correspond to extreme congestion in which the traffic conditions in the two lanes are very similar.

Now suppose we want to estimate the expected occupancy in lane 3 given the occupancy in lane 1, for example, to use the values from lane 1 to estimate those for lane 3 when the latter are missing due to detector malfunction. First, observe that although the relationship is clearly not globally linear, it is locally linear—over a small range of lane 1 values, the relationship between lane 3 and lane 1 is nearly linear, as is shown in Figure 14.26.

To conform with generic notation, let x and y denote occupancies in lanes 1 and 3, respectively, and suppose that we want to estimate the value of y corresponding to a value x_0. Local linearity suggests that we choose a "bandwidth" h (e.g., $h = .05$) and fit a linear relationship between y and x over the range $x_0 - h \leq x \leq x_0 + h$.

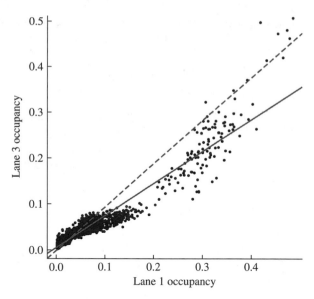

FIGURE 14.25 Occupancy in lane 3 versus that in lane 1. The dashed line is the line $y = x$, and the solid line is the least squares fit.

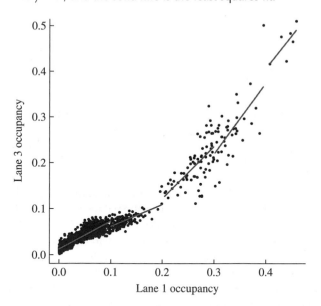

FIGURE 14.26 Linear relationships fit over the ranges $0 \leq x \leq .1$, $.1 < x \leq .2$, $.2 < x \leq .3$, $.3 < x \leq .4$, and $.4 < x \leq .5$.

This would amount to finding β_0 and β_1 to minimize

$$S(\beta_0, \beta_1) = \sum_{i=1}^{n}(y_i - \beta_0 - \beta_1 x_i)^2 w_h(x_i - x_0)$$

where the weight function $w_h(u)$ equals 1 for $-h \leq u \leq h$ and 0 elsewhere. The fitted value corresponding to x_0 would then be $\beta_0 + \beta_1 x_0$. For example, if $x_0 = .25$

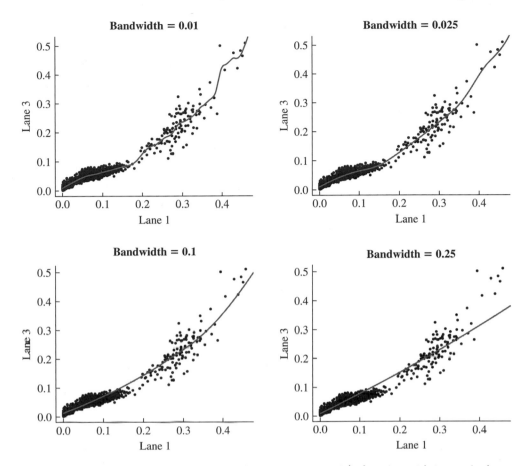

FIGURE 14.27   Local linear estimates using a Gaussian weight function with increasingly large bandwidths.

and $h = .05$, the fitted line is that shown over the region $.2 \le x \le .3$ in Figure 14.26, and the fitted value at $x_0 = .25$ is the height of the regression line at that point.

The weight function $w_h(u)$ is rectangular—it gives equal weight to all (y_i, x_i) pairs for which $x_0 - h \le x_i \le x_0 + h$. Rather than weight all pairs in this neighborhood equally, it is preferable to use weights that decay away from x_0. This can be accomplished by letting $w_h(u)$ be a probability density function with mean zero and standard deviation h, for example, a Gaussian density. The estimate can be computed on a dense grid of values of x_0, using at each point of the grid the weight function $w_h(x_i - x_0)$, which centers the density at x_0.

Results for four choices of the bandwidth h are shown in Figure 14.27. Notice that for the small value $h = .01$, the smoothed curve is quite wiggly, because for such a small bandwidth few points contribute to the fit. In contrast, the large value $h = .25$ produces a very smooth curve, but one that oversmooths and fails to track local trends. The intermediate value $h = .025$ appears to do best at tracking the local trend. Also note that the curves are continuous, since $S(\beta_0, \beta_1)$ is a continuous function of x_0.

Notice that the values of occupancy are not uniformly distributed but are much more dense for small values of occupancy than for large values. Thus a region of width

$h = .025$ centered on a low value of occupancy contains many more points than a region centered on a high value of occupancy. The smoothing is thus, in a sense, not uniform in occupancy. A common alternative to smoothing with a fixed bandwidth is to let the bandwidth depend on occupancy in such a way that a constant fraction, f, of points are contained within a bandwidth. Thus, if the fraction is $f = 0.10$, for example, the bandwidth, $h(x_0)$, corresponding to a value x_0, of the independent variable is such that 10% of the x values are in the interval $x_0 \pm h(x_0)$.

The bandwidth can often be reasonably chosen by visual examination of the smoothed scatter plots, as in Figure 14.27. In some circumstances, though, it is desirable to select the bandwidth automatically from the data. Cross-validation is a commonly used procedure; for choosing a fraction f the algorithm is as follows:

Specify a sequence of possible values of $f : f_1, f_2, \ldots, f_M$.
For each $k = 1, 2, \ldots, M$.
 For $i = 1, 2, \ldots, n$, leave out the data point (y_i, x_i), smooth the rest of the data using the bandwidth f_k, and use the result to predict y_i. Denote the predicted value by $\hat{y}_{(-i)}$.
 Compute the cross-validation score, $CV(f_k) = \sum_{i=1}^{n} (\hat{y}_{(-i)} - y_i)^2$.
Select the bandwidth, which minimizes $CV(f)$.

Figure 14.28 shows the results of cross-validated choice of a smoothing fraction f and a Gaussian weight function. The left panel shows the cross-validation score. It is high for very small values of f, since the smoothed curve is very wiggly because locally it depends on a small number of observations. The score is also high for large values of f, which lead to over-smoothing. The minimizing fraction is $f = 0.28$ and

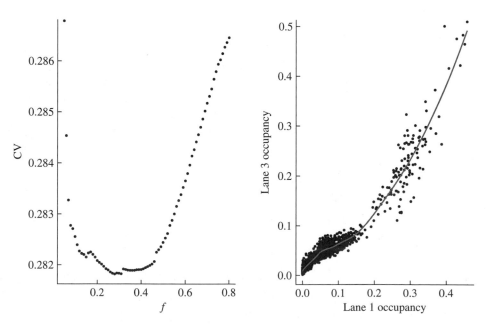

FIGURE **14.28** Left panel: cross-validation score as a function of f. Right panel: local linear fit for the value of the minimizer of f, $f = 0.28$.

the corresponding smoothed curve shown in the right panel of the figure apparently does a good job at estimating the local trend.

14.8 Concluding Remarks

We have developed theory and techniques only for linear least squares problems. If the unknown parameters enter into the prediction equation nonlinearly, the minimization cannot typically be done in closed form, and iterative methods are necessary. Also, expressions in closed form for the standard errors of the coefficients cannot usually be obtained; linearization is often used to obtain approximate standard errors. The bootstrap can also be used.

As has been mentioned, least squares estimates are not robust against outliers. There are robust methods for regression. The discussion of M estimates in Chapter 10 suggests minimizing

$$\sum_{i=1}^{n} \Psi(Y_i - \hat{Y}_i)$$

for a robust weight function, Ψ. Note that the least squares estimate corresponds to $\Psi(x) = x^2$. The choice $\Psi(x) = |x|$ gives the curve-fitting analogue of the median.

In some applications, a large number of independent variables are candidates for inclusion in the prediction equation. Various techniques for variable selection have been proposed, and research is still active in this area.

In simple linear regression, points with x values at the extremes of the data exert a large influence on the fitted line. In multiple regression, a similar phenomenon occurs, but is not so easily detectable usually. For this reason, several measures of "influence" have been proposed. Good software packages routinely flag influential observations.

The problem of errors introduced via calibration of instruments has not been fully discussed. Suppose, for example, that an instrument for measuring temperature is to be calibrated. Readings are taken at several known temperatures (the independent variables) and a functional relationship between the instrument readings (the dependent variable) and the temperatures is fit by the method of least squares. After this has been carried out, an unknown temperature is read by the instrument and is predicted using the fitted relationship. How do the errors in the estimates of the coefficients of the functional form propagate? That is, what is the uncertainty of the estimated temperature? This is an inverse problem, and its analysis is not completely straightforward.

14.9 Problems

1. Convert the following relationships into linear relationships by making transformations and defining new variables.

   a. $y = a/(b + cx)$
   b. $y = ae^{-bx}$
   c. $y = ab^x$

d. $y = x/(a + bx)$
e. $y = 1/(1 + e^{bx})$

2. Plot y versus x for the following pairs:

x	.34	1.38	−.65	.68	1.40	−.88	−.30	−1.18	.50	−1.75
y	.27	1.34	−.53	.35	1.28	−.98	−.72	−.81	.64	−1.59

a. Fit a line $y = a + bx$ by the method of least squares, and sketch it on the plot.
b. Fit a line $x = c + dy$ by the method of least squares, and sketch it on the plot.
c. Are the lines in parts (a) and (b) the same? If not, why not?

3. Suppose that $y_i = \mu + e_i$, where $i = 1, \ldots, n$ and the e_i are independent errors with mean zero and variance σ^2. Show that $\bar{y}$ is the least squares estimate of μ.

4. Consider a standard linear regression model in which the freshman GPA is modeled to depend linearly on high school GPA: $Y_i = \beta_0 + \beta_1 x_i + e_i, i = 1, 2 \ldots, n$. Suppose that different intercepts were to be allowed for females and males, and write the model as

$$Y_i = I_F(i)\beta_F + I_M(i)\beta_M + \beta_1 x_i + e_i$$

where $I_F(i)$ and $I_M(i)$ are indicator variables taking on values 0 and 1 according to whether the gender of the ith person is female or male. Give the form of the design matrix for such a model.

5. Three objects are located on a line at points $p_1 < p_2 < p_3$. These locations are not precisely known. A surveyor makes the following measurements:

a. He stands at the origin and measures the three distances from there to p_1, p_2, p_3. Let these measurements be denoted by Y_1, Y_2, Y_3.
b. He goes to p_1 and measures the distances from there to p_2 and p_3. Let these measurements be denoted by Y_4, Y_5.
c. He goes to p_2 and measures the distance from there to p_3. Denote this measurement by Y_6.

He thus makes six measurements in all, and they are all subject to error. In order to estimate the values p_1, p_2, p_3, he decides to combine all the measurements by the method of least squares. Using matrix notation, explain clearly how the least squares estimates would be calculated (you don't have to do the actual calculations).

6. Two objects of unknown weights w_1 and w_2 are weighed on an error-prone pan balance in the following way: (1) object 1 is weighed by itself, and the measurement is 3 g; (2) object 2 is weighed by itself, and the result is 3 g; (3) the difference of the weights (the weight of object 1 minus the weight of object 2) is measured by placing the objects in different pans, and the result is 1 g; (4) the sum of the weights is measured as 7g. The problem is to estimate the true weights of the objects from these measurements.

a. Set up a linear model, $\mathbf{Y} = \mathbf{X}\beta + \mathbf{e}$. (*Hint:* The entries of $\mathbf{X}$ are 0 and ± 1.)
b. Find the least squares estimates of w_1 and w_2.

c. Find the estimate of σ^2.

d. Find the estimated standard errors of the least squares estimates of part (b).

e. Estimate $w_1 - w_2$ and its standard error.

f. Test the null hypothesis H_0: $w_1 = w_2$.

7. (Weighted Least Squares) Suppose that in the model $y_i = \beta_0 + \beta_1 x_i + e_i$, the errors have mean zero and are independent, but $\text{Var}(e_i) = \rho_i^2 \sigma^2$, where the ρ_i are known constants, so the errors do not have equal variance. This situation arises when the y_i are averages of several observations at x_i; in this case, if y_i is an average of n_i independent observations, $\rho_i^2 = 1/n_i$ (why?). Because the variances are not equal, the theory developed in this chapter does not apply; intuitively, it seems that the observations with large variability should influence the estimates of β_0 and β_1 less than the observations with small variability.

The problem may be transformed as follows:

$$\rho_i^{-1} y_i = \rho_i^{-1}\beta_0 + \rho_i^{-1}\beta_1 x_i + \rho_i^{-1} e_i$$

or

$$z_i = u_i \beta_0 + v_i \beta_1 + \delta_i$$

where

$$u_i = \rho_i^{-1} \qquad v_i = \rho_i^{-1} x_i \qquad \delta_i = \rho_i^{-1} e_i$$

a. Show that the new model satisfies the assumptions of the standard statistical model.

b. Find the least squares estimates of β_0 and β_1.

c. Show that performing a least squares analysis on the new model, as was done in part (b), is equivalent to minimizing

$$\sum_{i=1}^{n} (y_i - \beta_0 - \beta_1 x_i)^2 \rho_i^{-2}$$

This is a weighted least squares criterion; the observations with large variances are weighted less.

d. Find the variances of the estimates of part (b).

8. (The QR Method) This problem outlines the basic ideas of an alternative method, the QR method, of finding the least squares estimate $\hat{\beta}$. An advantage of the method is that it does not include forming the matrix $\mathbf{X}^T \mathbf{X}$, a process that tends to increase rounding error. The essential ingredient of the method is that if $\mathbf{X}_{n \times p}$ has p linearly independent columns, it may be factored in the form

$$\underset{n \times p}{\mathbf{X}} = \underset{n \times p}{\mathbf{Q}} \ \underset{p \times p}{\mathbf{R}}$$

where the columns of $\mathbf{Q}$ are orthogonal ($\mathbf{Q}^T \mathbf{Q} = \mathbf{I}$) and $\mathbf{R}$ is upper-triangular ($r_{ij} = 0$, for $i > j$) and nonsingular. [For a discussion of this decomposition and its relationship to the Gram-Schmidt process, see Strang (1980).]

Show that $\hat{\beta} = (\mathbf{X}^T \mathbf{X})^{-1} \mathbf{X}^T \mathbf{Y}$ may also be expressed as $\hat{\beta} = \mathbf{R}^{-1} \mathbf{Q}^T \mathbf{Y}$, or $\mathbf{R}\hat{\beta} = \mathbf{Q}^T \mathbf{Y}$. Indicate how this last equation may be solved for $\hat{\beta}$ by back-substitution, using that $\mathbf{R}$ is upper-triangular, and show that it is thus unnecessary to invert $\mathbf{R}$.

9. (Cholesky Decomposition) This problem outlines the basic ideas of a popular and effective method of computing least squares estimates. Assuming that its inverse exists, $\mathbf{X}^T\mathbf{X}$ is a positive, definite matrix and may be factored as $\mathbf{X}^T\mathbf{X} = \mathbf{R}^T\mathbf{R}$, where $\mathbf{R}$ is an upper-triangular matrix. This factorization is called the **Cholesky decomposition**. Show that the least squares estimates can be found by solving the equations

$$\mathbf{R}^T\mathbf{v} = \mathbf{X}^T\mathbf{Y}$$
$$\mathbf{R}\hat{\beta} = \mathbf{v}$$

where $\mathbf{v}$ is appropriately defined. Show that these equations can be solved by back-substitution because $\mathbf{R}$ is upper-triangular, and that therefore it is not necessary to carry out any matrix inversions explicitly to find the least squares estimates.

10. Show that the least squares estimates of the slope and intercept of a line may be expressed as

$$\hat{\beta}_0 = \bar{y} - \hat{\beta}_1\bar{x}$$

and

$$\hat{\beta}_1 = \frac{\sum_{i=1}^{n}(x_i - \bar{x})(y_i - \bar{y})}{\sum_{i=1}^{n}(x_i - \bar{x})^2}$$

11. Show that if $\bar{x} = 0$, the estimated slope and intercept are uncorrelated under the assumptions of the standard statistical model.

12. Use the result of Problem 10 to show that the line fit by the method of least squares passes through the point $(\bar{x}, \bar{y})$.

13. Suppose that a line is fit by the method of least squares to n points, that the standard statistical model holds, and that we want to estimate the line at a new point, x_0. Denoting the value on the line by μ_0, the estimate is

$$\hat{\mu}_0 = \hat{\beta}_0 + \hat{\beta}_1 x_0$$

   a. Derive an expression for the variance of $\hat{\mu}_0$.
   b. Sketch the standard deviation of $\hat{\mu}_0$ as a function of $x_0 - \bar{x}$. The shape of the curve should be intuitively plausible.
   c. Derive a 95% confidence interval for $\mu_0 = \beta_0 + \beta_1 x_0$ under an assumption of normality.

14. Problem 13 dealt with how to form a confidence interval for the value of a line at a point x_0. Suppose that instead we want to predict the value of a new observation, Y_0, at x_0,

$$Y_0 = \beta_0 + \beta_1 x_0 + e_0$$

by the estimate

$$\hat{Y}_0 = \hat{\beta}_0 + \hat{\beta}_1 x_0$$

a. Find an expression for the variance of $\hat{Y}_0 - Y_0$, and compare it to the expression for the variance of $\hat{\mu}_0$ obtained in part (a) of Problem 13. Assume that e_0 is independent of the original observations and has the variance σ^2.

b. Assuming that e_0 is normally distributed, find the distribution of $\hat{Y}_0 - Y_0$. Use this result to find an interval I such that $P(Y_0 \in I) = 1 - \alpha$. This interval is called a $100(1 - \alpha)\%$ prediction interval.

15. Find the least squares estimate of β for fitting the line $y = \beta x$ to points (x_i, y_i), where $i = 1, \ldots, n$.

16. Consider fitting the curve $y = \beta_0 x + \beta_1 x^2$ to points (x_i, y_i), where $i = 1, \ldots, n$.

a. Use the matrix formalism to find expressions for the least squares estimates of β_0 and β_1.

b. Find an expression for the covariance matrix of the estimates.

17. This problem extends some of the material in Section 14.2.3. Let X and Y be random variables with

$$E(X) = \mu_x \qquad E(Y) = \mu_y$$
$$\text{Var}(X) = \sigma_x^2 \qquad \text{Var}(Y) = \sigma_y^2$$

$$\text{Cov}(X, Y) = \sigma_{xy}$$

Consider predicting Y from X as $\hat{Y} = \alpha + \beta X$, where α and β are chosen to minimize $E(Y - \hat{Y})^2$, the expected squared prediction error.

a. Show that the minimizing values of α and β are

$$\beta = \frac{\sigma_{xy}}{\sigma_x^2}$$
$$\alpha = \mu_y - \beta \mu_x$$

[*Hint:* $E(Y - \hat{Y})^2 = (EY - E\hat{Y})^2 + \text{Var}(Y - \hat{Y})$.]

b. Show that for this choice of α and β

$$\frac{\text{Var}(Y) - \text{Var}(Y - \hat{Y})}{\text{Var}(Y)} = r_{xy}^2$$

18. Suppose that

$$Y_i = \beta_0 + \beta_1 x_i + e_i, \qquad i = 1, \ldots, n$$

where the e_i are independent and normally distributed with mean zero and variance σ^2. Find the mle's of β_0 and β_1 and verify that they are the least squares estimates. (*Hint:* Under these assumptions, the Y_i are independent and normally distributed with means $\beta_0 + \beta_1 x_i$ and variance σ^2. Write the joint density function of the Y_i and thus the likelihood.)

19. a. Show that the vector of residuals is orthogonal to every column of **X**.

b. Use this result to show that the residuals sum to zero and thus the sum has expectation zero if the model contains an intercept term.

20. Assume that the columns of **X**, $\mathbf{X}_1, \ldots, \mathbf{X}_p$, are orthogonal; that is, $\mathbf{X}_i^T \mathbf{X}_j = 0$, for $i \neq j$. Show that the covariance matrix of the least squares estimates is diagonal.

21. Suppose that n points $x_1, \ldots, x_n$ are to be placed in the interval $[-1, 1]$ for fitting the model

$$Y_i = \beta_0 + \beta_1 x_i + \epsilon_i$$

where the ϵ_i are independent with common variance σ^2. How should the x_i be chosen in order to minimize $\text{Var}(\hat{\beta}_1)$?

22. Suppose that the relation of family income to consumption is linear. Of those families in the 90th percentile of income, what proportion would you expect to be at or above the 90th percentile of consumption: (a) exactly 50%, (b) less than 50%, (c) more than 50%? Justify your answers.

23. Suppose that grades on a midterm and a final have a correlation coefficient of .5 and both exams have an average score of 75 and a standard deviation of 10.

   a. If a student's score on the midterm is 95, what would you predict his score on the final to be?

   b. If a student scored 85 on the final, what would you guess that her score on the midterm was?

24. Suppose that the independent variables in a least squares problem are replaced by rescaled variables $u_{ij} = k_j x_{ij}$ (for example, centimeters are converted to meters.) Show that $\hat{Y}$ does not change. Does $\hat{\beta}$ change? (*Hint:* Express the new design matrix in terms of the old one.)

25. Suppose that each setting x_i of the independent variables in a simple least squares problem is duplicated, yielding two independent observations Y_{i_1}, Y_{i_2}. Is it true that the least squares estimates of the intercept and slope can be found by doing a regression of the mean responses of each pair of duplicates, $\overline{Y}_i = (Y_{i_1} + Y_{i_2})/2$ on the x_i? Why or why not?

26. Suppose that Z_1, Z_2, Z_3, Z_4 are random variables with $\text{Var}(Z_i) = 1$ and $\text{Cov}(Z_i, Z_j) = \rho$ for $i \neq j$. Use the matrix techniques we developed in Section 14.4.1 to show that $Z_1 + Z_2 + Z_3 + Z_4$ is uncorrelated with $Z_1 + Z_2 - Z_3 - Z_4$.

27. For the standard linear model of Section 14.4.2, show that

$$\sigma^2 I = \Sigma_{\hat{Y}\hat{Y}} + \Sigma_{\hat{e}\hat{e}}$$

Conclude that

$$n\sigma^2 = \sum_{i=1}^{n} \text{Var}(\hat{Y}_i) + \sum_{i=1}^{n} \text{Var}(\hat{e}_i)$$

28. Suppose that $X_1, \ldots, X_n$ are independent with mean μ_i and common variance σ^2. Let $Y = \sum_{i=1}^{n} a_i X_i$.

   a. Let $Z = \sum_{i=1}^{n} b_i X_i$. Use Theorem D of Section 14.4.1 to find $\text{Cov}(Y, Z)$.

   b. Use Theorem C of Section 14.4.1 to find $E(\sum_{i=1}^{n} \sum_{j=1}^{n} X_i X_j)$.

29. Assume that X_1 and X_2 are uncorrelated random variables with variance σ^2, and use matrix methods to show that $Y = X_1 + X_2$ and $Z = X_1 - X_2$ are uncorrelated. (*Hint:* Find Σ_{YZ}.)

30. Let $X_1, \ldots, X_n$ be random variables with $\text{Var}(X_i) = \sigma^2$ and $\text{Cov}(X_i, X_j) = \rho\sigma^2$, for $i \neq j$. Use matrix methods to find $\text{Var}(\overline{X})$.

31. Let Z be a random vector with 4 components and covariance matrix $\sigma^2 I$. Let $U = Z_1 + Z_2 + Z_3 + Z_4$ and $V = (Z_1 + Z_2) - (Z_3 + Z_4)$. Use matrix methods to find $\text{Cov}(U, V)$.

32. Let X be a random n-vector and let Y be a random vector with $Y_1 = X_1$, $Y_i = X_i - X_{i-1}, i = 1, 2, \ldots, n$.

    a. If the X_i are independent random variables with variances σ^2, find the covariance matrix of **Y.**

    b. If the Y_i are independent random variables with variances σ^2, find the covariance matrix of **X.**

33. a. Let $X \sim N(0, 1)$ and $E \sim N(0, 1)$ be independent, and let $Y = X + \beta E$. Show that

$$r_{xy} = \frac{1}{\sqrt{\beta^2 + 1}}$$

    b. Use the results of part (a) to generate bivariate samples (x_i, y_i) of size 20 with population correlation coefficients $-.9$, $-.5$, 0, .5, and .9, and compute the sample correlation coefficients.

    c. Have a partner generate scatterplots as in part (b) and then guess the correlation coefficients.

34. Generate a bivariate sample of size 50 as in Problem 33 with correlation coefficient .8. Find the estimated regression line and the residuals. Plot the residuals versus X and the residuals versus Y. Explain the appearance of the plots.

35. An investigator wants to use multiple regression to predict a variable, Y, from two other variables, X_1 and X_2. She proposes forming a new variable $X_3 = X_1 + X_2$ and using multiple regression to predict Y from the three X variables. Show that she will run into problems because the design matrix will not have full rank.

36. The file `bismuth` contains the transition pressure (bar) of the bismuth II–I transition as a function of temperature (°C) (see Example E in Section 14.2.2). Fit a linear relationship between pressure and temperature, examine the residuals, and comment.

37. Dissociation pressure for a reaction involving barium nitride was recorded as a function of temperature (Orcutt 1970). The second law of thermodynamics gives the approximate relationship

$$\ln(\text{pressure}) = A + \frac{B}{T}$$

where T is absolute temperature. From the data in the file `barium`, estimate A and B and their standard errors. Form approximate 95% confidence intervals for A and B. Examine the residuals and comment.

38. The file `sapphire` lists observed values of Young's modulus (g) measured at various temperatures (T) for sapphire rods (Ku 1969). Fit a linear relationship

$g = \beta_0 + \beta_1 t$, and form confidence intervals for the coefficients. Examine the residuals.

39. As part of a nuclear safeguards program, the contents of a tank are routinely measured. The determination of volume is made indirectly by measuring the difference in pressure at the top and at the bottom of the tank. The tank is cylindrical in shape, but its internal geometry is complicated by various pipes and agitator paddles. Without these complications, pressure and volume should have a linear relationship. To calibrate pressure with respect to volume, known quantities (x) of liquid are placed in the tank and pressure readings (y) are taken. The data in the file tankvolume are from Knafl et al. (1984). The units of volume are kiloliters and those of pressure are pascals.

   a. Plot pressure versus volume. Does the relationship appear linear?
   b. Calculate the linear regression of pressure on volume, and plot the residuals versus volume. What does the residual plot show?
   c. Try fitting pressure as a quadratic function of volume. What do you think of the fit?

40. The following data come from the calibration of a proving ring, a device for measuring force (Hockersmith and Ku 1969).

   a. Plot load versus deflection. Does the plot look linear?
   b. Fit deflection as a linear function of load, and plot the residuals versus load. Do the residuals show any systematic lack of fit?
   c. Fit deflection as a quadratic function of load, and estimate the coefficients and their standard errors. Plot the residuals. Does the fit look reasonable?

Load	Deflection		
	Run 1	Run 2	Run 3
10,000	68.32	68.35	68.30
20,000	136.78	136.68	136.80
30,000	204.98	205.02	204.98
40,000	273.85	273.85	273.80
50,000	342.70	342.63	342.63
60,000	411.30	411.35	411.28
70,000	480.65	480.60	480.63
80,000	549.85	549.85	549.83
90,000	619.00	619.02	619.10
100,000	688.70	688.62	688.58

41. The file chestnut contains the diameter (feet) at breast height (DBH) and the age (years) of 27 chestnut trees (Chapman and Demeritt 1936). Try fitting DBH as a linear function of age. Examine the residuals. Can you find a transformation of DBH and/or age that produces a more linear relationship?

42. The stopping distance (y) of an automobile on a certain road was studied as a function of velocity (Brownlee 1960). The data are listed in the following table.

Fit y and $\sqrt{y}$ as linear functions of velocity, and examine the residuals in each case. Which fit is better? Can you suggest any physical reason that explains why?

Velocity (mi/h)	Stopping Distance (ft)
20.5	15.4
20.5	13.3
30.5	33.9
40.5	73.1
48.8	113.0
57.8	142.6

43. Chang (1945) studied the rate of sedimentation of amoebic cysts in water, in attempting to develop methods of water purification. The following table gives the diameters of the cysts and the times required for the cysts to settle through $720\ \mu$m of still water at three temperatures. Each entry of the table is an average of several observations, the number of which is given in parentheses. Does the time required appear to be a linear or a quadratic function of diameter? Can you find a model that fits? How do the settling rates at the three temperatures compare? (See Problem 7.)

	Settling Times of Cysts (sec)		
Diameter (μm)	10°C	25°C	28°C
11.5	217.1 (2)	138.2 (1)	128.4 (2)
13.1	168.3 (3)	109.3 (3)	103.1 (4)
14.4	136.6 (11)	89.1 (13)	82.7 (11)
15.8	114.6 (17)	73.0 (11)	70.5 (18)
17.3	96.4 (8)	61.3 (6)	59.7 (6)
18.7	80.8 (5)	56.2 (4)	50.0 (4)
20.2	70.4 (2)	46.3 (1)	41.4 (2)

44. Cogswell (1973) studied a method of measuring resistance to breathing in children. The file `asthma` lists respiratory resistance and height (cm) for children with asthma and the file `cystfibr` contains results for children with cystic fibrosis. Is there a statistically significant relation between respiratory resistance and height in either group?

45. The file `reading` contains average reading scores of third-graders from several elementary schools on a standardized test in each of two successive years. Is there a "regression effect"?

46. Measurement of the concentration of small asbestos fibers is important in studies of environmental health issues and in setting and enforcing appropriate regulations. The concentrations of such fibers are measured most accurately by an electron microscope, but for practical reasons, optical microscopes must sometimes be used. Kiefer et al. (1987) compared measurements of asbestos fiber concentration from 30 airborne samples by a scanning electron microscope (SEM) and by a phase contrast microscope (PCM). The data are contained in the file `asbestos`. Study the relationship between the two measurements, taking the more accurate SEM measurements as the independent variable and the PCM measurements as the dependent variable.

47. Aerial survey methods are used to estimate the number of snow geese in their summer range areas west of Hudson's Bay in Canada. To obtain estimates, small aircraft fly over the range and, when a flock of snow geese is spotted, an experienced observer estimates the number of geese in the flock. To investigate the reliability of this method, an experiment in which an airplane carried two observers flew over 45 flocks, and each observer independently estimated the number of geese in the flock. Also, a photograph of the flock was taken so that an exact count of the number in the flock could be obtained (Weisberg 1985). The data are contained in the file `geese`.

   a. Draw scatterplots of observer counts, Y, versus photo count, x. Do these graphs suggest that a simple linear regression model might be appropriate?
   b. Calculate the linear regressions. What are the residual standard errors, what do they mean, and how do they compare? Do the fitted regressions appear to be different? Plot residuals and absolute values of residuals versus photo counts. Do the residuals indicate any systematic misfit? Does the residual variation appear to be constant?
   c. Repeat the above using the square root transformation on the counts. Does this transformation stabilize the variance?
   d. You have now computed the fits in two ways. How do they compare?
   e. Write a few sentences in answer to the questions, "How well do observers estimate the number of geese?" "How do the two observers compare?"

48. The volume, height, and diameter at 4.5 ft above ground level were measured for a sample of 31 black cherry trees in the Allegheny National Forest in Pennsylvania. The data were collected to provide a basis for determining an easy way of estimating the volume of a tree. Develop a model relating volume to height and diameter. The columns of the data matrix are diameter, height, and volume, in that order (Ryan, Joiner, & Ryan 1976). The data are contained in the file `treevolume`.

49. The file `flow-occ` contains data collected by loop detectors in all three lanes (see Section 14.7). Examine the relationship of flow in lane 3 versus that in lane 1. Make a scatterplot and fit a regression line. Does the linear relationship look accurate or is there some systematic misfit? Fit local linear relationships with several bandwidths. Identify a bandwidth that is too small and one that is too large. What bandwidth appears to provide a good balance between being too wiggly and being over-smooth?

50. The file `binary59683` contains measurements of the light of an astronomical source as a function of time. Time is in units of days (Julian date), and brightness is measured as "magnitude." According to this system of measurement, the brightest star has magnitude -1.4 and the faintest visible star has magnitude 6, so the larger the magnitude, the dimmer the light.

a. Plot magnitude versus time. Do you see any structure?
b. The object is actually an eclipsing binary system (two stars rotating around each other) with a period $P = 0.407528$ days. Define $s = t$ mod P and plot magnitude versus s. Can you qualitatively explain the shape of the "lightcurve" you see?
c. The lightcurve contains a lot of information about the nature of the binary system. Use local linear smoothing to estimate the underlying lightcurve.
d. Change the period slightly and see how the lightcurve changes. Do this several times. Can you propose a method to find an unknown period? About how accurately can the period be estimated from this data?

This data comes from the Hipparcos mission. More information, more lightcurves, and interactive demos can be found at http://www.rssd.esa.int/SA-general/Projects/Hipparcos/education.html.

51. The following table shows the monthly returns of stock in Disney, MacDonalds, Schlumberger, and Haliburton for January through May 1998. Fit a multiple regression to predict Disney returns from those of the other stocks. What is the standard deviation of the residuals? What is R^2?

Disney	MacDonalds	Schlumberger	Haliburton
0.08088	−0.01309	−0.08463	−0.13373
0.04737	0.15958	0.02884	0.03616
−0.04634	0.09966	0.00165	0.07919
0.16834	0.03125	0.09571	0.09227
−0.09082	0.06206	−0.05723	−0.13242

Next, using the regression equation you have just found, carry out the predictions for January through May of 1999 and compare to the actual data listed below. What is the standard deviation of the prediction error? How can the comparison with the results from 1998 be explained? Is a reasonable explanation that the fundamental nature of the relationships changed in the one year period?

Disney	MacDonalds	Schlumberger	Haliburton
0.1	0.02604	0.02695	0.00211
0.06629	0.07851	0.02362	−0.04
−0.11545	0.06732	0.23938	0.35526
0.02008	−0.06483	0.06127	0.10714
−0.08268	−0.09029	−0.05773	−0.02933

52. The file bodytemp contains normal body temperature readings (degrees Fahrenheit) and heart rates (beats per minute) of 65 males (coded by 1) and 65 females (coded by 2) from Shoemaker (1996).

a. For both males and females, make scatterplots of heart rate versus body temperature. Comment on the relationship or lack thereof.

b. Does the relationship for males appear to be the same as that for females? Examine this question graphically, by making a scatterplot showing both females and males and identifying females and males by different plotting symbols.

c. For the males, fit a linear regression to predict heart rate from temperature. Plot the residuals versus temperature and comment on whether the relationship is linear. Find the estimated slope and its standard error.

d. Repeat the above for females.

e. Test whether the slopes for males and females are equal. (*Hint:* Consider the difference of the slopes.)

f. Test whether the intercepts are equal.

53. Old Faithful geyser in Yellowstone National Park, Wyoming, derives its name from the regularity of its eruptions. The file `oldfaithful` contains measurements on eight successive days of the durations of the eruptions (in minutes) and the subsequent time interval before the next eruption. The park posts predicted eruption times for vistors. How well can the time until the next eruption be predicted by the duration of the current one?

a. Does the use of linear regression appear to be appropriate?

b. If the duration is 2 minutes, what would you predict the time until the next eruption to be? How can you quantify the accuracy of the prediction? Repeat this analysis for a duration of 4.5 minutes.

54. In 1970, Congress instituted a lottery for the military draft to support the unpopular war in Vietnam. All 366 possible birth dates were placed in plastic capsules in a rotating drum and were selected one by one. Eligible males born on the first day drawn were first in line to be drafted, etc. The results were criticized by some who claimed that government incompetency at running a fair lottery resulted in a tendency of men born later in the year being more likely to be drafted. Indeed, later investigation revealed that the birth dates were placed in the drum by month and were not thoroughly mixed. The columns of the file `1970lottery` are month, month number, day of the year, and draft number.

a. Plot draft number versus day number. Do you see any trend?

b. Plot the linear regression line on the scatterplot.

c. Plot a local linear smoothing on the scatterplot. Try varying the bandwidth.

55. When gasoline is pumped into the tank of an automobile, hydrocarbon vapors in the tank are forced out and into the atmosphere, producing a significant amount of air pollution. For this reason, vapor-recovery devices are often installed on gasoline pumps. It is difficult to test a recovery device in actual operation, because all that can be measured is the amount of vapor actually recovered and, by means of a "sniffer," whether *any* vapor escaped into the atmosphere. To estimate the efficiency of the device, it is thus necessary to estimate the total amount of vapor in the tank by using its relation to the values of variables that can actually be measured. In this exercise, you will try to develop such a predictive relationship using data that were obtained in a laboratory experiment. The file `gasvapor` contains recordings of the following variables: initial tank temperature ($°F$), temperature of the dispensed gasoline (°F), initial vapor pressure in the tank (psi), vapor pressure of the dispensed gasoline (psi), and emitted hydrocarbons (g). A prediction of emitted hydrocarbons is desired.

First, randomly select 40 observations and set them aside. You will develop a predictive relationship based on the remaining observations and then test its strength on the observations you have held out. (It is instructive to have each student in the class hold out the same 40 observations and then compare results.)

a. Look at the relationships among the variables by scatterplots. Comment on which relationships look strong. Based on this information, what variables would you conjecture will be important in the model? Do the plots suggest that transformations will be helpful? Do there appear to be any outliers?

b. Try fitting a few different models and select two that you think are the best.

c. Using these two models, predict the responses for the 40 observations you have held out and compare the predictions to the observed values by plotting predicted versus observed values, and by plotting prediction errors versus each of the independent variables. Summarize the strength of the prediction by the root mean square prediction error:

$$\text{RMSPE} = \sqrt{\frac{1}{40} \sum_{i=1}^{40} (Y_i - \hat{Y}_i)^2}$$

where Y_i is the ith observed value and $\hat{Y}_i$ is the predicted value.

56. Recordings of the levels of pollutants and various meteorological conditions are made hourly at several stations by the Los Angeles Pollution Control District. This agency attempts to construct mathematical/statistical models to predict pollution levels and to gain a better understanding of the complexities of air pollution. Obviously, very large quantities of data are collected and analyzed, but only a small set of data will be considered in this problem. The file `airpollution` contains the maximum level of an oxidant (a photochemical pollutant) and the morning averages of four meteorological variables: wind speed, temperature, humidity, and insolation (a measure of the amount of sunlight). The data cover 30 days during one summer.

a. Examine the relationship of oxidant level to each of the four meteorological variables and the relationships of the meteorological variables to each other. How well can the maximum level of oxidant be predicted from some or all of the meteorological variables? Which appear to be most important?

b. The standard statistical model used in this chapter assumes that the errors are random and independent of one another. In data that are collected over time, the error at any given time may well be correlated with the error from the preceding time. This phenomenon is called **serial correlation,** and in its presence the estimated standard errors of the coefficients developed in this chapter may be incorrect. The parameter estimates are still unbiased, however. (Why?) Can you detect serial correlation in the errors from your fits?

Common Distributions

Discrete Distributions

Binomial

$$p(k) = \binom{n}{k} p^k (1-p)^{n-k}, \qquad k = 0, 1, \ldots, n$$
$$E(X) = np$$
$$\text{Var}(X) = np(1-p)$$
$$M(t) = (1 - p + pe^t)^n$$

Geometric

$$p(k) = p(1-p)^{k-1}, \qquad k = 1, \ldots$$
$$E(X) = \frac{1}{p}$$
$$\text{Var}(X) = \frac{1-p}{p^2}$$
$$M(t) = \frac{e^t p}{1 - (1-p)e^t}$$

Negative Binomial

$$p(k) = \binom{k-1}{r-1} p^r (1-p)^{k-r}, \qquad k = r, r+1, \ldots$$
$$E(X) = \frac{r}{p}$$
$$\text{Var}(X) = \frac{r(1-p)}{p^2}$$
$$M(t) = \left(\frac{e^t p}{1 - (1-p)e^t} \right)^r$$

Poisson

$$p(k) = \frac{\lambda^k e^{-\lambda}}{k!}, \qquad k = 0, 1, \ldots$$

$$E(X) = \lambda$$

$$\mathrm{Var}(X) = \lambda$$

$$M(t) = e^{\lambda(e^t - 1)}$$

Continuous Distributions

Normal

$$f(x) = \frac{1}{\sigma\sqrt{2\pi}} e^{-\frac{1}{2\sigma^2}(x-\mu)^2}, \qquad -\infty < x < \infty$$

$$E(X) = \mu$$

$$\mathrm{Var}(X) = \sigma^2$$

$$M(t) = e^{\mu t} e^{\sigma^2 t^2 / 2}$$

Gamma

$$f(x) = \frac{\lambda^\alpha}{\Gamma(\alpha)} x^{\alpha-1} e^{-\lambda x}, \qquad x \geq 0$$

$$E(X) = \frac{\alpha}{\lambda}$$

$$\mathrm{Var}(X) = \frac{\alpha}{\lambda^2}$$

$$M(t) = \left(\frac{\lambda}{\lambda - t}\right)^\alpha, \qquad t < \lambda$$

Exponential (Special Case of Gamma with $\alpha = 1$)

Chi-Square with n Degrees of Freedom (Special Case of Gamma with $\alpha = n/2$, $\lambda = \frac{1}{2}$)

Uniform

$$f(x) = 1, \qquad 0 \leq x \leq 1$$

$$E(X) = \tfrac{1}{2}$$

$$\mathrm{Var}(X) = \tfrac{1}{12}$$

$$M(t) = \frac{e^t - 1}{t}$$

Beta

$$f(x) = \frac{\Gamma(a+b)}{\Gamma(a)\Gamma(b)} x^{a-1}(1-x)^{b-1}, \qquad 0 \le x \le 1$$

$$E(X) = \frac{a}{a+b}$$

$$\mathrm{Var}(X) = \frac{ab}{(a+b)^2(a+b+1)}$$

$M(t)$ is not useful.

Tables

TABLE 1 Binomial Probabilities

Tabulated values are $\sum_{x=0}^{k} p(x)$. (Computations are rounded off at the third decimal place.)

$n = 5$

k	.01	.05	.10	.20	.30	.40	.50	.60	.70	.80	.90	.95	.99
0	.951	.774	.590	.328	.168	.078	.031	.010	.002	.000	.000	.000	.000
1	.999	.977	.919	.737	.528	.337	.188	.087	.031	.007	.000	.000	.000
2	1.000	.999	.991	.942	.837	.683	.500	.317	.163	.058	.009	.001	.000
3	1.000	1.000	1.000	.993	.969	.913	.812	.663	.472	.263	.081	.023	.001
4	1.000	1.000	1.000	1.000	.998	.990	.969	.922	.832	.672	.410	.226	.049

$n = 10$

k	.01	.05	.10	.20	.30	.40	.50	.60	.70	.80	.90	.95	.99
0	.904	.599	.349	.107	.028	.006	.001	.000	.000	.000	.000	.000	.000
1	.996	.914	.736	.376	.149	.046	.011	.002	.000	.000	.000	.000	.000
2	1.000	.988	.930	.678	.383	.167	.055	.012	.002	.000	.000	.000	.000
3	1.000	.999	.987	.879	.650	.382	.172	.055	.011	.001	.000	.000	.000
4	1.000	1.000	.998	.967	.850	.633	.377	.166	.047	.006	.000	.000	.000
5	1.000	1.000	1.000	.994	.953	.834	.623	.367	.150	.033	.002	.000	.000
6	1.000	1.000	1.000	.999	.989	.945	.828	.618	.350	.121	.013	.001	.000
7	1.000	1.000	1.000	1.000	.998	.988	.945	.833	.617	.322	.070	.012	.000
8	1.000	1.000	1.000	1.000	1.000	.998	.989	.954	.851	.624	.264	.086	.004
9	1.000	1.000	1.000	1.000	1.000	1.000	.999	.994	.972	.893	.651	.401	.096

$n = 15$

k \ p	.01	.05	.10	.20	.30	.40	.50	.60	.70	.80	.90	.95	.99
0	.860	.463	.206	.035	.005	.000	.000	.000	.000	.000	.000	.000	.000
1	.990	.829	.549	.167	.035	.005	.000	.000	.000	.000	.000	.000	.000
2	1.000	.964	.816	.398	.127	.027	.004	.000	.000	.000	.000	.000	.000
3	1.000	.995	.944	.648	.297	.091	.018	.002	.000	.000	.000	.000	.000
4	1.000	.999	.987	.836	.515	.217	.059	.009	.001	.000	.000	.000	.000
5	1.000	1.000	.998	.939	.722	.403	.151	.034	.004	.000	.000	.000	.000
6	1.000	1.000	1.000	.982	.869	.610	.304	.095	.015	.001	.000	.000	.000
7	1.000	1.000	1.000	.996	.950	.787	.500	.213	.050	.004	.000	.000	.000
8	1.000	1.000	1.000	.999	.985	.905	.696	.390	.131	.018	.000	.000	.000
9	1.000	1.000	1.000	1.000	.996	.966	.849	.597	.278	.061	.002	.000	.000
10	1.000	1.000	1.000	1.000	.999	.991	.941	.783	.485	.164	.013	.001	.000
11	1.000	1.000	1.000	1.000	1.000	.998	.982	.909	.703	.352	.056	.005	.000
12	1.000	1.000	1.000	1.000	1.000	1.000	.996	.973	.873	.602	.184	.036	.000
13	1.000	1.000	1.000	1.000	1.000	1.000	1.000	.995	.965	.833	.451	.171	.010
14	1.000	1.000	1.000	1.000	1.000	1.000	1.000	1.000	.995	.965	.794	.537	.140

$n = 20$

k \ p	.01	.05	.10	.20	.30	.40	.50	.60	.70	.80	.90	.95	.99
0	.818	.358	.122	.002	.001	.000	.000	.000	.000	.000	.000	.000	.000
1	.983	.736	.392	.069	.008	.001	.000	.000	.000	.000	.000	.000	.000
2	.999	.925	.677	.206	.035	.004	.000	.000	.000	.000	.000	.000	.000
3	1.000	.984	.867	.411	.107	.016	.001	.000	.000	.000	.000	.000	.000
4	1.000	.997	.957	.630	.238	.051	.006	.000	.000	.000	.000	.000	.000
5	1.000	1.000	.989	.804	.416	.126	.021	.002	.000	.000	.000	.000	.000
6	1.000	1.000	.998	.913	.608	.250	.058	.006	.000	.000	.000	.000	.000
7	1.000	1.000	1.000	.968	.772	.416	.132	.021	.001	.000	.000	.000	.000
8	1.000	1.000	1.000	.990	.887	.596	.252	.057	.005	.000	.000	.000	.000
9	1.000	1.000	1.000	.997	.952	.755	.412	.128	.017	.001	.000	.000	.000
10	1.000	1.000	1.000	.999	.983	.872	.588	.245	.048	.003	.000	.000	.000
11	1.000	1.000	1.000	1.000	.995	.943	.748	.404	.113	.010	.000	.000	.000
12	1.000	1.000	1.000	1.000	.999	.979	.868	.584	.228	.032	.000	.000	.000
13	1.000	1.000	1.000	1.000	1.000	.994	.942	.750	.392	.087	.002	.000	.000
14	1.000	1.000	1.000	1.000	1.000	.998	.979	.874	.584	.196	.011	.000	.000
15	1.000	1.000	1.000	1.000	1.000	1.000	.994	.949	.762	.370	.043	.003	.000
16	1.000	1.000	1.000	1.000	1.000	1.000	.999	.984	.893	.589	.133	.016	.000
17	1.000	1.000	1.000	1.000	1.000	1.000	1.000	.996	.965	.794	.323	.075	.001
18	1.000	1.000	1.000	1.000	1.000	1.000	1.000	.999	.992	.931	.608	.264	.017
19	1.000	1.000	1.000	1.000	1.000	1.000	1.000	1.000	.999	.988	.878	.642	.182

$n = 25$

k \ p	.01	.05	.10	.20	.30	.40	.50	.60	.70	.80	.90	.95	.99
0	.778	.277	.072	.004	.000	.000	.000	.000	.000	.000	.000	.000	.000
1	.974	.642	.271	.027	.002	.000	.000	.000	.000	.000	.000	.000	.000
2	.998	.873	.537	.098	.009	.000	.000	.000	.000	.000	.000	.000	.000
3	1.000	.966	.764	.234	.033	.002	.000	.000	.000	.000	.000	.000	.000
4	1.000	.993	.902	.421	.090	.009	.000	.000	.000	.000	.000	.000	.000
5	1.000	.999	.967	.617	.193	.029	.002	.000	.000	.000	.000	.000	.000
6	1.000	1.000	.991	.780	.341	.074	.007	.000	.000	.000	.000	.000	.000
7	1.000	1.000	.998	.891	.512	.154	.022	.001	.000	.000	.000	.000	.000
8	1.000	1.000	1.000	.953	.677	.274	.054	.004	.000	.000	.000	.000	.000
9	1.000	1.000	1.000	.983	.811	.425	.115	.013	.000	.000	.000	.000	.000
10	1.000	1.000	1.000	.994	.902	.586	.212	.034	.002	.000	.000	.000	.000
11	1.000	1.000	1.000	.998	.956	.732	.345	.078	.006	.000	.000	.000	.000
12	1.000	1.000	1.000	1.000	.983	.846	.500	.154	.017	.000	.000	.000	.000
13	1.000	1.000	1.000	1.000	.994	.922	.655	.268	.044	.002	.000	.000	.000
14	1.000	1.000	1.000	1.000	.998	.966	.788	.414	.098	.006	.000	.000	.000
15	1.000	1.000	1.000	1.000	1.000	.987	.885	.575	.189	.017	.000	.000	.000
16	1.000	1.000	1.000	1.000	1.000	.996	.946	.726	.323	.047	.000	.000	.000
17	1.000	1.000	1.000	1.000	1.000	.999	.978	.846	.488	.109	.002	.000	.000
18	1.000	1.000	1.000	1.000	1.000	1.000	.993	.926	.659	.220	.009	.000	.000
19	1.000	1.000	1.000	1.000	1.000	1.000	.998	.971	.807	.383	.033	.001	.000
20	1.000	1.000	1.000	1.000	1.000	1.000	1.000	.991	.910	.579	.098	.007	.000
21	1.000	1.000	1.000	1.000	1.000	1.000	1.000	.998	.967	.766	.236	.034	.000
22	1.000	1.000	1.000	1.000	1.000	1.000	1.000	1.000	.991	.902	.463	.127	.002
23	1.000	1.000	1.000	1.000	1.000	1.000	1.000	1.000	.998	.973	.729	.358	.026
24	1.000	1.000	1.000	1.000	1.000	1.000	1.000	1.000	1.000	.996	.928	.723	.222

TABLE 2   Cumulative Normal Distribution—Values of *P* Corresponding to z_p for the Normal Curve

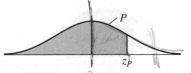

z is the standard normal variable. The value of *P* for $-z_p$ equals 1 minus the value of *P* for $+z_p$; for example, the *P* for –1.62 equals 1 – .9474 = .0526.

z_p	.00	.01	.02	.03	.04	.05	.06	.07	.08	.09
.0	.5000	.5040	.5080	.5120	.5160	.5199	.5239	.5279	.5319	.5359
.1	.5398	.5438	.5478	.5517	.5557	.5596	.5636	.5675	.5714	.5753
.2	.5793	.5832	.5871	.5910	.5948	.5987	.6026	.6064	.6103	.6141
.3	.6179	.6217	.6255	.6293	.6331	.6368	.6406	.6443	.6480	.6517
.4	.6554	.6591	.6628	.6664	.6700	.6736	.6772	.6808	.6844	.6879
.5	.6915	.6950	.6985	.7019	.7054	.7088	.7123	.7157	.7190	.7224
.6	.7257	.7291	.7324	.7357	.7389	.7422	.7454	.7486	.7517	.7549
.7	.7580	.7611	.7642	.7673	.7704	.7734	.7764	.7794	.7823	.7852
.8	.7881	.7910	.7939	.7967	.7995	.8023	.8051	.8078	.8106	.8133
.9	.8159	.8186	.8212	.8238	.8264	.8289	.8315	.8340	.8365	.8389
1.0	.8413	.8438	.8461	.8485	.8508	.8531	.8554	.8577	.8599	.8621
1.1	.8643	.8665	.8686	.8708	.8729	.8749	.8770	.8790	.8810	.8830
1.2	.8849	.8869	.8888	.8907	.8925	.8944	.8962	.8980	.8997	.9015
1.3	.9032	.9049	.9066	.9082	.9099	.9115	.9131	.9147	.9162	.9177
1.4	.9192	.9207	.9222	.9236	.9251	.9265	.9279	.9292	.9306	.9319
1.5	.9332	.9345	.9357	.9370	.9382	.9394	.9406	.9418	.9429	.9441
1.6	.9452	.9463	.9474	.9484	.9495	.9505	.9515	.9525	.9535	.9545
1.7	.9554	.9564	.9573	.9582	.9591	.9599	.9608	.9616	.9625	.9633
1.8	.9641	.9649	.9656	.9664	.9671	.9678	.9686	.9693	.9699	.9706
1.9	.9713	.9719	.9726	.9732	.9738	.9744	.9750	.9756	.9761	.9767
2.0	.9772	.9778	.9783	.9788	.9793	.9798	.9803	.9808	.9812	.9817
2.1	.9821	.9826	.9830	.9834	.9838	.9842	.9846	.9850	.9854	.9857
2.2	.9861	.9864	.9868	.9871	.9875	.9878	.9881	.9884	.9887	.9890
2.3	.9893	.9896	.9898	.9901	.9904	.9906	.9909	.9911	.9913	.9916
2.4	.9918	.9920	.9922	.9925	.9927	.9929	.9931	.9932	.9934	.9936
2.5	.9938	.9940	.9941	.9943	.9945	.9946	.9948	.9949	.9951	.9952
2.6	.9953	.9955	.9956	.9957	.9959	.9960	.9961	.9962	.9963	.9964
2.7	.9965	.9966	.9967	.9968	.9969	.9970	.9971	.9972	.9973	.9974
2.8	.9974	.9975	.9976	.9977	.9977	.9978	.9979	.9979	.9980	.9981
2.9	.9981	.9982	.9982	.9983	.9984	.9984	.9985	.9985	.9986	.9986
3.0	.9987	.9987	.9987	.9988	.9988	.9989	.9989	.9989	.9990	.9990
3.1	.9990	.9991	.9991	.9991	.9992	.9992	.9992	.9992	.9993	.9993
3.2	.9993	.9993	.9994	.9994	.9994	.9994	.9994	.9995	.9995	.9995
3.3	.9995	.9995	.9995	.9996	.9996	.9996	.9996	.9996	.9996	.9997
3.4	.9997	.9997	.9997	.9997	.9997	.9997	.9997	.9997	.9997	.9998

TABLE 3 Percentiles of the χ^2 Distribution—Values of χ_P^2 Corresponding to P

df	$\chi^2_{.005}$	$\chi^2_{.01}$	$\chi^2_{.025}$	$\chi^2_{.05}$	$\chi^2_{.10}$	$\chi^2_{.90}$	$\chi^2_{.95}$	$\chi^2_{.975}$	$\chi^2_{.99}$	$\chi^2_{.995}$
1	.000039	.00016	.00098	.0039	.0158	2.71	3.84	5.02	6.63	7.88
2	.0100	.0201	.0506	.1026	.2107	4.61	5.99	7.38	9.21	10.60
3	.0717	.115	.216	.352	.584	6.25	7.81	9.35	11.34	12.84
4	.207	.297	.484	.711	1.064	7.78	9.49	11.14	13.28	14.86
5	.412	.554	.831	1.15	1.61	9.24	11.07	12.83	15.09	16.75
6	.676	.872	1.24	1.64	2.20	10.64	12.59	14.45	16.81	18.55
7	.989	1.24	1.69	2.17	2.83	12.02	14.07	16.01	18.48	20.28
8	1.34	1.65	2.18	2.73	3.49	13.36	15.51	17.53	20.09	21.96
9	1.73	2.09	2.70	3.33	4.17	14.68	16.92	19.02	21.67	23.59
10	2.16	2.56	3.25	3.94	4.87	15.99	18.31	20.48	23.21	25.19
11	2.60	3.05	3.82	4.57	5.58	17.28	19.68	21.92	24.73	26.76
12	3.07	3.57	4.40	5.23	6.30	18.55	21.03	23.34	26.22	28.30
13	3.57	4.11	5.01	5.89	7.04	19.81	22.36	24.74	27.69	29.82
14	4.07	4.66	5.63	6.57	7.79	21.06	23.68	26.12	29.14	31.32
15	4.60	5.23	6.26	7.26	8.55	22.31	25.00	27.49	30.58	32.80
16	5.14	5.81	6.91	7.96	9.31	23.54	26.30	28.85	32.00	34.27
18	6.26	7.01	8.23	9.39	10.86	25.99	28.87	31.53	34.81	37.16
20	7.43	8.26	9.59	10.85	12.44	28.41	31.41	34.17	37.57	40.00
24	9.89	10.86	12.40	13.85	15.66	33.20	36.42	39.36	42.98	45.56
30	13.79	14.95	16.79	18.49	20.60	40.26	43.77	46.98	50.89	53.67
40	20.71	22.16	24.43	26.51	29.05	51.81	55.76	59.34	63.69	66.77
60	35.53	37.48	40.48	43.19	46.46	74.40	79.08	83.30	88.38	91.95
120	83.85	86.92	91.58	95.70	100.62	140.23	146.57	152.21	158.95	163.64

For large degrees of freedom,

$$\chi_P^2 = \tfrac{1}{2}(z_P + \sqrt{2v - 1})^2 \text{ approximately,}$$

where v = degrees of freedom and z_P is given in Table 2.

TABLE 4   Percentiles of the *t* Distribution

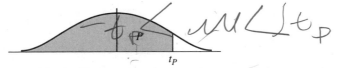

df	$t_{.60}$	$t_{.70}$	$t_{.80}$	$t_{.90}$	$t_{.95}$	$t_{.975}$	$t_{.99}$	$t_{.995}$
1	.325	.727	1.376	3.078	6.314	12.706	31.821	63.657
2	.289	.617	1.061	1.886	2.920	4.303	6.965	9.925
3	.277	.584	.978	1.638	2.353	3.182	4.541	5.841
4	.271	.569	.941	1.533	2.132	2.776	3.747	4.604
5	.267	.559	.920	1.476	2.015	2.571	3.365	4.032
6	.265	.553	.906	1.440	1.943	2.447	3.143	3.707
7	.263	.549	.896	1.415	1.895	2.365	2.998	3.499
8	.262	.546	.889	1.397	1.860	2.306	2.896	3.355
9	.261	.543	.883	1.383	1.833	2.262	2.821	3.250
10	.260	.542	.879	1.372	1.812	2.228	2.764	3.169
11	.260	.540	.876	1.363	1.796	2.201	2.718	3.106
12	.259	.539	.873	1.356	1.782	2.179	2.681	3.055
13	.259	.538	.870	1.350	1.771	2.160	2.650	3.012
14	.258	.537	.868	1.345	1.761	2.145	2.624	2.977
15	.258	.536	.866	1.341	1.753	2.131	2.602	2.947
16	.258	.535	.865	1.337	1.746	2.120	2.583	2.921
17	.257	.534	.863	1.333	1.740	2.110	2.567	2.898
18	.257	.534	.862	1.330	1.734	2.101	2.552	2.878
19	.257	.533	.861	1.328	1.729	2.093	2.539	2.861
20	.257	.533	.860	1.325	1.725	2.086	2.528	2.845
21	.257	.532	.859	1.323	1.721	2.080	2.518	2.831
22	.256	.532	.858	1.321	1.717	2.074	2.508	2.819
23	.256	.532	.858	1.319	1.714	2.069	2.500	2.807
24	.256	.531	.857	1.318	1.711	2.064	2.492	2.797
25	.256	.531	.856	1.316	1.708	2.060	2.485	2.787
26	.256	.531	.856	1.315	1.706	2.056	2.479	2.779
27	.256	.531	.855	1.314	1.703	2.052	2.473	2.771
28	.256	.530	.855	1.313	1.701	2.048	2.467	2.763
29	.256	.530	.854	1.311	1.699	2.045	2.462	2.756
30	.256	.530	.854	1.310	1.697	2.042	2.457	2.750
40	.255	.529	.851	1.303	1.684	2.021	2.423	2.704
60	.254	.527	.848	1.296	1.671	2.000	2.390	2.660
120	.254	.526	.845	1.289	1.658	1.980	2.358	2.617
∞	.253	.524	.842	1.282	1.645	1.960	2.326	2.576

TABLE 5 Percentiles of the F Distribution: $F_{.90}(n_1, n_2)$

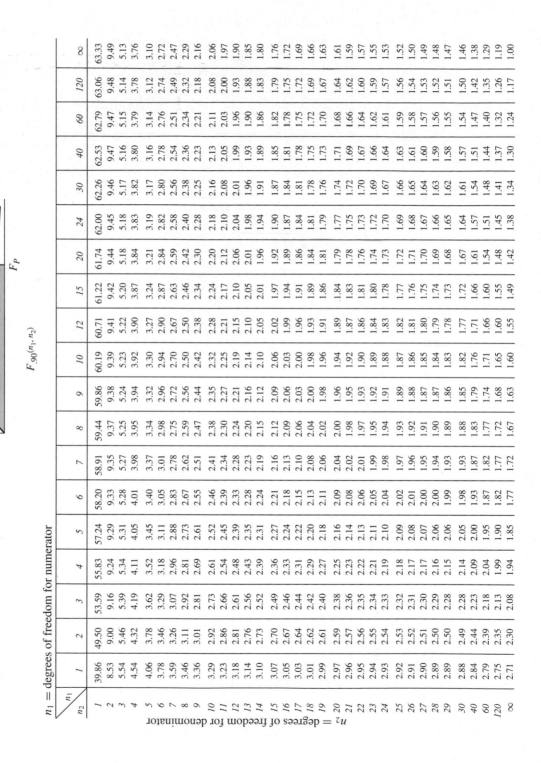

P

F_P

$F_{.90}(n_1, n_2)$

n_1 = degrees of freedom for numerator

n_2 = degrees of freedom for denominator

$n_2 \backslash n_1$	1	2	3	4	5	6	7	8	9	10	12	15	20	24	30	40	60	120	∞
1	39.86	49.50	53.59	55.83	57.24	58.20	58.91	59.44	59.86	60.19	60.71	61.22	61.74	62.00	62.26	62.53	62.79	63.06	63.33
2	8.53	9.00	9.16	9.24	9.29	9.33	9.35	9.37	9.38	9.39	9.41	9.42	9.44	9.45	9.46	9.47	9.47	9.48	9.49
3	5.54	5.46	5.39	5.34	5.31	5.28	5.27	5.25	5.24	5.23	5.22	5.20	5.18	5.18	5.17	5.16	5.15	5.14	5.13
4	4.54	4.32	4.19	4.11	4.05	4.01	3.98	3.95	3.94	3.92	3.90	3.87	3.84	3.83	3.82	3.80	3.79	3.78	3.76
5	4.06	3.78	3.62	3.52	3.45	3.40	3.37	3.34	3.32	3.30	3.27	3.24	3.21	3.19	3.17	3.16	3.14	3.12	3.10
6	3.78	3.46	3.29	3.18	3.11	3.05	3.01	2.98	2.96	2.94	2.90	2.87	2.84	2.82	2.80	2.78	2.76	2.74	2.72
7	3.59	3.26	3.07	2.96	2.88	2.83	2.78	2.75	2.72	2.70	2.67	2.63	2.59	2.58	2.56	2.54	2.51	2.49	2.47
8	3.46	3.11	2.92	2.81	2.73	2.67	2.62	2.59	2.56	2.50	2.50	2.46	2.42	2.40	2.38	2.36	2.34	2.32	2.29
9	3.36	3.01	2.81	2.69	2.61	2.55	2.51	2.47	2.44	2.42	2.38	2.34	2.30	2.28	2.25	2.23	2.21	2.18	2.16
10	3.29	2.92	2.73	2.61	2.52	2.46	2.41	2.38	2.35	2.32	2.28	2.24	2.20	2.18	2.16	2.13	2.11	2.08	2.06
11	3.23	2.86	2.66	2.54	2.45	2.39	2.34	2.30	2.27	2.25	2.21	2.17	2.12	2.10	2.08	2.05	2.03	2.00	1.97
12	3.18	2.81	2.61	2.48	2.39	2.33	2.28	2.24	2.21	2.19	2.15	2.10	2.06	2.04	2.01	1.99	1.96	1.93	1.90
13	3.14	2.76	2.56	2.43	2.35	2.28	2.23	2.20	2.16	2.14	2.10	2.05	2.01	1.98	1.96	1.93	1.90	1.88	1.85
14	3.10	2.73	2.52	2.39	2.31	2.24	2.19	2.15	2.12	2.10	2.05	2.01	1.96	1.94	1.91	1.89	1.86	1.83	1.80
15	3.07	2.70	2.49	2.36	2.27	2.21	2.16	2.12	2.09	2.06	2.02	1.97	1.92	1.90	1.87	1.85	1.82	1.79	1.76
16	3.05	2.67	2.46	2.33	2.24	2.18	2.13	2.09	2.06	2.03	1.99	1.94	1.89	1.87	1.84	1.81	1.78	1.75	1.72
17	3.03	2.64	2.44	2.31	2.22	2.15	2.10	2.06	2.03	2.00	1.96	1.91	1.86	1.84	1.81	1.78	1.75	1.72	1.69
18	3.01	2.62	2.42	2.29	2.20	2.13	2.08	2.04	2.00	1.98	1.93	1.89	1.84	1.81	1.78	1.75	1.72	1.69	1.66
19	2.99	2.61	2.40	2.27	2.18	2.11	2.06	2.02	1.98	1.96	1.91	1.86	1.81	1.79	1.76	1.73	1.70	1.67	1.63
20	2.97	2.59	2.38	2.25	2.16	2.09	2.04	2.00	1.96	1.94	1.89	1.84	1.79	1.77	1.74	1.71	1.68	1.64	1.61
21	2.96	2.57	2.36	2.23	2.14	2.08	2.02	1.98	1.95	1.92	1.87	1.83	1.78	1.75	1.72	1.69	1.66	1.62	1.59
22	2.95	2.56	2.35	2.22	2.13	2.06	2.01	1.97	1.93	1.90	1.86	1.81	1.76	1.73	1.70	1.67	1.64	1.60	1.57
23	2.94	2.55	2.34	2.21	2.11	2.05	1.99	1.95	1.92	1.89	1.84	1.80	1.74	1.72	1.69	1.66	1.62	1.59	1.55
24	2.93	2.54	2.33	2.19	2.10	2.04	1.98	1.94	1.91	1.88	1.83	1.78	1.73	1.70	1.67	1.64	1.61	1.57	1.53
25	2.92	2.53	2.32	2.18	2.09	2.02	1.97	1.93	1.89	1.87	1.82	1.77	1.72	1.69	1.66	1.63	1.59	1.56	1.52
26	2.91	2.52	2.31	2.17	2.08	2.01	1.96	1.92	1.88	1.86	1.81	1.76	1.71	1.68	1.65	1.61	1.58	1.54	1.50
27	2.90	2.51	2.30	2.17	2.07	2.00	1.95	1.91	1.87	1.85	1.80	1.75	1.70	1.67	1.64	1.60	1.57	1.53	1.49
28	2.89	2.50	2.29	2.16	2.06	2.00	1.94	1.90	1.87	1.84	1.79	1.74	1.69	1.66	1.63	1.59	1.56	1.52	1.48
29	2.89	2.50	2.28	2.15	2.06	1.99	1.93	1.89	1.86	1.83	1.78	1.73	1.68	1.65	1.62	1.58	1.55	1.51	1.47
30	2.88	2.49	2.28	2.14	2.05	1.98	1.93	1.88	1.85	1.82	1.77	1.72	1.67	1.64	1.61	1.57	1.54	1.50	1.46
40	2.84	2.44	2.23	2.09	2.00	1.93	1.87	1.83	1.79	1.76	1.71	1.66	1.61	1.57	1.54	1.51	1.47	1.42	1.38
60	2.79	2.39	2.18	2.04	1.95	1.87	1.82	1.77	1.74	1.71	1.66	1.60	1.54	1.51	1.48	1.44	1.40	1.35	1.29
120	2.75	2.35	2.13	1.99	1.90	1.82	1.77	1.72	1.68	1.65	1.60	1.55	1.48	1.45	1.41	1.37	1.32	1.26	1.19
∞	2.71	2.30	2.08	1.94	1.85	1.77	1.72	1.67	1.63	1.60	1.55	1.49	1.42	1.38	1.34	1.30	1.24	1.17	1.00

TABLE 5 Percentiles of the F Distribution: $F_{.95}(n_1, n_2)$ (Continued)

n_1 = degrees of freedom for numerator

n_2 = degrees of freedom for denominator

n_2 \ n_1	1	2	3	4	5	6	7	8	9	10	12	15	20	24	30	40	60	120	∞
1	161.4	199.5	215.7	224.6	230.2	234.0	236.8	238.9	240.5	241.9	243.9	245.9	248.0	249.1	250.1	251.1	252.2	253.3	254.3
2	18.51	19.00	19.16	19.25	19.30	19.33	19.35	19.37	19.38	19.40	19.41	19.43	19.45	19.45	19.46	19.47	19.48	19.49	19.50
3	10.13	9.55	9.28	9.12	9.01	8.94	8.89	8.85	8.81	8.79	8.74	8.70	8.66	8.64	8.62	8.59	8.57	8.55	8.53
4	7.71	6.94	6.59	6.39	6.26	6.16	6.09	6.04	6.00	5.96	5.91	5.86	5.80	5.77	5.75	5.72	5.69	5.66	5.63
5	6.61	5.79	5.41	5.19	5.05	4.95	4.88	4.82	4.77	4.74	4.68	4.62	4.56	4.53	4.50	4.46	4.43	4.40	4.36
6	5.99	5.14	4.76	4.53	4.39	4.28	4.21	4.15	4.10	4.06	4.00	3.94	3.87	3.84	3.81	3.77	3.74	3.70	3.67
7	5.59	4.74	4.35	4.12	3.97	3.87	3.79	3.73	3.68	3.64	3.57	3.51	3.44	3.41	3.38	3.34	3.30	3.27	3.23
8	5.32	4.46	4.07	3.84	3.69	3.58	3.50	3.44	3.39	3.35	3.28	3.22	3.15	3.12	3.08	3.04	3.01	2.97	2.93
9	5.12	4.26	3.86	3.63	3.48	3.37	3.29	3.23	3.18	3.14	3.07	3.01	2.94	2.90	2.86	2.83	2.79	2.75	2.71
10	4.96	4.10	3.71	3.48	3.33	3.22	3.14	3.07	3.02	2.98	2.91	2.85	2.77	2.74	2.70	2.66	2.62	2.58	2.54
11	4.84	3.98	3.59	3.36	3.20	3.09	3.01	2.95	2.90	2.85	2.79	2.72	2.65	2.61	2.57	2.53	2.49	2.45	2.40
12	4.75	3.89	3.49	3.26	3.11	3.00	2.91	2.85	2.80	2.75	2.69	2.62	2.54	2.51	2.47	2.43	2.38	2.34	2.30
13	4.67	3.81	3.41	3.18	3.03	2.92	2.83	2.77	2.71	2.67	2.60	2.53	2.46	2.42	2.38	2.34	2.30	2.25	2.21
14	4.60	3.74	3.34	3.11	2.96	2.85	2.76	2.70	2.65	2.60	2.53	2.46	2.39	2.35	2.31	2.27	2.22	2.18	2.13
15	4.54	3.68	3.29	3.06	2.90	2.79	2.71	2.64	2.59	2.54	2.48	2.40	2.33	2.29	2.25	2.20	2.16	2.11	2.07
16	4.49	3.63	3.24	3.01	2.85	2.74	2.66	2.59	2.54	2.49	2.42	2.35	2.28	2.24	2.19	2.15	2.11	2.06	2.01
17	4.45	3.59	3.20	2.96	2.81	2.70	2.61	2.55	2.49	2.45	2.38	2.31	2.23	2.19	2.15	2.10	2.06	2.01	1.96
18	4.41	3.55	3.16	2.93	2.77	2.66	2.58	2.51	2.46	2.41	2.34	2.27	2.19	2.15	2.11	2.06	2.02	1.97	1.92
19	4.38	3.52	3.13	2.90	2.74	2.63	2.54	2.48	2.42	2.38	2.31	2.23	2.16	2.11	2.07	2.03	1.98	1.93	1.88
20	4.35	3.49	3.10	2.87	2.71	2.60	2.51	2.45	2.39	2.35	2.28	2.20	2.12	2.08	2.04	1.99	1.95	1.90	1.84
21	4.32	3.47	3.07	2.84	2.68	2.57	2.49	2.42	2.37	2.32	2.25	2.18	2.10	2.05	2.01	1.96	1.92	1.87	1.81
22	4.30	3.44	3.05	2.82	2.66	2.55	2.46	2.40	2.34	2.30	2.23	2.15	2.07	2.03	1.98	1.94	1.89	1.84	1.78
23	4.28	3.42	3.03	2.80	2.64	2.53	2.44	2.37	2.32	2.27	2.20	2.13	2.05	2.01	1.96	1.91	1.86	1.81	1.76
24	4.26	3.40	3.01	2.78	2.62	2.51	2.42	2.36	2.30	2.25	2.18	2.11	2.03	1.98	1.94	1.89	1.84	1.79	1.73
25	4.24	3.39	2.99	2.76	2.60	2.49	2.40	2.34	2.28	2.24	2.16	2.09	2.01	1.96	1.92	1.87	1.82	1.77	1.71
26	4.23	3.37	2.98	2.74	2.59	2.47	2.39	2.32	2.27	2.22	2.15	2.07	1.99	1.95	1.90	1.85	1.80	1.75	1.69
27	4.21	3.35	2.96	2.73	2.57	2.46	2.37	2.31	2.25	2.20	2.13	2.06	1.97	1.93	1.88	1.84	1.79	1.73	1.67
28	4.20	3.34	2.95	2.71	2.56	2.45	2.36	2.29	2.24	2.19	2.12	2.04	1.96	1.91	1.87	1.82	1.77	1.71	1.65
29	4.18	3.33	2.93	2.70	2.55	2.43	2.35	2.28	2.22	2.18	2.10	2.03	1.94	1.90	1.85	1.81	1.75	1.70	1.64
30	4.17	3.32	2.92	2.69	2.53	2.42	2.33	2.27	2.21	2.16	2.09	2.01	1.93	1.89	1.84	1.79	1.74	1.68	1.62
40	4.08	3.23	2.84	2.61	2.45	2.34	2.25	2.18	2.12	2.08	2.00	1.92	1.84	1.79	1.74	1.69	1.64	1.58	1.51
60	4.00	3.15	2.76	2.53	2.37	2.25	2.17	2.10	2.04	1.99	1.92	1.84	1.75	1.70	1.65	1.59	1.53	1.47	1.39
120	3.92	3.07	2.68	2.45	2.29	2.17	2.09	2.02	1.96	1.91	1.83	1.75	1.66	1.61	1.55	1.50	1.43	1.35	1.25
∞	3.84	3.00	2.60	2.37	2.21	2.10	2.01	1.94	1.88	1.83	1.75	1.67	1.57	1.52	1.46	1.39	1.32	1.22	1.00

TABLE 5 Percentiles of the F Distribution: $F_{.975}(n_1, n_2)$ (Continued)

n_1 = degrees of freedom for numerator

n_2 \ n_1	1	2	3	4	5	6	7	8	9	10	12	15	20	24	30	40	60	120	∞
1	647.8	799.5	864.2	899.6	921.8	937.1	948.2	956.7	963.3	968.6	976.7	984.9	993.1	997.2	1001	1006	1010	1014	1018
2	38.51	39.00	39.17	39.25	39.30	39.33	39.36	39.37	39.39	39.40	39.41	39.43	39.45	39.46	39.46	39.47	39.48	39.49	39.50
3	17.44	16.04	15.44	15.10	14.88	14.73	14.62	14.54	14.47	14.42	14.34	14.25	14.17	14.12	14.08	14.04	13.99	13.95	13.90
4	12.22	10.65	9.98	9.60	9.36	9.20	9.07	8.98	8.90	8.84	8.75	8.66	8.56	8.51	8.46	8.41	8.36	8.31	8.26
5	10.01	8.43	7.76	7.39	7.15	6.98	6.85	6.76	6.68	6.62	6.52	6.43	6.33	6.28	6.23	6.18	6.12	6.07	6.02
6	8.81	7.26	6.60	6.23	5.99	5.82	5.70	5.60	5.52	5.46	5.37	5.27	5.17	5.12	5.07	5.01	4.96	4.90	4.85
7	8.07	6.54	5.89	5.52	5.29	5.12	4.99	4.90	4.82	4.76	4.67	4.57	4.47	4.42	4.36	4.31	4.25	4.20	4.14
8	7.57	6.06	5.42	5.05	4.82	4.65	4.53	4.43	4.36	4.30	4.20	4.10	4.00	3.95	3.89	3.84	3.78	3.73	3.67
9	7.21	5.71	5.08	4.72	4.48	4.32	4.20	4.10	4.03	3.96	3.87	3.77	3.67	3.61	3.56	3.51	3.45	3.39	3.33
10	6.94	5.46	4.83	4.47	4.24	4.07	3.95	3.85	3.78	3.72	3.62	3.52	3.42	3.37	3.31	3.26	3.20	3.14	3.08
11	6.72	5.26	4.63	4.28	4.04	3.88	3.76	3.66	3.59	3.53	3.43	3.33	3.23	3.17	3.12	3.06	3.00	2.94	2.88
12	6.55	5.10	4.47	4.12	3.89	3.73	3.61	3.51	3.44	3.37	3.28	3.18	3.07	3.02	2.96	2.91	2.85	2.79	2.72
13	6.41	4.97	4.35	4.00	3.77	3.60	3.48	3.39	3.31	3.25	3.15	3.05	2.95	2.89	2.84	2.78	2.72	2.66	2.60
14	6.30	4.86	4.24	3.89	3.66	3.50	3.38	3.29	3.21	3.15	3.05	2.95	2.84	2.79	2.73	2.67	2.61	2.55	2.49
15	6.20	4.77	4.15	3.80	3.58	3.41	3.29	3.20	3.12	3.06	2.96	2.86	2.76	2.70	2.64	2.59	2.52	2.46	2.40
16	6.12	4.69	4.08	3.73	3.50	3.34	3.22	3.12	3.05	2.99	2.89	2.79	2.68	2.63	2.57	2.51	2.45	2.38	2.32
17	6.04	4.62	4.01	3.66	3.44	3.28	3.16	3.06	2.98	2.92	2.82	2.72	2.62	2.56	2.50	2.44	2.38	2.32	2.25
18	5.98	4.56	3.95	3.61	3.38	3.22	3.10	3.01	2.93	2.87	2.77	2.67	2.56	2.50	2.44	2.38	2.32	2.26	2.19
19	5.92	4.51	3.90	3.56	3.33	3.17	3.05	2.96	2.88	2.82	2.72	2.62	2.51	2.45	2.39	2.33	2.27	2.20	2.13
20	5.87	4.46	3.86	3.51	3.29	3.13	3.01	2.91	2.84	2.77	2.68	2.57	2.46	2.41	2.35	2.29	2.22	2.16	2.09
21	5.83	4.42	3.82	3.48	3.25	3.09	2.97	2.87	2.80	2.73	2.64	2.53	2.42	2.37	2.31	2.25	2.18	2.11	2.04
22	5.79	4.38	3.78	3.44	3.22	3.05	2.93	2.84	2.76	2.70	2.60	2.50	2.39	2.33	2.27	2.21	2.14	2.08	2.00
23	5.75	4.35	3.75	3.41	3.18	3.02	2.90	2.81	2.73	2.67	2.57	2.47	2.36	2.30	2.24	2.18	2.11	2.04	1.97
24	5.72	4.32	3.72	3.38	3.15	2.99	2.87	2.78	2.70	2.64	2.54	2.44	2.33	2.27	2.21	2.15	2.08	2.01	1.94
25	5.69	4.29	3.69	3.35	3.13	2.97	2.85	2.75	2.68	2.61	2.51	2.41	2.30	2.24	2.18	2.12	2.05	1.98	1.91
26	5.66	4.27	3.67	3.33	3.10	2.94	2.82	2.73	2.65	2.59	2.49	2.39	2.28	2.22	2.16	2.09	2.03	1.95	1.88
27	5.63	4.24	3.65	3.31	3.08	2.92	2.80	2.71	2.63	2.57	2.47	2.36	2.25	2.19	2.13	2.07	2.00	1.93	1.85
28	5.61	4.22	3.63	3.29	3.06	2.90	2.78	2.69	2.61	2.55	2.45	2.34	2.23	2.17	2.11	2.05	1.98	1.91	1.83
29	5.59	4.20	3.61	3.27	3.04	2.88	2.76	2.67	2.59	2.53	2.43	2.32	2.21	2.15	2.09	2.03	1.96	1.89	1.81
30	5.57	4.18	3.59	3.25	3.03	2.87	2.75	2.65	2.57	2.51	2.41	2.31	2.20	2.14	2.07	2.01	1.94	1.87	1.79
40	5.42	4.05	3.46	3.13	2.90	2.74	2.62	2.53	2.45	2.39	2.29	2.18	2.07	2.01	1.94	1.88	1.80	1.72	1.64
60	5.29	3.93	3.34	3.01	2.79	2.63	2.51	2.41	2.33	2.27	2.17	2.06	1.94	1.88	1.82	1.74	1.67	1.58	1.48
120	5.15	3.80	3.23	2.89	2.67	2.52	2.39	2.30	2.22	2.16	2.05	1.94	1.82	1.76	1.69	1.61	1.53	1.43	1.31
∞	5.02	3.69	3.12	2.79	2.57	2.41	2.29	2.19	2.11	2.05	1.94	1.83	1.71	1.64	1.57	1.48	1.39	1.27	1.00

n_2 = degrees of freedom for denominator

TABLE 5 Percentiles of the F Distribution: $F_{.99}(n_1, n_2)$ (Continued)

n_1 = degrees of freedom for numerator

n_2	1	2	3	4	5	6	7	8	9	10	12	15	20	24	30	40	60	120	∞
1	4052	4999.5	5403	5625	5764	5859	5928	5982	6022	6056	6106	6157	6209	6235	6261	6287	6313	6339	6366
2	98.50	99.00	99.17	99.25	99.30	99.33	99.36	99.37	99.39	99.40	99.42	99.43	99.45	99.46	99.47	99.47	99.48	99.49	99.50
3	34.12	30.82	29.46	28.71	28.24	27.91	27.67	27.49	27.35	27.23	27.05	26.87	26.69	26.60	26.50	26.41	26.32	26.22	26.13
4	21.20	18.00	16.69	15.98	15.52	15.21	14.98	14.80	14.66	14.55	14.37	14.20	14.02	13.93	13.84	13.75	13.65	13.56	13.46
5	16.26	13.27	12.06	11.39	10.97	10.67	10.46	10.29	10.16	10.05	9.89	9.72	9.55	9.47	9.38	9.29	9.20	9.11	9.02
6	13.75	10.92	9.78	9.15	8.75	8.47	8.26	8.10	7.98	7.87	7.72	7.56	7.40	7.31	7.23	7.14	7.06	6.97	6.88
7	12.25	9.55	8.45	7.85	7.46	7.19	6.99	6.84	6.72	6.62	6.47	6.31	6.16	6.07	5.99	5.91	5.82	5.74	5.65
8	11.26	8.65	7.59	7.01	6.63	6.37	6.18	6.03	5.91	5.81	5.67	5.52	5.36	5.28	5.20	5.12	5.03	4.95	4.86
9	10.56	8.02	6.99	6.42	6.06	5.80	5.61	5.47	5.35	5.26	5.11	4.96	4.81	4.73	4.65	4.57	4.48	4.40	4.31
10	10.04	7.56	6.55	5.99	5.64	5.39	5.20	5.06	4.94	4.85	4.71	4.56	4.41	4.33	4.25	4.17	4.08	4.00	3.91
11	9.65	7.21	6.22	5.67	5.32	5.07	4.89	4.74	4.63	4.54	4.40	4.25	4.10	4.02	3.94	3.86	3.78	3.69	3.60
12	9.33	6.93	5.95	5.41	5.06	4.82	4.64	4.50	4.39	4.30	4.16	4.01	3.86	3.78	3.70	3.62	3.54	3.45	3.36
13	9.07	6.70	5.74	5.21	4.86	4.62	4.44	4.30	4.19	4.10	3.96	3.82	3.66	3.59	3.51	3.43	3.34	3.25	3.17
14	8.86	6.51	5.56	5.04	4.69	4.46	4.28	4.14	4.03	3.94	3.80	3.66	3.51	3.43	3.35	3.27	3.18	3.09	3.00
15	8.68	6.36	5.42	4.89	4.56	4.32	4.14	4.00	3.89	3.80	3.67	3.52	3.37	3.29	3.21	3.13	3.05	2.96	2.87
16	8.53	6.23	5.29	4.77	4.44	4.20	4.03	3.89	3.78	3.69	3.55	3.41	3.26	3.18	3.10	3.02	2.93	2.84	2.75
17	8.40	6.11	5.18	4.67	4.34	4.10	3.93	3.79	3.68	3.59	3.46	3.31	3.16	3.08	3.00	2.92	2.83	2.75	2.65
18	8.29	6.01	5.09	4.58	4.25	4.01	3.84	3.71	3.60	3.51	3.37	3.23	3.08	3.00	2.92	2.84	2.75	2.66	2.57
19	8.18	5.93	5.01	4.50	4.17	3.94	3.77	3.63	3.52	3.43	3.30	3.15	3.00	2.92	2.84	2.76	2.67	2.58	2.49
20	8.10	5.85	4.94	4.43	4.10	3.87	3.70	3.56	3.46	3.37	3.23	3.09	2.94	2.86	2.78	2.69	2.61	2.52	2.42
21	8.02	5.78	4.87	4.37	4.04	3.81	3.64	3.51	3.40	3.31	3.17	3.03	2.88	2.80	2.72	2.64	2.55	2.46	2.36
22	7.95	5.72	4.82	4.31	3.99	3.76	3.59	3.45	3.35	3.26	3.12	2.98	2.83	2.75	2.67	2.58	2.50	2.40	2.31
23	7.88	5.66	4.76	4.26	3.94	3.71	3.54	3.41	3.30	3.21	3.07	2.93	2.78	2.70	2.62	2.54	2.45	2.35	2.26
24	7.82	5.61	4.72	4.22	3.90	3.67	3.50	3.36	3.26	3.17	3.03	2.89	2.74	2.66	2.58	2.49	2.40	2.31	2.21
25	7.77	5.57	4.68	4.18	3.85	3.63	3.46	3.32	3.22	3.13	2.99	2.85	2.70	2.62	2.54	2.45	2.36	2.27	2.17
26	7.72	5.53	4.64	4.14	3.82	3.59	3.42	3.29	3.18	3.09	2.96	2.81	2.66	2.58	2.50	2.42	2.33	2.23	2.13
27	7.68	5.49	4.60	4.11	3.78	3.56	3.39	3.26	3.15	3.06	2.93	2.78	2.63	2.55	2.47	2.38	2.29	2.20	2.10
28	7.64	5.45	4.57	4.07	3.75	3.53	3.36	3.23	3.12	3.03	2.90	2.75	2.60	2.52	2.44	2.35	2.26	2.17	2.06
29	7.60	5.42	4.54	4.04	3.73	3.50	3.33	3.20	3.09	3.00	2.87	2.73	2.57	2.49	2.41	2.33	2.23	2.14	2.03
30	7.56	5.39	4.51	4.02	3.70	3.47	3.30	3.17	3.07	2.98	2.84	2.70	2.55	2.47	2.39	2.30	2.21	2.11	2.01
40	7.31	5.18	4.31	3.83	3.51	3.29	3.12	2.99	2.89	2.80	2.66	2.52	2.37	2.29	2.20	2.11	2.02	1.92	1.80
60	7.08	4.98	4.13	3.65	3.34	3.12	2.95	2.82	2.72	2.63	2.50	2.35	2.20	2.12	2.03	1.94	1.84	1.73	1.60
120	6.85	4.79	3.95	3.48	3.17	2.96	2.79	2.66	2.56	2.47	2.34	2.19	2.03	1.95	1.86	1.76	1.66	1.53	1.38
∞	6.63	4.61	3.78	3.32	3.02	2.80	2.64	2.51	2.41	2.32	2.18	2.04	1.88	1.79	1.70	1.59	1.47	1.32	1.00

n_2 = degrees of freedom for denominator

TABLE 6 Percentiles of the Studentized Range: $q_{.90}$

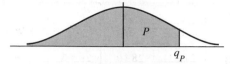

$q = w/s$ where w is the range of t observations, and v is the number of degrees of freedom associated with the standard deviation s.

v	2	3	4	5	6	7	8	9	10
1	8.93	13.44	16.36	18.49	20.15	21.51	22.64	23.62	24.48
2	4.13	5.73	6.77	7.54	8.14	8.63	9.05	9.41	9.72
3	3.33	4.47	5.20	5.74	6.16	6.51	6.81	7.06	7.29
4	3.01	3.98	4.59	5.03	5.39	5.68	5.93	6.14	6.33
5	2.85	3.72	4.26	4.66	4.98	5.24	5.46	5.65	5.82
6	2.75	3.56	4.07	4.44	4.73	4.97	5.17	5.34	5.50
7	2.68	3.45	3.93	4.28	4.55	4.78	4.97	5.14	5.28
8	2.63	3.37	3.83	4.17	4.43	4.65	4.83	4.99	5.13
9	2.59	3.32	3.76	4.08	4.34	4.54	4.72	4.87	5.01
10	2.56	3.27	3.70	4.02	4.26	4.47	4.64	4.78	4.91
11	2.54	3.23	3.66	3.96	4.20	4.40	4.57	4.71	4.84
12	2.52	3.20	3.62	3.92	4.16	4.35	4.51	4.65	4.78
13	2.50	3.18	3.59	3.88	4.12	4.30	4.46	4.60	4.72
14	2.49	3.16	3.56	3.85	4.08	4.27	4.42	4.56	4.68
15	2.48	3.14	3.54	3.83	4.05	4.23	4.39	4.52	4.64
16	2.47	3.12	3.52	3.80	4.03	4.21	4.36	4.49	4.61
17	2.46	3.11	3.50	3.78	4.00	4.18	4.33	4.46	4.58
18	2.45	3.10	3.49	3.77	3.98	4.16	4.31	4.44	4.55
19	2.45	3.09	3.47	3.75	3.97	4.14	4.29	4.42	4.53
20	2.44	3.08	3.46	3.74	3.95	4.12	4.27	4.40	4.51
24	2.42	3.05	3.42	3.69	3.90	4.07	4.21	4.34	4.44
30	2.40	3.02	3.39	3.65	3.85	4.02	4.16	4.28	4.38
40	2.38	2.99	3.35	3.60	3.80	3.96	4.10	4.21	4.32
60	2.36	2.96	3.31	3.56	3.75	3.91	4.04	4.16	4.25
120	2.34	2.93	3.28	3.52	3.71	3.86	3.99	4.10	4.19
∞	2.33	2.90	3.24	3.48	3.66	3.81	3.93	4.04	4.13

TABLE 6   Percentiles of the Studentized Range: $q_{.90}$ (Continued)

v \ t	11	12	13	14	15	16	17	18	19	20
1	25.24	25.92	26.54	27.10	27.62	28.10	28.54	28.96	29.35	29.71
2	10.01	10.26	10.49	10.70	10.89	11.07	11.24	11.39	11.54	11.68
3	7.49	7.67	7.83	7.98	8.12	8.25	8.37	8.48	8.58	8.68
4	6.49	6.65	6.78	6.91	7.02	7.13	7.23	7.33	7.41	7.50
5	5.97	6.10	6.22	6.34	6.44	6.54	6.63	6.71	6.79	6.86
6	5.64	5.76	5.87	5.98	6.07	6.16	6.25	6.32	6.40	6.47
7	5.41	5.53	5.64	5.74	5.83	5.91	5.99	6.06	6.13	6.19
8	5.25	5.36	5.46	5.56	5.64	5.72	5.80	5.87	5.93	6.00
9	5.13	5.23	5.33	5.42	5.51	5.58	5.66	5.72	5.79	5.85
10	5.03	5.13	5.23	5.32	5.40	5.47	5.54	5.61	5.67	5.73
11	4.95	5.05	5.15	5.23	5.31	5.38	5.45	5.51	5.57	5.63
12	4.89	4.99	5.08	5.16	5.24	5.31	5.37	5.44	5.49	5.55
13	4.83	4.93	5.02	5.10	5.18	5.25	5.31	5.37	5.43	5.48
14	4.79	4.88	4.97	5.05	5.12	5.19	5.26	5.32	5.37	5.43
15	4.75	4.84	4.93	5.01	5.08	5.15	5.21	5.27	5.32	5.38
16	4.71	4.81	4.89	4.97	5.04	5.11	5.17	5.23	5.28	5.33
17	4.68	4.77	4.86	4.93	5.01	5.07	5.13	5.19	5.24	5.30
18	4.65	4.75	4.83	4.90	4.98	5.04	5.10	5.16	5.21	5.26
19	4.63	4.72	4.80	4.88	4.95	5.01	5.07	5.13	5.18	5.23
20	4.61	4.70	4.78	4.85	4.92	4.99	5.05	5.10	5.16	5.20
24	4.54	4.63	4.71	4.78	4.85	4.91	4.97	5.02	5.07	5.12
30	4.47	4.56	4.64	4.71	4.77	4.83	4.89	4.94	4.99	5.03
40	4.41	4.49	4.56	4.63	4.69	4.75	4.81	4.86	4.90	4.95
60	4.34	4.42	4.49	4.56	4.62	4.67	4.73	4.78	4.82	4.86
120	4.28	4.35	4.42	4.48	4.54	4.60	4.65	4.69	4.74	4.78
∞	4.21	4.28	4.35	4.41	4.47	4.52	4.57	4.61	4.65	4.69

TABLE 6   Percentiles of the Studentized Range: $q_{.95}$ (Continued)

v \ t	2	3	4	5	6	7	8	9	10
1	17.97	26.98	32.82	37.08	40.41	43.12	45.40	47.36	49.07
2	6.08	8.33	9.80	10.88	11.74	12.44	13.03	13.54	13.99
3	4.50	5.91	6.82	7.50	8.04	8.48	8.85	9.18	9.46
4	3.93	5.04	5.76	6.29	6.71	7.05	7.35	7.60	7.83
5	3.64	4.60	5.22	5.67	6.03	6.33	6.58	6.80	6.99
6	3.46	4.34	4.90	5.30	5.63	5.90	6.12	6.32	6.49
7	3.34	4.16	4.68	5.06	5.36	5.61	5.82	6.00	6.16
8	3.26	4.04	4.53	4.89	5.17	5.40	5.60	5.77	5.92
9	3.20	3.95	4.41	4.76	5.02	5.24	5.43	5.59	5.74
10	3.15	3.88	4.33	4.65	4.91	5.12	5.30	5.46	5.60
11	3.11	3.82	4.26	4.57	4.82	5.03	5.20	5.35	5.49
12	3.08	3.77	4.20	4.51	4.75	4.95	5.12	5.27	5.39
13	3.06	3.73	4.15	4.45	4.69	4.88	5.05	5.19	5.32
14	3.03	3.70	4.11	4.41	4.64	4.83	4.99	5.13	5.25
15	3.01	3.67	4.08	4.37	4.59	4.78	4.94	5.08	5.20
16	3.00	3.65	4.05	4.33	4.56	4.74	4.90	5.03	5.15
17	2.98	3.63	4.02	4.30	4.52	4.70	4.86	4.99	5.11
18	2.97	3.61	4.00	4.28	4.49	4.67	4.82	4.96	5.07
19	2.96	3.59	3.98	4.25	4.47	4.65	4.79	4.92	5.04
20	2.95	3.58	3.96	4.23	4.45	4.62	4.77	4.90	5.01
24	2.92	3.53	3.90	4.17	4.37	4.54	4.68	4.81	4.92
30	2.89	3.49	3.85	4.10	4.30	4.46	4.60	4.72	4.82
40	2.86	3.44	3.79	4.04	4.23	4.39	4.52	4.63	4.73
60	2.83	3.40	3.74	3.98	4.16	4.31	4.44	4.55	4.65
120	2.80	3.36	3.68	3.92	4.10	4.24	4.36	4.47	4.56
∞	2.77	3.31	3.63	3.86	4.03	4.17	4.29	4.39	4.47

TABLE 6   Percentiles of the Studentized Range: $q_{.95}$ (Continued)

v \ t	11	12	13	14	15	16	17	18	19	20
1	50.59	51.96	53.20	54.33	55.36	56.32	57.22	58.04	58.83	59.56
2	14.39	14.75	15.08	15.38	15.65	15.91	16.14	16.37	16.57	16.77
3	9.72	9.95	10.15	10.35	10.52	10.69	10.84	10.98	11.11	11.24
4	8.03	8.21	8.37	8.52	8.66	8.79	8.91	9.03	9.13	9.23
5	7.17	7.32	7.47	7.60	7.72	7.83	7.93	8.03	8.12	8.21
6	6.65	6.79	6.92	7.03	7.14	7.24	7.34	7.43	7.51	7.59
7	6.30	6.43	6.55	6.66	6.76	6.85	6.94	7.02	7.10	7.17
8	6.05	6.18	6.29	6.39	6.48	6.57	6.65	6.73	6.80	6.87
9	5.87	5.98	6.09	6.19	6.28	6.36	6.44	6.51	6.58	6.64
10	5.72	5.83	5.93	6.03	6.11	6.19	6.27	6.34	6.40	6.47
11	5.61	5.71	5.81	5.90	5.98	6.06	6.13	6.20	6.27	6.33
12	5.51	5.61	5.71	5.80	5.88	5.95	6.02	6.09	6.15	6.21
13	5.43	5.53	5.63	5.71	5.79	5.86	5.93	5.99	6.05	6.11
14	5.36	5.46	5.55	5.64	5.71	5.79	5.85	5.91	5.97	6.03
15	5.31	5.40	5.49	5.57	5.65	5.72	5.78	5.85	5.90	5.96
16	5.26	5.35	5.44	5.52	5.59	5.66	5.73	5.79	5.84	5.90
17	5.21	5.31	5.39	5.47	5.54	5.61	5.67	5.73	5.79	5.84
18	5.17	5.27	5.35	5.43	5.50	5.57	5.63	5.69	5.74	5.79
19	5.14	5.23	5.31	5.39	5.46	5.53	5.59	5.65	5.70	5.75
20	5.11	5.20	5.28	5.36	5.43	5.49	5.55	5.61	5.66	5.71
24	5.01	5.10	5.18	5.25	5.32	5.38	5.44	5.49	5.55	5.59
30	4.92	5.00	5.08	5.15	5.21	5.27	5.33	5.38	5.43	5.47
40	4.82	4.90	4.98	5.04	5.11	5.16	5.22	5.27	5.31	5.36
60	4.73	4.81	4.88	4.94	5.00	5.06	5.11	5.15	5.20	5.24
120	4.64	4.71	4.78	4.84	4.90	4.95	5.00	5.04	5.09	5.13
∞	4.55	4.62	4.68	4.74	4.80	4.85	4.89	4.93	4.97	5.01

TABLE 6   Percentiles of the Studentized Range: $q_{.99}$ (Continued)

v＼t	2	3	4	5	6	7	8	9	10
1	90.03	135.0	164.3	185.6	202.2	215.8	227.2	237.0	245.6
2	14.04	19.02	22.29	24.72	26.63	28.20	29.53	30.68	31.69
3	8.26	10.62	12.17	13.33	14.24	15.00	15.64	16.20	16.69
4	6.51	8.12	9.17	9.96	10.58	11.10	11.55	11.93	12.27
5	5.70	6.98	7.80	8.42	8.91	9.32	9.67	9.97	10.24
6	5.24	6.33	7.03	7.56	7.97	8.32	8.61	8.87	9.10
7	4.95	5.92	6.54	7.01	7.37	7.68	7.94	8.17	8.37
8	4.75	5.64	6.20	6.62	6.96	7.24	7.47	7.68	7.86
9	4.60	5.43	5.96	6.35	6.66	6.91	7.13	7.33	7.49
10	4.48	5.27	5.77	6.14	6.43	6.67	6.87	7.05	7.21
11	4.39	5.15	5.62	5.97	6.25	6.48	6.67	6.84	6.99
12	4.32	5.05	5.50	5.84	6.10	6.32	6.51	6.67	6.81
13	4.26	4.96	5.40	5.73	5.98	6.19	6.37	6.53	6.67
14	4.21	4.89	5.32	5.63	5.88	6.08	6.26	6.41	6.54
15	4.17	4.84	5.25	5.56	5.80	5.99	6.16	6.31	6.44
16	4.13	4.79	5.19	5.49	5.72	5.92	6.08	6.22	6.35
17	4.10	4.74	5.14	5.43	5.66	5.85	6.01	6.15	6.27
18	4.07	4.70	5.09	5.38	5.60	5.79	5.94	6.08	6.20
19	4.05	4.67	5.05	5.33	5.55	5.73	5.89	6.02	6.14
20	4.02	4.64	5.02	5.29	5.51	5.69	5.84	5.97	6.09
24	3.96	4.55	4.91	5.17	5.37	5.54	5.69	5.81	5.92
30	3.89	4.45	4.80	5.05	5.24	5.40	5.54	5.65	5.76
40	3.82	4.37	4.70	4.93	5.11	5.26	5.39	5.50	5.60
60	3.76	4.28	4.59	4.82	4.99	5.13	5.25	5.36	5.45
120	3.70	4.20	4.50	4.71	4.87	5.01	5.12	5.21	5.30
∞	3.64	4.12	4.40	4.60	4.76	4.88	4.99	5.08	5.16

TABLE 6   Percentiles of the Studentized Range: $q_{.99}$ (Continued)

v \ t	11	12	13	14	15	16	17	18	19	20
1	253.2	260.0	266.2	271.8	277.0	281.8	286.3	290.4	294.3	298.0
2	32.59	33.40	34.13	34.81	35.43	36.00	36.53	37.03	37.50	37.95
3	17.13	17.53	17.89	18.22	18.52	18.81	19.07	19.32	19.55	19.77
4	12.57	12.84	13.09	13.32	13.53	13.73	13.91	14.08	14.24	14.40
5	10.48	10.70	10.89	11.08	11.24	11.40	11.55	11.68	11.81	11.93
6	9.30	9.48	9.65	9.81	9.95	10.08	10.21	10.32	10.43	10.54
7	8.55	8.71	8.86	9.00	9.12	9.24	9.35	9.46	9.55	9.65
8	8.03	8.18	8.31	8.44	8.55	8.66	8.76	8.85	8.94	9.03
9	7.65	7.78	7.91	8.03	8.13	8.23	8.33	8.41	8.49	8.57
10	7.36	7.49	7.60	7.71	7.81	7.91	7.99	8.08	8.15	8.23
11	7.13	7.25	7.36	7.46	7.56	7.65	7.73	7.81	7.88	7.95
12	6.94	7.06	7.17	7.26	7.36	7.44	7.52	7.59	7.66	7.73
13	6.79	6.90	7.01	7.10	7.19	7.27	7.35	7.42	7.48	7.55
14	6.66	6.77	6.87	6.96	7.05	7.13	7.20	7.27	7.33	7.39
15	6.55	6.66	6.76	6.84	6.93	7.00	7.07	7.14	7.20	7.26
16	6.46	6.56	6.66	6.74	6.82	6.90	6.97	7.03	7.09	7.15
17	6.38	6.48	6.57	6.66	6.73	6.81	6.87	6.94	7.00	7.05
18	6.31	6.41	6.50	6.58	6.65	6.73	6.79	6.85	6.91	6.97
19	6.25	6.34	6.43	6.51	6.58	6.65	6.72	6.78	6.84	6.89
20	6.19	6.28	6.37	6.45	6.52	6.59	6.65	6.71	6.77	6.82
24	6.02	6.11	6.19	6.26	6.33	6.39	6.45	6.51	6.56	6.61
30	5.85	5.93	6.01	6.08	6.14	6.20	6.26	6.31	6.36	6.41
40	5.69	5.76	5.83	5.90	5.96	6.02	6.07	6.12	6.16	6.21
60	5.53	5.60	5.67	5.73	5.78	5.84	5.89	5.93	5.97	6.01
120	5.37	5.44	5.50	5.56	5.61	5.66	5.71	5.75	5.79	5.83
∞	5.23	5.29	5.35	5.40	5.45	5.49	5.54	5.57	5.61	5.65

TABLE 7 Percentage Points of the Bonferroni t Statistic: $t_v^{\alpha/2k}$

$\alpha = .05$

v \ k	2	3	4	5	6	7	8	9	10	15	20	25	30	35	40	45	50
5	3.17	3.54	3.81	4.04	4.22	4.38	4.53	4.66	4.78	5.25	5.60	5.89	6.15	6.36	6.56	6.70	6.86
7	2.84	3.13	3.34	3.50	3.64	3.76	3.86	3.95	4.03	4.36	4.59	4.78	4.95	5.09	5.21	5.31	5.40
10	2.64	2.87	3.04	3.17	3.28	3.37	3.45	3.52	3.58	3.83	4.01	4.15	4.27	4.37	4.45	4.53	4.59
12	2.56	2.78	2.94	3.06	3.15	3.24	3.31	3.37	3.43	3.65	3.80	3.93	4.04	4.13	4.20	4.26	4.32
15	2.49	2.69	2.84	2.95	3.04	3.11	3.18	3.24	3.29	3.48	3.62	3.74	3.82	3.90	3.97	4.02	4.07
20	2.42	2.61	2.75	2.85	2.93	3.00	3.06	3.11	3.16	3.33	3.46	3.55	3.63	3.70	3.76	3.80	3.85
24	2.39	2.58	2.70	2.80	2.88	2.94	3.00	3.05	3.09	3.26	3.38	3.47	3.54	3.61	3.66	3.70	3.74
30	2.36	2.54	2.66	2.75	2.83	2.89	2.94	2.99	3.03	3.19	3.30	3.39	3.46	3.52	3.57	3.61	3.65
40	2.33	2.50	2.62	2.71	2.78	2.84	2.89	2.93	2.97	3.12	3.23	3.31	3.38	3.43	3.48	3.51	3.55
60	2.30	2.47	2.58	2.66	2.73	2.79	2.84	2.88	2.92	3.06	3.16	3.24	3.30	3.34	3.39	3.42	3.46
120	2.27	2.43	2.54	2.62	2.68	2.74	2.79	2.83	2.85	2.99	3.09	3.15	3.22	3.27	3.31	3.34	3.37
∞	2.24	2.39	2.50	2.58	2.64	2.69	2.74	2.77	2.81	2.94	3.02	3.09	3.15	3.19	3.23	3.26	3.29

$\alpha = .01$

v \ k	2	3	4	5	6	7	8	9	10	15	20	25	30	35	40	45	50
5	4.78	5.25	5.60	5.89	6.15	6.36	6.56	6.70	6.86	7.51	8.00	8.37	8.68	8.95	9.19	9.41	9.68
7	4.03	4.36	4.59	4.78	4.95	5.09	5.21	5.31	5.40	5.79	6.08	6.30	6.49	6.67	6.83	6.93	7.06
10	3.58	3.83	4.01	4.15	4.27	4.37	4.45	4.53	4.59	4.86	5.06	5.20	5.33	5.44	5.52	5.60	5.70
12	3.43	3.65	3.80	3.93	4.04	4.13	4.20	4.26	4.32	4.56	4.73	4.86	4.95	5.04	5.12	5.20	5.27
15	3.29	3.48	3.62	3.74	3.82	3.90	3.97	4.02	4.07	4.29	4.42	4.53	4.61	4.71	4.78	4.84	4.90
20	3.16	3.33	3.46	3.55	3.63	3.70	3.76	3.80	3.85	4.03	4.15	4.25	4.33	4.39	4.46	4.52	4.56
24	3.09	3.26	3.38	3.47	3.54	3.61	3.66	3.70	3.74	3.91	4.04	4.1	4.2	4.3	4.3	4.3	4.4
30	3.03	3.19	3.30	3.39	3.46	3.52	3.57	3.61	3.65	3.80	3.90	3.98	4.13	4.26	4.1	4.2	4.2
40	2.97	3.12	3.23	3.31	3.38	3.43	3.48	3.51	3.55	3.70	3.79	3.88	3.93	3.97	4.01	4.1	4.1
60	2.92	3.06	3.16	3.24	3.30	3.34	3.39	3.42	3.46	3.59	3.69	3.76	3.81	3.84	3.89	3.93	3.97
120	2.86	2.99	3.09	3.15	3.22	3.27	3.31	3.34	3.37	3.50	3.58	3.64	3.69	3.73	3.77	3.80	3.83
∞	2.81	2.94	3.02	3.09	3.15	3.19	3.23	3.26	3.29	3.40	3.48	3.54	3.59	3.63	3.66	3.69	3.72

TABLE 8 Critical Values of Smaller Rank Sum for the Wilcoxon Mann-Whitney Test

n_2	α for Two-Sided Test	α for One-Sided Test	n_1 (Smaller Sample)																			
			1	2	3	4	5	6	7	8	9	10	11	12	13	14	15	16	17	18	19	20
3	.20	.10		3	7																	
	.10	.05			6																	
	.05	.025																				
	.01	.005																				
4	.20	.10		3	7	13																
	.10	.05			6	11																
	.05	.025				10																
	.01	.005																				
5	.20	.10		4	8	14	20															
	.10	.05		3	7	12	19															
	.05	.025			6	11	17															
	.01	.005					15															
6	.20	.10		4	9	15	22	30														
	.10	.05		3	8	13	20	28														
	.05	.025			7	12	18	26														
	.01	.005				10	16	23														
7	.20	.10		4	10	16	23	32	41													
	.10	.05		3	8	14	21	29	39													
	.05	.025			7	13	20	27	36													
	.01	.005				10	16	24	32													
8	.20	.10		5	11	17	25	34	44	55												
	.10	.05		4	9	15	23	31	41	51												
	.05	.025		3	8	14	21	29	38	49												
	.01	.005				11	17	25	34	43												
9	.20	.10	1	5	11	19	27	36	46	58	70											
	.10	.05		4	*10	16	24	33	43	54	66											
	.05	.025		3	8	14	22	31	40	51	62											
	.01	.005			6	11	18	26	35	45	56											

(continued)

TABLE 8 Critical Values of Smaller Rank Sum for the Wilcoxon Mann-Whitney Test (Continued)

n_2	α for Two-Sided Test	α for One-Sided Test	n_1 (Smaller Sample) 1	2	3	4	5	6	7	8	9	10	11	12	13	14	15	16	17	18	19	20
10	.20	.10	1	6	12	20	28	38	49	60	73	87										
	.10	.05		4	10	17	26	35	45	56	69	82										
	.05	.025		3	9	15	23	32	42	53	65	78										
	.01	.005			6	12	19	27	37	47	58	71										
11	.20	.10	1	6	13	21	30	40	51	63	76	91	106									
	.10	.05		4	11	18	27	37	47	59	72	86	100									
	.05	.025		3	9	16	24	34	44	55	68	81	96									
	.01	.005			6	12	20	28	38	49	61	73	87									
12	.20	.10	1	7	14	22	32	42	54	66	80	94	110	127								
	.10	.05		5	11	19	28	38	49	62	75	89	104	120								
	.05	.025		4	10	17	26	35	46	58	71	84	99	115								
	.01	.005			7	13	21	30	40	51	63	76	90	105								
13	.20	.10	1	7	15	23	33	44	56	69	83	98	114	131	149							
	.10	.05		5	12	20	30	40	52	64	78	92	108	125	142							
	.05	.025		4	10	18	27	37	48	60	73	88	103	119	136							
	.01	.005			7	*13	22	31	41	53	65	79	93	109	125							
14	.20	.10	1	*8	16	25	35	46	59	72	86	102	118	136	154	174						
	.10	.05		*6	13	21	31	42	54	67	81	96	112	129	147	166						
	.05	.025		4	11	19	28	38	50	62	76	91	106	123	141	160						
	.01	.005			7	14	22	32	43	54	67	81	96	112	129	147						
15	.20	.10	1	8	16	26	37	48	61	75	90	106	123	141	159	179	200					
	.10	.05		6	13	22	33	44	56	69	84	99	116	133	152	171	192					
	.05	.025		4	11	20	29	40	52	65	79	94	110	127	145	164	184					
	.01	.005			8	15	23	33	44	56	69	84	99	115	133	151	171					
16	.20	.10	1	8	17	27	38	50	64	78	93	109	127	145	165	185	206	229				
	.10	.05		6	14	24	34	46	58	72	87	103	120	138	156	176	197	219				
	.05	.025		4	12	21	30	42	54	67	82	97	113	131	150	169	190	211				
	.01	.005			8	15	24	34	46	58	72	86	102	119	136	155	175	196				

TABLE 8   Critical Values of Smaller Rank Sum for the Wilcoxon Mann-Whitney Test (Continued)

n_2	α for Two-Sided Test	α for One-Sided Test	1	2	3	4	5	6	7	8	9	10	11	12	13	14	15	16	17	18	19	20
17	.20	.10	1	9	18	28	40	52	66	81	97	113	131	150	170	190	212	235	259			
	.10	.05		6	15	25	35	47	61	75	90	106	123	142	161	182	203	225	249			
	.05	.025		5	12	21	32	43	56	70	84	100	117	135	154	174	195	217	240			
	.01	.005			8	16	25	36	47	60	74	89	105	122	140	159	180	201	223			
18	.20	.10	1	9	19	30	42	55	69	84	100	117	135	155	175	196	218	242	266	291		
	.10	.05		7	15	26	37	49	63	77	93	110	127	146	166	187	208	231	255	280		
	.05	.025		5	13	22	33	45	58	72	87	103	121	139	158	179	200	222	246	270		
	.01	.005			8	16	26	37	49	62	76	92	108	125	144	163	184	206	228	252		
19	.20	.10	2	10	20	31	43	57	71	87	103	121	139	159	180	202	224	248	273	299	325	
	.10	.05	1	7	16	27	38	51	65	80	96	113	131	150	171	192	214	237	262	287	313	
	.05	.025		5	13	23	34	46	60	74	90	107	124	143	163	*183	205	228	252	277	303	
	.01	.005		3	9	17	27	38	50	64	78	94	111	129	*148	168	189	210	234	258	283	
20	.20	.10	2	10	21	32	45	59	74	90	107	125	144	164	185	207	230	255	280	306	333	361
	.10	.05	1	7	17	28	40	53	67	83	99	117	135	155	175	197	220	243	268	294	320	348
	.05	.025		5	14	24	35	48	62	77	93	110	128	147	167	188	210	234	258	283	309	337
	.01	.005		3	9	18	28	39	52	66	81	97	114	132	151	172	193	215	239	263	289	315

For larger values of n_1 and n_2, critical values are given to a good approximation by the formula:

$$\frac{n_1}{2}(n_1 + n_2 + 1) - z \left\{ \frac{n_1 n_2 (n_1 + n_2 + 1)}{12} \right\}^{1/2}$$

where $z = 1.28$ for $\alpha = .20$ (two-sided test)

$z = 1.64$ for $\alpha = .10$ (two-sided test)

$z = 1.96$ for $\alpha = .05$ (two-sided test)

$z = 2.58$ for $\alpha = .01$ (two-sided test)

* Values have been corrected to the values given by D. B. Owen, *Handbook of Statistical Tables*, copyright 1962, Addison-Wesley Publishing Co., Inc.

TABLE 9   Critical Values of $W_\alpha(n)$ for the Wilcoxon Signed-Ranks Test

W_α is the integer such that the probability that $W \le W_\alpha$ is closest to α. For example, for $n = 8$, $P(W \le 3) = .020$ and $P(W \le 4) = .027$; therefore, $W_{.025}(8) = 4$.

	α for One-Sided Test		
	.025	.01	.005
	α for Two-Sided Test		
n	.05	.02	.01
6	0	—	—
7	2	0	—
8	4	2	0
9	6	3	2
10	8	5	3
11	11	7	5
12	14	10	7
13	17	13	10
14	21	16	13
15	25	20	16
16	30	24	20
17	35	28	23
18	40	33	28
19	46	38	32
20	52	43	38
21	59	49	43
22	66	56	49
23	73	62	55
24	81	69	61
25	89	77	68

For large n,

$$W_P(n) = \frac{n(n+1)}{4} - z_{1-P}\sqrt{\frac{n(n+1)(2n+1)}{24}}$$

approximately, where z is given in Table 2.

Bibliography

Adguna, W., and Labuschagne, M. (2002). Genotype-environment interactions and phenotypic stability analysis of linseed in Ethiopia. *Plant Breeding,* 66–71.

Agresti, A. (1996). *An Introduction to Categorical Data Analysis.* Wiley.

Albert, J., and Bennett, J. (2003). *Curve Ball: Baseball, Statistics, and the Role of Chance in the Game.* Springer.

Allison, T., and Cicchetti, D. (1976). Sleep in mammals: ecological and constitutional correlates. *Science,* November 12, vol. 194, pp. 732–734.

Andrews, D., Bickel, P., Hampel, F., Huber, P., Rogers, W., and Tukey, J. (1972). *Robust Estimates of Location.* Princeton, N.J.: Princeton University Press.

Andrews, D., and Herzberg. (1985). *Data.* Springer-Verlag.

Anscombe, F. J. (1950). Sampling theory of the negative binomial and logarithmic series distributions. *Biometrika, 37,* 358–382.

Armitage, P. (1983). *Statistical Methods in Medical Research.* Boston: Blackwell.

Bailey, C., Cox, E., and Springer, J. (1978). High pressure liquid chromatographic determination of the intermediate/side reaction products in FD&C Red No. 2 and FD&C Yellow No. 5; Statistical analysis of instrument response. *J. Assoc. Offic. Anal. Chem., 61,* 1404–1414.

Barlow, R. E., Toland, R. H., and Freeman, T. (1984). A Bayesian analysis of stress-rupture life of Kevlar/epoxy spherical pressure vessels. In *Proceedings of the Canadian Conference in Applied Statistics.* T. D. Dwivedi (ed.). New York: Marcel-Dekker.

Barnothy, J. M. (1964). Development of young mice. In *Biological Effects of Magnetic Fields.* M. Barnothy, ed. New York: Plenum Press.

Beecher, H. K. (1959). *Measurement of Subjective Responses.* Oxford, England: Oxford University Press.

Beller, G., Smith, T., Abelmann, W., Haber, E., and Hood, W. (1971). Digitalis intoxication: A prospective clinical study with serum level correlations. *N. Eng. J. Med., 284,* 989–997.

Benjamin, J., and Cornell, C. (1970). *Probability, Statistics, and Decision for Civil Engineers.* New York: McGraw-Hill.

Bennett, C., and Franklin, N. (1954). *Statistical Analysis in Chemistry and the Chemical Industry.* New York: Wiley.

Berkson, J. (1966). Examination of randomness of alpha particle emissions. In *Research Papers in Statistics.* F. N. David (ed.). New York: Wiley.

Bernstein, P. (1998). *Against the Gods: the Remarkable Story of Risk.* Wiley.

Bevan, S., Kullberg, R., and Rice, J. (1979). An analysis of cell membrane noise. *Annals of Statistics, 7,* 237–257.

Bhattacharjee, C., Bradley, P., Smith, M., Scally A., and Wilson, B. (2000). Do animals bite more during a full moon? Retrospective observational analysis. *British Medical Journal, 321,* 1559–1561.

Bickel, P., Chen, C., Kwon, J., Rice, J., van Zwet, E., and Varaiya, P. (2004). *Measuring Traffic. Berkeley Department of Statistics Technical Report 664.*

Bickel, P., and Doksum, K. (1977). *Mathematical Statistics: Basic Ideas and Selected Topics.* Oakland, Calif.: Holden-Day.

Bickel, P., and Doksum, K. (2001). *Mathematical Statistics: Basic Ideas and Selected Topics.* Prentice-Hall.

Bickel, P., and O'Connell, J. W. (1975). Is there a sex bias in graduate admissions? *Science, 187,* 398–404.

Bishop, Y., Fienberg, S., and Holland, P. (1975). *Discrete Multivariate Analysis: Theory and Practice.* Cambridge, Mass.: MIT Press.

Bjerkdal, T. (1960). Acquisition of resistance in guinea pigs infected with different doses of virulent tubercle bacilli. *Amer. J. Hygiene, 72,* 130–148.

Bliss, C., and Fisher, R. A. (1953). Fitting the negative binomial distribution to biological data. *Biometrics, 9,* 174–200.

Box, G. E. P., and Cox, D. R. (1964). An analysis of transformations (with discussion). *J. Royal Stat. Soc., Series B 26,* 211–246.

Box, G. E. P., and Tiao, G. C. (1973). *Bayesian Inference in Statistical Analysis.* Reading, Mass.: Addison-Wesley.

Box, G. E. P., Hunter, W. G., and Hunter, J. S. (1978). *Statistics for Experimenters.* New York: Wiley.

Brownlee, K. A. (1960). *Statistical Theory and Methodology in Science and Engineering.* New York: Wiley.

Brunk, H. D. (1975). *An Introduction to Mathematical Statistics.* Gardena, Calif.: Xerox.

Burr, I. (1974). *Applied Statistical Methods.* New York: Academic Press.

Campbell, J. A., and Pelletier, O. (1962). Determination of niacin (niacinamide) in cereal products. *J. Assoc. Offic. Anal. Chem., 45,* 449–453.

Carey, J. R., Liedo, P., Orozco, D., and Vaupel, J. W. (1992). Slowing of mortality rates at older ages in large medfly cohorts. *Science, 258,* 457–461.

Chambers, J., Cleveland, W., Kleiner, B., and Tukey, P. (1983). *Graphical Methods for Data Analysis.* Boston: Duxbury.

Chang, A. E., et al. (1979). Delta-9-Tetrahydrocannibol as an antiemetic in cancer patients receiving high-dose methotrexate. *The Science of Medical Marijuana Ann. Internal Med., 91,* 819–824.

Chang, S. L. (1945). Sedimentation in water and the specific gravity of cysts of *Entamoeba histolytica. Amer. J. Hygiene, 41,* 156–163.

Chapman, H., and Demeritt, D. (1936). *Elements of Forest Mensuration.* Nashville, Tenn: Williams Press.

Chernoff, H., and Lehman, E. (1954). The use of maximum likelihood estimates in tests for goodness of fit. *Annals of Math. Stat., 23,* 315–345.

Clancy, V. J. (1947). Empirical distributions in chemistry. *Nature, 159,* 340.

Cleveland, W., Graedel, T., Kleiner, B., and Warner, J. (1974). Sunday and workday variations in photochemical air pollutants in New Jersey and New York. *Science, 186,* 1037–1038.

Cobb, L., Thomas, G., Dillard, D., Merendino, J., and Bruce, R. (1959). An evaluation of internal mammary artery ligation by a double blind technique. *N. Eng. J. Med., 260,* 1115–1118.

Cochran, W. G. (1977). *Sampling Techniques.* New York: Wiley.

Cogswell, J. J. (1973). Forced oscillation technique for determination of resistance to breathing in children. *Arch. Dis. Child., 48,* 259–266.

Converse, P., and Traugott, M. (1986). Assessing the accuracy of polls and surveys. *Science, 234,* 1094–1098.

Cook, R. D., and Weisberg, S. (1982). *Residuals and Influence in Regression.* New York: Chapman and Hall.

Cramer, H. (1946). *Mathematical Methods of Statistics.* Princeton, N.J.: Princeton University Press.

Cummings, K. M., Giovino, G., Sciandra, R., Koenigsberg, M., and Emont, S. (1987). Physician advice to quit smoking: Who gets it and who doesn't? *Am. J. Prev. Med., 3,* 69–75.

Czitrom, V., and Reece, J. (1997). *Statistical Case Studies for Process Improvement.* SIAM-ASA, 87–103.

Dahiya, R., and Gurland, J. (1972). Pearson chi-squared test of fit with random intervals. *Biometrika, 59,* 147–153.

Dahlquist, G., and Bjorck, A. (1974). *Numerical Methods.* Englewood Cliffs, N.J.: Prentice-Hall.

David, H. (1981). *Order Statistics.* New York: Wiley.

Davies, O. (1960). *The Design and Analysis of Industrial Experiments.* London: Oliver and Boyd.

De Forina, M., Armanino, C., Lanteri, S., and Tiscornia, E. (1983). Classification of olive oils from their fatty acid composition. In *Food Research and Data Analysis,* 189–214. H. Martens and H. Russwurm Jr., (eds.). London: Applied Science Publishers.

DeHoff, R., and Rhines, F. (eds.) (1968). *Quantitative Microscopy.* New York: McGraw-Hill.

Deming, W. (1960). *Sample Design in Business Research.* New York: Wiley.

Diamond, G., and Forrester, J. (1979). Analysis of probability as an aid in the clinical diagnosis of coronary-artery disease. *New Eng. J. Med., 300,* 1350–1358.

Doksum, K., and Sievers, G. (1976). Plotting with confidence. Graphical comparisons of two populations. *Biometrika, 63,* 421–434.

Dongarra, J. (1979). *LINPACK Users' Guide.* Philadelphia: SIAM.

Donoho, A., Donoho, D., and Gasko, M. (1986). *MacSpin: Dynamic Data Display.* Belmont, Calif.: Wadsworth.

Dorfman, D. (1978). The Cyril Burt question: New findings. *Science, 201,* 1177–1186.

Dowdall, J. A. (1974). Women's attitudes toward employment and family roles. *Soc. Anal., 35,* 251–262.

Draper, N., and Smith, H. (1981). *Applied Regression Analysis.* New York: Wiley.

The Economist (2002). All in the Mind. February 21.

The Economist (2002). Try It and See. February 28.

Eddy, D. M. (1982). Probabilistic reasoning in clinical medicine: Problems and opportunities. In *Judgment under Uncertainty: Heuristics and Biases,* 249–267. D. Kahneman, P. Slovic, and A. Tversky (eds.). Cambridge University Press.

Edwards, W., Lindman, H., and Savage, L. J. (1963). Bayesian statistical inference for psychological research. *Psych. Rev., 70,* 193–242.

Efron, B., and Tibshirani, R. (1993). *An Introduction to the Bootstrap.* New York: Chapman and Hall.

Evans, D. (1953). Experimental evidence concerning contagious distributions in ecology. *Biometrika, 40,* 186–211.

Fechter, J. V., and Porter, L. G. (1979). Kitchen range energy consumption. Prepared for Office of Conservation, U. S. Department of Energy, NBSIR 78-1556 (Washington, D.C.).

Ferguson, T. S. (1967). *Mathematical Statistics: A Decision Theoretic Approach.* New York: Academic Press.

Filliben, J. (1975). The probability plot correlation coefficient test for normality. *Technometrics, 17,* 111–117.

Finkner, A. (1950). Methods of sampling for estimating commercial peach production in North Carolina. *North Carolina Agricultural Experiment Station Technical Bulletin, 91.*

Fisher, R. A. (1936). Has Mendel's work been rediscovered? *Annals of Science, 1,* 115–137.

Fisher, R. A. (1958). *Statistical Methods for Research Workers.* New York: Hafner.

Freedman, D., Pisani, R., and Purves, R. (1978). *Statistics.* New York: Norton.

Gardner, M. (1976). Mathematical games. *Scientific American, 234,* 119–123.

Gastwirth, J. (1987). The statistical precision of medical screening procedures. *Statistical Science, 3,* 213–222.

Geissler, A. (1889). Beiträge zur Frage des Geschlechtsverhältnisses der Gebornen. *Z. K. Sachs. Stat. Bur., 35,* 1–24.

Gerlough, D., and Schuhl, A. (1955). *Use of Poisson Distribution in Highway Traffic.* Eno Foundation for Highway Traffic Control.

Glass, D., and Hall, J. (1954). A study of intergeneration changes in status. In *Social Mobility in Britain,* D. Glass (ed.). Glencoe, Ill.: Free Press.

Gosset, W. S. (1931). The Lanarkshire milk experiment. *Biometrika, 23,* 398.

Grace, N., Muench, H., and Chalmers, T. (1966). The present status of shunts for portal hypertension in cirrhosis. *Gastroenterology, 50,* 684–691.

Haberman, S. (1978). *Analysis of Qualitative Data.* New York: Academic Press.

Hampson, R., and Walker, R. (1961). Vapor pressures of platinum, iridium, and rhodium. *J. Res. Nat. Bur. Stand., 65 A,* 289–295.

Hanley, J. A., and Shapiro, S. H. (1994). Sexual activity and the lifespan of male fruitflies: A dataset that gets attention. *J. Stat. Edu., 2*(1).

Harbaugh, J., Doveton, J., and Davis, J. (1977). *Probability Methods in Oil Exploration.* New York: Wiley.

Hartley, H. O., and Ross, A. (1954). Unbiased ratio estimates. *Nature, 174,* 270–271.

Heckman, M. (1960). Flame photometric determination of calcium in animal feeds. *J. Assoc. Offic. Anal. Chem., 43,* 337–340.

Hennekens, C., Drolette, M., Jesse, M., Davies, J., and Hutchison, G. (1976). Coffee drinking and death due to coronary heart disease. *N. Eng. J. Med., 294,* 633–636.

Herson, J. (1976). An investigation of the relative efficiency of least-squares prediction to conventional probability sampling plans. *J. Amer. Stat. Assoc. 71,* 700–703. From the National Center for Health Statistics Hospital Discharge Survey (January 1968).

Hill, R. A., and Barton, R. A. (2005). Red enhances human performance in contests. *Nature, 435,* 293.

Hoaglin, D. (1980). A Poissoness plot. *Amer. Stat., 34,* 146–149.

Hoaglin, D., Mosteller, F., and Tukey, J. (1983). *Understanding Robust and Exploratory Data Analysis.* New York: Wiley.

Hockersmith, T., and Ku, H. (1969). Uncertainties associated with proving ring calibration. In *Precision Measurement and Calibration,* H. Ku (ed). U.S. National Bureau of Standards Special Publication 300, Vol. I (Washington, D.C.).

Hollander, M., and Wolfe, D. (1973). *Nonparametric Statistical Methods.* New York: Wiley.

Hopper, J. H., and Seeman, E. (1994). The bone density of female twins discordant for tobacco use. *N. Eng. J. Med., 330,* 387–392.

Horvitz, D., Shah, B., and Simmons, W. (1967). The unrelated randomized response model. *Proc. Soc. Stat. Sect. Amer. Stat.,* 65–72.

Houck, J. C. (1970). Temperature coefficient of the bismuth I–II transition pressure. *J. Res. Nat. Bur. Stand., 74 A,* 51–54.

Huber, P. (1981). *Robust Statistics.* New York: Wiley.

Huie, R. E., and Herron, J. T. (1972). Rates of reaction of atomic oxygen III, spiropentane, cyclopentane, cyclohexane, and cycloheptane. *J. Res. Nat. Bur. Stand., 74 A,* 77–80.

Johnson, S., and Johnson, R. (1972). Tonsillectomy history in Hodgkin's disease. *N. Eng. J. Med., 287,* 1122–1125.

Joiner, B. (1981). Lurking variables: Some examples. *Amer. Stat., 35,* 227–233.

Kacprzak, J., and Chvojka, R. (1976). Determination of methyl mercury in fish by flameless atomic absorption spectroscopy and comparison with an acid digestion method for total mercury. *J. Assoc. Offic. Anal. Chem., 59,* 153–157.

Kamal, A. A., Eldamaty, S. E., and Faris, R. (1991). Blood lead level of Cairo traffic policemen. *Science of the Total Environment, 105,* 165–170.

Kiefer, M. J., Buchan, R. M., Keefe, T. J., and Blehm, K. D. (1987). A predictive model for determining asbestos concentrations for fibers less than five micrometers in length. *Environmental Research, 43,* 31–38.

Kirchoefer, R. (1979). Semiautomated method for the analysis of chlopheniramine maleate tables: Collaborative study. *J. Assoc. Offic. Anal. Chem., 62,* 1197–1120.

Kiser, C. V., and Schaefer, N. L. (1949). Demographic characteristics of women in Who's Who. *Milbank Memorial Fund Quarterly, 27,* 422.

Kish, L. (1965). *Survey Sampling.* New York: Wiley.

Knafl, G., Spiegelman, C., Sacks, J., and Ylvisaker, D. (1984). Nonparametric calibration. *Technometrics, 26,* 233–241.

Ku, H. (1969). *Precision Measurement and Calibration.* National Bureau of Standards Special Publication 300 (Washington, D.C.).

Ku, H. (1981). Personal communication.

Lagakos, S., and Mosteller, F. (1981). FD&C Red No. 40 experiments. *J. Nat. Canc. Instit., 66,* 197–213.

Lawson, C. L., and Hanson, R. J. (1974). *Solving Least Squares Problems.* Englewood Cliffs, N.J.: Prentice-Hall.

Lazarsfeld, P., Berelson, B., and Gaudet, H. (1948). *The People's Choice: How the Voter Makes Up His Mind in a Presidential Election.* New York: Columbia University Press.

Le Cam, L., and Neyman, J. (eds.) (1967). *Proceedings of the Fifth Berkeley Symposium on Mathematical Statistics and Probability. Volume V: Weather Modification.* Berkeley: University of California Press.

Lehmann, E. (1975). *Nonparametrics: Statistical Methods Based on Ranks.* Oakland, Calif.: Holden-Day.

Lehmann, E., and Casella, G. (1998). *Theory of Point Estimation.* Springer.

Lehmann, E. L. (1983). *Theory of Point Estimation.* New York: Wiley.

Levine, J. D., Gordon, N. C., and Fields, H. L. (1978). The mechanism of placebo analgesia. *Lancet,* 654–657.

Levine, P. H. (1973). An acute effect of cigarette smoking on platelet function. *Circulation, 48,* 619–623.

Lin, S.-L., Sutton, V., and Quarashi, M. (1979). Equivalence of microbiological and hydroxylamine methods of analysis for ampicillin dosage forms. *J. Assoc. Offic. Anal. Chem., 62,* 989–997.

Lynd, R. S., and Lynd, H. M. (1956). *Middletown: A Study in Modern American Culture.* New York: Harcourt-Brace.

MacFarquhar, L. (2004). The pollster. *The New Yorker.* October 18.

Malkiel, B. G. (2004). *A Random Walk Down Wall Street: Completely Revised and Updated Eighth Edition.* W. W. Norton & Company.

Marshall, C. G., Ogden, D. C., and Colquhoun, D. (1990). The actions of suxamethonium (succinyldicholine) as an agonist and channel blocker at the nicotinic receptor of frog muscle. *Journal of Physiology, 428,* 155–174.

Martin, H., Gudzinowicz, B., and Fanger, H. (1975). *Normal Values in Clinical Chemistry.* New York: Marcel-Dekker.

McCool, J. (1979). Analysis of single classification experiments based on censored samples from the two-parameter Weibull distribution. *J. Stat. Planning and Inference, 3,* 39–68.

McNish, A. (1962). The speed of light. *IRE Trans. on Instrumentation, 11,* 138–148.

Mecklenburg, R. S., Benson, E. A., Benson, J. W., Fredlung, P. N., Guinn, T., Metz, R. J., Nielson, R. L., and Sannar, C. A. (1984). Acute complications associated with insulin pump therapy: Report of experience with 161 patients. *J. Amer. Med. Assoc.* 252(23), 3265–3269.

Miller, R. (1981). *Simultaneous Statistical Inference.* New York: Springer-Verlag.

Mittleman, M. A., Maclure, M., Tofler, G. H., et al. (1993). Triggering of acute myocardial infarction by heavy exertion. *N. Eng. J. Med., 329,* 1677–1683.

Morton, A. Q. (1978). *Literary Detection.* New York: Scribner's.

Natrella, M. (1963). *Experimental Statistics.* National Bureau of Standards Handbook 91 (Washington, D.C.).

Olsen, A., Simpson, J., and Eden, J. (1975). A Bayesian analysis of a multiplicative treatment effect in weather modification. *Technometrics, 17,* 161–166.

Orcutt, R. H. (1970). Generation of controlled low pressures of nitrogen by means of dissociation equilibria. *J. Res. Nat. Bur. Stand., 74 A,* 45–49.

Overfield, T., and Klauber, M. R. (1980). Prevalence of tuberculosis in Eskimos having blood group B gene. *Hum. Bio., 52,* 87–92.

Partridge, L., and Marion, M. (1981). Sexual activity and the lifespan of male fruitflies. *Nature, 294,* 580–581.

Pearson, E. S., and Wishart, J. (ed.) (1958). *Student's Collected Works.* Cambridge, England: Cambridge University Press.

Pearson, E., D'Agostino, R., and Bowman, K. (1977). Tests for departure from normality: Comparison of powers. *Biometrika, 64,* 231–246.

Pearson, K., and Hartley, H. (1966). *Biometrika Tables for Statisticians.* Cambridge, England: Cambridge University Press.

Peck, R., Casella, G., Cobb, G., Hoerl, R., Nolan, D., Starbuck, R., and Stern, H. (2005). *Statistics: A Guide to the Unknown.* Duxbury.

Perry, L., Van Dyke, R., and Theye, R. (1974). Sympathoadrenal and hemodynamic effects of isoflurane, halothane, and cyclopropane in dogs. *Anesthesiology, 40,* 465–470.

Phillips, D. P., and King, E. W. (1988). Death takes a holiday: Mortality surrounding major social occasions. *Lancet, 2,* 728–732.

Phillips, D. P., and Smith, D. G. (1990). Postponement of death until symbolically meaningful occasions. *J. Amer. Med. Assoc., 263,* 1947–1961.

Plato, C., Rucknagel, D., and Gerschowitz, H. (1964). Studies of the distribution of glucose-6-phosphate dehydrogenase deficiency, thalassemia, and other genetic traits in the coastal and mountain villages of Cyprus. *Amer. J. Human Genetics, 16,* 267–283.

Preston–Thomas, H., Turnbull, G., Green, E., Dauphinee, T., and Kalra, S. (1960). *Can. Jour. Phys., 38,* 824–852.

Quenouille, M. (1956). Notes on bias in estimation. *Biometrika, 43,* 353–360.

Raftery, A., and Zeh, J. (1993). Estimation of bowhead whale, *Balaena mysticetus,* population size. In *Case Studies in Bayesian Statistics.* C. Gatsonis, J. Hodges, R. Kass, and N. Singpurwalla, (eds.). *Springer Lecture Notes in Statistics, 83,* 163–240.

Ratcliff, J. (1957). New surgery for ailing hearts. *Reader's Digest, 71,* 70–73.

Redelmeier, D. A., and Tibshirani, R. J. (1997). Association between cellular-telephone calls and motor vehicle collisions. *N. Eng. J. Med., 336,* 453–458.

Redelmeier, D. A., and Tibshirani, R. J. (1997). Is using a car phone like driving drunk? *Chance Magazine, 10*(2), 5–9.

Rice, J. R. (1983). *Numerical Methods, Software, and Analysis.* New York: McGraw-Hill.

Robson, G. (1929). Monograph of the recent cephalopoda, part I. London: British Museum.

Rosen, B., and Jerdee, T. (1974). Influence of sex role stereotypes on personnel decisions. *J. Appl. Psych., 59,* 9–14.

Rosner, B. (2006). *Fundamentals of Biostatistics.* Duxbury.

Rudemo, H. (1982). Empirical choice of histograms and kernel density estimators. *Scand. J. Stat., 9,* 65–78.

Ryan, T., and Joiner, B. (unpublished ms.). Normal probability plots and tests for normality. Pennsylvania State Univ.

Ryan, T., Joiner, B., and Ryan, B. (1976). *Minitab Student Handbook.* Boston, Mass.: Duxbury.

Sachs, R. K., van den Engh, G., Trask, B., Yokota, H., and Hearst, J. E. (1995). A random-walk/giant-loop model for interphase chromosome. *Proceedings of the National Academy of Sciences,* USA, *92,* 2710–2714.

Schachter, A. (1959). *The Psychology of Affiliation.* Stanford, Calif.: Stanford University Press.

Scheffe, H. (1959). *The Analysis of Variance.* New York: Wiley.

Scott, D. W. (1992). *Multivariate Density Estimation: Theory and Practice.* Wiley.

Shoemaker, A. L. (1996). What's normal? Temperature, gender, and heart rate. *J. Stat. Edu., 3*(2).

Simiu, E., and Filliben, J. (1975). Statistical analysis of extreme winds. *Nat. Bur. Stand. Tech.* Note No. 868 (Washington, D.C.).

Simpson, J., Olsen, A., and Eden, J. (1975). A Bayesian analysis of a multiplicative treatment effect in weather modification. *Technometrics, 17,* 161–166.

Smith, D. G., Prentice, R., Thompson, D. J., and Hermann, W. L. (1975). Association of exogeneous estrogen and endometrial carcinoma. *N. Eng. J. Med., 293,* 1164–1167.

Snyder, R. G. (1961). The sex ratio of offspring of pilots of high performance military aircraft. *Hum. Biol., 3,* 1–10.

Stanley, W., and Walton, D. (1961). Trifluoperazine ("Stelazine"). A controlled clinical trial in chronic schizophrenia. *J. Mental Sci., 107,* 250–257.

Steel, E., Small, J., Leigh, S., and Filliben, J. (1980). Statistical considerations in the preparation of chrysotile filter standard reference materials. NBS Technical Report (Washington, D.C.).

Steering Committee of the Physicians' Health Study Research Group (1989). Final report on the aspirin component of the ongoing physicians' health study. *N. Eng. J. Med., 32*(3), 129–135.

Stigler, S. M. (1977). Do robust estimates work with real data? *Annals of Statistics, 5,* 1055–1098.

Strang, G. (1980). *Linear Algebra and Its Applications.* New York: Academic Press.

Student (1907). On the error of counting with a haemacytometer. *Biometrika, 5,* 351.

Subotičanec, K., Folnegović-Šmalc, V., Turčin, R., Meštrović, B., and Buzina, R. (1986). Plasma levels and urinary vitamin C excretion in schizophrenic patients. *Hum. Nut.: Clin. Nut., 40C,* 421–428.

Taguma, Y., Kitamoto, Y., Furaki, G., Ueda, H., Monma, H., Ishisaki, M., Takahahsi, H., Sekino, H., and Sasaki, Y. (1985). Effects of catopril on heavy proteinurea in axotemic diabetes. *N. Eng. J. Med., 313*(26), 1617–1620.

Tanur, J., Mosteller, F., Kruskal, W., Link, R., Pieters, R., and Rising, G. (1972). *Statistics: A Guide to the Unknown.* Oakland, Calif.: Holden-Day.

Thomas, H. A. (1948). Frequency of minor floods. *J. Boston Soc. Civil Engineers, 34,* 425–442.

Tukey, J. (1977). *Exploratory Data Analysis.* Reading, Mass.: Addison-Wesley.

Tversky, A., and Kahneman, D. (1974). Judgement under uncertainty: Heuristics and biases in judgements reveal some heuristics of thinking under uncertainty. *Science, 185,* 1124–1131.

Udias, A., and Rice, J. (1975). Statistical analysis of microearthquake activity near San Andreas Geophysical Observatory, Hollister, California. *Bulletin of the Seismological Society of America, 65,* 809–828.

Van Atta, C., and Chen, W. (1968). Correlation measurements in grid turbulence using digital harmonic analysis. *J. Fluid Mech., 34,* 497–515.

Veitch, J., and Wilks, A. (1985). A characterization of Arctic undersea noise. *J. Acoust. Soc. Amer., 77,* 989–999.

Velleman, P., and Hoaglin, D. (1981). *Applications, Basics, and Computing of Exploratory Data Analysis.* Boston, Mass.: Duxbury.

Vianna, N., Greenwald, P., and Davies, J. (1971). Tonsillectomy and Hodgkin's disease: The lymphoid tissue barrier. *Lancet, 1,* 431–432.

Warner, S. (1965). Randomized response: A survey technique for eliminating evasive answer bias. *J. Amer. Stat. Assoc., 60,* 63–69.

Weindling, S. (1977). Statistics report: Math 80B.

Weindruch et al from Sleuth.

Weisburg, S. (1980). *Applied Linear Regression.* New York: Wiley.

Welch, W. J. (1987). Rerandomizing the median in matched-pairs designs. *Biometrika, 74,* 609–614.

White, J., Riethof, M., and Kushnir, I. (1960). Estimation of microcrystalline wax in beeswax. *J. Assoc. Offic. Anal. Chem., 43,* 781–790.

Wilk, M., and Gnanadesikan, R. (1968). Probability plotting methods for the analysis of data. *Biometrika, 55,* 1–17.

Williams, W. (1978). How bad can "good" data really be? *Amer. Stat., 32,* 61–67.

Wilson, E. B. (1952). *An Introduction to Scientific Research.* New York: McGraw-Hill.

Wood, L. A. (1972). Modulus of natural rubber crosslinked by dicumyl peroxide. I. Experimental observations. *J. Res. Nat. Bur. Stand., 76A,* 51–59.

Woodward, P. (1948). A statistical theory of cascade multiplication. *Proc. Camb. Phil. Soc., 44,* 404–412.

Woolf, B. (1955). On estimating the relation between blood group and disease. *Annals of Hum. Genetics, 19,* 251–253.

Yates, F. (1960). *Sampling Methods for Censuses and Surveys.* New York: Hafner.

Yip, P., Chao, A., and Chiu, C. (2000). Seasonal variation in suicides: Diminished or vanished. Experience from England and Wales, 1982–1996. *The British Journal of Psychiatry, 177,* 366–369.

Yokota, H., van den Engh, G., Hearst, J. E., Sachs, R. K., and Trask, B. (1995). Evidence for the organization of chromatin in megabase pair-sized loops arranged along a random walk path in the human G0/G1 interphase nucleus. *J. Cell Biol., 130,* 1239–1249.

Youden, J. (1972). Enduring values. *Technometrics, 14,* 1–11.

Youden, W. J. (1962). *Experimentation and Measurement*. Washington, D.C.: National Science Teachers Association.

Youden, W. J., (1974). *Risk, Choice and Prediction*. Boston, Mass.: Duxbury.

$= \#\,of\,ways$

$13 \cdot \binom{4}{1} = 13 \cdot 6 \cdot \# \binom{4}{3} = 13 \cdot 6 \cdot \#\circ 4$

Answers to Selected Problems

Following are answers to those odd-numbered problems for which a short answer can be given. No proofs, graphs, or extensive data analysis are given.

Chapter 1

1. a. $\Omega = \{hhh,\ hht,\ htt,\ hth,\ ttt,\ tth,\ thh,\ tht\}$
   b. $A = \{hhh,\ hht,\ hth,\ thh\}$
      $B = \{hht,\ hhh\}$
      $C = \{hht,\ htt,\ ttt,\ tht\}$
   c. $A^c = \{htt,\ ttt,\ tth,\ tht\}$
      $A \cap B = \{hht,\ hhh\}$
      $A \cup C = \{hhh,\ hht,\ hth,\ thh,\ htt,\ ttt,\ tht\}$

3. $\Omega = \{rrr,\ rrg,\ rrw,\ rwg,\ rgw,\ rgr,\ rwr,\ rgg,\ ggr,\ ggw,\ grr,\ grw,$
   $gwr,\ grg,\ gwg,\ wrr,\ wgg,\ wrg,\ wgr\}$

5. $\Omega = (A \cap B)^c \cap (A \cup B)$ **9.** Not 50%   **11.** $7 \times 6 \times 5 \times 4/10^4$

13. a. $10(4^5 - 4)/\binom{52}{5}$ **b.** $13 \times 48/\binom{52}{5}$ **c.** $13 \times 12 \times 4 \times 6/\binom{52}{5}$

15. 72   **19. a.** $5 \times 3 \times 2 \times 2/\binom{12}{4}$ **b.** $240/\binom{12}{5}$

21. $\frac{8}{32}$ **23.** $n(n-1)$ **25.** 6

27. $26 \times 25 \times 24 \times 23 \times 22/26^5$ **29.** $\binom{10}{2}/\binom{47}{2}$ **31.** $6^2 \times 5^2 \times 4^2 \times 3^2 \times 2^2$

33. $7 \times 6 \times 5 \times 4 \times 3/7^5$ **37.** 210

39. a. $21!/26!$ **b.** 1.818×10^7

41. a. $[\binom{7}{2} + \binom{8}{2} + \binom{9}{2}]/\binom{24}{2}$ **b.** $\binom{7}{2}/\binom{24}{2}$

43. $\binom{10}{3\,3\,4}$ **47. a.** $11/45$ **b.** $6/11$

49. a. $4/7$ **b.** $3/7$ **51.** $2/5$

53. 0.35

55. a. .48, .70   **b.** .064, .614, .322

57. 2/3

59. a. 2/3   **b.** 5/6

61. .86

63. 1/3   **69.** Yes

73. $\sum_{j=k}^{n} \binom{n}{j} p^j (1-p)^{n-j}$ **75.** $p^3 - 2p^2 + 1$; .597   **77.** 14

79. a. $P(aa) = 1/4$, $P(Aa) = 1/2$, $P(AA) = 1/4$
   b. 2/3
   c. $P(aa) = p/6$, $P(Aa) = 1/3 + p/6$, $P(AA) = 2/3 - p/3$
   d. $p_c = [(1 - p/4)(2/3)]/(1 - p/6)$

Chapter 2

3. $p(1) = .1$, $p(2) = .2$, $p(3) = .4$, $p(4) = .1$, $p(5) = .2$

7. $F(x) = \begin{cases} 0 & x < 0 \\ 1 - p & 0 \le x < 1 \\ 1 & x \ge 1 \end{cases}$ **9.** $p < .5$ **11.** $[(n+1)p]$

13. a. .0130   **b.** .2517   **15.** 3 of 5

17. $P(X = k) = p(1 - p)^k, k = 0, 1, \ldots$ **19.** $F(n) = 1 - (1 - p)^n$

23. $\binom{k+r-1}{r} p^r (1-p)^k$ **25. a.** .9987   **b.** 9×10^{-7}

27. $p(k) = 100^k e^{-100}/k!$, approximately   **29.** $P(X \le 4) = .532104$

31. a. .28   **b.** 20.79 min   **33.** $f(x) = \alpha\beta x^{\beta-1} \exp(-\alpha x^\beta)$ **37.** 2/3

39. b. $f(x) = [\pi(1 + x^2)]^{-1}$, $-\infty < x < \infty$ **c.** 3.08

41. $-\log(1/4)/\lambda$, $-\log(3/4)/\lambda$ **43.** $f(x) = 4\lambda\pi x^2 \exp(-4\lambda\pi x^3/3)$

45. a. $1 - e^{-1}$ **b.** $e^{-.5} - e^{-1.5}$ **c.** 46.1

53. a. 0.3085   **b.** 0.8351   **c.** 21.5

55. $c = 1.96\sigma$ **59.** $f(x) = x^{-1/2}/2$

61. $(\lambda/c)^\alpha t^{\alpha-1} \exp(-\lambda t/c)/\Gamma(\alpha)$ **63.** $[\pi(1 + x^2)]^{-1}$

65. $X = [-1 + 2\sqrt{1/4 - \alpha(1/2 - \alpha/4 - U)}]/\alpha$, where U is uniform

67. a. $f(x) = (\beta/\alpha^\beta)x^{\beta-1} \exp(-(x/\alpha)^\beta)$

69. $f(x) = (\lambda/3)(3/4\pi)^{1/3} x^{-2/3} \exp(-\lambda(3x/4\pi)^{1/3})$

Chapter 3

1. a. $p_1 = .19$, $p_2 = .32$, $p_3 = .31$, $p_4 = .18$, for both X and Y
   b. $p(1|1) = .526$, $p(2|1) = .263$, $p(3|1) = .105$, $p(4|1) = .105$, for both
   X and Y

3. Multinomial, $n = 10$, $p_1 = p_2 = p_3 = 1/3$

7. $f_{XY}(x, y) = \alpha\beta \exp[-\alpha x - \beta y]$; $f_x(x) = \alpha \exp[-\alpha x]$, $f_Y(y) = \beta \exp[-\beta y]$

9. a. $f_X(x) = 3(1 - x^2)/4$, $-1 \le x \le 1$, $f_Y(y) = 3\sqrt{1 - y}/2$, $0 \le y \le 1$
   b. $f_{X|Y}(x|y) = 1/(2\sqrt{1 - y})$, $f_{Y|X}(y|x) = 1/(1 - x^2)$

11. $5/36 + \log(2)/6$

13. $p(0) = 1/2$, $p(1) = p(2) = 1/4$

15. a. $c = 3/2\pi$ **c.** $\dfrac{2\sqrt{2} - 1}{2\sqrt{2}}$

   d. $f_Y(y) = \frac{3}{4}(1 - y^2)$, $-1 \le y \le 1$
       $f_X(x) = \frac{3}{4}(1 - x^2)$, $-1 \le x \le 1$
       X and Y are not independent.

   e. $f_{Y|X}(y|x) = \dfrac{\sqrt{1 - x^2 - y^2}}{\pi(1 - x^2)}$

      $f_{X|Y}(x|y) = \dfrac{\sqrt{1 - x^2 - y^2}}{\pi(1 - y^2)}$

17. b. $f_X(x) = 1 - |x|$, $-1 \le x \le 1$; $f_Y(y) = 1 - |y|$, $-1 \le y \le 1$
   c. $f_{X|Y}(x|y) = 1/(2 - 2|y|)$, $1 - |y| \le x \le 1 + |y|$
      $f_{Y|X}(y|x) = 1/(2 - 2|x|)$, $1 - |x| \le y \le 1 + |x|$

19. a. $\beta/(\alpha + \beta)$ **b.** $\beta/(2\alpha + \beta)$

23. Binomial (m, pr)

29. $h(x, y) = \lambda\mu e^{-\lambda x} e^{-\mu y}[1 + \alpha(1 - 2e^{-\lambda x})(1 - 2e^{-\mu y})]$

33. a. $f_{\Theta|N}(\theta|n) = n(n + 1)\theta(1 - \theta)^{n-1}$

43. $f_S(s) = s$ for $0 \le s \le 1$ and $= 2 - s$ for $1 \le s \le 2$

49. $\lambda e^{-\lambda S/2} - \lambda e^{-\lambda S}$ **53.** $5/9$

55. $f_{XY}(x, y) = \frac{1}{2\pi}(x^2 + y^2)^{-1/2}$, $x^2 + y^2 \le 1$

57. $x_1 = y_1$; $x_2 = -y_1 + y_2$

61. $f_{UV}(u, v) = \dfrac{1}{bd} f_{XY}\left(\dfrac{u - a}{b}, \dfrac{v - c}{d}\right)$

63. a. $f_{UV}(u, v) = \dfrac{1}{2} f_{XY}\left(\dfrac{u + v}{2}, \dfrac{u - v}{2}\right)$ where $U = X + Y$, $V = X - Y$

   b. $f_{UV}(u, v) = \dfrac{1}{2|v|} f_{XY}((uv)^{1/2}, (u/v)^{1/2})$ where $U = XY$, $V = X/Y$

67. $f(t) = n(n - 1)\lambda[\exp(-(n - 1)\lambda t) - \exp(-n\lambda t)]$

69. $n\beta v^{\beta-1}\alpha^{-\beta}\exp(-n(v/\alpha)^{\beta})$ **71.** $1-\gamma^n$

75. Let $U = X_{(i)}$, $V = X_{(j)}$

$$f_{UV}(u, v) = \frac{n!}{(i-1)!(j-i-1)!(n-j)!}$$
$$\times [F(u)]^{i-1} f(u)[F(v) - F(u)]^{j-i-1} f(v)[1 - F(v)]^{n-j}$$

77. $n(1-x)^{n-1}$ **79.** Exponential (λ)

81. a. $n/(n+1)$ **b.** $(n-1)/(n+1)$

Chapter 4

3. $E(X) = 3.1$; $\text{Var}(X) = 1.49$

5. $E(X) = \alpha/3$; $\text{Var}(X) = 1/3 - \alpha^2/9$

7. a. $E(X) = 5/8$
b. $p_Y(0) = 1/2$, $p_Y(1) = 3/8$, $p_Y(4) = 1/8$, $E(Y) = 7/8$
c. $E(X^2) = 7/8$ **d.** $\text{Var}(X) = 31/64$

9. That value of n such that $s \sum_{k=n}^{\infty} p(k) > c \sum_{k=1}^{n-1} p(k)$ and $s \sum_{k=n+1}^{\infty} p(k) < c \sum_{k=1}^{n} p(k)$

15. It makes no difference.

17. a. $E(X_{(k)}) = k/(n+1)$
b. $\text{Var}(X_{(k)}) = k(n-k+1)/[(n+1)^2(n+2)]$

19. $1/(n+1)$ **21.** $1/3$ **23.** $2/\lambda^2$ (square), $1/\lambda^2$ (rectangle)

25. $2\alpha(\alpha+1)/\lambda^2$ **27.** 1 **31.** no

35. r/p **37.** $p > (1/k)^{1/k}$

39. a. 4606 **b.** $10,000$

41. The expected number of occurrences is 4.62. Using Markov's inequality, the chance of 100 or more occurrences is less than 0.0462, so you should be surprised.

45. $\text{Cov}(N_i, N_j) = -np_i p_j$

47. $\text{Cov}(X, Z) = -\sigma_X^2$; $\text{Corr}(X, Z) = -\dfrac{\sigma_X}{(\sigma_X^2 + \sigma_Y^2)^{1/2}}$

49. b. $\alpha = \sigma_Y^2/(\sigma_Y^2 + \sigma_X^2)$
c. $(X+Y)/2$ is better when $1/3 < \sigma_X^2/\sigma_Y^2 < 3$.

51. $\pi_i = n^{-1}$ for the optimal portfolio. If each individual return has standard deviation σ, the standard deviation of the return from this portfolio is $\sigma/\sqrt{n}$. If the entire investment is in one security, the standard deviation of the return is σ.

55. $E(T) = n(n+1)\mu/2$; $\text{Var}(T) = n(n+1)(2n+1)\sigma^2/6$

57. $\sigma_X^2\sigma_Y^2 + \mu_X^2\sigma_Y^2 + \mu_Y^2\sigma_X^2$

61. a. $\text{Cov}(x, Y) = 1/36$; $\text{Corr}(X, Y) = 1/2$
  b. $E(X|Y) = Y/2$, $E(Y|X) = (X + 1)/2$
  c. If $Z = E(X|Y)$, the density of Z is $f_Z(z) = 8z$, $0 \le z \le 1/2$
     If $Z = E(Y|X)$, the density of Z is $f_Z(z) = 8(1 - z)$, $1/2 \le z \le 1$
  d. $\hat{Y} = \frac{1}{2} + \frac{1}{2}X$; the mean squared prediction error is $1/24$
  e. $\hat{Y} = \frac{1}{2} + \frac{1}{2}X$; the mean squared prediction error is $1/24$

63. a. $\text{Cov}(X, Y) = -.0085$; $\rho_{XY} = -.1256$
  b. $E(Y|X) = (6X^2 + 8X + 3)/[4(3X^2 + 3X + 1)]$

65. In the claim that $E(T|N = n) = nE(X)$ **67.** $3/2$, $1/6$

71. $p_{Y|X}(y|x)$ is hypergeometric. $E(Y|X = x) = mx/n$

73. $np(1 + p)$ **75. a.** $1/2\lambda$;   **b.** $5/12\lambda^2$

77. $E(X|Y) = Y/2$, $E(Y|X) = X + 1$

79. $M(t) = \frac{1}{2} + \frac{3}{8}e^t + \frac{1}{8}e^{2t}$ **81.** $M(t) = 1 - p + pe^t$

85. $M(t) = e^t p/[1 - (1 - p)e^t]$; $E(X) = 1/p$; $\text{Var}(X) = (1 - p)/p^2$

87. Same p **93.** Exponential

99. b. $E[g(X)] \approx \log \mu - \sigma^2/2\mu^2$; $\text{Var}[g(X)] \approx \sigma^2/\mu^2$

101. $E(Y) \approx \sqrt{\lambda} - 1/(8\sqrt{\lambda})$; $\text{Var}(Y) \approx 1/4$ **103.** .0628 mm

Chapter 5

3. .0228   **13.** $N(0, 150{,}000)$; most likely to be where he started

15. $p = .017$ **17.** $n = 96$

21. b. $\text{Var}(\hat{I}(f)) = \frac{1}{n}\left[\int_a^b \frac{f^2(x)}{g(x)}dx - I^2(f)\right]$

29. Let $Z_n = n(U_{(n)} - 1)$. Then $P(Z_n \le z) \to e^z$, $-1 \le z \le 0$

Chapter 6

3. $c = .17$ **9.** $E(S^2) = \sigma^2$; $\text{Var}(S^2) = 2\sigma^4/(n - 1)$

Chapter 7

1. $p(1.5) = 1/5$, $p(2) = 1/10$, $p(2.5) = 1/10$, $p(3) = 1/5$, $p(4.5) = 1/10$,
  $p(5) = 1/5$, $p(6) = 1/10$; $E(\overline{X}) = 17/5$; $\text{Var}(\overline{X}) = 2.34$

3. d, f, h   **7.** $n = 319$, ignoring the fpc

9. $\text{SE} = .026$. CI: $(.05, .15)$ **11. a.** 6 samples.   **b.** Yes

15. b.

n	Δ_1	Δ_2
20	211.6	86.8
40	145.6	59.7
80	96.9	39.8

17. no

19. 1.28, 1.645

21. The sample size should be multiplied by 4.

29. a. $\hat{Q} = \dfrac{R - t(1 - p)}{p}$, where t = probability of answering yes to unrelated question

c. $\mathrm{Var}(\hat{Q}) = r(1 - r)/(np^2)$, where $r = P(\text{yes}) = qp + t(1 - p)$

31. $n = 395$

33. The sample size for each survey should be 1250.

35. a. $\overline{X} = 98.04$

b. $s^2 \dfrac{N - 1}{N} = 133.64, \dfrac{s^2}{n}\left(1 - \dfrac{n}{N}\right) = 5.28$

c. 98.04 ± 4.50 and $196{,}080 \pm 9008$

37. a. $\alpha + \beta = 1$

b. $\alpha = \dfrac{\sigma^2_{\overline{X}_2}}{\sigma^2_{\overline{X}_1} + \sigma^2_{\overline{X}_2}} \qquad \beta = \dfrac{\sigma^2_{\overline{X}_1}}{\sigma^2_{\overline{X}_1} + \sigma^2_{\overline{X}_2}}$

39. Choose n such that $p = 1 - \dfrac{(N - k)(N - k - 1)\cdots(N - n + k - 1)}{N(N - 1)\cdots(N - k + 1)}$, which can be done by a recursive computation; $n = 581$

41. b. $\dfrac{N^2}{n}\left(\sigma^2_A + \sigma^2_B - 2\rho\sigma_A\sigma_B\right)$

c. The proposed method has smaller variance if $\rho > \dfrac{\sigma^2_B}{2\sigma_A\sigma_B}$.

d. The ratio estimate is biased. The approximate variance of the ratio estimate is greater if $\dfrac{\mu_A}{\mu_B} > 1$.

43. $R = \dfrac{\overline{V}}{\overline{O}} = .73, s_R = .02, .73 \pm .04$

47. The bias is .96 for $n = 64$ and .39 for $n = 128$.

49. Ignoring the fpc,

a. $R = 31.25;$

b. $s_R = .835; 31.25 \pm 1.637;$

c. $T = 10^7; 10^7 \pm 5, 228, 153;$

d. $s_{T_R} = 266{,}400$, which is much better.

53. a. For optimal allocation, the sample sizes are 10, 18, 17, 19, 12, 9, 15. For proportional allocation they are 20, 23, 19, 17, 8, 6, 7.

b. $\mathrm{Var}(\overline{X}_{SO}) = 2.90, \mathrm{Var}(\overline{X}_{sp}) = 3.4, \mathrm{Var}(\overline{X}_{srs}) = 6.2$

55. a. $\frac{1}{6}\overline{X}_H + \frac{5}{6}\overline{X}_L$
   b. 0.68
   c. No, the standard error would be 0.87.
   d. No, the standard error would be 0.71.

57. $p(2.2) = 1/6, \ p(2.8) = 1/3, \ p(3.8) = 1/6, \ p(4.4) = 1/3; \ E(\overline{X}_s) = 3.4;$
   $\text{Var}(\overline{X}_s) = .72$

61. a. $w_1 + w_2 + w_3 = 0$ and $w_1 + 2w_2 + 3w_3 = 1$
   b. $w_1 = -1/2, \ w_2 = 0, \ w_3 = 1/2$

Chapter 8

3. For concentration (1),
   a. $\hat{\lambda} = .6825;$ **b.** $.6825 \pm .081;$
   c. There are not gross differences between observed and expected counts.

5. a. $\hat{\theta} = 1/3$ **b.** $\text{Lik}(\theta) = \theta(1 - \theta)^2$
   c. $\hat{\theta} = 1/3$ **d.** $\beta(2, 3)$

7. a. $\hat{p} = 1/\overline{X}$ **b.** $\tilde{p} = 1/\overline{X}$
   c. $\text{Var}(\tilde{p}) \approx p^2(1 - p)/n$
   d. The posterior distribution is $\beta(2, k)$; the posterior mean is $2/(k + 2)$.

13. $P(|\hat{\alpha}| > .5) \approx .1489$

17. b. $\hat{\alpha} = n(8\Sigma_{i=1}^{n} X_i^2 - 2n)^{-1} - 1/2$

   c. $\dfrac{\Gamma'(2\alpha)}{\Gamma(2\alpha)} - \dfrac{\Gamma'(\alpha)}{\Gamma(\alpha)} + \dfrac{1}{2n} \sum_{i=1}^{n} \log[X_i(1 - X_i)] = 0$

   d. $\left(2n \left[\dfrac{\Gamma''(\alpha)\Gamma(\alpha) - \Gamma'(\alpha)^2}{\Gamma(\alpha)^2} - \dfrac{2\Gamma''(2\alpha)\Gamma(2\alpha) - \Gamma'(2\alpha)^2}{\Gamma(2\alpha)^2} \right] \right)^{-1}$

19. a. $\hat{\sigma} = \sqrt{n^{-1}\Sigma_{i=1}^{n}(X_i - \mu)^2}$ **b.** $\hat{\mu} = \overline{X}$ **c.** no

21. a. $\overline{X} - 1$ **b.** $\min(X_1, X_2, \ldots, X_n)$ **c.** $\min(X_1, X_2, \ldots, X_n)$

23. Method of moments estimate is 1775. MLE is 888.

27. Let T be the time of the first failure.

   a. $\dfrac{5}{\tau} \exp\left(-\dfrac{5t}{\tau}\right)$ **b.** $\hat{\tau} = 5T$

   c. $\hat{\tau} \sim \exp\left(\dfrac{1}{\tau}\right)$ **d.** $\sigma_{\hat{\tau}} = \tau$

31. a. $\theta(1 - \theta)^6$ **b.** $\hat{\theta} = 1/7$

33. Let q be the .95 quantile of the t distribution with $n - 1$ df; $c = qs_{\overline{X}}$.

41. For α the relative efficiency is approximately .444; for λ it is approximately .823.

47. a. $\hat{\theta} = \overline{X}/(\overline{X} - x_0)$
   b. $\tilde{\theta} = n/(\Sigma \log X_i - n \log x_0)$
   c. $\text{Var}(\tilde{\theta}) \approx \theta^2/n$

49. a. Let $\hat{p}$ be the proportion of the n events that go forward. Then $\hat{\alpha} = 4\hat{p} - 2$.
   b. $\text{Var}(\hat{\alpha}) = (2 - \alpha)(2 + \alpha)/n$

53. a. $\hat{\theta} = 2\overline{X}$; $E(\hat{\theta}) = \theta$; $\text{Var}(\hat{\theta}) = \theta^2/3n$
   b. $\tilde{\theta} = \max(X_1, X_2, \ldots, X_n)$
   c. $E(\tilde{\theta}) = n\theta/(n + 1)$; bias $= -\theta/(n + 1)$; $\text{Var}(\hat{\theta}) = n\theta^2/(n + 2)(n + 1)^2$;
   $\text{MSE} = 2\theta^2/(n + 1)(n + 2)$
   d. $\theta^* = (n + 1)\tilde{\theta}/n$

55. a. Let n_1, n_2, n_3, n_4 denote the counts. The mle of θ is the positive root of the equation

$$(n_1 + n_2 + n_3 + n_4)\theta^2 - (n_1 - 2n_2 - 2n_3 - n_4)\theta - 2n_4 = 0$$

The asymptotic variance is $\text{Var}(\hat{\theta}) = 2(2 + \theta)(1 - \theta)\theta/(n_1 + n_2 + n_3 + n_4)$
$(1 + \theta)$. For these data, $\hat{\theta} = .0357$ and $s_{\hat{\theta}} = .0057$.
   b. An approximate 95% confidence interval is $.0357 \pm .0112$.

57. a. s^2 is unbiased.   **b.** $\hat{\sigma}^2$ has smaller MSE.   **c.** $\rho = 1/(n + 1)$

59. b. $\hat{\alpha} = (n_1 + n_2 - n_3)/(n_1 + n_2 + n_3)$ if this quantity is positive and 0 otherwise.

63. In case (1) the posterior is $\beta(4, 98)$ and the posterior mean is 0.039. In case (2) the posterior is $\beta(3.5, 102)$ and the posterior mean is 0.033. The posterior for case (2) rises more steeply and falls off more rapidly than that of case (1).

65. $\mu_0 = 16.25$, $\xi_0 = 80$ **71.** $\displaystyle\prod_{i=1}^{n} (1 + X_i)$

73. $\displaystyle\sum_{i=1}^{n} X_i^2$

Chapter 9

1. a. $\alpha = .002$ **b.** power $= .349$

3. a. $\alpha = .046$ **5.** F, F, F, F, F, F, F, T

7. Reject when $\sum X_i > c$. Since under H_0, $\sum X_i$ follows a Poisson distribution with parameter $n\lambda$, c can be chosen so that $P(\sum X_i > c | H_0) = \alpha$.

9. For $\alpha = .10$, the test rejects for $\overline{X} > 2.56$, and the power is .2981. For $\alpha = .01$, the test rejects for $\overline{X} > 4.66$, and the power is .0571.

17. a. $LR = \frac{\sigma_1}{\sigma_0} \exp\left[\frac{1}{2}x^2\left(\frac{1}{\sigma_1^2} - \frac{1}{\sigma_0^2}\right)\right]$. A level α test rejects for $X^2 > \sigma_0^2 \chi_1^2(\alpha)$.

   b. Reject for $\sum_{i=1}^{n} X_i^2 > \sigma_0^2 \chi_n^2(\alpha)$ **c.** Yes

19. a. $X < 2/3$ **b.** Reject for large values of X
   c. Reject for $X > \sqrt{1 - \alpha}$ **d.** $1 - (1 - \alpha)^{3/2}$

21. a. Reject for $X > 1$; power $= 1/2$
  b. Significance level $= \alpha$, power $= \alpha/2$
  c. Significance level $= \alpha$, power $= \alpha/2$
  d. Reject when $(1 - \alpha)/2 \leq X \leq (1 + \alpha)/2$
  e. For $\alpha > 0$, the rejection region is not uniquely determined.
  f. The rejection region is not uniquely determined.

23. yes   **25.** $-2 \log \Lambda = 54.6$. Strongly rejects   **27.** ≥ 12.02

29. yes   **31.** 2.6×10^{-1}, 9.8×10^{-3}, 3×10^{-4}, 7×10^{-7}

33. $-2 \log \Lambda$ and X^2 are both approximately 2.93. $.05 < p < .10$; not significant for Chinese and Japanese; both $\approx .3$.

35. $X^2 = .0067$ with 1 df and $p \approx .90$. The model fits well.

37. $X^2 = 79$ with 11 df and $p \approx 0$. The accidents are not uniformly distributed, apparently varying seasonally with the greatest number in November–January and the fewest in March–June. There is also an increased incidence in the summer months, July–August.

39. $\chi^2 = 85.5$ with 9 df, and thus provides overwhelming evidence against the null hypothesis of constant rate.

41. Let $\hat{p}_i = X_i/n_i$ and $\hat{p} = \sum X_i / \sum n_i$. Then

$$\Lambda = \frac{\hat{p}^{\Sigma n_i \hat{p}_i} (1 - \hat{p})^{\Sigma n_i (1 - \hat{p}_i)}}{\prod \hat{p}_i^{n_i \hat{p}_i} (1 - \hat{p}_i)^{n_i (1 - \hat{p}_i)}}$$

and

$$-2 \log \Lambda \approx \sum \frac{(X_i - n_i \hat{p})^2}{n_i \hat{p}(1 - \hat{p})}$$

is approximately distributed as χ^2_{m-1} under H_0.

43. a. 9207 heads out of 17950 tosses is not consistent with the null hypothesis of 17950 independent Bernoulli trials with probability .5 of heads. ($X^2 = 11.99$ with 1 df).
  b. The data are not consistent with the model ($X^2 = 21.57$ with 5 df, $p \approx .001$).
  c. A chi-square test gives $X^2 = 8.74$ with 4 df and $p \approx .07$. Again, the model looks doubtful.

45. The binomial model does not fit the data ($X^2 = 110.5$ with 11 df). Relative to the binomial model, there are too many families with very small and very large numbers of boys. The model might fail because the probability of a male child differs from family to family.

51. The horizontal bands are due to identical data values.

57. The tails decrease less rapidly than do those of a normal probability distribution, causing the normal probability plot to deviate from a straight line at the ends by curving below the line on the left and above the line on the right.

59. The rootogram shows no systematic deviation.

Chapter 10

3. $q_{.25} \approx 63.4$; $q_{.5} \approx 63.6$; $q_{.75} \approx 63.8$

7. Differences are about 50 days for the weakest, 150 days for the median. Can't tell for the strongest.

9. Bias $\approx -\dfrac{1}{2n} \dfrac{F(x)}{1 - F(x)}$, which is large for large x.

11. $h(t) = \alpha \beta t^{\beta-1}$

13. The uniform distribution on $[0, 1]$ is an example.

15. $h(t) = (24 - t)^{-1}$. It increases from 0 to 24. It is more likely after 5 hours.

23. $(n + 1)\left(\dfrac{k + 1}{n + 1} - p\right) X_{(k)} + (n + 1)\left(p - \dfrac{k}{n + 1}\right) X_{(k+1)}$

29. b. $\approx .018$ **c.** ≈ 18 **d.** $\approx 2.4 \times 10^{-19}$

31. a. n^n

    b.

x	1/3	5/3	2	7/3	8/3	3	10/3	11/3
$p(x)$	1/27	3/27	3/27	3/27	8/27	3/27	3/27	3/27

33. The mean and standard deviation

37. Median $= 14.57$, $\bar{x} = 14.58$, $\bar{x}_{.10} = 14.59$, $\bar{x}_{.20} = 14.59$; $s = .78$, $\text{IQR}/1.35 = .74$, $\text{MAD}/.65 = .82$

41. The interval $(X_{(r)}, X_{(s)})$ covers x_p with probability $\sum_{i=r}^{s-1} \binom{n}{i} p^i (1 - p)^{n-i}$.

Chapter 11

7. Throughout. For example, all are used in the assertion that $\text{Var}(\overline{X} - \overline{Y}) = \sigma^2(n^{-1} + m^{-1})$. All are used in Theorem A and Corollary A. Independence is used in the expression for the likelihood.

11. Use the test statistic

$$t = \frac{(\overline{X} - \overline{Y}) - \Delta}{s_p\sqrt{\dfrac{1}{n} + \dfrac{1}{m}}}$$

13. The power of the sign test is .35, and the power of the normal theory test is .46.

15. $n = 768$

19. a. $\sqrt{2}$ **b.** $\overline{Y} - \overline{X} > 2.33$ **c.** 0.17

    d. Yes   **e.** $\overline{Y} - \overline{X} > 2.78$; power $= 0.11$

21. a. A pooled t test gives a p-value of .053.

    b. The p-value from the Mann-Whitney test is .064.

    c. The sample sizes are small and normal probability plots suggest skewness, so the Mann-Whitney test is more appropriate.

25. a. No   **b.** No   **c.** Yes, yes

27.

w	0	1	2	3	4	5	6	7	8	9	10
$p(w)$	.0625	.0625	.0625	.125	.125	.125	.125	.125	.0625	.0625	.0625

31. $E\hat{\pi} = 1/2$; $\text{Var}(\hat{\pi}) = \dfrac{1}{12}\dfrac{m+n+1}{mn}$, which is smallest when $m = n$.

33. Let $\theta = \sigma_X^2/\sigma_Y^2$ and $\hat{\theta} = s_X^2/s_Y^2$.

    a. For H_1: $\theta > 1$, reject if $\hat{\theta} > F_{n-1,m-1}(\alpha)$. For H_2: $\theta \neq 1$, reject if $\hat{\theta} > F_{n-1,m-1}(\alpha/2)$ or $\hat{\theta} < F_{n-1,m-1}(1 - \alpha/2)$.

    b. A $100(1 - \alpha)\%$ confidence interval for θ is

$$\left[\frac{\hat{\theta}}{F_{n-1,m-1}(\alpha/2)}, \frac{\hat{\theta}}{F_{n-1,m-1}(1 - \alpha/2)} \right].$$

    c. $\hat{\theta} = .60$. The p-value for a two-sided test is .42. A 95% confidence interval for θ is $(.13, 2.16)$.

37. a. For each patient, compute a difference score (after − before), and compare the difference scores of the treatment and control by a signed rank test or a paired t test. A signed rank test gives for Ward A $W_+ = 36$, $p = .124$ and for Ward B $W_+ = 22$, $p = .205$.

    b. To compare the two wards, use a two-sample t test or a Mann-Whitney test on difference scores. Using a Mann-Whitney test, there is strong evidence that the stelazine group in Ward A improved more than the stelazine group in Ward B $(p = .02)$ and weaker evidence that the placebo group improved more in Ward A than in Ward B $(p = .09)$.

45. a. For example, for 1957 by a Wilcoxon signed rank test there is no evidence that seeding is effective $(p = .73)$. For this and other years, it appears that the gain in seeding over not seeding may be greatest when rainfall in the unseeded area is low.

    b. Randomization guards against possibly confounding the effect of seeding with cyclical weather patterns. Pairing is effective if rainfall on successive days is positively correlated; in these data, the correlation is weak.

47. a. To test for an effect of seeding, compare the differences (target − control) to each other by a two-sample t test or a Mann-Whitney test. A Mann-Whitney test gives a p-value of .73.

    b. The square root transformation makes the distribution of the data less skewed.

    c. Using a control area is effective if the correlation between the target and control areas is large enough that the standard deviation of the difference (target − control) is smaller than the standard deviation of the target rainfalls. This was indeed the case.

49. 95% CI: (8.9, 13.1). Null hypothesis is overwhelmingly rejected.

51. The durations of the bottle-fed are typically much longer. Because the distribution is very skewed with some large outliers, a nonparametric test is preferable. The p-value from a signed-rank test is 0.012.

53. The lettuce leaf cigarettes were controls to ensure that the effects of the experiment were due to tobacco specifically, not just due to smoking a lit cigarette. The unlit cigarettes were controls to ensure that the effects were due to lit tobacco, not just unlit tobacco.

Chapter 12

11.

	A	B	C	D
I	2	3	4	5
II	3	4	5	6
III	4	5	6	7

17. $\hat{\alpha}_i = \overline{Y}_{i..} - \overline{Y}...$
$\hat{\beta}_j = \overline{Y}_{.j.} - \overline{Y}...$
$\hat{\delta}_{ij} = \overline{Y}_{ij.} - \overline{Y}_{i..} - \overline{Y}_{.j.} + \overline{Y}...$
$\hat{\mu} = \overline{Y}...$

19. $Y_{ijkl} = \mu + \alpha_i + \beta_j + \gamma_k + \delta_{ij} + \upsilon_{jk} + \rho_{ik} + \phi_{ijk} + \epsilon_{ijkl}$
The main effects α_i, β_j, γ_k satisfy constraints of the form $\sum \alpha_i = 0$. The two-factor interactions, δ, υ, and ρ, satisfy constraints of the form $\sum_i \delta_{ij} = \sum_j \delta_{ij} = 0$. The three-factor interactions, ϕ_{ijk}, sum to zero over each subscript.

21. A graphical display suggests that Group IV may have a higher infestation rate than the other groups, but the F test only gives a p-value of .12 ($F_{3,16} = 2.27$). The Kruskal–Wallis test results in $K = 6.2$ with a p-value of .10 (3 df).

23. For the male rats, both dose and light are significant (LH increases with dose and is higher in normal light), and there is an indication of interaction ($p = .07$) (the difference in LH production between normal and constant light increases with dose), summarized in the following anova table:

Source	df	SS	MS
Dose	4	545549	136387
Light	1	242189	242189
Interaction	4	55099	13775
Error	50	301055	6021

The variability is not constant from cell to cell but is proportional to the mean. When the data are analyzed on a log scale, the cell variability is stabilized and the interaction disappears. The effects of light and dose are still clear.

25. The following anova table shows that none of the main effects or interactions are significant:

Source	df	SS	MS
Position	9	83.84	9.32
Bar	4	46.04	11.51
Interaction	36	334.36	9.29
Error	50	448.00	8.96

There are some odd things about the data. The first reading is almost always larger than the second, suggesting that the measurement procedure changed somehow between the first and second measurements. One notable exception to this is position 7 on bar 50, which looks anomalous.

27.

Source	df	SS	MS
Species	2	836131	418066
Error	131	446758	3410
Total	133	1282889	

The variance increases with the mean and is stabilized by a square root transformation. The Bonferroni method shows that there are significant differences between all the species.

29.

Source	Df	Sum Sq	Mean Sq	F value	p-value
Furnace	2	4.1089	2.0544	1.4460	0.26159
Wafer.Type	2	5.8756	2.9378	2.0678	0.15547
Furnace x Wafer.Type	4	21.3489	5.3372	3.7566	0.02162
Residuals	18	25.5733	1.4207		

Only interactions are significant. Lines are not parallel in the interaction plot, in which the relationship of thickness of external wafers to furnaces appears quite different than that of the other two wafer types.

33. a. N/R50 and R/R50   **b.** N/R50 and lopro   **c.** N/R50 and N/R40

Chapter 13

1. $X^2 = 5.10$ with 1 df; $p < .025$

3. For the ABO group there is a significant association ($X^2 = 15.37$ with 6 df, $p = .02$), due largely to the higher than average incidence of moderate-advanced TB in B. For the MN group there is no significant association ($X^2 = 4.73$ with 4 df, $p = .32$).

5. $X^2 = 6.03$ with 7 df and $p = .54$, so there is no convincing evidence of a relationship.

7. $X^2 = 12.18$ with 6 df and $p = .06$. It appears that psychology majors do a bit worse and biology majors a bit better than average.

9. In this aspect of her style, Jane Austen was not consistent. *Sense and Sensibility* and *Emma* do not differ significantly from each other ($X^2 = 6.17$ with 5 df and $p = .30$), but *Sanditon* I differs from them, *and* not being followed by *I* less frequently and *the* not being preceded by *on* more frequently ($X^2 = 23.29$ with 10 df and $p = .01$). *Sanditon* I and II were not consistent ($X^2 = 17.77$, df $= 5$, $p < .01$), largely due to the different incidences of *and* followed by *I*.

11. a. In both cases the statistic is

$$-2 \log \Lambda = 2 \sum_i \sum_j O_{ij} \log(O_{ij}/E_{ij})$$

b. $-2 \log \Lambda = 12.59$
c. $-2 \log \Lambda = 16.52$

13. Arrange a table with the number of children of an older sister as rows and the number of children of her younger sister as columns.

a. $H_o: \pi_{ij} = \pi_{i.}\pi_{.j}$. This is the usual test for independence, with

$$X^2 = \sum_{ij} (n_{ij} - n_{i.}n_{.j}/n_{..})^2/(n_{i.}n_{.j}/n_{..})$$

b. $H_0: \sum_{i \neq j} \pi_{ij} = \sum_{j \neq i} \pi_{ji}$ is equivalent to $H_0: \pi_{ij} = \pi_{ji}$. The test statistic is

$$X^2 = \sum_{i \neq j} (n_{ij} - (n_{ij} + n_{ji})/2)^2/((n_{ij} + n_{ji})/2)$$

which follows a χ_2^2 distribution under H_0. The null hypotheses of (a) and (b) are not equivalent. For example, if the younger sister had exactly the same number of children as the older, (a) would be false and (b) would be true.

15. For males, $X^2 = 13.39$, df $= 4$, $p = .01$. For females, $X^2 = 4.47$, df $= 4$, $p = .35$. We would conclude that for males the incidence was especially high in Control I and especially low in Medium Dose and that there was no evidence of a difference in incidence rates among females.

17. There is clear evidence of different rates of ulcers for A and O in both London and Manchester ($X^2 = 43.4$ and 5.52 with 1 df respectively). Comparing London A to Manchester A, we see that the incidence rate is higher in Manchester ($X^2 = 91$, df $= 1$), whereas the incidence rate is higher for London O than for Manchester O ($X^2 = 204$, df $= 1$).

19. $p = .01$ **21.** $\hat{\Delta} = 3.77$

23. McNemar's test gives a chi-square statistic equal to 28.5. Comparing this to the chi-square distribution with 1 df, the result is highly significant: heavy exertion is associated with myocardial infarction. This design is similar to the cell phone study in that each subject acts as his own control.

25. a. The total incidence of myocardial infarction (MCI) is reduced by aspirin ($X^2 = 26.4$ with 1 df). The odds ratio is 0.58, which is a considerable reduction in risk due to aspirin. The incidences of fatal and nonfatal are both significantly reduced as well ($X^2 = 6.2, 20.43, \mathrm{df} = 1$). There is no indication that among those having MCI, the fatality rate was reduced (p-value $= 0.32$). The difference in the incidence of strokes was not statistically significant, $X^2 = 1.67, \mathrm{df} = 1$.

b. There is no evidence that total cardiovascular mortality is decreased by aspirin, but the reduction in mortality due to myocardial infarction is significant.

27. The death penalty was given in 13% of the cases in which the victim was white and the defendant was not. In all other cases the death penalty was given only 5–6% of the time. A chi-square test of independence yields a statistic equal to 15.9 with 3 df, so the p-value is 0.001. Whether such a test is valid is debatable. The use of the test could be criticized on the grounds that these are all the data there are for the years 1993–97, the numbers speak for themeselves, and there is no plausible probability model on which to base probability calculations, like p-values. The use of the test could be defended by arguing that for a table with these row and column marginal totals, it would be very unlikely that there would be such variation of the proportions between rows if only chance were at work.

29. It depends on how the sampling is done. If the number of males and females are determined prior to the sample being drawn, a test of homogeneity would be appropriate. If only the total sample size are fixed, a test of independence would be appropriate. Management won't care, because the qualitative nature of the conclusion would be the same in either case.

Chapter 14

1. b. $\log y = \log a - bx$. Let $u = \log y$ and $v = \log a$.

d. $y^{-1} = ax^{-1} + b$. Let $u = y^{-1}$ and $v = x^{-1}$.

5. This can be set up as a least squares problem with the parameter vector $\beta = (p_1, p_2, p_3)^T$ and the design matrix

$$X = \begin{pmatrix} 1 & 0 & 0 \\ 0 & 1 & 0 \\ 0 & 0 & 1 \\ -1 & 1 & 0 \\ -1 & 0 & 1 \\ 0 & -1 & 1 \end{pmatrix}$$

The least squares estimate is $\hat{\beta} = (X^T X)^{-1} X^T Y$. This gives, for example,

$$\hat{p}_1 = \frac{1}{2}Y_1 + \frac{1}{4}Y_2 + \frac{1}{4}Y_3 + \frac{1}{4}Y_4 + \frac{1}{4}Y_5$$

13. a. $\mathrm{Var}(\hat{\mu}_0) = \sigma^2 \left[\frac{1}{n} + \frac{(x_0 - \bar{x})^2}{\sum (x_i - \bar{x})^2} \right]$

c. $\hat{\mu}_0 \pm s_{\hat{u}_0} t_{n-2}(\alpha/2)$, where

$$s_{\hat{\mu}_0} = s \left[\frac{1}{n} + \frac{(x_0 - \bar{x})^2}{\sum(x_i - \bar{x})^2} \right]^{1/2}$$

15. $\hat{\beta} = \left(\sum x_i y_i \right) / \left(\sum x_i^2 \right)$

21. Place half the x_i at -1 and half at $+1$.

23. a. 85　　　**b.** 80　　　　　　**25.** true

31. $\text{Cov}(U, V) = 0$

37. $\hat{A} = 18.18, s_{\hat{A}} = .14; 18.18 \pm .29$
$\hat{B} = -2.126 \times 10^4, s_{\hat{B}} = 1.33 \times 10^2; -2.126 \times 10^4 \pm 2.72 \times 10^2$

39. Neither a linear nor a quadratic function fits the data.

41. One possibility is DBH versus the square root of age.

43. A physical argument suggests that settling times should be inversely proportional to the squared diameter; empirically, such a fit looks reasonable. Using the model $T = \beta_0 + \beta_1/D^2$ and weighted least squares, we find (standard errors listed in parentheses)

	10	25	28
$\hat{\beta}_0$	$-.403(1.59)$	$1.48 \ (2.50)$	$2.25 \ (2.08)$
$\hat{\beta}_1$	$28672 \ (371)$	$18152 \ (573)$	$16919 \ (474)$

From the table we see that the intercept can be taken to be 0.

51. For 1998, RSS $= .016$. For the 1999 predictions, RSS $= .055$, which is much larger. The predicted values for 1999 appear unrelated to the observed values. The poor performance in 1999 of the predictions formed from the 1998 data is due to over-fitting—4 parameters were estimated from 5 data points.

53. a. There appear to be two regimes corresponding to durations less than or greater than 3 min, and it is best to fit separate linear regressions to each regime.
b. For a duration of 2 min the prediction would be 54.3 min. The standard error of this fitted value is 1.04 min. But there are two parts to the prediction error: the error of the fitted value and the variability of a new observation around its expected value. This latter is measured by the standard deviation of the residuals, 5.9 min. For a duration of 4.5 min, the prediction is 80.3 min. The standard error of this prediction is 1.09 min and the residual standard deviation is 6.7 min. A 95% prediction interval is (67 min, 94 min). See problems 13 and 14.

Author Index

Applications Index

Subject Index

Credits

(This page constitutes an extension of the copyright page.)

We have made every effort to trace the ownership of all copyrighted material and to secure permission from copyright holders. In the event of any question arising as to the use of any material, we will be happy to make the necessary corrections in future printings.

Page 52, Figure 2-10 taken from Figure 5 of "The Action of Suxamethicin as an Agonist and Channel Blocker at the Nicotinic Receptor of Frog Muscle," by C. G. Marshall, D. C. Ogden, and D. Colguhoun, *Journal of Physiology,* 1990, 162. Reprinted with permission.

Page 54, Figure 2-12 taken from Figure 4 of "Statistical Analysis of Microearthquake Activity near San Andreas Geophysical Observatory, Hollister, California," by A. Udias and J. Rice, *Bulletin of the American Seismological Society,* 1975, *65,* 809–828. Reprinted with permission.

Page 56, Figures 2-14 and 2-15 taken from Figures 5 and 11 of "A Characterization of Arctic Undersea Noise," by J. Veitch and A. Wilks, *J. Acoust. Soc. of Amer.,* 1985, *77,* 989–999. Reprinted with the permission of the American Institute of Physics.

Page 57, Figure 2-16 taken from Figure 2 of "Correlation Measurements in Grid Turbulence Using Digital Harmonic Analysis," by C. Van Atta and W. Chen, *Journal of Fluid Mechanics,* 1968, *34,* 497–515. Copyright © 1968, Cambridge University Press. Reprinted with permission.

Page 92, Figure 3-14 taken from Figure 5 of "Correlation Measurements in Grid Turbulence Using Digital Harmonic Analysis," by C. Van Atta and W. Chen, *Journal of Fluid Mechanics,* 1968, *34,* 497–515. Copyright © 1968, Cambridge University Press. Reprinted with permission.

Page 136, Figure 4-3 taken from Figure 2 of "Enduring Values," by J. Youden, *Technomeasurement,* 1972, *14,* 1–11. Copyright 1972 the American Statistical Association. Reprinted with permission.

Page 137, Figure 4-4 taken from Figure 1 of "The Speed of Light," by A. McNish, *IRE Trans. on Instrum.,* 1962, *11,* 138–148. Copyright © 1962 IEEE. Reprinted with permission.

Page 145, Figure 4-6 taken from "Against the Gods: the Remarkable Story of Risk," by Peter L. Bernstein, 1996, 254. Wiley and Sons.

Page 258, Figure 8-1 taken from Figure 10 of "An Analysis of Cell Membrane Noise," by S. Bevan, R. Kullberg, and J. Rice, *Annals of Statistics,* 1979, *7,* 237–257. Reprinted with permission.

Page 259, Figure 8-2 taken from Figure 6 of *Proceedings of the Fifth Berkeley Symposium on Mathematical Statistics and Probability, Volume V: Weather Modification,* by L. LeCam and J. Neyman (Eds.), 1967. Reprinted with the permission of the University of California Press.

Page 304, Figure 8-11 taken from Figure 1 of "Sampling Theory of the Negative Binomial and Logarithmic Series Distributions," by F. J. Anscombe, *Biometrika,* 1950, *37,* 358–382. Reprinted with the permission of the Biometrika Trustees.

Page 562, Figure 14-19 reproduced from *Statistics,* 2nd ed., by David Freedman, Robert Pisani, and Roger Purves with the permission of W. W. Norton & Company, Inc. Copyright © 1991 by David Freedman, Robert Pisani, and Roger Purves.